Soft Computing for Risk Evaluation and Management

Studies in Fuzziness and Soft Computing

Editor-in-chief
Prof. Janusz Kacprzyk
Systems Research Institute
Polish Academy of Sciences
ul. Newelska 6
01-447 Warsaw, Poland
E-mail: kacprzyk@ibspan.waw.pl
http://www.springer.de/cgi-bin/search_book.pl?series=2941

Further volumes of this series can be found at our homepage.

Vol. 55. J. N. Mordeson, D. S. Malik and S.-C. Cheng
Fuzzy Mathematics in Medicine, 2000
ISBN 3-7908-1325-7

Vol. 56. L. Polkowski, S. Tsumoto and T. Y. Lin (Eds.)
Rough Set Methods and Applications, 2000
ISBN 3-7908-1328-1

Vol. 57. V. Novák and I. Perfilieva (Eds.)
Discovering the World with Fuzzy Logic, 2001
ISBN 3-7908-1330-3

Vol. 58. D. S. Malik and J. N. Mordeson
Fuzzy Discrete Structures, 2000
ISBN 3-7908-1335-4

Vol. 59. T. Furuhashi, S. Tano and H.-A. Jacobsen (Eds.)
Deep Fusion of Computational and Symbolic Processing, 2001
ISBN 3-7908-1339-7

Vol. 60. K. J. Cios (Ed.)
Medical Data Mining and Knowledge Discovery, 2001
ISBN 3-7908-1340-0

Vol. 61. D. Driankov and A. Saffiotti (Eds.)
Fuzzy Logic Techniques for Autonomous Vehicle Navigation, 2001
ISBN 3-7908-1341-9

Vol. 62. N. Baba and L. C. Jain (Eds.)
Computational Intelligence in Games, 2001
ISBN 3-7908-1348-6

Vol. 63. O. Castillo and P. Melin
Soft Computing for Control of Non-Linear Dynamical Systems, 2001
ISBN 3-7908-1349-4

Vol. 64. I. Nishizaki and M. Sakawa
Fuzzy and Multiobjective Games for Conflict Resolution, 2001
ISBN 3-7908-1341-9

Vol. 65. E. Orłowska and A. Szalas (Eds.)
Relational Methods for Computer Science Applications, 2001
ISBN 3-7908-1365-6

Vol. 66. R. J. Howlett and L. C. Jain (Eds.)
Radial Basis Function Networks 1, 2001
ISBN 3-7908-1367-2

Vol. 67. R. J. Howlett and L. C. Jain (Eds.)
Radial Basis Function Networks 2, 2001
ISBN 3-7908-1368-0

Vol. 68. A. Kandel, M. Last and H. Bunke (Eds.)
Data Mining and Computational Intelligence, 2001
ISBN 3-7908-1371-0

Vol. 69. A. Piegat
Fuzzy Modeling and Control, 2001
ISBN 3-7908-1385-0

Vol. 70. W. Pedrycz (Ed.)
Granular Computing, 2001
ISBN 3-7908-1387-7

Vol. 71. K. Leiviskä (Ed.)
Industrial Applications of Soft Computing, 2001
ISBN 3-7908-1388-5

Vol. 72. M. Mareš
Fuzzy Cooperative Games, 2001
ISBN 3-7908-1392-3

Vol. 73. Y. Yoshida (Ed.)
Dynamical Aspects in Fuzzy Decision, 2001
ISBN 3-7908-1397-4

Vol. 74. H.-N. Teodorescu, L. C. Jain and A. Kandel (Eds.)
Hardware Implementation of Intelligent Systems, 2001
ISBN 3-7908-1399-0

Vol. 75. V. Loia and S. Sessa (Eds.)
Soft Computing Agents, 2001
ISBN 3-7908-1404-0

Da Ruan · Janusz Kacprzyk
Mario Fedrizzi
Editors

Soft Computing for Risk Evaluation and Management

Applications in Technology, Environment and Finance

With 155 Figures
and 55 Tables

Physica-Verlag
A Springer-Verlag Company

Dr. Da Ruan
Belgian Nuclear Research Centre
Boeretang 200
2400 Mol
Belgium
druan@sckcen.be

Professor Janusz Kacprzyk
Polish Academy of Sciences
Systems Research Institute
ul. Newelska 6
01-447 Warsaw
Poland
kacprzyk@ibspan.waw.pl

Professor Mario Fedrizzi
Universita degli Studi di Trento
Dipartimento di Informatica e Studi Aziendali
Via Inama 5–7
38100 Trento
Italy
fedrizzi@cs.unitn.it

ISSN 1434-9922
ISBN 978-3-662-00348-0 ISBN 978-3-7908-1814-7 (eBook)
DOI 10.1007/978-3-7908-1814-7

Cataloging-in-Publication Data applied for
Die Deutsche Bibliothek – CIP-Einheitsaufnahme
Soft computing for risk evaluation and management: applications in technology, environment and finance; with 55 tables / Da Ruan ... ed. – Heidelberg; New York: Physica-Verl., 2001
(Studies in fuzziness and soft computing; Vol. 76)
ISBN 978-3-662-00348-0

Physica-Verlag Heidelberg New York
a member of BertelsmannSpringer Science+Business Media GmbH

Softcover reprint of the hardcover 1st edition 2001

Hardcover Design: Erich Kirchner, Heidelberg

SPIN 10837849 88/2202-5 4 3 2 1 0 – Printed on acid-free paper

Preface

This volume is a result of a deep conviction of a larger and larger community of scholars, researchers and practitioners – including the editors of this volume – that the combination of two recent "hot topics", i.e. the analysis of *risk*, and its evaluation and management, on the one hand, and *soft computing* (including as its main elements fuzzy logic, various probabilistic-type reasonings, neural networks, evolutionary computation, etc.), on the other hand, can be very beneficial.

Basically, risk is omnipresent, involved in virtually all problems people in diverse areas face in their activities, and without taking it into account it is impossible to find adequate models and solutions. This concerns not only fields that obviously involve risk as a relevant, and "natural" element, exemplified by finance. Also, risk is a crucial element of problems in other areas, maybe less visible and evident to general public. This concerns as, e.g., environmental problems, energy systems, man-machine systems, technological systems, etc.

In all those problems there is much uncertainty concerning data and relations between variables, and human testimonies and perceptions play a crucial role. Moreover, the resulting problems may become complex that may prohibit the use of traditional, "hard" tools and techniques for their formulation and solution. Tools and techniques of broadly perceived *soft computing* may come here to the rescue, and the purpose of this volume is to show how they can be employed for the formulation and solution of problems in which risk plays a crucial role, we have to evaluated it, and finally manage.

The volume starts with some introductory sections that provide the reader with some foundations that may facilitate the reading and use of papers included in the next parts. These papers discuss mainly issues related to decision making under uncertain, risky and imprecise information, testing of imprecise hypotheses, and some elements of Dempster – Shafer's theory of evidence.

The second part is concerned with modern tools and techniques for the analysis of evidence and data that is a point of departure of any problem analysis. The papers in this part deal with broadly perceived data mining and knowledge discovery, in particular the use of evidence theory, the discovery of rules and associations, the use of neuro-fuzzy approaches for time series analysis, and tools for preference analysis and classification.

The third part deals with various issues related to risk evaluation and management in broadly perceived decision support and control systems. The consecutive papers are concerned with the analysis of human reliability through the use of linguistic terms and relations, the analysis of concurrent real-time distributed software systems, and issues related to safety in control systems.

The fourth part is devoted to risk in energy and environmental systems. A comprehensive exposition of the use of soft (mainly fuzzy) tools for the analysis of price forecasting and energy contracting in energy markets, monitoring of energy and heat supply systems, nuclear waste management, sea water pollution, etc. All these analyses clearly involve risk as a relevant factor.

The fifth part of the volume is concerned with risk in finance. The main issues discussed deal with capital budgeting with fuzzy cash flows, project evaluation, various risk measures, financial intermediation, creditworthiness, options and management of a their portfolio.

The editors think that the volume will be of use for a wide audience of researchers, scholars and practitioners in diverse areas of science and technology. We wish to thank all the authors for their valuable contributions. Our thanks and gratitude is also for the Springer-Verlag group (Physica-Verlag, Heidelberg and New York) – notably Dr. Martina Bihn and Mrs. Gabriele Keidel – for their support and consideration.

Mol, Belgium — Da Ruan
Warsaw, Poland — Janusz Kacprzyk
Trento, Italy — Mario Fedrizzi

Contents

3. RISK IN DECISION SUPPORT AND CONTROL SYSTEMS

4. RISK IN ENERGY AND ENVIRONMENTAL SYSTEMS

1.

BASIC ISSUES

Decision Making under Uncertainty with Ordinal Linguistic Data

Thomas Whalen
Professor of Decision Sciences
Georgia State University, Atlanta, GA 30303-3083
whalen@gsu.edu

Abstract

Decision Science has long provided prescriptive methods for choice decisions (in the tradition of Von Neumann and Morgenstern) based on interval scaled (dis)utility consequence measures and ratio scaled subjective probability measures. Over the past three decades, soft computing has provided alternative approaches with less stringent scale assumptions. After a brief introduction, this paper presents a description of five ordinal methods for decision making under uncertainty in the context of linguistic data: Possibilistic Decisionmaking, Revised Possibilistic Decisionmaking, Commensurate L-Fuzzy Risk Minimization, Fuzzy Relational Ordinal Risk Minimization, and Quadratic Ordinal Psychophysical Optimization. Finally, these five techniques are illustrated using a single example.

Decision Analysis

A decision analysis problem arises when we must choose between two or more alternative actions whose outcomes depend on which one of a collection of possible states of nature is the actual state. If it is possible to assign a well-defined subjective (or objective) probability distribution to the states of nature and to assign a numerical utility score measured on an interval scale to the outcome of each state-action pair, then we can compute the expected utility of each action and pick the action whose expected utility is greatest or disutility is least. Probabilities measured on a ratio scale are multiplied times utilities measured on an interval scale; adding the results for a given action gives expected utility measured on the same interval scale as the individual utilities.

While this method makes no guarantees for any one decision, financial portfolio theory and the law of large numbers imply that a lifetime of making decisions by this rule has a very high probability of generating a higher lifetime utility than using any other rule when the scale assumptions are met.

From a standpoint of computational theory of perception or "computing with words," this method begins by converting the natural perceptions of how usual or unusual the possible states are into probability estimates on a ratio scale, and converting the natural perceptions of how acceptable or unacceptable the possible outcomes of state-action pairs are into utility numbers on an interval scale using Von Neuman-Morgenstern [1947] utility theory or some similar method.

A direct generalization of this is to use fuzzy numbers for probabilities and utilities, computing a fuzzy expected utility for each action by the extension principle of fuzzy mathematics. Since natural perceptions of usuality and acceptability are more likely to be in the form of words than numbers, the fuzzy method has the advantage of a more natural representation. [Watson et al, 1979]

Ordinal Methods

However, sometimes it is not reasonable to assume that the perceived usuality of the states of nature can be converted to a ratio scale of probability, even a fuzzy one, without excessive distortion. Similarly, sometimes it is not possible to assign utility scores on an interval scale, crisp or fuzzy, to adequately represent the perceived acceptability or unacceptability of outcomes. In such a case, it is necessary to rely on the ordinal properties of the perceived usuality and acceptability.

One method, widely used in practice even if frowned upon in theory, is to ignore the uncertainty about the states of nature, and base the decision entirely on the one state that is the most usual or likely one. Obviously, the appropriateness of this approach is directly proportional to the degree to which usuality is in fact concentrated in one state of nature.

Another ordinal method represents the attractiveness of each alternative action by the least desirable outcome for that action. The alternative selected is the one for which this worst-case value is most acceptable or least unacceptable.

The above two ordinal methods consider either the maximal usuality alone or the maximal disutility alone. The remainder of this paper is concerned with methods that try to capture the extremes of both utility and possbility within the confines of ordinal calculation.

Possibilistic Decisionmaking

Yager [1979] proposed a method of "possibilistic decisionmaking" which represented utility as membership in the fuzzy set of good outcomes and usuality as membership in the fuzzy set of possible states of nature. The potential utility of a state-action outcome is its membership in the fuzzy set of outcomes that are both possible and good, calculated as the minimum of the membership of the outcome in "good" and the state in "possible." The attractiveness of an action is then taken as the greatest potential utility of any of its possible outcomes. The choice of action is made by taking the alternative whose attractiveness is greatest.

Revised Possibilistic Decisionmaking

Whalen [1984] pointed out that Yager's approach treated choice nodes and chance nodes the same way, maximizing in both cases, and suggested an algorithm known as "revised possibilistic decisionmaking" that measured the disutility of each state-action outcome rather than its utility. The threat of a particular state-action outcome is its membership in the fuzzy set of outcomes that are both possible and bad, found by minimization; the risk of an action is the greatest threat of its possible outcomes. The choice of action is made by taking the alternative whose risk is least.

Both of these varieties of possibilistic decisionmaking were introduced assuming that set memberships follow a complete weak order, represented without loss of generality by numbers in the unit interval. Thus, the nuances of natural perceptions of likelihood and acceptability were rapidly lost.

Commensurate L-Fuzzy Risk Minimization

The next stage in generalization is the use of L-fuzzy (variously rendered "lattice fuzzy" or "linguistic fuzzy") sets [Goguen, 1967] for bad outcomes and possible states, with set memberships defined on an incompletely ordered abstract lattice. If all memberships are measured on a common lattice, it is still possible in principle to take the minimum of the membership of a state-action outcome in the L-fuzzy set of bad outcomes and the membership of the corresponding state in the L-fuzzy set of possible states. But since the ordering is incomplete, it is not always possible to find an explicit minimum or maximum. Furthermore, it may not be meaningful to compare the membership of a state in the set of usual states with the membership of an outcome in the set of bad outcomes, since the two are qualitatively so different.

The Commensurate L-Fuzzy Risk Minimization technique deals with this by placing usuality set memberships on one incompletely ordered lattice and disutility set memberships on a separate incompletely ordered lattice. The version presented in [Whalen & Bronn, 1988] uses a (disutility, possibility) ordered pair to specify the membership of a state-action outcome in the set of outcomes that are both possible and bad. The risk of an alternative action is found by symbolically maximizing these pairs similarly to the way the previous method maximizes unresolved minima; the selected action is the one for which this symbolic structure is least risky. Recognizing that this process is not highly decisive, the method successively falls back to the L-fuzzy approach with a common incomplete lattice, and the revised possibilistic approach with a single complete weak order of membership grades.

Fuzzy Relational Ordinal Risk Minimization

Whalen & Schott [2000] have suggested an approach to commensurate ordinal decision making based on recent insights in knowledge granulation, computing with words, fuzzy relations as fuzzy x-y graphs, and second order fuzzy sets. In this approach, the disutility of a state-action outcome is represented by a linguistic variable. The base values of this linguistic variable may be well-defined utility numbers, or they may be a purely abstract ordered set. The usuality of a state is similarly represented by a linguistic variable defined on a base variable of degrees of possibility.

The **second order fuzzy threat of a state-action pair** $\widetilde{\widetilde{T(A,s)}}$ is s second order fuzzy set of disutilities. The membership of a disutility base value x in this set is itself a (first-ordet) fuzzy set of possibilities $\widetilde{\mu_{\widetilde{\widetilde{T(A,s)}}}(x)}$. The membership of a possibility base value p in the fuzzy possibility of disutility x (that is, its membership in the fuzzy set of possibilities that the disutility if (A,s) is x) is denoted $\mu_{\widetilde{\mu_{\widetilde{\widetilde{T(A,s)}}}(x)}}(p)$. The second-order mambership function is defined by the Cartesian minimum of the fuzzy disutility of the state-action pair and the fuzzy possibility of the state: the value of $\mu_{\widetilde{\mu_{\widetilde{\widetilde{T(A,s)}}}(x)}}(p)$ is the mionimum of the membership of x in the fuzzy disutility of the state-action pair (A,s), $\widetilde{d(A,s)}(x)$ and the membership of p in the fuzzy possibility of state p, $\widetilde{\pi(s)}(p)$

$$\mu_{\widetilde{\mu_{\widetilde{\widetilde{T(A,s)}}}(x)}}(p) = \min\left\{ \begin{array}{c} \mu_{\widetilde{d(A,s)}}(x), \\ \mu_{\widetilde{\pi(s)}}(p) \end{array} \right\}$$

The **second order fuzzy risk profile of an action,** $\widetilde{\widetilde{rA}}$ is a second order fuzzy set of disutilities formed by the union of the fuzzy relations corresponding to the threat of all the state-action pairs for that action. The possibility that the disutility of a given action is x is a fuzzy set of possibilities formed as the union of the fuzzy possibility that each possible outcome has disutility x. This is equivalent to a fuzzy x-y graph for each action; the knowledge granules making up this graph have fuzzy coordinates defined by the disutility (x) and usuality (y) of the possible outcomes.

$$\mu_{\widetilde{\mu_{\widetilde{\widetilde{rA}}}(x)}}(p) = \max_{s}\left\{\mu_{\widetilde{\mu_{\widetilde{\widetilde{rA}}}(x)}}(p)\right\}$$

The **first order fuzzy risk profile of an action,** $\widetilde{rA}$ is created from the action's second order risk profile using a Sugeno integral. The first order possibility that the disutility of an action is x is the maximum over possibility grades (0 to 1) of the minimum of each possibility grade with its membership in the second order possibility that the disutility of the action is x.

$$\mu_{\widetilde{rA}}(x) = \max_{p}\left\{\min(p, \mu_{\widetilde{\mu_{\widetilde{\widetilde{rA}}}(x)}}(p)\right\}$$

The **linguistic risk assessment** is the final stage in the process, found by converting the first order fuzzy risk profile of each alternative action back into words. This allows the user to exercise judgment as to what action should be chosen and what rhetorical argument to use to justify that choice. Wenstop's [1980] classic approach to linguistic approximation does not work well for this purpose because the possibility distributions tend to be multimodal, so a method based on linear integer programming is used instead. Future research will evaluate the effectiveness of genetic algorithms and other methods for linguistic approximation in the context of risk minimization using computing with words.

Quadratic Ordinal Psychophysical Optimization

Whalen and Wang [2000] apply quadratic programming to an ordinal interpretation of linguistic probability and utility terms. This interpretation incorporates a well-established finding of psychophysics: the degree to which stimuli must differ physically to be discerned perceptually is proportional to the magnitude for the stimuli according to a logarithmic law of human (and animal) perception.

If the utility of each state-action pair is known only roughly, for instance as "good," "fair," or "poor," we will represent this ordinally; each fair utility is greater than any poor utility and less than any good utility.

Over a century of research in the field of psychophysics indicates a very strong tendency for human perception to operate on a logarithmic scale, in which the "just noticeable difference" between stimuli is a constant proportion of the magnitude of the stimulus rather than a fixed incremental amount. Based on this, we require that each "fair" utility is greater than or equal to a fixed constant called the distinguishability ratio times any "poor" utility, and each "good" utility is greater than or equal to the same fixed constant times any "fair" utility. The distinguishability ratio is a generalization of the psychophysical concept of a decibel.

To represent the fact that two quantities with the same rough description need not be identical as long as their difference is not psychologically significant, the quantitative representation of each "fair" utility must be less than or equal to the distinguishability constant times that of every other "fair" utility, and similarly among "good" and "poor." This automatically entails that each utility in a class is also greater than or equal to any other utility divided by the distinguishability ratio.

If the probability of each state is specified numerically, then the overall utility of an action is the sum of the unknown numeric representations of the utility of each of the action's state-action pairs, weighted by the numerical probability weights. This is a linear function with linear inequality constraints; thus, it is possible to find the maximum and minimum overall utility for each action by linear programming. More importantly, we can find the minimum and maximum difference between the overall utility of two actions. Alternative A dominates Alternative B if max{utility(B)-utiility(A)}<0; it is not sufficient that max{utility(A)}>max{utiility(B)} and min{utility(A)}>min{utiility(B)}, but Alternative A can dominate alternative B even if max{utiility(B)}> min{utility(A).

A dominated action can be removed from consideration, leaving a short list of nondominated actions. These can be re-analyzed in several ways. One can use a finer grid of linguistic terms like "very low" or "upper medium" and repeat the analysis. One can also introduce additional inequalities into the linear programming formulation. For example, there may be two actions that both generate "low" utility in a particular state, but further introspection and/or economic analysis may indicate one is discernibly lower than the other. Finally, one can move to methods that are more decisive than the ordinal ones, but which require stronger assumptions.

If the utility of each state-action pair is specified numerically but the probability of the different attributes is only specified roughly, the linear programming problem is very similar to the one described above.

If utility is known only as good, fair, poor and probability is known only as low, medium, high, finding the maximum and minimum of the difference between two alternative actions, and thus identifying dominated alternatives, becomes a problem in quadratic programming. The difference in overall utility of between action i and action j is a weighted sum,

$$Dij = \sum_k (Uik*Pk - Ujk*Pk),$$

in which each term is the product of two variables, the utility of action i in state k and the unknown probability, where these variables U•• and P• are subject to a set of linear inequality constraints. If the maximum value of the quadratic function Dij, subject to the linear constraints, is negative then action j dominates action i.

Example

To illustrate the four ordinal approaches discussed above, we will look at a simple abstract decision problem in which utility is granulated to good, fair, and poor, and usuality is granulated to usual, plausible, and rare. If there are just three states of nature, assessed as a usual state, a plausible state, and a rare state, there are six possible risk profiles without ties in utility. In the illustrative example, we will compare six alternative actions, one for each of these risk profiles.

Table 1: Linguistic Utility & Probability

actions:	Usual State	Plausible State	Rare State
A1	good	fair	poor
A2	good	poor	fair
A3	fair	good	poor
A4	fair	poor	good
A5	poor	good	fair
A6	poor	fair	good

Possibilistic Decisionmaking

To illustrate Yager's possibilistic decisionmaking, it is necessary to place these quantities in a complete ranking. For simplicity, assume that the membership of "good" in the fuzzy set of good outcomes is equal to the membership of "usual" in the fuzzy set of possible states, and similarly for fair and plausible and for poor and rare. Then the ordinal calculations to assess the overall attractiveness of the six actions are as follows:

Table 2: Possibilistic Decisionmaking

	Usual State	Plausible State	Rare State	Overall Utility
A1	min(good, usual) = good	min(fair, plausible) = fair	min(poor, rare) = rare	max(good, fair, rare) = good
A2	min(good, usual) = good	min(poor, plausible) = poor	min(fair, rare) = rare	max(good, poor, rare) = good
A3	min(fair, usual) = fair	min(good, plausible) = plausible	min(poor, rare) = rare	max(fair, plausible, rare) = fair
A4	min(fair, usual) = fair	min(poor, plausible) = poor	min(good, rare) = rare	max(fair, poor, rare) = fair
A5	min(poor, usual) = poor	min(good, plausible) = plausible	min(fair, rare) = rare	max(poorr, plausible, rare) = plausible = fair
A6	min(poor, usual) = poor	min(fair, plausible) = fair	min(good, rare) = rare	max(poor, fair, rare) = fair

If we use the second-highest term as a tie-breaker, the method ranks the six actions from highest to lowest utility in the order listed, with A1 being most attractive, except that the last two actions are still tied for last place.

Revised Possibilistic Decisionmaking

To illustrate the revised possibilistic decisionmaking algorithm, convert good, medium, and poor utility into low, medium, and high disutility respectively and assume that the membership of "high" in the fuzzy set of bad outcomes is equal to the membership of "usual" in the fuzzy set of possible states, and similarly for medium and plausible and for good and rare. The ordinal calculations to assess the overall attractiveness of the six actions is shown in the next table.

Ranking actions by attractiveness from lowest to highest disutility, again using the second place as a tie-breaker, we have A1 through A3 tied for best, followed by A4, A5, A6.

Table 3: Revised Possibilistic Decisionmaking

	Usual State	Plausible State	Rare State	Overall Risk
A1	min(low, usual) = low	min(medium, Plausible) = medium	min(high, rare) = rare	max(low, medium, rare) = medium
A2	min(low, usual) = low	min(high, plausible) = plausible	min(mediu, rare) = rare	max(low, plausible, rare) = plausible = medium
A3	min(medium, usual) = medium	min(low, plausible) = low	min(high, rare) = rare	max(medium, low, rare) = medium
A4	min(medium, usual) = medium	min(high, plausible) = plausible	min(low, Rare) = rare	max(medium, plausible, rare) = medium
A5	min(high, usual) = high	min(low plausible) = low	Min(mediu, rare) = rare	max(high, low, rare) = high
A6	min(high, usual) = high	min(medium, Plausible) = medium	min(low, rare) = rare	max(high, medium, rare) = high

Commensurate L-Fuzzy Risk Minimization

For this method, we assume that low, medium, high disutility and usual, plausible, rare usuality are on two separate scales; see Table 4.

In this case, actions A1, A2, and A3 are difficult to rank among themselves without additional information, but all three are discernibly better than A4, while A5 and A6 are the two least attractive alternatives.

Fuzzy Relational Ordinal Risk Minimization

Suppose that low, medium and high disunity are represented by the following vectors giving the compatibility of utility base variable scores with membership grades shown in Table 5. Also suppose that rare, plausible, and usual have the same vector representations, respectively, though of course applying to possibility base values rather than disutility. Then the three second-order fuzzy sets representing the three threats associated with action A2 are as shown in Tables 6, 7, and 8.

Table 4: Commensurate L-Fuzzy Risk Minimization:

	Usual State	Plausible State	Rare State	Overall Risk
A1	(low, usual)	(medium, plausible)	(high, rare)	(low, usual) or (medium, plausible) or (high, rare)
A2	(low, usual)	(high, plausible)	(medium, rare)	(low, usual) or (high, plausible)
A3	(medium, usual)	(low, plausible)	(high, rare)	(medium, usual) or (high, rare)
A4	(medium, usual)	(high, plausible)	(low, rare)	(medium, usual) or (high, plausible)
A5	(high, usual)	(low, plausible)	(medium, rare)	(high, usual)
A6	high, usual)	(medium, plausible)	(low, rare)	(high, usual)

Table 5: Membership Functions for Fuzzy Disutilities

1	0.75	0.5	0.25	0	0	0	0	0	low
0	0.25	0.5	0.75	1	0.75	0.5	0.25	0	medium
0	0	0	0	0	0.25	0.5	0.75	1	high
0	**0.125**	**0.250**	**0.375**	**0.500**	**0.625**	**0.750**	**0.875**	**1**	**disutilities**

Table 6: Second-Order Fuzzy Threat of (low, usual)

Poss	Disutility:								
	0.000	**0.125**	**0.250**	**0.375**	**0.500**	**0.625**	**0.750**	**0.875**	**1.000**
0.000	0	0	0	0	0	0	0	0	0
0.125	0	0	0	0	0	0	0	0	0
0.250	0	0	0	0	0	0	0	0	0
0.375	0	0	0	0	0	0	0	0	0
0.500	0	0	0	0	0	0	0	0	0
0.625	0.25	0.25	0.25	0.25	0	0	0	0	0
0.750	0.5	0.5	0.5	0.25	0	0	0	0	0
0.875	0.75	0.75	0.5	0.25	0	0	0	0	0
1.000	1	0.75	0.5	0.25	0	0	0	0	0

to the fuzzy set of possibilities comaptible with disutility .625 in the risk profile of Action A2. This is bcause:

•possibility .250 belongs 0% to the fuzzy set of possibilities compatible with disutility .625 in the fuzzy threat of (low, usual) associated with the usual state and action A2 (Table 6),

•possibility .250 belongs 25% to the fuzzy set of possibilities compatible with disutility .625 in the fuzzy threat of (high, plausible) associated with the plausible state and action A2 (Table 7),

•possibility .250 belongs 50% to the fuzzy set of possibilities compatible with disutility .625 in the fuzzy threat of (medium, rare) associated with the rare state and action A2 (Table 8),

The first-order risk profile of action A2, found by Sugeno integral, is shown in Table 10/

Table 10: First-Order Fuzzy Risk Profile of Action A2:

	Disutility:								
	0.000	0.125	0.250	0.375	0.500	0.625	0.750	0.875	1.000
Poss.	1.000	0.750	0.500	0.250	0.250	0.250	0.500	0.625	0.625

Figure 1: Sugeno Integral for Membership of .625 in Action A2's First-Order Fuzzy Risk Profile

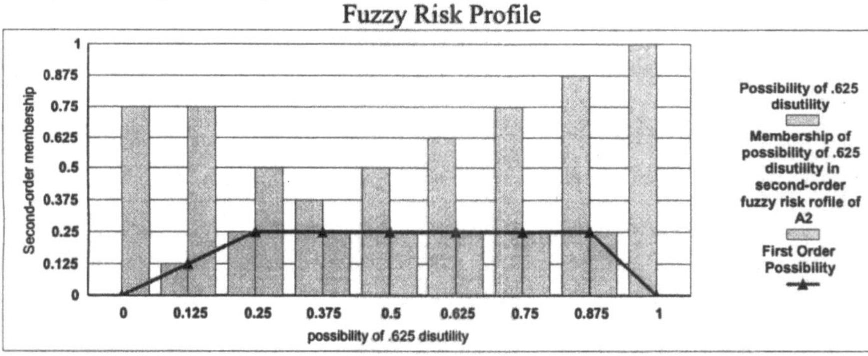

Figure 1 illustrated the calculation that gives .25 as the membership of disutility .625 in the first-order risk profile for action A2.

This value was converted back into words by integer linear programming. The lexicon consisted of ten fuzzy sets: low, lower medium, medium, upper medium, high, plus five more labeled possibly low through possibly high and having (subnormal) memberships equal to one half times the memberships in the corresponding sets in the first five. The best fit to the risk profile of action A2 according to this method turned out to be "low or high." A more precise

Table 7: Second-Order Fuzzy Threat of (high, plausible)

Poss	Disutility:								
	0.000	**0.125**	**0.250**	**0.375**	**0.500**	**0.625**	**0.750**	**0.875**	**1.000**
0.000	0	0	0	0	0	0	0	0	0
0.125	0	0	0	0	0	0.25	0.25	0.25	0.25
0.250	0	0	0	0	0	0.25	0.5	0.5	0.5
0.375	0	0	0	0	0	0.25	0.5	0.75	0.75
0.500	0	0	0	0	0	0.25	0.5	0.75	1
0.625	0	0	0	0	0	0.25	0.5	0.75	0.75
0.750	0	0	0	0	0	0.25	0.5	0.5	0.5
0.875	0	0	0	0	0	0.25	0.25	0.25	0.25
1.000	0	0	0	0	0	0	0	0	0

Table 8: Second-Order Fuzzy Threat of (medium, rare)

Poss	Disutility:								
	0.000	**0.125**	**0.250**	**0.375**	**0.500**	**0.625**	**0.750**	**0.875**	**1.000**
0.000	0	0.25	0.5	0.75	1	0.75	0.5	0.25	0
0.125	0	0.25	0.5	0.75	0.75	0.75	0.5	0.25	0
0.250	0	0.25	0.5	0.5	0.5	0.5	0.5	0.25	0
0.375	0	0.25	0.25	0.25	0.25	0.25	0.25	0.25	0
0.500	0	0	0	0	0	0	0	0	0
0.625	0	0	0	0	0	0	0	0	0
0.750	0	0	0	0	0	0	0	0	0
0.875	0	0	0	0	0	0	0	0	0
1.000	0	0	0	0	0	0	0	0	0

The second-order fuzzy risk profile of action A2 is represented by the union of these three second-order fuzzy sets:

Table 9: Second-Order Fuzzy Risk Profile of Action A2

Poss	Disutility:								
	0.000	**0.125**	**0.250**	**0.375**	**0.500**	**0.625**	**0.750**	**0.875**	**1.000**
0.000	0	0.25	0.5	0.75	1	0.75	0.5	0.25	0
0.125	0	0.25	0.5	0.75	0.75	0.75	0.5	0.25	0.25
0.250	0	0.25	0.5	0.5	0.5	0.5	0.5	0.5	0.5
0.375	0	0.25	0.25	0.25	0.25	0.25	0.25	0.75	0.75
0.500	0	0	0	0	0	0.25	0.5	0.75	1
0.625	0.25	0.25	0.25	0.25	0	0.25	0.5	0.75	0.75
0.750	0.5	0.5	0.5	0.25	0	0.25	0.5	0.5	0.5
0.875	0.75	0.75	0.5	0.25	0	0.25	0.25	0.25	0.25
1.000	1	0.75	0.5	0.25	0	0	0	0	0

As an illustrative example, the entry 0.75 in the row for .250 possibility and column for .625 disutility in Table 9 means that the possibility .250 belongs 50%

linguistic approximator is under development, but this suffices for an initial demonstration.

The linguistic approximations to the risk profiles of the six alternative actions are as follows:

A1: low or possibly (lower medium or upper medium)
A2: low or high
A3: medium or possibly low
A4: medium or possibly low
A5: high or low
A4: high or possibly (lower medium or upper medium)

Action A1 seems to be indicated by this output; however, the user has more flexibility to construct and evaluate alternative rhetorical arguments for and against various action, combining the output of the method with other information. Thus, computing with words is well suited here to a decision support application rather than an automated decision mechanism.

Quadratic Ordinal Psychophysical Optimization

Table 11 shows the maximum difference between the expected utility of the action corresponding to the row label and the expected utility of the action corresponding to the column label. Each entry in the table is the result of a separate quadratic optimization over the numeric state probabilities and state-action utilities subject to linear ordinal constraints.

Table 11: Maximum Quadratic Expected Utility Differences

	A1	A2	A3	A4	A5	A6
A1		0.275	0.95548	0.95438	0.9607	0.9614
A2	0.19556		0.95546	0.95435	0.9592	0.9599
A3	-0.04	0.055		0.30622	0.766	0.7655
A4	-0.15093	-0.12	0.15396		0.7635	0.7644
A5	-0.44287	-0.44	0.03606	0.18		0.1812
A6	-0.59363	-0.51	-0.002	0.00179	0.006	

Table 12 shows the ranges of numeric values that are assigned to the linguistic terms for utility (good, fair, and poor) and probabilities (usual, plausible, and rare) when the objective function is to maximize the quantity (expected utility of action A5) - (expected utility of action A3). Table 13 shows the assignment of numerical values to each linguistic probability and state-action pair to optimize the same objective function. Together, these two tables how that the objective function is maximized when:

- the probability assigned to the "plausible" state is as small as possible,
- the spread within "good" values is as large as possible, and
- the spread between "fair" and "poor" utilitiesis as small as possible.

This maximizes the difference in utility between A5 and A3 by bringing the utility of A3 to 0.16210 and simultaneously bringing the utility of A5 to 0.19816. Since the resulting difference, 0.03606, is positive, action A3 does not dominate action A5; despite the overall impression that A3 is better, the ambiguity inherent in the ordinal linguistic terms allows for an interpretation in which A5 could conceivably be slightly better.

Table 12:
Maxima and Minima

maxgood	1
mingood	0.8
maxfair	0.015625
minfair	0.0125
maxpoor	0.01
minpoor	0.01
maxusual	1
minusual	0.8
maxplaus	0.64
minplaus	0.016535
maxrare	0.01125
minrare	0.01

Table 13: Probabilities and Utilities

	Usual	Plausible	Rare	Expected Utility
prob	0.8	0.19	0.01	
A1	0.8	0.015625	0.01	0.64307
A2	0.8	0.01	0.015625	0.64206
A3	0.0125	0.8	0.01	0.1621
A4	0.01563	0.01	0.8	0.0224
A5	0.01	1	0.015625	0.19816
A6	0.01	0.0125	0.999428	0.02037

Figure 2 shows the lattice structure of dominance relations between the six actions created by quadratic ordinal psychophysical optimization.

Figure 2: Dominance Relations Between Actions

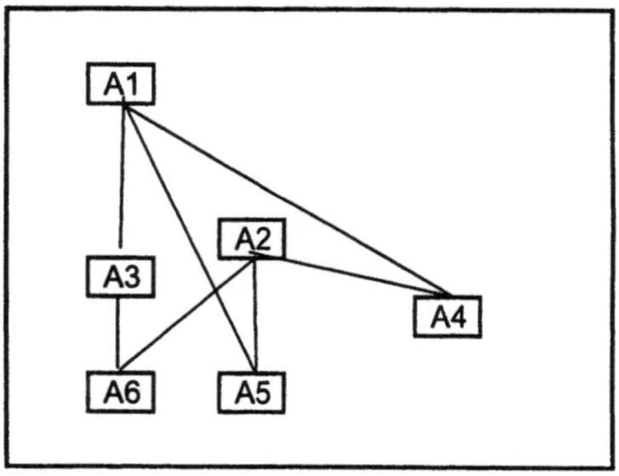

References

Goguen, J. A., "L-Fuzzy Sets,", J. Math. Analysis and Application 18, 1967.

Von Neumann, J. and Morgenstern, O., THE THEORY OF GAMES AND ECONOMIC BEHAVIOR, Wiley, New York [1947].

Watson, S., Weiss, J., and Donnell, M., "Fuzzy decision analysis", IEEE Transactions on Systems, Man, and Cybernetics, Vol. SMC-9, No. 1 [1979].

Wenstop, Fred, "Quantitative analysis with linguistic values." Fuzzy Sets and Systems 4, p. 99-115 [1980]

Whalen, T., "Decision Making Under Uncertainty With Various Assumptions About Available Information", IEEE Transactions on Systems, Man, and Cybernetics, Vol. SMC-14, No. 6 [1984a].

Whalen, T & Bronn, C., "Essentials of Decision Making Under Generalized Uncertainty," in J. Kacprzyk & M. Fedrizzi (eds.) Combining Fuzzy Imprecision with Probabilistic Uncertainty in Decision Making, Springer-Verlag, 1988.

Whalen, T & Wang, G., "Optimizing Ordinal Utility Using Quadratic Programming and Pshcyophysical Theory," Proceedings, 2000 North American Fuzzy Information Processing Association Conference, pages 307-310, 2000.

Whalen, T & Schott, B., "Decision Risk Minimization by Computational Semiotics," Proceeedings, 2000 Joint Conference on Information Sciences, 2000.

Yager, R., "Possibilistic Decisionmaking," IEEE Transactions on Systems, Man, and Cybernetics, Vol SMC-9, July 1979.

Modeling Risk in Sequential Decision Making with Interval Probabilities

Kurt J. Engemann
Iona College
New Rochelle, NY 10801
kengemann@iona.edu

Holmes E. Miller
Muhlenberg College
Allentown, PA 18104
homiller@muhlenberg.edu

Ronald R. Yager
Iona College
New Rochelle, NY 10801
yager@panix.com

Abstract

There are some drawbacks in attempting to apply the classical approach to risk analysis. We propose a new method to analyze scenarios involving a sequence of decisions which have imprecise information regarding event likelihoods. The decision maker's knowledge and degree of optimism are included as key elements in the analysis. An example in risk management in presented.

1. Introduction

The initial step of any risk analysis is understanding what events can occur. The decision maker must decide how to respond to these events. In [1] Miller and Engemann identify options such as: avoiding risks; changing the likelihood of occurrence of an event; reducing the losses if an event does occur, and assigning the risk to others. Because investment decisions address organizational needs and because the risk components of these investments lead to consequences with business implications, business managers must be involved from the inception to provide direction regarding threats, events, controls, and the appropriate level of risk they are willing to incur.

Using probabilities can be problematic when data is incomplete or unavailable for specific events, or when the analysis requires incorporating the decision maker's degree of optimism or pessimism. Therefore, alternative methods are required which consider both the nature of the information used to evaluate the alternatives, and the behavioral characteristics of the decision makers themselves. Many events associated with risk either have not occurred or if they have, have occurred with such infrequency that obtaining accurate probability estimates is problematic.

2. Interval Probabilities

In many decision making problems the information available regarding events is not precise, but is in the form of interval probabilities. We shall illustrate a new method designed to address this issue. Consider the situation in which, rather than having a precise probability, p_j, for state of nature S_j, we have a range of possible values, $[B_j, T_j]$, such that $B_j \le p_j \le T_j$.

We note that a required condition for the existence of a consistent set of probability intervals is $\Sigma_j B_j \le 1 \le \Sigma_j T_j$. Given a consistent set of probability intervals, a refinement can be made to narrow the ranges of the probabilities. To illustrate this, consider the case in which we have two events with interval probabilities: $.5 \le p_1 \le 1$ and $.2 \le p_2 \le .7$. Note that since the minimal value for p_2 is 0.2 and the total of p_1 and p_2 must be one, p_1 will never exceed 0.8 for any valid probability distribution. In a similar way since the minimal value for p_1 is 0.5, we see that the value for p_2 will never exceed 0.5 in any valid probability distribution. The implication of the above example is that we may provide tighter bounds for interval probabilities without losing any information about possible probability distributions. More generally, given $B_j \le p_j \le T_j$ is a consistent set of interval probabilities, then the associated refined set of bounds is $L_j \le p_j \le U_j$ such that:

$$L_j = \mathrm{Max}(B_j, 1 - \Sigma_{i \ne j} T_i)$$

$$U_j = \mathrm{Min}(T_j, 1 - \Sigma_{i \ne j} B_i).$$

We note the following properties:

1. $[L_j, U_j] \subseteq [B_j, T_j]$ for all j
2. Any $p_j \in [B_j, T_j]$ also satisfies $p_j \in [L_j, U_j]$.

Determining the expected value for each alternative A_i requires selecting a probability distribution, $P = (p_1, p_2, \ldots, p_n)$, where $p_j \in [L_j, U_j]$ and $\Sigma_j p_j = 1$. It should be emphasized that the interval uncertainty manifests itself in an uncertainty with respect to the value of the probabilities. The problem is complicated by the fact that the set of valid probability distributions is continuously infinite.

The valuation of alternative A_i can be uniquely determined if we incorporate the disposition (level of optimism) of the decision maker. As in the case of uncertainty, the disposition of the decision maker is a key element. The evaluation of the A_i can be obtained for particular types of decision attitude, as shown by Yager and Kreinovich [2].

In the case of a neutral decision maker the valuation of A_i can be expressed as:

$$\mathrm{Val}(A_i) = \Sigma_j C_{ij} \hat{p}_j$$

where C_{ij} is the payoff using alterative A_i if S_j is the state of nature, and where $\hat{p}_j$ is a probability defined such that:

$$\hat{p}_j = L_j + \frac{\Delta_j}{\sum_{i=1}^{n} \Delta_i}\left(1 - \sum_{i=1}^{n} L_i\right)$$

where $\Delta_j = U_j - L_j$, the spread associated with P_j. Letting $L = \Sigma_i L_i$, the total lower probability, and $\Delta = \Sigma_i \Delta_i$, we get:

$$\hat{p}_j = L_j + \frac{\Delta_j}{\Delta}(1 - L)$$

Note the $\hat{p}_j$ are determined without regard to the payoff since the decision maker is risk neutral. It can be shown that the $\hat{p}_j$ essentially are a probability distribution thus, $Val(A_i)$ is an expected value.

In the case of the optimistic decision maker, we find the valid probability distribution which gives the maximal expected value. Essentially, this becomes a problem of finding the valid probability distribution that assigns as much as possible of the probabilities to S_j's that have high payoffs. In the following we shall, without loss of generality, assume the C_{ij} for A_i have been indexed in descending order, $C_{ir} \geq C_{is}$ if $r < s$. In this case

$$Val_{opt}(A_i) = \Sigma_j C_{ij} \tilde{p}_j$$

where $\tilde{p}_j = L_j + \beta_j$. The β_j are determined as follows:

1. Let $L = \Sigma_j L_j$
2. If $L = 1$, set all $\beta_j = 0$ and exit
3. Calculate $Sum(j) = L + \Sigma_{i=1}^{j} \Delta_i$ and let j^* be the minimal j such that $sum(j) \geq 1$. Then
   i. $\beta_j = \Delta_j$ for $j < j^*$
   ii. $\beta_j = 0$ for $j > j^*$
   iii. $\beta_j = 1 - Sum(j^* - 1)$ for $j = j^*$.

In the case of the pessimistic decision maker, we find the valid probability distribution which gives the minimal expected value. Essentially, this becomes a problem of finding the valid probability distribution that assigns as much as possible of the probabilities to S_j's that have low payoffs. Again, we shall assume that the C_{ij} for A_i have been indexed in descending order. In this case

$$Val_{pes}(A_i) = \Sigma_j C_{ij} \ddot{p}_j$$

when $\ddot{p}_j = L_j + \alpha_j$. The α_j are determined as follows:

1. Let $L = \Sigma_j L_j$
2. If $L = 1$, set all $\alpha_j = 0$ and exit
3. Calculate $Sum(j) = L + \Sigma_{i=j}^{n} \Delta_i$ and let j^* be the maximal j such that $sum(j) \geq 1$. Then
   i. $\alpha_j = \Delta_j$ for $j > j^*$
   ii. $\alpha_j = 0$ for $j < j^*$
   iii. $\alpha_j = 1 - Sum(j^* + 1)$ for $j = j^*$.

Example: The following example illustrates the methodology to determine the valuation of an alternative, A_1. We are given the following payoffs (costs in this example) for the states of nature S_j:

	S_1	S_2	S_3	S_4
A_1	250	150	350	50

Assume that the decision maker evaluates the event likelihoods as intervals as follows:

$p_1 \in [.3, 1]$ "likely" $\quad$ $p_2 \in [0, 0.1]$ "very unlikely"

$p_3 \in [0, .2]$ "could happen" $\quad$ $p_4 \in [0, 1.0]$ "unknown"

From this we can determine the following refined probability intervals:

$p_1 \in [.3, 1.0] \quad p_2 \in [0, .1] \quad p_3 \in [0, 0.2] \quad p_4 \in [0, 0.7]$

Here we see:

$L_1 = .3 \quad L_2 = 0 \quad L_3 = 0 \quad L_4 = 0 \quad L = 0.3$

$\Delta_1 = 0.7 \quad \Delta_2 = 0.1 \quad \Delta_3 = 0.2 \quad \Delta_4 = 0.7 \quad \Delta = 1.7.$

For the risk neutral decision maker we determine the probabilities using:

$$\hat{p}_j = L_j + \frac{\Delta_j}{\Delta}(1 - L).$$

Substituting from above we obtain:

$$\hat{p}_j = L_j + .412\Delta_j$$

Hence: $\hat{p}_1 = 0.59, \hat{p}_2 = 0.04, \hat{p}_3 = 0.08, \hat{p}_4 = 029.$

Using these risk neutral probabilities probabilities we get the valuation for action A_1:

$$Val(A_1) = (0.59)250 + (0.04)150 + (0.08)350 + (0.29)50 = 196.$$

Consider an optimistic decision. The ordered payoffs for alternative A_1 are: [50 150 250 350]. Remember in this example the payoffs are costs, so smaller is better. In this case the associated probabilities are found using:

$$\tilde{p}_4 = L_4 + \beta_4 = 0 + \beta_4$$

$$\tilde{p}_2 = L_2 + \beta_2 = 0 + \beta_2$$

$$\tilde{p}_1 = L_1 + \beta_1 = .3 + \beta_1$$

$$\tilde{p}_3 = L_3 + \beta_3 = 0 + \beta_3$$

Since $L = 0.3$ we have $1 - 0.3 = 0.7$ to allocate to the β_j values. For an optimistic decision maker, we start at the top. Since $\Delta_4 = 0.7$, we allocate this to β_1 and this uses up all our available weight. Hence, for alternative A_1: $\tilde{p}_4 = 0.7$, $\tilde{p}_2 = 0$, $\tilde{p}_1 = 0.3$, $\tilde{p}_3 = 0$. Using this we get:

$$Val(A_1) = (0.7)50 + 0(150) + (0.3)(250) + (0)(350) = 110.$$

Consider a pessimistic decision maker. The ordered payoffs for alternative A_1 are: [50 150 250 350]. In this case the associated probabilities are found using:

$$\ddot{p}_4 = L_4 + \beta_4 = 0.0 + \beta_4$$
$$\ddot{p}_2 = L_2 + \beta_2 = 0.0 + \beta_2$$
$$\ddot{p}_1 = L_1 + \beta_1 = 0.3 + \beta_1$$
$$\ddot{p}_3 = L_3 + \beta_3 = 0.0 + \beta_3$$

Here we start allocating the available probability weight of 0.7 from the bottom upward and obtain ; $\ddot{p}_3 = .2$, $\ddot{p}_1 = .8$, $\ddot{p}_2 = 0$, $\ddot{p}_4 = 0$. In this case the valuation of alternative 1 is: $Val(A_1) = (0.2)350 + (0.8)\,250 = 270$.

Engemann and Yager [3] provide a unification and generalization of the above procedures for a decision maker with any level of optimism. The probabilities to be used in the evaluation of the alternatives can be obtained iteratively from:

$$\hat{p}_j = L_j + \min\left(\frac{w_j(\Delta_j/\Delta)}{\sum_{k=1}^{n} w_k(\Delta_k/\Delta)}(1\text{-}L),\ \Delta_j\right)$$

where the w_j are found using the maximum entropy approach of O'Hagan [4]. Table 1 displays the results of applying the Engemann and Yager [3] algorithm to the entire range of optimism of the decision maker. A pessimist has a level of optimism of 0, a risk neutral decision maker has a level of optimism of .5, and an optimist has a level of optimism of 1. It is seen in Table 1 that the same probabilities and alternative valuations determined above for the risk neutral, optimistic and pessimistic decision makers are also obtained using this more general approach. These results are also displayed graphically in Figures 1 and 2.

TABLE 1:
PROBABILITIES OF EVENTS AND EXPECTED VALUES FOR LEVELS OF OPTIMISM

LEVEL OF OPTIMISM

	0.00	0.10	0.20	0.30	0.40	0.50	0.60	0.70	0.80	0.90	1.00
S4	0.00	0.03	0.07	0.13	0.21	0.29	0.38	0.47	0.56	0.64	0.70
S2	0.00	0.02	0.02	0.03	0.04	0.04	0.04	0.04	0.03	0.02	0.00
S1	0.80	0.76	0.70	0.66	0.63	0.59	0.53	0.47	0.40	0.34	0.30
S3	0.20	0.20	0.20	0.17	0.12	0.08	0.05	0.03	0.01	0.00	0.00
EV	270	263	253	238	217	196	176	156	137	120	110

FIGURE 1:
PROBABILITIES OF EVENTS FOR VARIOUS LEVELS OF OPTIMISM

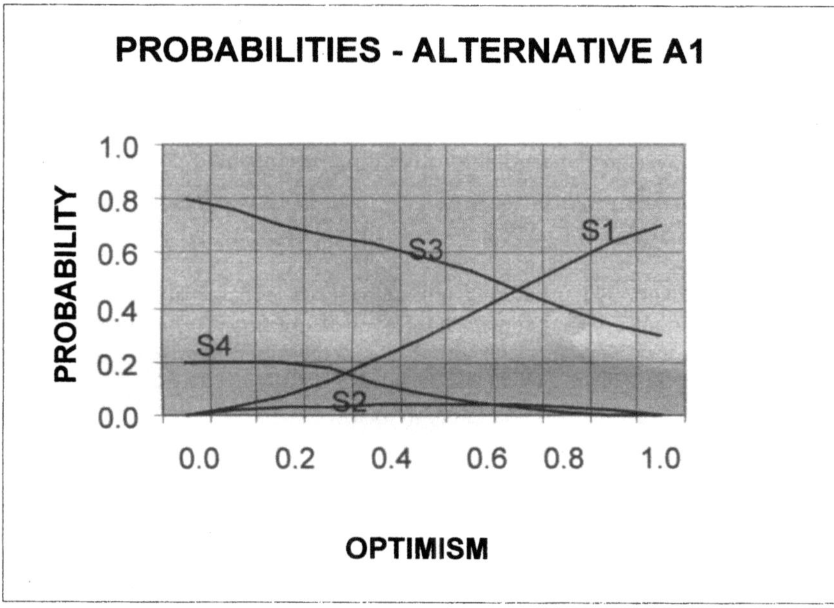

FIGURE 2:
EXPECTED VALUE FOR VARIOUS LEVELS OF OPTIMISM

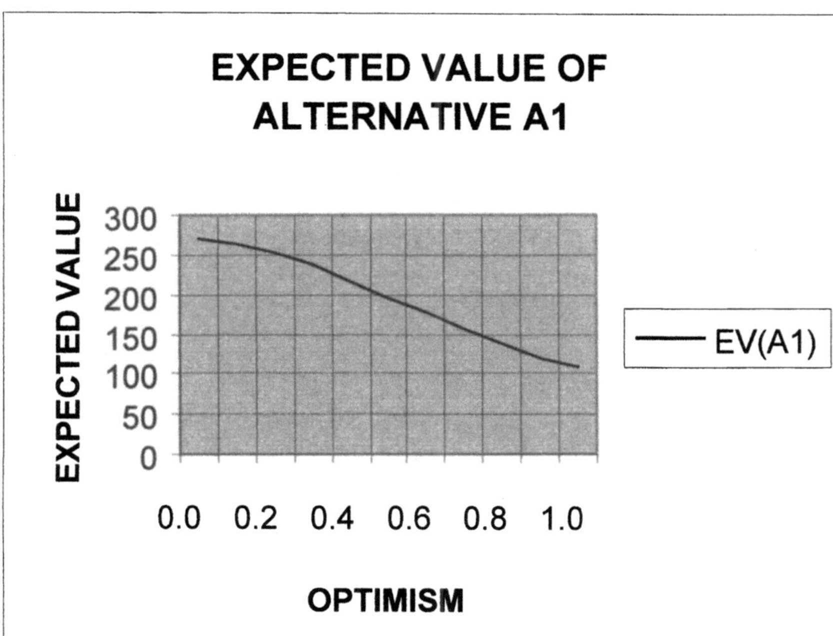

3. A Risk Management Example

In developing contingency plans for catastrophic events, many organizations must decide on appropriate strategies for backing up data processing operations. Two alternatives are Hot Sites and Cold Sites. Hot Sites are located off premises and contain a complement of hardware, telecommunications lines, software, and other supplies such that processing immediately can be moved from the primary site to the Hot Site. Cold Sites also are located off premises and contain linkages that facilitate equipment and telecommunications being installed. The Cold Site, however, must be upgraded before processing can begin. Naturally, each option has its advantages: The Hot Site option allows for immediate processing but at a greater cost; the Cold Site is less costly but - if an event occurs - may result in higher losses.

The following example illustrates a simplified view of the decisions an organization faces regarding whether to choose a Hot Site or a Cold Site strategy. For our example, which is grounded in an actual study that the authors were involved in, we assume only one type of catastrophic event can occur -- a fire. The process is a multistage decision process. Stage 1 involves the decision regarding the strategic choice -- namely, the Hot Site or the Cold Site. Given this choice, one of three states of nature can occur: No fire, a minor fire, or a major fire. We shall assume both the minor and major fires necessitate moving to the Hot or Cold Site. At the new site, the organization faces two other strategic choices: use a "Transaction" strategy, which favors customers whose business consists primarily of processing transactions, or use a "Relationship" strategy, which favors customers whose business consists primarily of relationship products. For example, in a financial services setting transaction customers would be those needing funds transfer processing, and relationship customers, would be those needing investment and financial consulting services. After the "Relationship" or "Transaction" strategy is chosen, we define four possible outcomes: The firm can suffer losses from transaction customers but not relationship customers; the firm can suffer losses from relationship customers but not transaction customers; the firm can suffer losses from both transaction customers and from relation customers, and the firm can suffer no losses from either type of customer. For short, we call these outcomes, "T," "R," "B," and "N." As is discussed below, assuming "not" is unrealistic, but may be necessary in developing point probability estimates; such assumptions need not be made when using interval probabilities. Figure 3 presents a decision tree than treats this problem using estimates for losses and point estimates for probabilities.

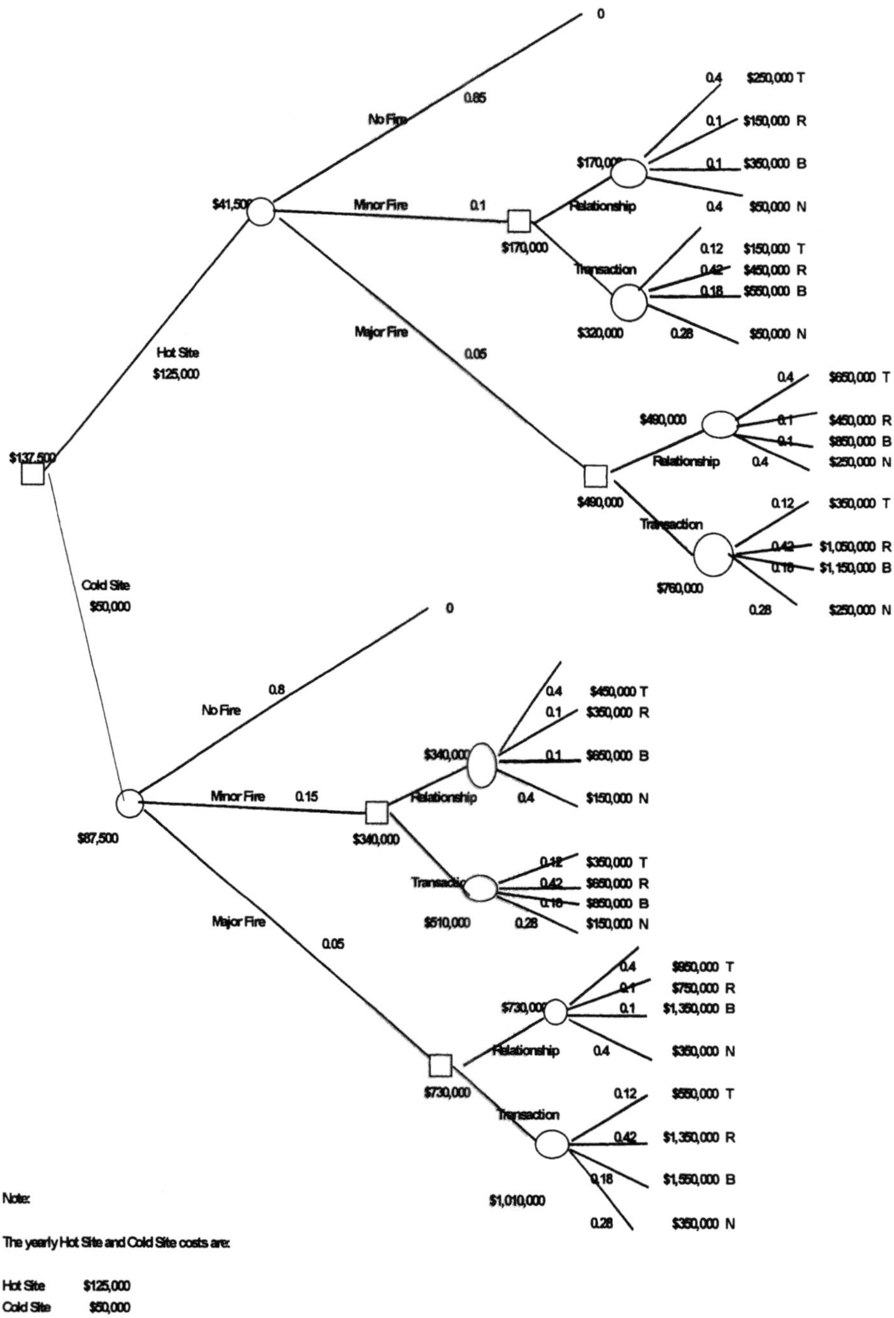

FIGURE 3:
DECISION TREE FOR HOT SITE-COLD SITE DECISION: BASE CASE

Analyzing Figure 3 indicates the optimal strategy is to choose the Cold Site option. If there is a minor fire, choose the Relationship strategy and if there is a major fire, also choose the relationship strategy. In developing the data to support this analysis, both loss estimates and probability estimates are necessary. The loss estimates appear in Table 2 and include customer losses due to lost business, losses incorporating implementing the Hot or Cold Site strategies, and direct losses due to the fire.

TABLE 2: CALCULATION OF LOSSES

Losses -- Includes lost business, fire damage, variable processing costs					
Strategy	T	R	B	N	
Relationship	$250,000	$150,000	$350,000	$50,000	Hot Site - Minor Fire
Transaction	$150,000	$450,000	$550,000	$50,000	Hot Site - Minor Fire
Relationship	$650,000	$450,000	$850,000	$250,000	Hot Site - Major Fire
Transaction	$350,000	$1,050,000	$1,150,000	$250,000	Hot Site - Major Fire
Relationship	$450,000	$350,000	$650,000	$150,000	Cold Site - Minor Fire
Transaction	$350,000	$650,000	$850,000	$150,000	Cold Site - Minor Fire
Relationship	$950,000	$750,000	$1,350,000	$350,000	Cold Site - Major Fire
Transaction	$550,000	$1,350,000	$1,550,000	$350,000	Cold Site - Major Fire

Probability estimates appear in Table 3. Using the probabilities for a Relationship loss and a Transaction loss in Table 3, probabilities of each of the four events described above can be calculated using elementary probability theory and an assumption of independence of events.

TABLE 3: CALCULATION OF POINT PROBABILITIES

	Marginal Probabilities of Losses for Various Events	
Strategy	Transaction Losses	Relationship Losses
Transaction	0.3	0.6
Relationship	0.5	0.2

	Joint Probabilities of Losses for Various Events			
Strategy	T	R	B	N
Transaction	0.12	0.42	0.18	0.28
Relationship	0.4	0.1	0.1	0.4

(Assume probabilities are the same for minor and major fires and for Hot Site and Cold Site alternatives)

As is given in the above table, to develop probability estimates we assume that the probability of a transaction Loss is exactly 0.3 when a transaction strategy is followed and 0.5 when a relationship strategy is followed. Similarly, we assume the relationship loss in each of these cases is 0.6 and 0.2 respectively. How do we know this? In practice, an analyst would collect data, make a best estimate, and perhaps perform a sensitivity analysis to see how the solution would change with changes in these probabilities.

Realistically, point estimates for events such as the above are hard to obtain. Indeed, even the events themselves as defined are unrealistic. For example, to obtain probabilities of events as in the example above, one defines events as "Transaction & Not Relationship," which implies there are transaction losses but no relationship losses. A more realistic phrasing might be "Primarily Transaction and some Relationship losses." But what does this mean and how might one obtain the requisite probabilities?

One alternative is to express the probabilities qualitatively or as a quantitative range. Table 4 contains an example of the events (as they have been redefined) in the above problem.

TABLE 4: DEFINITIONS FOR INTERVAL PROBABILITIES

FOR RELATIONSHIP STRATEGY	Verbal description of likelihood	Probability Interval
Primarily Transaction Losses	Likely	0.3 - 1.0
Primarily Relationship Losses	Very unlikely; no more than 1 in 10	0.0 - 0.1
Both Transaction and Relationship	Could happen, no more than 1 in 5	0.0 - 0.20
Minimal Losses	No real information	0.0 - 1.0
FOR TRANSACTION STRATEGY		
Primarily Transaction Losses	Somewhat unlikely	0.1 - 0.2
Primarily Relationship Losses	Most Likely	0.4 - 1.0
Both Transaction and Relationship	Might happen	0.0 -0.5
Minimal Losses	Might happen	0.0 -0.5

In addition of defining interval probabilities for these events, interval probabilities can be defined for the various magnitudes of fire. We will use the following probability intervals:

Fire Magnitude	Verbal description	Probability Interval
None	By far the most likely	0.5 - 1.0
Minor	Could happen every 5-10 years	0.1 - 0.2
Major	Very unlikely	0.0 - 0.1

Using the methodology described by Engemann and Yager [3] we can analyze the sequence of decisions outlined in the decision tree as a function of the degree of optimism of the decision-maker.

Tables 5, 6, 7 and 8 present the resulting probabilities and valuations for Relationship and Transaction Strategies given a Minor Fire and Major Fire and Utilizing a Hot Site.

TABLE 5:
PROBABILITIES OF EVENTS AND EXPECTED VALUES FOR LEVELS OF OPTIMISM
(Relationship Strategy – Minor Fire – Hot Site)

LEVEL OF OPTIMISM

	0.00	0.10	0.20	0.30	0.40	0.50	0.60	0.70	0.80	0.90	1.00
N	0.00	0.03	0.07	0.13	0.21	0.29	0.38	0.47	0.56	0.64	0.70
R	0.00	0.02	0.02	0.03	0.04	0.04	0.04	0.04	0.03	0.02	0.00
T	0.80	0.76	0.70	0.66	0.63	0.59	0.53	0.47	0.40	0.34	0.30
B	0.20	0.20	0.20	0.17	0.12	0.08	0.05	0.03	0.01	0.00	0.00
EV	270	263	253	238	217	196	176	156	137	120	110

TABLE 6:
PROBABILITIES OF EVENTS AND EXPECTED VALUES FOR LEVELS OF OPTIMISM
(Transaction Strategy – Minor Fire – Hot Site)

LEVEL OF OPTIMISM

	0.00	0.10	0.20	0.30	0.40	0.50	0.60	0.70	0.80	0.90	1.00
N	0.00	0.01	0.02	0.06	0.10	0.16	0.22	0.30	0.37	0.45	0.50
T	0.10	0.10	0.11	0.12	0.13	0.13	0.13	0.14	0.13	0.12	0.10
R	0.40	0.49	0.54	0.56	0.56	0.56	0.54	0.51	0.47	0.43	0.40
B	0.50	0.40	0.33	0.27	0.21	0.16	0.11	0.06	0.03	0.01	0.00
EV	470	456	439	418	393	364	331	297	264	235	220

TABLE 7:
PROBABILITIES OF EVENTS AND EXPECTED VALUES FOR LEVELS OF OPTIMISM
(Relationship Strategy – Major Fire – Hot Site)

LEVEL OF OPTIMISM

	0.00	0.10	0.20	0.30	0.40	0.50	0.60	0.70	0.80	0.90	1.00
N	0.00	0.03	0.07	0.13	0.21	0.29	0.38	0.47	0.56	0.64	0.70
R	0.00	0.02	0.02	0.03	0.04	0.04	0.04	0.04	0.03	0.02	0.00
T	0.80	0.76	0.70	0.66	0.63	0.59	0.53	0.47	0.40	0.34	0.30
B	0.20	0.20	0.20	0.17	0.12	0.08	0.05	0.03	0.01	0.00	0.00
EV	690	676	656	627	585	543	502	461	424	390	370

TABLE 8:
PROBABILITIES OF EVENTS AND EXPECTED VALUES FOR LEVELS OF OPTIMISM
(Transaction Strategy – Major Fire – Hot Site)

LEVEL OF OPTIMISM

	0.00	0.10	0.20	0.30	0.40	0.50	0.60	0.70	0.80	0.90	1.00
N	0.00	0.01	0.02	0.06	0.10	0.16	0.22	0.30	0.37	0.45	0.50
T	0.10	0.10	0.11	0.12	0.13	0.13	0.13	0.14	0.13	0.12	0.10
R	0.40	0.49	0.54	0.56	0.56	0.56	0.54	0.51	0.47	0.43	0.40
B	0.50	0.40	0.33	0.27	0.21	0.16	0.11	0.06	0.03	0.01	0.00
EV	1030	1012	985	948	902	849	789	725	662	608	580

Note from the above results that a Relationship Strategy is preferred for both Minor and Major Fires given the use of a Hot Site for all levels of optimism of the decision maker. Given these results, we apply the same method to determine the valuation of the Hot Site alternative. Table 9 displays the results and adds the fixed cost of $125,00 for the hot site. Likewise Table 10 displays the results of the same analysis on the Cold Site alternative with the $50,000 fixed cost added. Figure 4 illustrates a Hot Site is preferred if the decision maker's optimism is less than .4, and if the decision maker's optimism is greater than .4 a Cold Site is preferred.

TABLE 9:
PROBABILITIES OF EVENTS AND EXPECTED VALUES FOR LEVELS OF OPTIMISM
(Hot Site)

LEVEL OF OPTIMISM

HOT SITE	0.00	0.10	0.20	0.30	0.40	0.50	0.60	0.70	0.80	0.90	1.00
NO FIRE	0.70	0.72	0.74	0.75	0.78	0.80	0.82	0.84	0.86	0.88	0.90
MINOR FIRE	0.20	0.18	0.16	0.15	0.15	0.15	0.14	0.14	0.13	0.12	0.10
MAJOR FIRE	0.10	0.10	0.10	0.10	0.07	0.05	0.03	0.02	0.01	0.00	0.00
	0.00	0.00	0.00	0.00	0.00	0.00	0.00	0.00	0.00	0.00	0.00
EV	123	114	106	96	74	57	42	31	22	15	11
FIXED COST	125	125	125	125	125	125	125	125	125	125	125
TOTAL COST	248	239	231	221	199	182	167	156	147	140	136

TABLE 10:
PROBABILITIES OF EVENTS AND EXPECTED VALUES FOR LEVELS OF OPTIMISM
(Cold Site)

LEVEL OF OPTIMISM

COLD SITE	0.00	0.10	0.20	0.30	0.40	0.50	0.60	0.70	0.80	0.90	1.00
NO FIRE	0.70	0.72	0.74	0.75	0.78	0.80	0.82	0.84	0.86	0.88	0.90
MINOR FIRE	0.20	0.18	0.16	0.15	0.15	0.15	0.14	0.14	0.13	0.12	0.10
MAJOR FIRE	0.10	0.10	0.10	0.10	0.07	0.05	0.03	0.02	0.01	0.00	0.00
	0.00	0.00	0.00	0.00	0.00	0.00	0.00	0.00	0.00	0.00	0.00
EV	201	186	172	156	124	96	74	56	42	31	24
FIXED COST	75	75	75	75	75	75	75	75	75	75	75
TOTAL COST	276	261	247	231	199	171	149	131	117	106	99

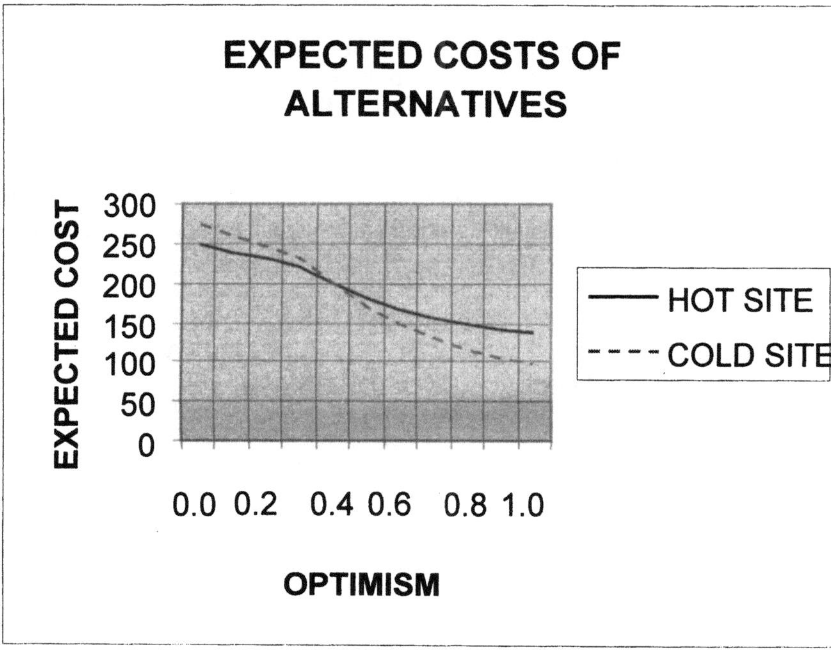

FIGURE 4:
SENSITIVITY OF HOT SITE-COLD SITE DECISION TO OPTIMISM

4. Conclusion

We have introduced a new method to analyze sequential decisions involving imprecise information regarding event likelihoods. A key ingredient in the methodology is the decision maker's knowledge and degree of optimism. The methodology facilitates sensitivity analysis in this environment.

5. References

[1]. Miller, H. and Engemann, K., "A Methodology for Managing Information-based Risk," Information Resources Management Journal, 9:2; 17-24, 1996.

[2]. Yager, R. and Kreinovich, V., "Decision Making Under Interval Probabilities," International Journal of Approximate Reasoning, 22, 195-215, 1999.

[3]. Engemann, K., and Yager, R., "On a General Approach to Decision Making Using Interval Probabilities," Technical Report #MII-1950, Machine Intelligence Institute, Iona College, New Rochelle, NY, 1999.

[4]. O'Hagan, M., "Using Maximum Entropy-ordered Weighted Averaging to Construct a Fuzzy Neuron," Proceedings of the 24th Annual IEEE Asilomar Conference on Signals, Systems and Computers, Pacific Grove, CA, 618-623, 1990.

Fuzzy Risk Analysis Vs. Probability Risk Analysis *

Huang Chongfu[1,2]

[1]Institute of Resources Science, Beijing Normal University
[2] Key Laboratory of Environmental Change and Natural Disaster
The Ministry of Education of China
No.19 Xinjiekouwai Street, Beijing 100875, China
hchongfu@bnu.edu.cn

In this chapter, we discuss the essential difference between fuzzy risk analysis and probability risk analysis. Then, we use the method of the information distribution to improve probability estimate in the probability risk analysis, and we develop the method to calculate the fuzzy risk with respect to the possibility-probability. The benefit of fuzzy risk assessment is that the new result saves more information for risk management.

1 Risk Analysis

Hardly a week goes by in our society without news of yet another hazard. Earthquake occurring, flooding, nuclear leaking out, food contaminated, environment pollution, air disaster, banking crisis, business bankrupt, traffic accident, all of them are terrible things. If we could predict these adverse events, it would become much easier to avoid the loss from them. However, in some degree, it is almost impossible. Any adverse event with uncertainty is related to the terminology "risk."

Everyone is constantly subjected to many risks, both as individuals and as members of various societal groups. There are voluntary risks which individuals elect to assume, such as those due to personal smoking, and involuntary risks which individuals do not elect to assume, such as those due to being forced to nuclear waste or flood. There are very few situations which create no risk to individuals or members of various societal groups.

Since we celebrated the first Earth Day in 1970, we have made considerable progress in the quality of our environment – air, water, land, and other natural resources. And yet myriad new concerns have emerged: second-hand tobacco smoke, cellular phones, and global warming, to name just a few. Concerns about industrial accidents, chemical and food safety, and natural disasters also have intensified. The heightened public sensitivities to multiple

* Project Supported by National Natural Science Foundation of China, No.49971001

environmental hazards, along with perceptions that the hazards are getting out of government control, spawned a number of public "right to know" and "right to know more" movements. The abundance of information that surfaced, however, proved to be more baffling than enlightening – not really surprising given the degree of scientific uncertainty, the divers interpretations of the same data by experts, and the billions of dollars at stake [15].

Terminologies "risk assessment" and "risk management" frequently appear in the risk literatures.

Risk assessment provides a systematic framework, grounded in scientific principles, to understand diverse risks. *Risk management* is the process of evaluation and, if necessary, controlling source of exposure (i.e., potential receptors) and risk (as hazardous materials). Sound environmental risk management, whether corporate or regulatory, means weighing many different attributes of a decision and developing alternatives. In risk assessment and management, we look at a situation or scenario and ask these types of questions: what can go wrong and why, how likely is it, how bad can it be, and what can we do about it? Risk, then, is a function of the natural of the hazard, accessibility or avenue of contact (exposure potential), characteristics of the exposed populations (receptors), the likelihood of occurrence, and the magnitude of exposure and consequences, as well as public values.

Table 1 presents a classical overview of safety, health, and ecological risk assessments including major process steps.

Table 1 Classical overview of three major types of risk assessments

Step	Safety	Human health	Ecological/environment
Step 1	Hazard identification	Data analysis/ hazard identification	Problem formulation (hazard screening)
Step 2	Probability/frenquency estimation of causes	Exposure assessment	Exposure assessment
Step 3	Consequence analysis	Dose-response or toxicity assessment	Toxicity effects assessment
Step 4	Risk evaluation	Risk characterization	Risk Characterization

Risk criteria provide an effective frame of reference for prioritizing problems, allocating resources, and reducing risks. However, there is a crucial limitation to realizing the full potential of risk assessment. That is the paucity of professionals who have the broad training and the perspective needed to transcend the "cookbook" approach and focus on the critical issues. Many believe that risk expressions are an attempt to obfuscate responses to a simple question:"Is it safe or unsafe?" In fact, it is very difficult to answer the question due to another limitation of risk assessment that no broad consensus on the purpose, the approach, or the result; inadequate data, speculative and myopic nature of assumptions.

The scientific information provided by risk assessment is but one input to the process. Other criteria include politics, economics, competing risks, and equity and other social concerns. Although risk assessment is rooted in the science, how useful its results are to risk management depends on the questions it is designed to answer, how it is conducted, and the way it is structured. Unfortunately, too many risk assessments prove to be of little value to risk managers because of inadequate planning.

Risk assessment and ***risk analysis*** are often used synonymously, as in this chapter, but risk analysis is sometimes used broadly to include risk management aspects as well.

2 Probability Risk Analysis and Fuzzy Risk Analysis

The traditional approach used to model risky choicemaking situations is to describe choices involving risk in term of their underlying probability distributions and associated utilities. In many literatures, risk is regarded as a product of severity of loss and likelihood of loss and the likelihood is measured by probability.

Intuitively, risk exists when loss is possible and its financial impact is significant. This linguistic definition captures a property of risk that eludes definition in terms of mathematical formulas. In general, risks are evaluated qualitatively rather than analyzed quantitatively [17]. In fact, as Jablonowski [12] stated, in the real wold, the possibility and financial significance of loss cannot be defined with precision.

It may be argued that only the statistical risk involves a measure of the probability and severity of adverse impacts [11].

In fact, the probabilistic method is an extremely effective tool to study risk only in the case when a great amount of data can be collected. However, even to issues with respect to statistical risk, the available data are insufficient to permit estimating reliably the frequencies of release of risk agents or other characteristics of concern.

Safety risk analysis tend to be much more probabilistic, incorporating the likelihood of initiating events, as well as the likelihood of exposures and the range of consequences. Human health and ecological risk assessments are typically deterministic and use single-point estimate where valuable information remains hidden from risk managers.

Traditionally, the major challenge in risk analysis is considered to find a scientific approach to estimate some probability distribution. It is true, if and only if the risks in a system are statistical risks. However, in many risk systems, randomness is just one of risk natures. Risk essence is based on all risk natures. For studying the essence, firstly, we view the aim of risk analysis and the situation we usually meet for a practical system.

It may be argued that, the aim of risk analysis is to describe or understand some states of a system serving for risk management to reduce or control loss. The states are relative to adverse outcomes. Therefore, it is necessary for risk analysis to show the relations among states, time, inputs and so on. A probability distribution is just a relation between events and probabilities of occurrence, where an event and a probability value can be regarded as a state and an input, respectively. For many systems, it is impossible to precisely calculate the relation, and we face the problem of imprecise probability. In other words, the aim of risk analysis is to answer how and why an adverse outcome produces. Based on the view, it may be argued that, **risk essence is dynamics of adverse outcomes**. In fact, a risk system could be studied with some state equations if we can find them.

In many cases, it is difficult to obtain the state equations and all data we need, and it may be unnecessary to study the equations over. Probabilistic methods simplify the procedure. However, it isn't reasonable to replace risk analysis by probability analysis.

When we study a risk system using a probabilistic method, usually, it is difficult to judge if a hypothesis of probability distribution is suitable, and sometimes we may meet the small-sample problem, where the data is too scanty to make a decision in any classical approach. It means that it is difficult to obtain a precise relation between events and probabilities of occurrence.

In the decision theory, we refer risk to a three-dimensional concept involving the following natures:

(N1) Adverse outcome for individuals;

(N2) Uncertainty over the occurrence, timing, site, or magnitude of that adverse outcome;

(N3) Complexity to show precisely by a state equation or a probability distribution.

Clearly, due to N3, we know that, risk is a complex phenomenon. When the complex nature is ignored, the concept of risk may reduced to be probability-risk, which means that we can find a probability distribution, obeying a statistical law, to show the risk phenomenon properly. As stated above, it is impossible to obtain a precise probability estimator in many cases.

The concept of fuzzy risk is not strange to us. However, most of us may consider the concept as a copy of simple expansion such as fuzzy group, fuzzy ring or fuzzy reliability. In the attitude, we can do noting to promote risk science with fuzzy set theory.

Risk, as a natural or societal phenomenon, is neither precise nor fuzzy. Risk is expressed in terms of the probability-risk only when we know that a risk phenomenon can be studied by a probability method.

Fuzzy risk is an engineering concept which emphasizes showing. For risk, existence differs from showing. Similarly, as we know well that, mass is a pure physical concept, and weight is one in engineering to show mass. Therefore, roughly speaking, **fuzzy risk can be defined as an approximate representation to show risk with fuzzy theory and techniques.** In general, a fuzzy risk is a fuzzy relation between loss events and concerning factors.

3 A Survey of Fuzzy Risk Analysis

Since the publication of the first paper on fuzzy sets by Zadeh [19] in 1965, there were many researchers [1,3,5–8] who use fuzzy methods to risk problems.

The oldest approach to fuzzy risk analysis is based on the premise that one can provide the input of natural language estimate as probability of failure, severity of loss and reliability of the estimate. A typical technique, suggested by Schmucker [18], is combining the fuzzy risk of subsystem to calculate the fuzzy risk of the entire system.

Let A be a fuzzy event such as "Johe may lose around 3000 US$", and B be another fuzzy event such as "Mary may lose around 2000 US$." Where "around" is a fuzzy concept. Suppose we have known the probabilities of A and B occurring, which are $P(A)$ and $P(B)$. More general, the probabilities can be represented by natural language or fuzzy numbers. Then, using the combining model, we can calculate $P(A \cup B)$.

Recent years, the premise is unnecessary. A method has been proposed by Machias and Skikos [16] for the computation of the fuzzy risk index of wind site, where the installation of wind energy conversion systems seems to be very possible in the near future. The proposed method is based on a heuristic rules matrix, which is a relation among linguistic terms of wind speed, frequency and risk value. The matrix which is defined by the authors shows the way that the two sets of independent information for the wind speed and its frequency of occurrence will be combined to yield the combined risk level in a classical risk analysis approach. According to these of the extreme wind events and the matrix, one can calculate membership values and risk index results. The major advantage of the method is that, in the cases where the classical one computes the same risk value, however these sites have completely different risk when fuzzy method is applied. The proposed approach presents a better sensitivity in ranking the wind site according to their risk index.

A fuzzy formalism [12] opens the way for computerization of the process of risk analysis, using a variety of AI techniques. Where, risk is defined, with respect to any probability(p)/loss(l) combination, as:

$$< \text{RISK}_{p,l} >=< \text{POSSIBLE}_{p,l} > \cap < \text{SIGNIFICANT FINANCIAL IMPACT}_{p,l} >$$

where $<>$ denotes a fuzzy set, and $\cap$ is the symbol for the set intersection. Here, the intersection of the fuzzy sets *possible* and *significant financial impact* is defined by multiplying the two. The components of risk are combined to produce the definition of risk in probability-loss space and the results appear as a three-dimensional membership function. Based on the definition, contour lines, such as those on a topographic map, can be employed to display membership of risk. Considering the determinations of probabilities of loss themselves being imprecise and fuzzy and combining the fuzziness of the definition of risk with the fuzziness inherent in the measurement of probabilities associated with exposure to accidental loss, the author has given something called as a fuzzy risk profile. A neural network model suggested to learn the fuzzy risk profiles to help select insurance policy limits for novel exposures.

For dealing with decision-making problems in which the decision maker has a vague and incomplete information about results and external factors, basic decision rules based on fuzzy risk intervals are developed by Delgado, Verdegay and Vila [4]. Where, from the point of view of semantic, a trapezoidal membership function is to be considered as a fuzzy risk interval.

All of these models can be regarded as fuzzy classifiers in risk space. The fuzzy risk index model of Machias and Skikos divides speed-frenquency space of wind into risk and no risk zone in some degree. The fuzzy formalism model of Jablonowski defined a fuzzy set in probability-loss space. And, the fuzzy risk interval model of Delgado and *et al.* just replace classical intervals by fuzzy numbers. A classical interval is a hard classifier and a fuzzy number is a soft classifier

These fuzzy models open ways for computerization of the process of risk analysis, using a variety of classical fuzzy techniques. However, they have not touched the essence of risk, to say nothing of both of risk and fuzzy.

A fuzzy mathematical model, which differs from above ones, has been suggested by Huang [13] to assess risk of urban natural hazard when there is not enough information. The model takes account of the comprehensive analyses on single hazard factor, hazard-effected body and urban whole system. It proves that fuzzy approach is the best choice to deal with the relationships between earthquake and disaster. The success from the model testify that, fuzzy set theory is an ideal method of incorporation inherent imprecision into risk analysis because our intuitions of risk do not translate well into precise mathematical formulas. The model is grounded in the *principle of information diffusion* [2,9] by which we can change the crisp observations of a given sample into fuzzy sets so that the information gaps caused by incompleteness can be filled, in some degree. Then a more accurate probability estimate for risk analysis can be obtained. Recent studies show that the method of information distribution from the principle can also be employed to calculate the fuzzy risk with respect to the so-called possibility-probability.

4 Principle of Information Diffusion

To help the readers who never touch the principle of information diffusion, in this section, we outline the principle and give a simple model called the *method of information distribution.*

4.1 Defintion

Suppose X represents a set of observations $x_1, x_2, \cdots, x_n$ on a given universe $U = \{u\}$.

Definition 1 A mapping μ_x from $X \times U$ to [0,1]:

$$\begin{aligned} \mu_x : X \times U &\to [0,1] \\ (x,u) &\mapsto \mu_x(u), \forall (x,u) \in X \times U \end{aligned}$$

is called information diffusion of X on U, if and only if it satisfies:
(1) For $x \in X$, $\exists u_0 \in U$ such that $|u_0 - x| = \inf\{|u - x| \; |u \in U\}$ and $\mu_x(u_0) = \sup\{\mu_x(u)|u \in U\}$;
(2) For every $x \in X$, $\mu_x(u)$ is a convex function about u;

$\mu_x(u)$ is called an *information diffusion function* of X on U.

Definition 2 If a relation R is constructed by X with an operation γ directly, then R is called a *non-diffusion estimator*, which is written as $R = \gamma(X)$.

Definition 3 If a relation R is constructed by X with an information diffusion function μ_x and an operation β, then R is called an *information diffusion estimator*, which is written as $R = \beta \circ \mu_x(X)$.

The principle of the information diffusion is an assertion that **there must exist a reasonable information diffusion function μ_x to change an observation x into many parts to partly fill the gaps caused by incomplete data and improve non-diffusion estimator.**

The principle of information diffusion holds, at the least, in the case of estimating a *probability density function* if we take information diffusion functions which satisfy the conditions as same as that in the kernel estimator, and use the formulae which are employed in the kernel estimator.

The simplest information diffusion function is the distribution function with respect to the information distribution.

4.2 Information Distribution

Let $X = \{x_1, \cdots, x_n\}$ be a sample observed from an experiment, and $U = \{u_1, u_2, \cdots, u_m\}$ be the discrete universe of X. This experiment has a number (possibly infinite) of *outcomes*. The *population* of the experiment is the collection of all possible outcomes. We assume that X is a *random sample*, to guarantee that every element of the population has the same chance of being included. We also assume that $x_i (i = 1, 2, \cdots, n)$ are *independent and identical distribution (i.i.d.)*. x_i is called a sample point.

Definition 4 A mapping from $X \times U$ to [0,1]

$$\begin{aligned} \mu : X \times U &\to [0,1] \\ (x,u) &\mapsto \mu(x,u), \forall (x,u) \in X \times U \end{aligned}$$

is called an *information distribution* of X on U, if $\mu(x,u)$ has the following properties :

(1) $\forall x \in X$, if $\exists u \in U$, such that $x = u$, then $\mu(x,u) = 1$, i.e., μ is reflexive;
(2) For $x \in X$, $\forall u', u'' \in U$, if $||u' - x|| \le ||u'' - x||$ then $\mu(x,u') \ge \mu(x,u'')$, i.e., μ is decreasing when $\| x - u \|$ is increasing;
(3) $\sum_{j=1}^{m} \mu(x_i, u_j) = 1$, $i = 1, 2, \cdots, n$, i.e., conservation of information.

In Definition 4, $u_j, j = 1, 2, \cdots, m$ are called the *controlling points*. μ is called a *distribution function* of X on U. We say that observation x_i gives information, gain at $q_{ij} = \mu(x_i, u_j)$, to controlling point u_j. q_{ij} is called *distributed information* on u_j from x_i. U is also called the *framework space*. An information distribution hangs a sample on a chosen framework space.

Let $Q_j = \sum_{i=1}^{n} q_{ij}, j = 1, 2, \cdots, m$. We say that sample X provides information in the total gain Q_j to controlling point u_j. Q_j is also called the *total distributed information* to controlling point u_j. $Q = (Q_1, Q_2, \cdots, Q_m)$ is called the *primary information distribution* of X on U.

Obviously, for x, the mapping $\mu(x,u)$ defines a fuzzy set of U. Sometimes, we use $\mu_x(u)$ to represent the fuzzy set.

The simplest distribution function is the 1-dimension linear-information-distribution.

Definition 5 Let $X = \{x_1, x_2, \cdots, x_n\}$ be a 1-dimension random sample, i.e., $x_I \in \mathbb{R}$, and $U = \{u_1, u_2, \cdots, u_m\}$ be the chosen framework space with $u_j - u_{j-1} \equiv \Delta$, $j = 2, 3, \cdots, m$. For any $x \in X$, and $u \in U$, the following formula is called *1-dimensional linear-information-distribution*:

$$\mu(x,u) = \begin{cases} 1 - \frac{|x-u|}{\Delta}, & \text{for } |x-u| \le \Delta, \\ 0, & \text{otherwise.} \end{cases} \tag{1}$$

Where, Δ is called step length. In short, this μ is called the *linear distribution*.

5 Fuzzy Model to Improve Probability Estimate

5.1 Benefit from Linear Distribution

We use an example to show the benefit of linear distribution for probability distribution estimation serving for risk assessment.

When we only have a given sample, without any assumption on the sample, the simplest method to estimate a probability distribution is the histogram.

A *histogram* is a graph of grouped (binned) data in which the number of values in each bin is represented by the area of a rectangular box.

A *frequency histogram* is a bar graph constructed in such a way that the area of each bar is proportional to the number of observations in the category that it represents. We read the actual number of observations per class from a frequency histogram.

A *relative frequency histogram* is a bar graph constructed in such a way that the area of each bar is proportional to the fraction of observations in the category that it represent.

According to the definition of the probability density function, we can develop a relative frequency histogram by dividing the width of the bin to be an estimator about probability distribution.

Example 1 We have 24 seismic records observed in China from 1900 to 1975 with magnitudes shown in Table 2 and epiceptral intensity $I_0 = VIII$.

Table 2. Magnitude records with epiceptral intensity $I_0 = VIII$

i	1	2	3	4	5	6	7	8	9	10	11	12
x_i	6.4	6.4	6.0	6.0	7.0	6.25	5.75	5.8	6.5	5.8	6.0	6.5
i	13	14	15	16	17	18	19	20	21	22	23	24
x_i	6.3	6.0	6.0	6.0	6.8	6.4	6.5	6.6	6.5	6.0	5.5	6.0

Given an origin $x_0 = 5.4$, a bin width $h = 0.4$, for positive integers $m = 5$, we can construct the histogram on the intervals:

$$I_1 = [5.4, 5.8[,\ I_2 = [5.8, 6.2[,\ I_3 = [6.2, 6.6[,\ I_4 = [6.6, 7.0[,\ I_5 = [7.0, 7.4[$$

We now count the observations falling into each interval. The observations that fall into interval $I_1 = [5.4, 5.8[$ are $x_{23} = 5.5$ and $x_7 = 5.75$. That is, there are two observations, which fall into this interval. We do the same for each interval. The result of these counts are recorded in Table 3 in the column labeled frequency.

Table 3. Frequency distribution of the magnitude records

Interval	Midpoint	Frequency
$[5.4, 5.8[$	5.6	2
$[5.8, 6.2[$	6.0	10
$[6.2, 6.6[$	6.4	9
$[6.6, 7.0[$	6.8	2
$[7.0, 7.4[$	7.2	1

From this frequency table, we can construct the frequency histogram. The histogram, shown in Fig. 1, consists of five vertical bars each of width 0.4, each centered at its respective interval mark. The height of each bar is equal to the interval frequency, the number of observations per interval.

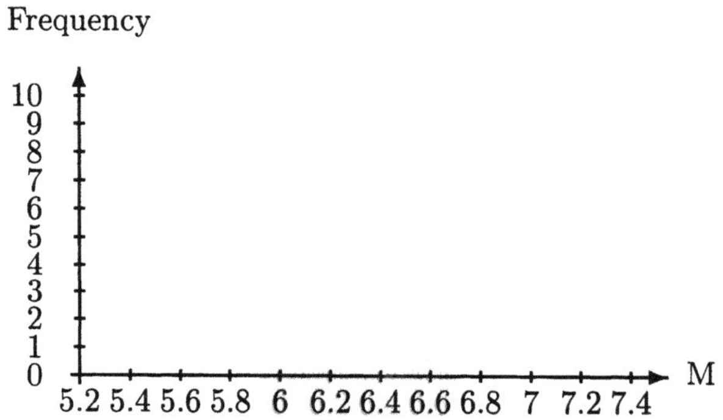

Fig. 1 Frequency histogram for a sample of observations on magnitude.

Dividing the number of observations per interval by the total number 24 of observations, we obtain a relative frequency histogram shown in Fig. 2. Note that the vertical axes is labeled 0 to 1.

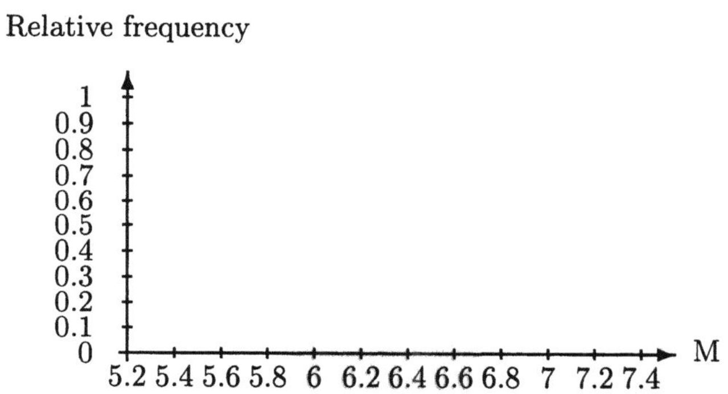

Fig. 2 Relative frequency histogram on magnitude.

We use $u_1, u_2, \cdots, u_m$ to denote the midpoints of the intervals in the histogram. They are the standard points of the intervals. Dividing the relative frequency per interval by the bin width, we write the result by $\hat{p}(u_i)$, $i = 1, 2, \cdots, m$. Plotting $(u_i, \hat{p}(u_i))$, $i = 1, 2, \cdots, m$ and linking them, we obtain an estimate of probability distribution $p(x)$ of the population. For simplicity, $(u_i, \hat{p}(u_i))$, $i = 1, 2, \cdots, m$ and the curve produced from them are often considered to be same.

Definition 6 Let $X = \{x_1, x_2, \cdots, x_n\}$ be a given sample drawn from a population with probability distribution function (pdf) $p(x)$ in terms of the density function. Given an *origin* x_0 and a *bin width* h, we define the *bins* of the histogram to be the intervals $[x_0 + mh, x_0 + (m + 1)h[$ for positive and negative integers m. The intervals are chosen to be closed on the left and open on the right.

$$\hat{p}(x) = \frac{1}{nh}(\text{number of } x_i \text{ in the same bin as } x). \tag{2}$$

is called a *histogram estimate* (HE) of $p(x)$.

From the data in Table 2, we can obtain a histogram estimate of $p(x)$ that an earthquake in magnitude x causes the damage measured by epiceptral intensity $I_0 = VIII$. The estimate is shown in Fig. 3.

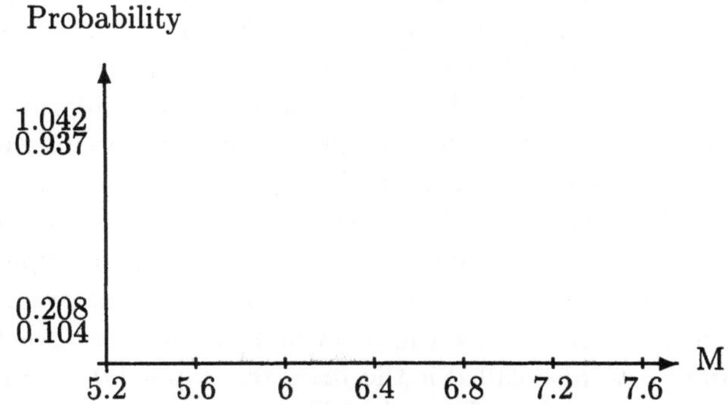

Fig. 3 Histogram estimate of the probability distribution to $I_0 = VIII$.

We know that, if the intervals in Fig. 1 are combined into a large one [5.4,7.4[, the resulting histogram becomes one bar, which cannot show any gradient. In other words, if the intervals in a histogram are too large, we can learn nothing from the given sample. In fact, the larger the intervals are, the rougher the estimation of probability distribution is. Oppositly, if we take smaller intervals, what will happen?

In Fig. 1, the domain is divided into five intervals. Now, we divide it into six intervals

$$I_1 = [5.40, 5.73[, \quad I_2 = [5.73, 6.07[, \quad I_3 = [6.07, 6.40[,$$

$$I_4 = [6.40, 6.73[, \quad I_5 = [6.73, 7.07[, \quad I_6 = [7.07, 7.40[,$$

and obtain a correspending histogram shown in Fig. 4.

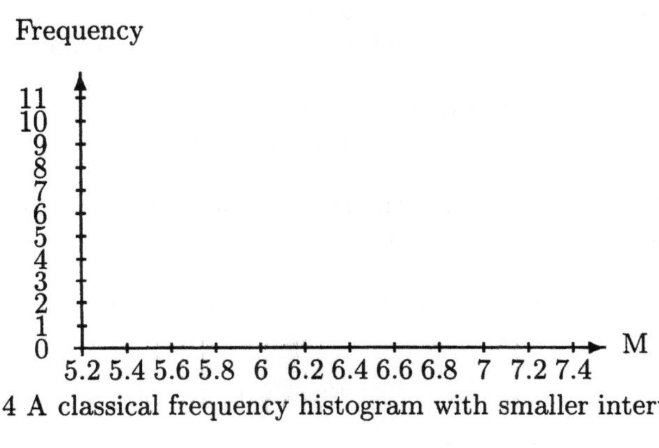

Fig. 4 A classical frequency histogram with smaller intervals.

It is waving with two crests (at I_2, I_4) and one trough (at I_3), which also cannot show any statistical law.

The reason is that, in the model of classical histogram, all observations falling into a same interval are considered to play a same role. It neglects the difference between them. In fact, the records may occupy different positions in the interval. For example, both of $x_1 = 6.4$ and $x_9 = 6.5$ belong to interval $I_4 = [6.40, 6.73[$, but their positions in I_4 are different. The former is located at the boundary, and the latter is near the center of the interval. Neglecting the position difference implies that we throw away some information.

If the sample size is large, by the central limit theorem, the neglected information is insignificant. However, for a small sample, we have to pay attention to each observation. Any observation may be near the center or boundary of an interval. A little random disturbance in the experiment may make the observations near the boundaries move other intervals. The information showing position is called the *transition information.* Due to fuzziness of a small sample, we also called it the *fuzzy transition information.*

It is easy to pick up the information, if we change observations as singletons into general fuzzy sets. The mechanism of the performance is to use fuzzy bounds instead of crisp bounds of the intervals in the classical histogram. The new histogram is called the soft frequency histogram.

Definition 7 Suppose a classical frequency histogram be constructed on m intervals $I_1, I_2, \cdots, I_m$ with width h. Let u_j be the center point of interval I_j. We use $U = \{u_1, u_2, \cdots, u_m\}$ to be the framework space with step length $\Delta = h$. Let

$\widetilde{H}(x)$ = total provided information to controlling point in the same interval as x.

which is called a *soft frequency histogram* of X on $\mathbb{R}$.

When we have got Q of X on U, the soft histogram, in fact, is

$$\forall x \in I_j, \quad \widetilde{H}(x) = Q_j,$$

Example 2 Let us remove the classical histogram in Fig. 4 into a soft histogram. Where, X is given in Table 2, $n = 24$, $m = 6$, $\Delta = h = (7.4 - 5.4)/6 = 0.3\dot{3}$, and the corresponding set of controlling points is

$$\begin{aligned} U &= \{u_1, u_2, u_3, u_4, u_5, u_6\} \\ &= \{5.57, 5.90, 6.24, 6.57, 6.90, 7.24\}. \end{aligned}$$

Using the linear distribution formula in (1), we obtain all distributed information q_{ij} shown in Table 4. For instance, for $x_3 = 6$, because

$$|x_3 - u_2| = |6 - 5.9| = 0.1 < \Delta, \text{ and } |x_3 - u_3| = |6 - 6.24| = 0.24 < \Delta,$$

and

$$|x_3 - u_1|, |x_3 - u_4|, |x_3 - u_5|, |x_3 - u_6| > \Delta,$$

we have

$$q_{32} = \mu(x_3, u_2) = 1 - \frac{|x_3 - u_2|}{\Delta} = 1 - \frac{|6 - 5.9|}{0.33} = 1 - \frac{0.1}{0.33} = 0.697,$$

$$q_{33} = \mu(x_3, u_3) = 1 - \frac{|x_3 - u_3|}{\Delta} = 1 - \frac{|6 - 6.24|}{0.33} = 1 - \frac{0.24}{0.33} = 0.273,$$

$$q_{3j} = \mu(x_3, u_j) = 0, j = 1, 4, 5, 6.$$

Table 4 Distributed information on u_j from x_i

q_{ij}	u_1 5.57	u_2 5.90	u_3 6.24	u_4 6.57	u_5 6.90	u_6 7.24
$x_1 = 6.4$	0	0	0.515	0.485	0	0
$x_2 = 6.4$	0	0	0.515	0.485	0	0
$x_3 = 6.0$	0	0.697	0.273	0	0	0
$x_4 = 6.0$	0	0.697	0.273	0	0	0
$x_5 = 7.0$	0	0	0	0	0.697	0.273
$x_6 = 6.25$	0	0	0.970	0.030	0	0
$x_7 = 5.75$	0.455	0.545	0	0	0	0
$x_8 = 5.8$	0.303	0.697	0	0	0	0
$x_9 = 6.5$	0	0	0.212	0.788	0	0
$x_{10} = 5.8$	0.303	0.697	0	0	0	0
$x_{11} = 6.0$	0	0.697	0.273	0	0	0
$x_{12} = 6.5$	0	0	0.212	0.788	0	0
$x_{13} = 6.3$	0	0	0.818	0.182	0	0
$x_{14} = 6.0$	0	0.697	0.273	0	0	0
$x_{15} = 6.0$	0	0.697	0.273	0	0	0
$x_{16} = 6.0$	0	0.697	0.273	0	0	0
$x_{17} = 6.8$	0	0	0	0.303	0.697	0
$x_{18} = 6.4$	0	0	0.515	0.485	0	0
$x_{19} = 6.5$	0	0	0.212	0.788	0	0
$x_{20} = 6.6$	0	0	0	0.909	0.091	0
$x_{21} = 6.5$	0	0	0.212	0.788	0	0
$x_{22} = 6.0$	0	0.697	0.273	0	0	0
$x_{23} = 5.5$	0.788	0	0	0	0	0
$x_{24} = 6.0$	0	0.697	0.273	0	0	0
$\sum_{i=1}^{24} q_{ij}$	1.848	7.515	6.364	6.030	1.485	0.273

Then, the primary information distribution of X on U is

$$\begin{aligned} Q &= (Q_1, Q_2, Q_3, Q_4, Q_5, Q_6) \\ &= (1.848, 7.515, 6.364, 6.030, 1.485, 0.273). \end{aligned}$$

Hence we obtain a soft frequency histogram shown in Fig. 5.

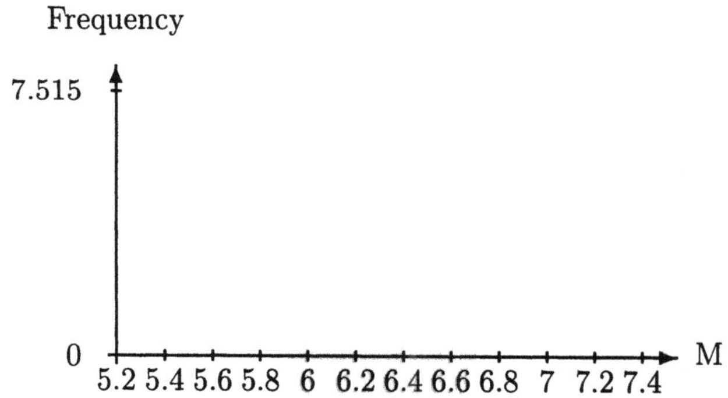

Fig. 5 A soft frequency histogram on magnitude.

Obviously, the new one is more regular than the classical one shown in Fig. 4 because the bars show some statistical law. It will be useful to estimate a probability distribution with the following expression.

Definition 8 Let $X = \{x_1, x_2, \cdots, x_n\}$ be a given sample drawn from a population with pdf $p(x)$. Given m intervals $I_1, I_2, \cdots, I_m$ with width Δ. Let u_j of the center point of I_j be a controlling point. If the distributed information is calculated by linear distribution,

$$\tilde{p}(x) = \frac{1}{n\Delta}(\text{total provided information to controlling point in the same interval as } x). \quad (3)$$

is called a *soft histogram estimate* (SHE) of $p(x)$.

By three typical simulation experiments for the estimation of probability distribution, we confirm that 1-dimensional linear-information-distribution can improve a classical histogram estimate in raising 23% work efficiency. In other words, using a small sample, less 23% size than one used in the histogram method, we can obtain a same accurate result by the method of information distribution. For example, if a statistician needs 30 observations to estimate risk by HE, then a fuzzy engineer only needs 23 observations (because 30-30×23% ≈ 23) to do it by SHE.

6 Fuzzy Model to Calculate Fuzzy Risk

To represent imprecise probability with a fuzzy relation, in this section, we introduce concept of possibility-probability distribution [10]. Then, we develop the linear distribution to calculate a possibility-probability distribution.

6.1 Possibility-Probability Distribution

Definition 9 Let x be an adverse event, $p(x)$ be probability of occurrence for x, and X be space of adverse events. Probability distribution $P = \{p(x)|x \in X\}$ is called the *probability-risk*.

Definition 10 A is called a *fuzzy set* on universe of discourse $U = \{u\}$ if and only if A is a mathematical mapping:

$$\begin{aligned} A: U &\to [0,1] \\ u &\mapsto \mu_A(u) \end{aligned}$$

$\mu_A(u)$ is called the *membership function* of fuzzy set A.

Definition 11 Let A be a fuzzy set. For $u \in U$, $\mu_A(u)$ is called *possibility* of the object x satisfies fuzzy concept A.

We use $\pi_A(u)$ to denote possibility of u relating to A. Obviously, $\forall u \in U$, $\pi_A(u) = \mu_A(u)$. $\pi_A(u)$ is called a *possibility distribution*.

Definition 12 Let $X = \{x\}$ be space of adverse events, $P = \{p\}$ be universe of discourse of probability, and $\pi_x(p)$ be possibility of that probability of x occurring is p. $\Pi_{X,P} = \{\pi_x(p)|x \in X, p \in P\}$ is called a *possibility-probability distribution.*

6.2 Calculating a Possibility-Probability Distribution

The linear distribution can be developed to calculate a possibility-probability distribution on $I = \{I_j | I_j = [u_j - \Delta/2, u_j + \Delta/2[, u_j \in U\}$, where intervals I_j are same as ones in the histogram.

We let

$$P = \{p_k | k = 0, 1, 2, \cdots, n\} = \{0/n, 1/n, 2/n, \cdots, n/n\}$$

be universe of discourse of probability.

For interval I_j, we use $\pi_{I_j}(p_k), k = 0, 2, \cdots, n$ to represent its possibility-probability.

Extending the information distribution method, we can calculate $\pi_{I_j}(p_k)$.

Firstly, we define an interior set and an outer set.

Definition 13 Let $X_1 = \{x_i | x_i \in X$ and $\exists u \in I_j$ such that $x_i = u\}$ and $X_2 = X \setminus X_1$. X_1 is called an *interior set* about I_j, and X_2 an *outer set* about the interval.

Let S be an index set such that $\forall s \in S$ then $x_s \in X_1$, and $\{x_s | s \in S\} = X_1$. Let T be one for X_2, that is, $\{x_t | t \in T\} = X_2$. S and T are called an interior and an outer index set, respectively.

Going a step further, we suppose there are $n1$ observations in X_1.

Then, for $\forall x_i \in X$, if $x_i \in X_1$ we say that it loses information, by gain at $1 - q_{ij}$, to other interval, we use $q_{ij}^- = 1 - q_{ij}$ to represent the loss; if $x_i \in X_2$ we say that it gives information, by gain at q_{ij}, to I_j, we use q_{ij}^+ to represent the addition.

Because there are $n1$ observations belong to jth interval I_j, $n1/n$ can be as an estimator of probability that adverse event x occurs in I_j. Hence, the possibility of the probability of occurrence can be defined as 1, that is,

$$\pi_{I_j}(n1/n) = 1.$$

However, if $x_i \in X_1$, the loss information q_{ij}^- to u_{j-1} or u_{j+1} implicates that x_i may leave interval I_j when there is a disturbance in the random experiment. The gain, q_{ij}^-, is the possibility of that x_i may leave I_j. When one of x_s , $s \in S$, has left I_j, we know that $P\{x \in I_j\} = (n1 - 1)/n$. Any one of them can make it true. Therefore, the possibility of $P\{x \in I_j\} = (n1-1)/n$ is

$$\pi_{I_j}(\frac{n1 - 1}{n}) = \bigvee_{s \in S} q_{sj}^-.$$

Now, the information distribution method has been extended to calculate th possibility of that x_i may leave I_j.

If two of them leave the interval, we obtain $P\{x \in I_j\} = (n1 - 2)/n$. According to properties of possibility, both of x_{s_1}, x_{s_2} leave I_j, the possibility is $q_{s_1 j}^- \wedge q_{s_2 j}^-$. Considering all pairing observations in I_j, we obtain the possibility of that probability of adverse event x is $(n1 - 2)/n$ as

$$\pi_{I_j}(\frac{n1 - 2}{n}) = \bigvee_{s_1, s_2 \in S, s_1 \neq s_2} (q_{s_1 j}^- \wedge q_{s_2 j}^-).$$

In other side, the observations in X_2 may move to I_j when there is a disturbance in the random experiment. $\forall x_t \in X_2$, it may move to I_j in possibility q_{tj}^+. Therefore, the possibility of $P\{x \in I_j\} = (n1 + 1)/n$ is

$$\pi_{I_j}(\frac{n1 + 1}{n}) = \bigvee_{t \in T} q_{tj}^+.$$

Now, the information distribution method has been extended to calculate the possibility of that x_t may move into I_j.

If two of elements in X_2 go into I_j, we obtain $P\{x \in I_j\} = (n1 + 2)/n$. Its possibility is

$$\bigvee_{t_1, t_2 \in T, t_1 \neq t_2} (q_{t_1 j}^+ \wedge q_{t_2 j}^+).$$

Hence, when there are $n1$ observations $\{x_s | s \in S\}$ in interval I_j, we can obtain a formula to calculate a possibility-probability distribution as

$$\pi_{I_j}(p) = \begin{cases} \bigwedge_{s \in S} q_{sj}^-, \ p = p_0; \\ \cdots \\ \bigvee_{s_1,s_2,s_3 \in S, s_1 \neq s_2 \neq s_3} (q_{s_1j}^- \wedge q_{s_2j}^- \wedge q_{s_3j}^-), \ p = p_{n1-3}; \\ \bigvee_{s_1,s_2 \in S, s_1 \neq s_2} (q_{s_1j}^- \wedge q_{s_2j}^-), \ p = p_{n1-2}; \\ \bigvee_{s \in S} q_{sj}^-, \ p = p_{n1-1}; \\ 1, \ p = p_{n1}; \\ \bigvee_{t \in T} q_{tj}^+, \ p = p_{n1+1}; \\ \bigvee_{t_1,t_2 \in T, t_1 \neq t_2} (q_{t_1j}^+ \wedge q_{t_2j}^+), \ p = p_{n1+2}; \\ \bigvee_{t_1,t_2,t_3 \in T, t_1 \neq t_2 \neq t_3} (q_{t_1j}^+ \wedge q_{t_2j}^+ \wedge q_{t_3j}^+), \ p = p_{n1+3}; \\ \cdots \\ \bigwedge_{t \in T} q_{tj}^+, \ p = p_n. \end{cases} \quad (4)$$

Example 3 We calculate fuzzy risk of flood in Changsha county of Hunan province in China where only 9 observations are available. We employ the loss ratio, called the disaster index, to represent flood disaster. Suppose that l is the value of loss in properties, and S is the total value of the properties. The disaster index related to lose is then defined by

$$x = \frac{l}{S}.$$

Table 5 shows flood affecting agriculture to the county, where S is seeded area; l is seeded area affected by flood; and x is disaster index.

Table 5 Flood affecting agriculture in Changsha

Year	1979	1980	1981	1982	1983	1984	1985	1987	1988
S^*	5581	5229	5186	5143	4887	4958	4966	4985	4999
l^*	56	199	205	112	636	242	198	1247	916
x	0.010	0.038	0.040	0.022	0.130	0.049	0.040	0.250	0.183

* in 1000 mu

We obtain

$$\begin{aligned} X &= \{x_1, x_2, \cdots, x_9\} \\ &= \{0.010, 0.038, 0.040, 0.022, 0.130, \\ &= 0.049, 0.040, 0.250, 0.183\}. \end{aligned}$$

The sample is too small to provide sufficient information for uncovering the relation between flood events and probabilities precisely. We call it an incomplete data set.

Here, the task of risk assessment is to calculate a fuzzy relation to represent the probability of exceeding the disaster index. Let the discrete universe of disaster index be

$$U = \{u_1, u_2, \cdots, u_{10}\} = \{0.05, 0.15, \cdots, 0.95\},$$

and universe of discourse of probability be

$$\begin{aligned}P &= \{p_0, p_1, p_2, p_3, p_4, p_5, p_6, p_7, p_8, p_9\}\\ &= \{0, 0.11, 0.22, 0.33, 0.44, 0.56, 0.67, 0.78, 0.89, 1\}.\end{aligned}$$

For $I_1 = [0, 0.1)$, $I_2 = [0.1, 0.2)$, $I_3 = [0.2, 0.3)$, using the formula (4), we obtain the possibility-probability distribution in flood disaster index as shown in Table 6.

Table 6 Possibility-probability distribution in flood

$\pi_{I_j}(p)$	0	0.11	0.22	0.33	0.44	0.56	0.67	0.78
I_1	0.01	0.10	0.10	0.12	0.28	0.40	1	0.20
I_2	0.20	0.33	1	0	0	0	0	0
I_3	0	1	0.33	0	0	0	0	0

For $I_4 = [0.3, 0.4), \cdots, I_{10} = [0.9, 1)$, we obtain

$$\pi_{I_j}(p) = \begin{cases} 1, & \text{for } p = 0; \\ 0, & \text{otherwise.} \end{cases}$$

$\pi_{I_j}(p)$, $j = 1, 2, \cdots, 10$, indicate that $P(\xi \in I)$ is a fuzzy number if we employ incomplete data set X to estimate it. For I_j, we denote the corresponding fuzzy number as A_j, i.e.,

$$A_j = \sum_{i=0}^{9} \frac{\pi_{I_j}(p_i)}{p_i}.$$

Then, we have

$$A_1 = \frac{0.01}{0} + \frac{0.10}{0.11} + \frac{0.10}{0.22} + \frac{0.12}{0.33} + \frac{0.28}{0.44} + \frac{0.40}{0.56} + \frac{1}{0.67} + \frac{0.20}{0.78},$$

$$A_2 = \frac{0.20}{0} + \frac{0.33}{0.11} + \frac{1}{0.22},$$

$$A_3 = \frac{1}{0.11} + \frac{0.33}{0.22},$$

$$A_4 = A_5 = \cdots = A_{10} = \frac{1}{0}.$$

Therefore, the probability of exceeding losses is a fuzzy number instead of one value if we employ X to estimate it. Obviously, $P(\xi \geq 0.3) = 0$, and $P(\xi \geq 0) = 1$. However, $P(\xi \geq 0.2)$ and $P(\xi \geq 0.1)$ are fuzzy numbers:

$$P(\xi \geq 0.2) = A_3 = \frac{1}{0.11} + \frac{0.33}{0.22},$$

$$P(\xi \geq 0.1) = A_2 + A_3 = \frac{0.2}{0.11} + \frac{0.33}{0.22} + \frac{1}{0.33} + \frac{0.33}{0.44}.$$

7 Risk Management Based on Fuzzy Risk

Let O be a risk system with n adverse outcomes and g be an alternative to reduce the loss. For example, in 50 years, a building in reinforced concrete frame can be considered as an earthquake-risk system with 4 adverse outcomes measured by earthquake magnitude, $m_1 = 5, m_2 = 6, m_3 = 7, m_4 = 8$. When we construct the building, one of the alternatives is to increase the ductility of the joints and strengthen the frame in order to meet the requirements of earthquake resistance. The increasing means that we have to spend more money. The addition is called the alternative cost, denoted by c. If we do not take any measure, suppose the loss is l_0. Taking the alternative, we lose l when an adverse outcome, in m, occurs. Then, the benefit output from the alternative is $b = (l_0 - l) - c$. The larger the benefit output is, the better the alternative is. In general, the benefit output of an alternative is determined by the adverse outcome that really occurs in the future, and the relation between the benefit output and the adverse outcome is fuzzy. To be easy for understanding, we suppose that the relation is a mathematical crisp mapping f from adverse outcome space, M, to benefit output space, B,

$$\begin{aligned} f : M &\to B \\ m &\mapsto f(m) \end{aligned}$$

In general, an alternative g defines a mapping f, and a set of alternatives $G = \{g_1, g_2, \cdots, g_n\}$ defines a set of mappings $F = \{f_1, f_2, \cdots, f_n\}$.

If we can precisely calculate the probability distribution of the adverse outcomes, it is easy to choice the best alternative. Let $p(m)$ be the probability distribution we obtained, and $f(m)$ be the benefit output from the alternative we take. Then the expect value of the benefit output is

$$Ef = \int_M f(m)p(m)dm. \tag{5}$$

Obviously, the best alternative is the one whose expect value is the largest among the alternatives.

Now, we turn to the case where possibility-probability distributions are used to represent fuzzy risks,

$$\Pi_{M,P} = \{\pi_m(p) | m \in M, p \in P\},$$

where M and P are the universes of the adverse outcome and concerning probability, respectively.

In the case, the expect value of the benefit output must be a fuzzy number, replacing the crisp value Ef in (5). It depends on the fuzzy probability distribution $\underset{\sim}{p}(m)$ defined by $\pi_m(p)$. We can calculate the fuzzy expect value by formula (6).

$$\underset{\sim}{E}f = \int_M f(m)\underset{\sim}{p}(m)dm. \tag{6}$$

Mathematically speaking, we can employ so-called α-levels tool to run the formula.

$\forall\alpha \in [0,1]$, we let

$$\underline{p}_\alpha(m) = \inf\{p|p \in P, \pi_m(p) \geq \alpha\},$$

$$\overline{p}_\alpha(m) = \sup\{p|p \in P, \pi_m(p) \geq \alpha\},$$

then, the set $p_\alpha(m) = \{p|p \in P, \pi_m(p) \geq \alpha\} = [\underline{p}_\alpha(m), \overline{p}_\alpha(m)]$ is a finite closed interval, called α-level.

Let

$$\begin{cases} \underline{E}_\alpha f = \int_M f(m)\underline{p}_\alpha(m)dm, \\ \overline{E}_\alpha f = \int_M f(m)\overline{p}_\alpha(m)dm. \end{cases}$$

Hence, the α-levels of the fuzzy expect value are

$$E_\alpha f = [\underline{E}_\alpha f, \overline{E}_\alpha f].$$

Then, according to the formula of decomposition[14] of fuzzy set in terms of its α-levels, we can obtain the fuzzy expect value

$$\underset{\sim}{E}f = \bigcup_{\alpha\in[0,1]} \alpha E_\alpha f.$$

Ordering the fuzzy expect values corresponding to all alternatives, we can know which alternative is the best one. The simplest criterion for the ordering is the center of gravity of $\underset{\sim}{E}f$, calculated by

$$CG(\mu_{\underset{\sim}{E}_f}(x)) = \frac{\int_{-\infty}^{+\infty} x\mu_{\underset{\sim}{E}_f}(x)dx}{\int_{-\infty}^{+\infty} \mu_{\underset{\sim}{E}_f}(x)dx}.$$

We notice that the operator which changes $\underset{\sim}{p}(m)$ into $\underset{\sim}{E}f$ is linear. Therefore, when we order the alternatives according to the center of gravity, the procedure can be simplified to avoid α-levels computation.

In the case, firstly, we calculate the center of gravity of $\underset{\sim}{p}(m)$, which is

$$c(m) = \frac{\int_P p\pi_m(p)dp}{\int_P \pi_m(p)dp},$$

then, the expect value, for the ordering, based on $\underset{\sim}{p}(m)$ is

$$Ef = \int_M f(m)c(m)dm. \tag{7}$$

Example 4 Let $G = \{g_1, g_2\}$ be the given set of alternatives to reduce the flood loss in Changsha county. Let $U = [0,1]$ be the universe of disaster index and $B = [0, 10]$ be the universe the benefit output, measured in million US dollar. Then, we suppose G defines the following mappings, from U to B,

$$g_1 : f_1(u) = -1 + 11u;$$
$$g_2 : f_2(u) = \begin{cases} -0.5 + 21u, & \text{for } 0 \le u \le 0.5; \\ 20 - 20u, & \text{for } 0.5 \le u \le 1. \end{cases}$$

as shown in Fig. 6.

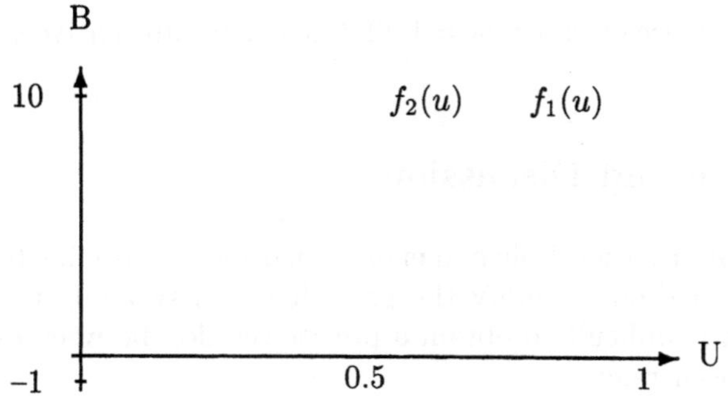

Fig. 6 Benefit-output functions corresponding to two alternatives

From Table 6, we can get the possibility-probability distribution in flood in Changsha county. Approximately, we let $\pi_{u_j}(p) = \pi_{I_j}(p),\ u_j \in U$. Then,

$$c(u_j) = \frac{\int_P p\pi_{u_j}(p)dp}{\int_P \pi_{u_j}(p)dp} = \frac{\sum_{i=1}^{9} p_i \pi_{u_j}(p_i)}{\sum_{i=1}^{9} \pi_{u_j}(p_i)}.$$

Therefore,

$$c(u_1) = \frac{0\times0.01+0.11\times0.1+0.22\times0.1+0.33\times0.12+0.44\times0.28+0.56\times0.4+0.67\times1+0.78\times0.2}{0.01+0.1+0.1+0.12+0.28+0.4+1+0.2}$$
$$= 0.56,$$
$$c(u_2) = \frac{0\times0.2+0.11\times0.33+0.22\times1}{0.2+0.33+1}$$
$$= 0.17,$$
$$c(u_3) = \frac{0.11\times1+0.22\times0.33}{1+0.33}$$
$$= 0.14,$$
$$c(u_4) = \cdots = c(u_{10}) = 0.$$

According to formula in (7), we can calculate the expect values for the two alternative,

$$g : Ef = \int_U f(u)c(u)du = \sum_{i=1}^{10} c(u_i)f(u_i) = 0.56f(u_1) + 0.17f(u_2) + 0.14f(u_3).$$

From the definition of f_1, we know

$$\begin{aligned} f_1(u_1) &= f_1(0.05) = -1 + 11 \times 0.05 = -0.45, \\ f_1(u_2) &= f_1(0.15) = -1 + 11 \times 0.15 = 0.65, \\ f_1(u_3) &= f_1(0.25) = -1 + 11 \times 0.25 = 1.75. \end{aligned}$$

Then,

$$Ef_1 = 0.56 \times (-0.45) + 0.17 \times 0.65 + 0.14 \times 1.75 = 0.1.$$

With the same way, we obtain $Ef_2 = 1.42$. Obviously, alternative g_2 is better than g_1.

8 Conclusion and Discussion

Risk, as a natural or societal phenomenon, is neither precise nor fuzzy.

Probabilistic methods simplify the procedure of risk analysis. However, in many cases, it is difficult to obtain a precise relation between events and probabilities of occurrence.

In this chapter, we review the contemporary efforts in fuzzy risk analysis and study the natures of risk. Then, we put forth a general definition of fuzzy risk and introduce concept of possibility-probability risk. A model to calculate the possibility-probability risk has been constructed in this chapter, which can improve probability-risk judgement results from the histogram method with a little observations.

An expect value, in benefit, of an alternative can be calculated based on a fuzzy assessment. It can show more information when we use a fuzzy number to represent the expect value.

As well-known, the public knowing fuzzy almost owe the success to fuzzy control, however it is just the use of fuzzy logic as the technology underlying the development of human-friendly machines, and it has not involved matters of the foundations of fuzzy logic. Recently, fuzzy information granulation, discussed by Prof. Zadeh [20], seems to be a very important issue for a restructuring of the foundations of fuzzy logic. Of particular importance is the need for a better understanding of decision-making in an environment generalized imprecision, uncertainty and partial truth [21].

Fuzzy risk analysis connects with more imprecision in the context of risk assessment and management for environment, health, and safety. The works in fuzzy risks must promote the study of the foundations of fuzzy logic.

References

[1] C.B. Brown, A fuzzy safety measure, *J. Engineering Mechanics* **105** (1979), 855-872.

[2] Chongfu Huang and Da Ruan, Information diffusion principle and application in fuzzy neuron, *Fuzzy Logic Foundations and Industrial Applications* (edited by Da Ruan, Kluwer Academic Publishers, Massachusetts,1996), 165-189.

[3] D.P. Clement, *Fuzzy Ratings for Computer Security Evaluation,* Ph.D. Dissertation, University of California at Berkeley (1977).

[4] M. Delgado, J.L. Verdegay and M.A. Vila, A model for linguistic partial information in decision-making problems, *International Journal of Intelligent Systems,* **9** (1994), 365-378.

[5] W.M. Dong and et al., Fuzzy computation in risk and decision analysis, *Civil Engineering Systems* **2** (1986), 201-208.

[6] A.O. Esogbue and et al., On the application of fuzzy sets theory to the optimal flood control problem arising in water resources systems, *Fuzzy Sets and Systems* **48** (1992), 155-172.

[7] F.C. Hadipriono, A rule-based fuzzy logic deduction technique for damage assessment of protective structures, *Fuzzy Sets and Systems* **44** (1991), 459-468.

[8] L.J. Hoffman, E.H. Michelmen and D.P. Clements, SEURAT–Security evaluation and analysis using fuzzy metrics, *Proc. of the 1978 National Computer Conference* (AFIPS Press, Montvale, New Jersey, 1978) **47**, 531-540.

[9] Huang Chongfu, Principle of information diffusion, *Fuzzy Sets and Systems,* **91** (1)(1997), 69-90.

[10] Huang Chongfu, Concepts and methods of fuzzy risk analysis, *Proceedings of the First China-Japan Conference on Risk Assessment and Management,* Beijing, China, November, 1998, 12-23.

[11] Huang Chongfu and Shi Peijun, Fuzzy risk and calculation, *Proceedings of 18th International Conference of the North American Fuzzy Information Processing Society,* (1999), 90-94

[12] M. Jablonowski, Fuzzy risk analysis: using AI system, *AI Expert,* **9**(12) (1994), 34-37.

[13] Huang Chongfu, Fuzzy risk assessment of urban natural hazards, *Fuzzy Sets and Systems,* **83** (1996), 271-282

[14] E.E. Kerre, *Fuzzy Sets and Approximate Reasoning* (Xian Jiaotong University Press, Xian, China, 1999).

[15] R. V. Kolluru, S. M. Bartell, R.M., Pitblado, and R. S. Stricoff, *Risk Assessment and Management Handbook for Environmental, Health, and Safety Professionals* (McGraw-Hill, New York, 1996)

[16] A.V. Machias and G.D. Skikos, Fuzzy risk index of wind sites, *IEEE Trans. Energy Conversion,* **7**(4) (1992), 638-643.

[17] A.P. Sage and E.B. White,Methodologies for risk and hazard assessment: a survey and status report, *IEEE Trans. Systems, Man, and Cybernetics,* **SMC-10**(8)(1980), 425-446.

[18] K.J. Schmucker, *Fuzzy Sets, Natural Language Computations, and Risk Analysis* (Computer Science Press, Rockvill, Maryland, 1984)

[19] L.A. Zadeh, Fuzzy sets, *Information and Control,* **8**(3) (1965), 338-353.

[20] L.A. Zadeh, Toward a theory of fuzzy information granulation and its centrality in human reasoning and fuzzy logic, *Fuzzy Sets and Systems* **90**(2)(1971), 111-127.

[21] L.A. Zadeh, Toward a restructuring of the foundations of fuzzy logic (FL), *Proceedings of FUZZ-IEEE'98,* Anchorage, USA, May, 1998, 1676-1677.

Soft Methods in Hypotheses Testing

Przemysław Grzegorzewski and Olgierd Hryniewicz

Systems Research Institute, Polish Academy of Sciences,
ul. Newelska 6, 01-447 Warsaw, Poland,

Abstract

The paper is devoted to nonclassical approaches to hypotheses testing. We consider testing statistical hypotheses in fuzzy environment, i.e. tests with vague data, tests for fuzzy hypotheses and tests for fuzzy hypotheses with vague data. We also present a possibilistic interpretation of statistical tests.
Key words: hypothesis testing, fuzzy data, fuzzy test, fuzzy hypotheses, possibility distribution.

1 Introduction

Nearly all aspects of human cognition is related with stating and veryfying hypotheses. Testing hypotheses is also one of the primary purposes of statistical inference. One meets statistical tests in such areas as regression analysis, analysis of variance, statistical quality control, risk theory etc.

In classical statistics all model parameters, especially data and hypotheses under study should be precise. However, in real life we meet very often vague data, like "about twenty", "more or less between nine and ten", "rather greater than one hundred", etc. Moreover, sometimes we are quite satisfied in verifying fuzzy hypotheses like "the mean m is about 40" instead of the crisp one "the mean is exactly 40" ($m = 40$) or "the frequency p is rather less than 5%" instead of "$p < 0.05$".

The problem of testing hypotheses in fuzzy environment was considered by many authors. For the review paper we refer the reader to Grzegorzewski and Hryniewicz (1997). Since then some new interesting approaches have appeared. Moreover, discussions on the nonstandard interpretation of the classical hypothesis testing (i.e. with crisp data and crisp hypotheses) have also appeared. Therefore it seems desirable to summarize them. In Sec. 2 we recall basic notions of the classical theory of hypotheses testing. Then, in Sec. 3, we sketch how to model vagueness met in statistics applying fuzzy set theory. Sec. 4 is devoted to testing hypotheses with vague data. Next we consider tests for veryfying fuzzy hypotheses (Sec. 5). We also show how to cope both with fuzzy hypotheses and vague data (Sec. 6). In Sec 7 we discuss some methods for the possibilistic interpretation of statistical tests.

2 Classical approach

Assume that the investigated phenomenon is described by a probability distribution P_θ which belongs to a family of distributions $\mathcal{P} = \{P_\theta : \theta \in \Theta\}$. We consider the null hypothesis $H : \theta \in \Theta_H$ concerning the parameter θ, with the alternative hypothesis $K : \theta \in \Theta_K$, where Θ_H and Θ_K are subsets of Θ such that $\Theta_H \cap \Theta_K = \emptyset$. We assume that if θ were known one would also know whether or not the hypothesis is true.

In the hypothesis testing problem we observe a random sample $V = (V_1, \ldots, V_n)$, and this observation can lead to one of two possible decisions: either to reject H (and to accept K), or to not reject H (usually identified with accepting H). Traditionally, rejection of H is denoted by one and acceptance of H by zero. Hence, a decision rule, called a *statistical test*, can be defined as a function $\varphi : \mathcal{X} \to \{0,1\}$, where $\mathcal{X}$ is a sample space. Each statistical test divides the sample space $\mathcal{X}$ into two exclusive subsets: $\{v \in \mathcal{X} : \varphi(v) = 0\}$ – the set of the acceptance of H, called an *acceptance region*, and $\mathcal{K} = \{v \in \mathcal{X} : \varphi(v) = 1\}$ – the set of the rejection of H which is also called a *critical region*. In practice, we compute a certain test statistic $T(V)$ (i.e. a function of the observations), then we find a critical region $\mathcal{K}$, and finally we reject the considered hypothesis if $T(V) \in \mathcal{K}$ or accept it otherwise. A typical statistical test has a following form:

$$\varphi(V) = \begin{cases} 1 & \text{if } \; T(V) \in \mathcal{K}, \\ 0 & \text{otherwise.} \end{cases} \tag{1}$$

The critical region depends on a preselected upper bound of the probability of the type I error (i.e. rejecting H when it is actually true), called a *significance level* δ . Thus we have

$$P\{\varphi(V) = 1 \mid H\} \leq \delta. \tag{2}$$

It is worth mention that while testing hypotheses one may also commit type II error, i.e. to accept H when it is actually false. For more details concerning the traditional theory of testing statistical hypotheses we refer the reader to Lehmann (1986).

3 Modelling vagueness

Now let $X_1, \ldots, X_n$ denote a fuzzy sample which is a fuzzy perception of the usual random sample $V_1, \ldots, V_n$, from the population with the distribution

P_θ. Assume that each X_i is a fuzzy number, i.e. X_i is a normal, fuzzy convex and bounded fuzzy subset of the real line $\mathcal{R}$ with an upper semicontinuous membership function $\mu_{X_i} : \mathcal{R} \to [0,1]$ (see, e.g., Dubois and Prade, 1980). A space of all fuzzy numbers will be denoted by $\mathcal{FN}(\mathcal{R})$. Of course, $\mathcal{FN}(\mathcal{R}) \subset \mathcal{F}(\mathcal{R})$, where $\mathcal{F}(\mathcal{R})$ denotes a space of all fuzzy sets on the real line. A useful tool for dealing with fuzzy numbers are their α–cuts. The α–cut, $\alpha \in (0,1]$, of a fuzzy number X with its membership function μ_X is a crisp set defined as

$$X_\alpha = \{t \in \mathcal{R} : \mu_X(t) \geq \alpha\}. \tag{3}$$

Every α–cut of a fuzzy number is a closed interval. A following notation will be useful below: $X_\alpha = [X_\alpha^L, X_\alpha^U]$, where

$$X_\alpha^L = \inf\{t \in \mathcal{R} : \mu_X(t) \geq \alpha\}, \tag{4}$$
$$X_\alpha^U = \sup\{t \in \mathcal{R} : \mu_X(t) \geq \alpha\}. \tag{5}$$

The precise definitions of a fuzzy random variable and fuzzy sample can be found in Kwakernaak (1978, 1979), Kruse (1982), Puri and Ralescu (1986). In this paper we adopt the approach similar to those of Kwakernaak and Kruse.

Suppose a random experiment is described as usual by a probability space $(\Omega, \mathcal{A}, P)$, where Ω is a set of all possible outcomes of the experiment, $\mathcal{A}$ is a σ–algebra of subsets of Ω (the set of all possible events) and P is a probability measure

Definition 1 *A mapping $X : \Omega \to \mathcal{FN}(\mathcal{R})$ is called a fuzzy random variable if it satisfies the following properties:*

(a) $\{X_\alpha(\omega) : \alpha \in [0,1]\}$ is a set representation of $X(\omega)$ for all $\omega \in \Omega$,

(b) for each $\alpha \in [0,1]$ both $X_\alpha^L = X_\alpha^L(\omega) = \inf X_\alpha(\omega)$ and $X_\alpha^U = X_\alpha^U(\omega) = \sup X_\alpha(\omega)$, are usual real-valued random variables on $(\Omega, \mathcal{A}, P)$.

Thus a fuzzy random variable X is considered as a perception of an unknown usual random variable $V : \Omega \to \mathcal{R}$, called an *original* of X. Let χ denote a set of all possible originals of X. If only vague data are available, it is of course impossible to show which of the possible originals is the true one. Therefore we can define a fuzzy set on χ, with a membership function $\nu : \chi \to [0,1]$ given as follows:

$$\nu(V) = \inf\{\mu_{X(\omega)}(V(\omega)) : \omega \in \Omega\}, \tag{6}$$

which corresponds to the grade of acceptability that a fixed random variable V is the original of the fuzzy random variable in question (see Kruse and Meyer, 1987).

In the presence of fuzzy data we cannot observe the parameter θ directly but only its vague image. Moreover, we cannot estimate θ better than to its fuzzy perception $\Lambda(\theta)$ defined as

$$\mu_{\Lambda(\theta)}(t) = \sup\{\nu(V_1,\dots,V_n) : (V_1,\dots,V_n) \in \chi^n, \theta(V_1) = t\}, \quad t \in \mathcal{R}, \quad (7)$$

where χ^n is a set of all possible originals of that fuzzy random sample with membership function

$$\nu(V_1,\dots,V_n) = \min_{i=1,\dots,n} \inf\{\mu_{X_i(\omega)}(V_i(\omega)) : \omega \in \Omega\}. \quad (8)$$

One can easily obtain α–cuts of $\Lambda(\theta)$:

$$\begin{aligned}(\Lambda(\theta))_\alpha = \{t \in \mathcal{R} : \exists (V_1,\dots,V_n) \in \chi^n, \theta(V_1) = t, \text{ such that} \\ V_i(\omega) \in (X_i(\omega))_\alpha \text{ for } \omega \in \Omega \text{ and for } i = 1,\dots,n\}.\end{aligned} \quad (9)$$

For more information we refer the reader to Kruse and Meyer (1987).

4 Testing hypotheses with vague data

Let $X_1,\dots,X_n$ denote a fuzzy sample, i.e. a fuzzy perception of the usual random sample $V_1,\dots,V_n$, from the population with the distribution P_θ. Let δ be a given number from the interval $(0,1)$. Grzegorzewski (2000a) proposed a definition of so called *fuzzy test* for vague data and a method of constructing such fuzzy tests.

Definition 2 *A function $\varphi : (\mathcal{FN}(\mathcal{R}))^n \to \mathcal{F}(\{0,1\})$ is called a fuzzy test for the hypothesis H, on the significance level δ, if*

$$\sup_{\alpha\in[0,1]} P\{\omega \in \Omega : \varphi_\alpha(X_1(\omega),\dots,X_n(\omega)) \subseteq \{1\} \,|H\} \le \delta, \quad (10)$$

where φ_α is the α–level set of $\varphi(X_1,\dots,X_n)$.

The fuzzy test can be regarded as a family of classical tests $\{\varphi_\alpha : \alpha \in (0,1]\}$. So the significance level of fuzzy test is the upper bound of type I error for the whole family $\{\varphi_\alpha : \alpha \in (0,1]\}$.

It is well known that in the classical theory of hypothesis testing there is an equivalence between the totality of parameter values for which the null hypothesis is accepted and the structure of the confidence intervals (see Lehmann, 1986). Namely, a family of acceptance regions for a statistical test concerning the parameter θ, on the significance level δ, is equivalent to a certain family of confidence intervals for that parameter, on the confidence level $1-\delta$. This observation is a base of the general method of constructing fuzzy test for fuzzy data.

First let us consider a following simple null hypothesis: $H : \theta = \theta_0$, against the composite two-sided alternative: $K : \theta \neq \theta_0$. Suppose we know a two-sided symmetrical confidence interval $[\pi_1, \pi_2]$ for θ, on a confidence level $1-\delta$, where obviously $\pi_1 = \pi_1(V_1, \ldots, V_n; \frac{\delta}{2})$ and $\pi_2 = \pi_2(V_1, \ldots, V_n; \frac{\delta}{2})$. Then the following theorem is true:

Theorem 3 *Let $X_1, \ldots, X_n$ be a fuzzy random sample, where $X_i \in \mathcal{FN}(\mathcal{R})$ for $i = 1, \ldots, n$, and let $\delta \in (0,1)$. Define $\Pi_\alpha = [\Pi_\alpha^L, \Pi_\alpha^U]$ for all $\alpha \in (0,1]$ as*

$$\begin{aligned} \Pi_\alpha^L = \Pi_\alpha^L(X_1, \ldots, X_n; \tfrac{\delta}{2}) = \inf \{ t \in \mathcal{R} : \forall i \in \{1, \ldots, n\} \\ \exists x_i \in (X_i)_\alpha \quad \textit{such that} \quad \pi_1(x_1, \ldots x_n; \tfrac{\delta}{2}) \leq t \}, \end{aligned} \tag{11}$$

$$\begin{aligned} \Pi_\alpha^U = \Pi_\alpha^U(X_1, \ldots, X_n; \tfrac{\delta}{2}) = \sup \{ t \in \mathcal{R} : \forall i \in \{1, \ldots, n\} \\ \exists x_i \in (X_i)_\alpha \quad \textit{such that} \quad \pi_2(x_1, \ldots, x_n; \tfrac{\delta}{2}) \geq t \}. \end{aligned} \tag{12}$$

Then a function $\varphi : (\mathcal{FN}(\mathcal{R}))^n \to \mathcal{F}(\{0,1\})$ with following $\alpha-$cuts

$$\varphi_\alpha(X_1, \ldots, X_n) = \begin{cases} \{0\} & \textit{if} \quad \theta_0 \in (\Pi_\alpha \setminus (\neg\Pi)_\alpha), \\ \{1\} & \textit{if} \quad \theta_0 \in ((\neg\Pi)_\alpha \setminus \Pi_\alpha), \\ \{0,1\} & \textit{if} \quad \theta_0 \in (\Pi_\alpha \cap (\neg\Pi)_\alpha), \\ \emptyset & \textit{if} \quad \theta_0 \notin (\Pi_\alpha \cup (\neg\Pi)_\alpha), \end{cases} \tag{13}$$

is a fuzzy test for $H : \theta = \theta_0$, against $K : \theta \neq \theta_0$, on the significance level δ.

Similarly we may obtain fuzzy tests for one-sided hypotheses. In particular, let us consider a null hypothesis $H : \theta \leq \theta_0$, against $K : \theta > \theta_0$. It is known that there is one-to-one correspondence between the acceptance region of the test designated for testing these hypotheses on the significance level δ and one-sided confidence interval $[\pi_1, +\infty)$ for the parameter θ on the confidence level $1-\delta$, where $\pi_1 = \pi_1(V_1, \ldots, V_n; \delta)$. Thus the following theorem may be proved:

Theorem 4 *Let $X_1, \ldots, X_n$ be a fuzzy random sample, where $X_i \in \mathcal{FN}(\mathcal{R})$ for $i = 1, \ldots, n$, and let $\delta \in (0,1)$. Then a function $\varphi : (\mathcal{FN}(\mathcal{R}))^n \to \mathcal{F}(\{0,1\})$ with $\alpha-$cuts given by (13), where $\Pi_\alpha = [\Pi_\alpha^L, \infty)$ for all $\alpha \in (0,1]$ and*

$$\begin{aligned} \Pi_\alpha^L = \Pi_\alpha^L(X_1, \ldots, X_n; \delta) = \inf \{ t \in \mathcal{R} : \forall i \in \{1, \ldots, n\} \\ \exists x_i \in (X_i)_\alpha \quad \textit{such that} \quad \pi_1(x_1, \ldots, x_n; \delta) \leq t \} \end{aligned} \tag{14}$$

is a fuzzy test for $H : \theta \leq \theta_0$, against $K : \theta > \theta_0$, on the significance level δ.

Now let us consider a hypothesis $H : \theta \geq \theta_0$, against the alternative $K : \theta < \theta_0$. One may show one-to-one correspondence between the acceptance region of the test designated for testing these hypotheses on the significance level δ and one-sided confidence interval $(-\infty, \pi_2]$ for the parameter θ on the confidence level $1-\delta$, where $\pi_2 = \pi_2(V_1, \ldots, V_n; \delta)$. So we can state a following theorem:

Theorem 5 *Let $X_1, \ldots, X_n$ be a fuzzy random sample, where $X_i \in \mathcal{FN}(\mathcal{R})$ for $i = 1, \ldots, n$, and let $\delta \in (0,1)$. Then a function $\varphi : (\mathcal{FN}(\mathcal{R}))^n \to \mathcal{F}(\{0,1\})$ with $\alpha-$cuts given by (13), where $\Pi_\alpha = (-\infty, \overline{\Pi}_\alpha^U]$ for all $\alpha \in (0,1]$ and*

$$\overline{\Pi}_\alpha^U = \overline{\Pi}_\alpha^U(X_1, \ldots, X_n; \delta) = \sup\{t \in \mathcal{R} : \forall i \in \{1, \ldots, n\} \\ \exists x_i \in (X_i)_\alpha \quad \text{such that} \quad \overline{\pi}_2(x_1, \ldots, x_n; \delta) \geq t\} \tag{15}$$

is a fuzzy test for $H : \theta \geq \theta_0$, against $K : \theta < \theta_0$, on the significance level δ.

It is not difficult to show a following lemma:

Lemma 6 *A membership function of the fuzzy test considered above for the hypothesis H against K is*

$$\mu_\varphi(t) = \mu_\Pi(\theta_0) I_{\{0\}}(t) + \mu_{\neg\Pi}(\theta_0) I_{\{1\}}(t) \\ = \mu_\Pi(\theta_0) I_{\{0\}}(t) + (1 - \mu_\Pi(\theta_0)) I_{\{1\}}(t), \qquad t \in \{0,1\}, \tag{16}$$

where Π is a fuzzy acceptance region depending on the considered hypotheses.

It is easily seen that our fuzzy test, contrary to the classical crisp test, does not lead to a binary decision – to accept or to reject the null hypothesis – but to a fuzzy decision: we may get $\varphi = 1/0 + 0/1$ which indicates that we should accept H, or $\varphi = 0/0 + 1/1$ which means that H should be rejected, but we may also get $\varphi = \mu_0/0 + (1 - \mu_0)/1$, where $\mu_0 \in (0,1)$, which can be interpreted as a degree of conviction that we should accept (μ_0) or reject ($1-\mu_0$) the hypothesis H. Thus, in situation when μ_0 is neither 0 nor 1, a user must decide whether to reject or to accept given hypothesis actually (however value μ_0 would support his decision). The problem of defuzzification of a fuzzy test is considered by Grzegorzewski (1998).

Hence it is also seen that the fuzzy test is well defined because if the data are not fuzzy but crisp, i.e. $X_1 = V_1, \ldots, X_n = V_n$, our fuzzy test reduces to the usual (i.e. crisp) statistical test.

Example

Suppose we have a fuzzy perception $X_1, \ldots, X_n$ of a random sample from the normal distribution $N(m, \sigma)$ and we want to verify a null hypothesis concerning the mean $H : m = m_0$ against the alternative hypothesis $K : m \neq m_0$. According to (11), (12) and (13) we can verify our hypothesis H against K using a fuzzy test $\varphi : (\mathcal{FN}(\mathcal{R}))^n \to \mathcal{F}(\{0,1\})$ with following $\alpha-$cuts

$$\varphi_\alpha(X_1, \ldots, X_n) = \begin{cases} \{0\}, & \text{if} \quad m_0 \in (\Pi_\alpha \setminus (\neg\Pi)_\alpha), \\ \{1\}, & \text{if} \quad m_0 \in ((\neg\Pi)_\alpha \setminus \Pi_\alpha), \\ \{0,1\}, & \text{if} \quad m_0 \in (\Pi_\alpha \cap (\neg\Pi)_\alpha), \\ \emptyset, & \text{if} \quad m_0 \notin (\Pi_\alpha \cup (\neg\Pi)_\alpha), \end{cases} \tag{17}$$

where

$$\Pi_\alpha = \left(\frac{1}{n} \sum_{i=1}^{n} (X_i)_\alpha^L - u_{1-\delta/2} \frac{\sigma}{\sqrt{n}}, \frac{1}{n} \sum_{i=1}^{n} (X_i)_\alpha^U + u_{1-\delta/2} \frac{\sigma}{\sqrt{n}} \right). \tag{18}$$

After simple calculations we get a following test

$$\varphi(X_1, \ldots, X_n) = \begin{cases} 1/0 + 0/1, & \text{if } m_0 \in \Pi_{\alpha=1}, \\ 0/0 + 1/1, & \text{if } m_0 \notin \Pi_{\alpha=0}, \\ \mu_\Pi(m_0)/0 + (1 - \mu_\Pi(m_0))/1, & \text{otherwise.} \end{cases} \tag{19}$$

5 Testing fuzzy hypotheses

Suppose that we consider a fuzzy null hypothesis H against fuzzy alternative K described by their membership functions $\mu_H : \Theta \to [0,1]$ and $\mu_K : \Theta \to [0,1]$, respectively. These membership functions give the degree to which θ belongs to each hypothesis. In particular, $\mu_H(\theta) = 1$ means that θ completely belongs to H, whereas $\mu_H(\theta) = 0$ indicates that θ is not at all a member of H.

In order to verify such fuzzily formulated hypotheses Arnold (1996) generalized the definition of probabilities of type I and type II errors. Let $M(\phi, \cdot) : \Theta \to [0,1]$ denote a power function of a given test ϕ. Then the quantities

$$\delta(\phi) = \sup_{\theta \in \Theta_H} \{(\mu_H(\theta) - \mu_K(\theta)) M(\phi, \theta)\} \tag{20}$$

$$\beta(\phi) = \sup_{\theta \in \Theta_K} \{(\mu_K(\theta) - \mu_H(\theta))(1 - M(\phi, \theta))\}, \tag{21}$$

where $\Theta_H = \{\theta \in \Theta : \mu_H(\theta) > \mu_K(\theta)\}$ and $\Theta_K = \{\theta \in \Theta : \mu_H(\theta) < \mu_K(\theta)\}$, are generalized probabilities of type I and type II errors, respectively. It is easily seen that in the traditional situation where $H \cap K = \emptyset$, (20) and (21) reduce to $\delta(\phi) = \sup_{\theta \in H} M(\phi, \theta)$ and $\beta(\phi) = \sup_{\theta \in K}(1 - M(\phi, \theta))$ which are classical probabilities of type I and type II errors, respectively.

Hence according to the classical approach to hypotheses testing (see 2) a test ϕ is said to be a test on significance level δ, $\delta \in (0,1)$, if the inequality $\delta(\phi) \le \delta$ holds. In his papers Arnold (1996, 1998) showed some examples how to construct a test of significance δ for particular fuzzy hypotheses.

A similar approach to veryfying fuzzy hypotheses was given by Taheri and Behboodian (1999). They proposed another definitions for probability of type I and type II errors. Let

$$\Delta_1 = \int_\Theta \mu_H(\theta) d\theta, \tag{22}$$

$$\Delta_2 = \int_\Theta \mu_K(\theta) d\theta, \tag{23}$$

where the integral should be substituted by summation in a discrete case. Taheri and Behboodian restrict their considerations to situations where $\Delta_1 < \infty$ and $\Delta_2 < \infty$ only. Then they define the probability of type I error as

$$\delta(\phi) = \frac{1}{\Delta_1} \int\limits_{\Theta} \mu_H(\theta) M(\phi, \theta) d\theta \tag{24}$$

and the probability of type II error as

$$\beta(\phi) = \frac{1}{\Delta_2} \int\limits_{\Theta} \mu_K(\theta)(1 - M(\phi, \theta)) d\theta. \tag{25}$$

It is not difficult to see that in the case of crisp simple hypotheses (24) and (25) reduce to ordinary probabilities of type I and type I error, respectively. Similarly as in Arnold's approach a test ϕ is said to be a test on significance level δ, $\delta \in (0, 1)$, if $\delta(\phi) \leq \delta$.

As it is seen Taheri and Behboodian approach leads to natural fuzzy generalization of simple hypotheses. And actually Taheri and Behboodian proved the Neyman-Pearson lemma for testing fuzzy hypotheses. Unfortunately this method seems to be limited to fuzzy simple hypotheses only, while the method due to Arnold is more general and enables to construct both one-sided as two-sided hypotheses.

It should be stressed that both in Arnold's approach and Taheri and Behboodian's approach the final result of the testing procedure is crisp. It is due to modifications of the definitions of the probabilities of type I and type II errors. All difficulties caused by fuzziness of the hypotheses were overcome by these very modifications. They are called "modifications" but actually they are quite new structures. However, one may look for the other method of testing fuzzy hypotheses which would be more "soft" and more natural in fuzzy environment in a sense that it would rather utilize classical tools through the extension principle than construct new structures. Such a method we show below. There are some similarities between it and the method of Watanabe and Imaizumi (1993).

Suppose that we consider a null hypothesis $H : \theta \in \Theta_H$ against an alternative hypothesis $K : \theta \in \Theta_K = \neg\Theta_H$, where Θ_H and Θ_K are fuzzy subsets of the parameter space Θ, with membership functions $\mu_H : \Theta \to [0, 1]$ and $\mu_K(x) = 1 - \mu_H(x)$, respectively. We'll transform such a fuzzily formulated problem into a family of crisp testing problems. We distinguish three basic situations:

- if $\text{supp}\mu_H$ is bounded then the problem of testing $H : \theta \in \Theta_H$ against $K : \theta \in \Theta_K = \neg\Theta_H$ can be transformed to a following family of crisp testing problems

$$\{H_{\theta_0} : \theta = \theta_0 \quad \text{vs.} \quad K_{\theta_0} : \theta \neq \theta_0, \text{ where } \theta_0 \in \text{supp}\mu_H\}; \tag{26}$$

- if $\text{supp}\mu_H$ is bounded only from the right then the problem of testing $H : \theta \in \Theta_H$ against $K : \theta \in \Theta_K = \neg\Theta_H$ can be transformed to a following family of crisp testing problems

$$\{H_{\theta_0} : \theta = \theta_0 \quad \text{vs.} \quad K_{\theta_0} : \theta > \theta_0, \text{ where } \theta_0 \in \text{supp}\mu_H\}\,; \tag{27}$$

- if $\text{supp}\mu_H$ is bounded only from the left then the problem of testing $H : \theta \in \Theta_H$ against $K : \theta \in \Theta_K = \neg\Theta_H$ can be transformed to a following family of crisp testing problems

$$\{H_{\theta_0} : \theta = \theta_0 \quad \text{vs.} \quad K_{\theta_0} : \theta < \theta_0, \text{ where } \theta_0 \in \text{supp}\mu_H\}\,. \tag{28}$$

Now let $\{\phi_{\theta_0} : \mathcal{X} \to \{0,1\}, \text{ where } \theta_0 \in \text{supp}\mu_H\}$ denote a family of tests on significance level δ for veryfying crisp hypotheses (26), (27) or (28). Then we get a following definition:

Definition 7 *A function $\psi : \mathcal{X} \to \mathcal{F}(\{0,1\})$ such that*

$$\mu_\psi(0) = \begin{cases} \sup\limits_{\theta_0 \in supp\mu_H \,:\, \phi_{\theta_0}(x)=0} \mu_H(\theta_0) & \text{if } \{\theta_0 \in supp\mu_H : \phi_{\theta_0}(x) = 0\} \neq \varnothing \\ 0 & \text{if } \{\theta_0 \in supp\mu_H : \phi_{\theta_0}(x) = 0\} = \varnothing \end{cases} \tag{29}$$
$$\mu_\psi(1) = 1 - \mu_\psi(0)$$

is called a fuzzy test for veryfying fuzzy hypotheses $H : \theta \in \Theta_H$ against $K : \theta \in \Theta_K = \neg\Theta_H$ on significance level δ.

Therefore our fuzzy test for fuzzy hypotheses does not always lead to binary decisions – to accept or to reject the null hypothesis – but to a fuzzy decision: we may get $\psi = 1/0 + 0/1$ which indicates that we should accept H, or $\psi = 0/0 + 1/1$ which means that H should be rejected, but we may also get $\psi = \xi/0 + (1-\xi)/1$, where $\xi \in (0,1)$, which can be interpreted as a degree of conviction that we should accept (ξ) or reject ($1-\xi$) the hypothesis H. It is easily seen that the interpretation of the output of this test resembles that of the fuzzy test for vague data described in Sec. 4. It is also seen that our fuzzy test for fuzzy hypotheses is well defined because if the hypotheses are not fuzzy but crisp then a fuzzy test ψ reduces to the usual statistical test.

6 Testing fuzzy hypotheses with vague data

In this section we describe a method of testing fuzzy hypotheses with vague data which can be used both for one-sided and two-sided alternatives (see Grzegorzewski, 2000b). We restrict our considerations to hypotheses on fuzzy perceptions of the parameter under study (see Sec. 3). It is known that in the space of fuzzy numbers statements of the type: "greater than" or "less than" have no sense in general, because there is no linear order in that space. Thus any time one wants to verify hypothesis that a fuzzy perception is, e.g., "greater" than given fuzzy number, he has to explain what does it mean, i.e. to specify precisely the way of ordering fuzzy values.

In the present section we consider a hypothesis $H : \Lambda(\theta) = \Lambda_0$ that a fuzzy perception $\Lambda(\theta)$ of a parameter θ is equal to Λ_0 against one-sided and two-sided alternatives constructed using the necessity index of strict dominance (*NSD*) due to Dubois and Prade (1983). Let us recall that for any fuzzy numbers A and B with membership functions μ_A and μ_B, respectively, we can evaluate the degree of necessity to which the relation $A > B$ is fulfilled

$$Ness(A > B) = 1 - \sup_{x,y:x \leq y} \min\{\mu_A(x), \mu_B(y)\}. \tag{30}$$

Dubois and Prade proposed also the possibility of strict dominance index and other indices. However, we decided to use *NSD* index because of its natural interpretation and effectiveness in solving real-life problems (see, e.g. Hryniewicz 1992, 1994).

Let us begin with the problem of testing the null hypothesis $H : \Lambda(\theta) = \Lambda_0$ against the one-sided alternative $K : Ness(\Lambda(\theta) > \Lambda_0) \geq \xi$, where ξ is a fixed number from the interval $[0, 1]$. Although we have both fuzzy data and fuzzy hypotheses the order based on *NSD* leads to a very simple statistical test. It can be shown that it is enough to consider only one $\alpha-$level in order to check whether the relation $Ness(\Lambda(\theta) > \Lambda_0) \geq \xi$ holds (see Grzegorzewski, 2000b). This conclusion makes the starting point for our test construction. A following theorems holds

Theorem 8 *Let $X_1, \ldots, X_n$, denote a fuzzy random sample, where $X_i \in \mathcal{FN}(\mathcal{R})$ for $i = 1, \ldots, n$, from the distribution with unknown real parameter θ and let $\xi \in [0, 1]$. Let $\Lambda(\theta) \in \mathcal{FN}(\mathcal{R})$ denote a fuzzy perception of θ and let $[\pi_1, +\infty)$ be the upper one-sided confidence interval for the parameter θ on the confidence level $1 - \delta$. Then a function $\phi : (\mathcal{FN}(\mathcal{R}))^n \to \{0, 1\}$ such that*

$$\phi(X_1, \ldots, X_n) = \begin{cases} 1 & \text{if } (\Lambda_0)^U_{1-\xi} < \Pi^L_{1-\xi} \\ 0 & \text{otherwise,} \end{cases} \tag{31}$$

where $\Pi^L_{1-\xi} = \Pi^L_{1-\xi}(X_1, \ldots, X_n; \delta)$ is given by (14), is a test for hypothesis $H : \Lambda(\theta) = \Lambda_0$ against one-sided alternative $K : Ness(\Lambda(\theta) > \Lambda_0) \geq \xi$, on the significance level δ.

Similarly, using one-to-one correspondence between the acceptance region of the test for the hypothesis $H : \theta = \theta_0$ against $K : \theta < \theta_0$ on the significance level δ and one-sided confidence interval $(-\infty, \pi_2]$ for the parameter θ on the confidence level $1 - \delta$, where $\pi_2 = \pi_2(V_1, \ldots, V_n; \delta)$, we get a test for the opposite one-sided alternative fuzzy hypothesis.

Theorem 9 *Let $X_1, \ldots, X_n$, denote a fuzzy random sample, where $X_i \in \mathcal{FN}(\mathcal{R})$ for $i = 1, \ldots, n$, from the distribution with unknown real parameter θ and let $\xi \in [0, 1]$. Let $\Lambda(\theta) \in \mathcal{FN}(\mathcal{R})$ denote a fuzzy perception of θ and let $(-\infty, \pi_2]$ be lower one-sided confidence interval for the parameter θ on the confidence level $1 - \delta$. Then a function $\phi : (\mathcal{FN}(\mathcal{R}))^n \to \{0, 1\}$ such that*

$$\phi(X_1, \ldots, X_n) = \begin{cases} 1 & \text{if } (\Lambda_0)^L_{1-\xi} > \Pi^U_{1-\xi} \\ 0 & \text{otherwise,} \end{cases} \tag{32}$$

where $\Pi^U_{1-\xi} = \Pi^U_{1-\xi}(X_1, \ldots, X_n; \delta)$ is given by (15), is a test for hypothesis $H : \Lambda(\theta) = \Lambda_0$ against one-sided alternative $K : Ness(\Lambda_0 > \Lambda(\theta)) \geq \xi$, on the significance level δ.

We can also use *NSD* index for testing our null hypothesis against two-sided alternative. Firstly we will define a following relation

Definition 10 *Let $X, Y \in \mathcal{FN}(\mathcal{R})$ and let $\xi \in [0, 1]$. Then*

$$Ness(X \neq Y) \geq \xi \Leftrightarrow (Ness(X > Y) \geq \xi \quad or \quad Ness(Y > X) \geq \xi) . \tag{33}$$

Now, keeping in mind that there is one-to-one correspondence between the acceptance region of the test for the hypothesis $H : \theta = \theta_0$ against $K : \theta \neq \theta_0$ on the significance level δ and two-sided confidence interval $[\pi_1, \pi_2]$ for the parameter θ on the confidence level $1 - \delta$, where $\pi_1 = \pi_1(V_1, \ldots, V_n; \frac{\delta}{2})$ $\pi_2 = \pi_2(V_1, \ldots, V_n; \frac{\delta}{2})$, we get:

Theorem 11 *Let $X_1, \ldots, X_n$, denote a fuzzy random sample, where $X_i \in \mathcal{FN}(\mathcal{R})$ for $i = 1, \ldots, n$, from the distribution with unknown real parameter θ and let $\xi \in [0, 1]$. Let $\Lambda(\theta) \in \mathcal{FN}(\mathcal{R})$ denote a fuzzy perception of θ and let $[\pi_1, \pi_2]$ be two-sided confidence interval for the parameter θ on the confidence level $1 - \delta$. Then a function $\phi : (\mathcal{FN}(\mathcal{R}))^n \to \{0, 1\}$ such that*

$$\phi(X_1, \ldots, X_n) = \begin{cases} 1 & \text{if } (\Lambda_0)^U_{1-\xi} < \Pi^L_{1-\xi} \quad or \quad (\Lambda_0)^L_{1-\xi} > \Pi^U_{1-\xi} \\ 0 & \text{otherwise,} \end{cases} \tag{34}$$

where $\Pi^L_{1-\xi} = \Pi^L_{1-\xi}(X_1, \ldots, X_n; \frac{\delta}{2})$ and $\Pi^U_{1-\xi} = \Pi^U_{1-\xi}(X_1, \ldots, X_n; \frac{\delta}{2})$ are given by (11) and (12), respectively, is a test for hypothesis $H : \Lambda(\theta) = \Lambda_0$ against one-sided alternative $K : Ness(\Lambda_0 \neq \Lambda(\theta)) \geq \xi$, on the significance level δ.

The tests shown above are well defined, because if we use crisp data instead of fuzzy observations and if we replace fuzzy hypotheses by crisp ones our tests reduce to the classical tests of significance. These tests are also very simple in use. Although we consider fuzziness both in data and in hypotheses the output of these test is crisp, i.e. our tests lead to precisely described decision: to rejection or to acceptance of the hypothesis under study. Thus they do not require any defuzzification method, which is also their advantage. Of course, one may construct similar tests for other indices, like the possibility index of strict dominance, etc.

7 Possibilistic interpretation of statistical tests

7.1 The use of possibility measures for the evaluation of the results of statistical tests

One of the most difficult problems that faces a practitioner is the interpretation of the results of statistical tests. The reasons for this problem stem from the fact that the null and the alternative hypotheses are not symmetric. It means, that the result of the test depends upon which hypothesis is considered as the null hypothesis, and which as the alternative one. Thus, it may happen that on a given significance level δ we cannot neither reject the null hypothesis H vs. the alternative K, nor the alternative hypothesis K (treated as a new null hypothesis) vs. H (being a new alternative hypothesis). The problem becomes very serious when both considered hypotheses are "close". For example, when we test the null hypothesis $H : \theta \leq \theta_0$ against the alternative $K : \theta > \theta_0$. As the result of this asymmetry a decision maker is very often advised not to reject the null hypothesis despite the fact that the test data apparently support the alternative. This makes him confused, especially when the decision has to be made only once.

To overcome this problem Hryniewicz (2000a) proposed a new possibilistic interpretation of the results of statistical tests. For this interpretation a well known concept of the observed test size (significance) p has been used. The observed test size p is the minimum value of the significance level δ for which the null hypothesis H has to be rejected for the given observed value of the test statistics T. In other words, if the chosen value of the significance level δ is greater than the value of p, then the null hypothesis H is rejected. Otherwise, we do not have any important reason to reject it. The concept of the significance p is widely used by statisticians despite the fact that it does not have a clear frequency interpretation. Loosely speaking, the value of p represents the measure how reliably the test data support the null hypothesis. There exists a consensus among statisticians, that the values of p not smaller than $0,05$ indicate that there is no reasons to reject the verified hypothesis.

Let us assume that our statistical decision problem is described, as usually, by setting two alternative hypotheses $H : \theta \in \Theta_H$ and $K : \theta \in \Theta_K$. Without the loss of generality we assume that $H : \theta \leq \theta_0$ and $K : \theta > \theta_1$, where $\theta_1 \geq \theta_0$. Hryniewicz (2000a) proposes to consider these two hypotheses *separately*. First, he analyses only the null hypothesis H. For the observed value of the test statistics $t = T(v_1, \ldots, v_n)$ he finds the observed test size p_H for this hypothesis. The value of observed test size p_H shows how the observed data support the null hypothesis. When this value is relatively large we may say that the observed data strongly support H. Otherwise, we should say that the data do not sufficiently support H. It is worthwhile to note that in the latter case we do not claim that the data support K. In the considered case of the null hypothesis $H : \theta \leq \theta_0$ the observed test size is defined as

$$p_H = p_H(\theta_0) = \inf_{\delta} \{\theta_0 \leq \pi_1(v_1, \ldots, v_n; \delta)\} \qquad (35)$$

The same can be done for the alternative hypothesis K, so we can find for this hypothesis the observed test size p_K. When $\Theta_H \cup \Theta_K = \Theta$ we have $p_K = 1 - p_H$. Thus, $p_K(\theta_1) = 1 - p_H(\theta_1)$.

Let us denote by 1 a situation when we decide that the data do not support the considered hypothesis, and by 0 a situation when we decide to accept the hypothesis. According to Hryniewicz (2000a) we propose to evaluate the null hypothesis H by a fuzzy subset $\widetilde{H}$ of $\{0, 1\}$ with the following membership function

$$\mu_{p_H}(x) = \begin{cases} \min[1, 2p_H] & \text{if } x = 0 \\ \min[1, 2(1 - p_H)] & \text{if } x = 1 \end{cases} \qquad (36)$$

which may be interpreted as a *possibility distribution* (see Dubois and Prade, 1997) of H. It is worthy to note that $\sup(\mu_{p_H}(0), \mu_{p_H}(1)) = 1$, and $\mu_{p_H}(1) = 1$ indicates that it is *plausible* that the hypothesis H is not true. On the other hand, when $\mu_{p_H}(0) = 1$ we wouldn't be surprised if H was true. It is necessary to stress here that the values of $\mu_{p_H}(x)$ do not represent the probabilities that H is false or true, but only a *possibility distribution* of the correctness of alternative decisions with respect to this hypothesis.

The same can be done for the alternative hypothesis K. We may evaluate the alternative hypothesis K by a fuzzy subset $\widetilde{K}$ of $\{0, 1\}$ with the following membership function

$$\mu_{p_K}(x) = \begin{cases} \min[1, 2p_K] & \text{if } x = 0 \\ \min[1, 2(1 - p_K)] & \text{if } x = 1 \end{cases} \qquad (37)$$

which may be interpreted as a *possibility distribution* of the correctness of alternative decisions with respect to K.

To choose an appropriate decision, i.e. to choose either H or K Hryniewicz (2000a) proposes to use four measures of possibility defined by Dubois and Prade (1983). First measure is called the *Possibility of Dominance*, and for two fuzzy sets A and B is defined as

$$PD = Poss(A \geq B) = \sup_{x,y: x \geq y} \min\{\mu_A(x), \mu_B(y)\}. \qquad (38)$$

PD is the measure for a possibility that the set A is not dominated by the set B. In the considered problem of testing hypotheses we have (see Hryniewicz, 2000a)

$$PD = Poss\left(\widetilde{H} \geq \widetilde{K}\right) = \max\{\mu_{p_H}(0), \mu_{p_K}(1)\}, \qquad (39)$$

and PD represents a *possibility* that choosing H over K is not a worse solution.

Second measure is called the *Possibility of Strict Dominance*, and for two fuzzy sets A and B is defined as

$$PSD = Poss\,(A > B) = \sup_{x} \inf_{y:y \geq x} \min \{\mu_A(x), 1 - \mu_B(y)\}. \tag{40}$$

PSD is the measure for a possibility that the set A strictly dominates the set B. In the considered problem of testing hypotheses we have

$$PSD = Poss\left(\widetilde{H} > \widetilde{K}\right) = \min \left\{\mu_{p_H}(0), 1 - \mu_{p_K}(0)\right\}, \tag{41}$$

and PSD represents a *possibility* that choosing H over K is a correct decision.

Third measure is called the *Necessity of Dominance*, and for two fuzzy sets A and B is defined as

$$\begin{aligned} ND &= Ness\,(A \geq B) = \\ &= \inf_{x} \sup_{y:y \leq x} \max \{1 - \mu_A(x), \mu_B(y)\} \end{aligned} \tag{42}$$

and represents a *necessity* that the set A dominates the set B. In the considered problem of testing hypotheses we have

$$ND = Ness\left(\widetilde{H} \geq \widetilde{K}\right) = \max \left\{1 - \mu_{p_H}(1), \mu_{p_K}(1)\right\}, \tag{43}$$

and ND represents a *necessity* of choosing H over K.

Fourth measure is called the *Necessity of Strict Dominance*, and for two fuzzy sets A and B is defined as

$$\begin{aligned} &NSD = Ness\,(A > B) = 1 - \sup_{x,y:x \leq y} \min \{\mu_A(x), \mu_B(y)\} = \\ &= 1 - Poss\,(B \geq A) \end{aligned} \tag{44}$$

and represents a *necessity* that the set A strictly dominates the set B (see also 30). In the considered problem of testing hypotheses we have

$$NSD = Ness\left(\widetilde{H} > \widetilde{K}\right) = 1 - \max \left\{\mu_{p_H}(1), \mu_{p_K}(0)\right\}, \tag{45}$$

and NSD represents a strict *necessity* of choosing H over K.

Close examinations of the proposed measures reveals that for all possibility distributions defined by (36) and (37) holds the following relation

$$PD \geq ND \geq PSD \geq NSD. \tag{46}$$

It means that according to the practical situation we can choose the appropriate measure of the correctness of our decision. If the choice between H

and K leads to serious consequences we should choose the NSD measure. In such a case $p_H > 0,5$ is required to have $NSD > 0$. When these consequences are not so serious we may choose the PSD measure. In that case $PSD > 0$ when $p_K < 0,5$, i.e. when there is no strong evidence that the alternative hypothesis is true. Finally, the PD measure gives us the information of the possibility that choosing H over K is not a wrong decision. It is also possible to use the ND measure instead of PD, especially when we expect a slightly stronger evidence that choosing H is allowable.

The proposed method allows to assign possibility measures for the results of classical statistical tests. It assigns numerical values for grades of possibility or necessity of choosing null hypothesis against the alternative one.

7.2 Possibilistic interpretation of statistical tests in the case of fuzzy data

When we observe imprecise data the problem of the interpretation of statistical tests becomes more complicated than in the classical case of crisp data. In the case of precise data, on the given significance level δ, we obtain a binary result of a statistical test: either to reject or to accept the null hypothesis. In the case of fuzzy data, the statistical tests described in the second section of this paper may indicate a fuzzy decision. In such a case, the statistical test does not provide a decision-maker with a univocal solution. The simple method for making an appropriate decision, i.e. for choosing this hypothesis that seems to be a correct one, has been proposed in Hryniewicz (2000b).

To deal with this problem Hryniewicz proposes to extend the results presented in the previous subsection of this paper to the case of fuzzy data. As previously, we assume that a fuzzy random sample $X_1, \ldots, X_n$ is observed. When the test statistics for the considered hypothesis is known for the case of crisp data, we can fuzzify it using the well known extension principle. Then, for this fuzzy test statistics we can define fuzzy confidence intervals applying the methodology proposed in Grzegorzewski (2000a).

As previously, we assume that we test the null hypothesis $H : \theta \leq \theta_0$, against the alternative $K : \theta > \theta_1$. As previously, we consider these two hypothesis separately. When the test data are fuzzy, for each α-cut ($\alpha \in [0,1]$) we introduce the following quantity

$$p_{H,\alpha} = p_{H,\alpha}(\theta_0) = \inf_{\delta} \left\{ \theta_0 \leq \Pi_\alpha^L(X_1, \ldots, X_n; \delta) \right\}, \tag{47}$$

where $\Pi_\alpha^L = \Pi_\alpha^L(X_1, \ldots, X_n; \delta)$ is the lower bound of the α-cut of the one-sided fuzzy confidence interval given by (14) .

For the alternative hypothesis the analogous quantity is defined as

$$p_{K,\alpha} = p_{K,\alpha}(\theta_1) = \inf_{\delta} \left\{ \theta_0 > \Pi_\alpha^U(X_1, \ldots, X_n; \delta) \right\}, \tag{48}$$

where $\Pi_\alpha^U = \Pi_\alpha^U(X_1, \ldots, X_n; \delta)$ is the upper bound of the α-cut of the one-sided fuzzy confidence interval given by (15) . It is easy to show that for each $\alpha_1 \geq \alpha_2$ we have $\Pi_{\alpha_1}^L \geq \Pi_{\alpha_2}^L$,and $\Pi_{\alpha_1}^U \leq \Pi_{\alpha_2}^U$. Thus, we have $p_{H,\alpha_1} \geq p_{H,\alpha_2}$ and $p_{K,\alpha_1} \geq p_{K,\alpha_2}$. Taking this into account, we can define the *fuzzy observed test size (fuzzy significance)* for the null hypothesis as

$$\widetilde{p}_H = \sup\left\{\alpha I_{[p_{H,1},p_{H,\alpha}]}; \alpha \in (0,1]\right\}. \tag{49}$$

and the *fuzzy observed test size (fuzzy significance)* for the alternative hypothesis as

$$\widetilde{p}_K = \sup\left\{\alpha I_{[p_{K,1},p_{K,\alpha}]}; \alpha \in (0,1]\right\}. \tag{50}$$

Having the fuzzy versions of the observed test sizes for both hypotheses we can define the fuzzy equivalents of the possibility and necessity indices presented in the previous subsection. It is worth of noting, that each of these indices can be described as a certain function $IND(p_H, p_K)$, where IND stands for PD, PSD, ND, or NSD, respectively. Using the extension principle, we can define the fuzzy counterparts of these indices $\widetilde{IND}$ using the following α-cuts:

$$(IND)_\alpha = \{IND(p_H, p_K) : p_H \in [p_{H,1}, p_{H,\alpha}],\ p_K \in [p_{K,1}, p_{K,\alpha}]\}. \tag{51}$$

When a decision maker needs an advice presented in a non-fuzzy form, we can defuzzify indices $\widetilde{IND}$ using, for example, Yager's F_1 index defined as

$$F_1 = \frac{\int_0^1 x_{IND}\mu(x_{IND})\, dx_{IND}}{\int_0^1 \mu(x_{IND})\, dx_{IND}} \tag{52}$$

where $\mu(x_{IND})$ is the membership function of the considered possibility (or necessity) index. The values of the defuzzified possibility (or necessity) indices may interpreted in exactly the same way as in the case of crisp data. Without the defuzzification, this interpretation is not, unfortunately, so straightforward.

8 Conclusions

Classical approach to hypotheses testing requires crisp hypotheses and crisp data. However, this assumption appears too rigid for many real-life problems, especially while dealing with imprecise data and linguistic terms. To relax this rigidity fuzzy models were incorporated into statistics. This idea seems to be promising and important for practitioners, but at the same time arises many new problems. It is due to fact that in fuzzy statistics one meets to sources of uncertainty: randomness and imprecision.

This paper together with the previous review presented by Grzegorzewski and Hryniewicz (1997) shows the state of art and some new ideas of hypotheses testing in fuzzy environment. Diversity of approaches indicates that we are yet in the initial stage in constructing the general theory and the commonly accepted methodology has not been worked out. However, it is seen that although many problems are still open, some progress have also been done.

References

Arnold B. F. (1996), *An approach to fuzzy hypothesis testing*, Metrika **44**, 119–126.

Arnold B. F. (1998), *Testing fuzzy hypotheses with crisp data*, Fuzzy Sets and Systems **94**, 323–333.

Dubois D., Prade H. (1980), *Fuzzy Sets and Systems: Theory and Applications*, Academic Press, New York.

Dubois, D., Prade H. (1983), *Ranking fuzzy numbers in the setting of possibility theory*, Information Sciences **30**, 184–244.

Dubois, D., Prade H. (1997), *Qualitative possibility theory and its applications to reasoning and decision under uncertainty*, Belgian Journal of Operations Research, Statistics and Computer Science **37**, 5–28.

Grzegorzewski P. (1998), *Fuzzy tests – defuzzification and randomization*, Fuzzy Sets and Systems (to appear).

Grzegorzewski, P. (2000a), *Testing statistical hypotheses with vague data*, Fuzzy Sets and Systems **112**, 501–510.

Grzegorzewski, P. (2000b), *Testing fuzzy hypotheses with vague data* (submitted).

Grzegorzewski, P., Hryniewicz O. (1997), *Testing hypotheses in fuzzy environment*, Mathware and Soft Computing **4**, 203–217.

Hryniewicz, O. (1994), *Statistical decisions with imprecise data and requirements*, In: R. Kulikowski, K. Szkatula, J. Kacprzyk (Eds.), *Systems Analysis and Decision Support in Economics and Technology*, Omnitech Press, Warsaw, 135–143.

Hryniewicz, O. (2000a), *Possibilistic interpretation of the results of statistical tests*, Proc. of the Eight International Conference IPMU, Madrid, vol. I, 215–219.

Hryniewicz, O. (2000b), *Possibilistic interpretation of fuzzy statistical tests* (submitted for publication).

Kruse R. (1982), *The strong law of large numbers for fuzzy random variables*, Inform. Sci. **28**, 233–241.

Kruse R., Meyer K. D. (1987), *Statistics with Vague Data*, D. Riedel Publishing Company.

Kwakernaak H. (1978), *Fuzzy random variables, Part I: Definitions and theorems*, Inform. Sci. **15**, 1–15.

Kwakernaak H. (1978), *Fuzzy random variables, Part II: Algorithms and examples for the discrete case*, Inform. Sci. **17**, 253–278.

Lehmann E. L. (1986), *Testing Statistical Hypotheses*, Wiley, New York, 2nd ed.

Puri M. L., Ralescu D. A. (1986), *Fuzzy random variables*, J. Math. Anal. Appl. **114**, 409–422.

Taheri S.M., Behboodian J. (1999), *Neyman–Pearson lemma for fuzzy hypotheses testing*, Metrika **49**, 3–17.

Watanabe N., Imaizumi T. (1993), *A fuzzy statistical test of fuzzy hypotheses*, Fuzzy Sets and Systems **53**, 167–178.

Extensions and Applications of Evidence Theory

Di Cai

Department of Computing Science
University of Glasgow
Glasgow G12 8RZ
Scotland, UK
caid@dcs.gla.ac.uk

Although the Dempster-Shafer theory of evidence fits in handling both imprecision and uncertainty very effectively, a large amount of researches have shown its extensions to be needed. The belief function can not handle the issue of comparisons. The orthogonal sum indeed plays a main role in evidential reasoning, but it has been criticised when it is used for combining two largely conflicting pieces of evidence. Also, it can only be used to combine evidence coming from the same frame of discernment.

This work discusses these problems and extends the D-S theory, emphasizing applications to knowledge discovery in dstributed databases.

1 Introduction

Uncertainty pervades a tremendous amount of information in the world. This uncertainty can take many forms. For instance, consider the problems of incomplete or inconsistent data, and the problems of heuristics or general principles, that have exceptions. Similarly, consider the uncertainty associated with locating useful information in a large repository. The problems of uncertainty become even more complex when merging different kinds of information from different sources and of differing reliabilities.

The Dempster-Shafer theory of evidence can deal with both imprecision and uncertainty very well. Distributed databases allow us to integrate data from different sources which have not previously been combined. The Dempster-Shafer theory of evidence and evidential reasoning are particularly suited to the integration of distributed databases. Evidential functions are suited to represent evidence from different sources. Evidential reasoning is carried out by the well-known orthogonal sum.

In the Dempster-Shafer theory, evidence is represented by evidential functions. Particularly, a belief degree is represented by a belief function. However, the belief functions do not handle the issue of comparisons. This problem initiates some proposals to extend the belief functions, and some proposals obtained some results but being not acceptable. We discuss these relevant problems and present our extensions to solve the issue of comparisons.

The orthogonal sum has been criticised when it is used for combining two largely conflicting pieces of evidence as the result obtained is controversial. This problem initiates some proposals to modify the D-S combination rule as well (Yager 1987, Hau & Rashyap 1990). After analysing these proposals, we present a appropriate modified combination method.

Also, the Dempster-Shafer combination operation can only be used to combine evidence coming from the same frame of discernment. We propose an extension of this operation, the orthogonal *product* operation, which can be used for combining evidence coming from different frames of discernment.

In this work we emphasize applications to knowledge discovery and linguistic summaries. The ability to summarize data provides an important method for getting a grasp of the meaning of a large collection of data. It enables humans to help understand the environment in a manner amenable to future useful manipulation. The key idea behind "linguistic summaries for distributed databases" is to use linguistic terms or natural language in presenting information and knowledge which are hidden in distributed databases. For the definition of linguistic summaries, see the other chapter "*Data Mining Based on Evidence Theory*" in this book.

This chapter is organised as follows. Section 1 extends the belief functions to deal with the issue of comparisons. Section 2 discusses orthogonal sum and presents its modifying approach. Section 3 introduces orthogonal product operation which can combine evidence coming from different frames of discernment.

2 Extending the Dempster-Shafer Theory of Evidence

As we know, evidence theory is a generalization of probability theory. In conventional probability theory, we can easily compute the probabilities of comparisons of two independent probability distributions. For example, we can work out the probability that one of the associated random variables is *less than* or *equal to* the other. But in standard D-S theory the definitions of the belief and plausibility functions (*bel* and *pls*) do not handle comparisons like this. When we use mass functions to represent uncertain and imprecise information instead of probability distributions, and conduct applications such as linguistic summary and knowledge extraction, it is necessary to extend the definitions of *bel* and *pls* functions to handle the comparison of two independent basic probability assignments.

Lee (1992) and Lim (1996) tried to extend the definitions. But according to their definitions, the blood types for Jim and his wife are impossible to be equal even if the blood type for each of them is A or B. Clearly, this is not acceptable within an evidential reasoning framework. A more acceptable result would be that the equal possibility is $1/2$.

It is interesting to extend definitions of the belief function and plausibility function and solve the Lee-Lim problem above.

2.1 The Extension for the Issue of Comparison

Since a mass function can be viewed as the probability distribution on the power set of a domain, the idea of comparison of probability distribution of independent random variables can be used to compute the *bel* and *pls* values of the various comparisons of the probability distributions on the power set of a domain.

Definition 2.1 Let X, Y be two independent random set variables which mass functions are $m_X, m_Y : 2^{\Theta} \to [0,1]$, respectively. Then,
(1) $bel(X = Y) = \sum_{U \subseteq \Theta} m_X(U) m_Y(U)$.
(2) $pls(X = Y) = \sum_{U,V \subseteq \Theta; U \cap V \neq \emptyset} m_X(U) m_Y(V)$
$= 1 - \sum_{U,V \subseteq \Theta; U \cap V = \emptyset} m_X(U) m_Y(V)$.

The definition of $bel(X = Y)$ is quite intuitive. Suppose that $f(U, V) = m_X(U) m_Y(V)$ is a joint mass function between two independent random set variables X, Y in the same way as a joint probability distribution between two independent random variables defined in probability theory. Then, $\{(U, U) | U \in 2^{\Theta}\}$ is used for the event $(X = Y)$. So $bel(X = Y)$ is the sum of the products $m_X(U) m_Y(U)$ of the degrees of support from two different sources represented by random set variables X and Y for every subset U in a domain Θ.

Plausibility $pls(X = Y)$ represents the sum of the products of the degrees of support from two different sources represented by random set variables X and Y for any two subsets U, V when they have common elements in a domain Θ.

In order to give an acceptable result for the blood types for Jim and Kim, we introduce the following definition.

Definition 2.2 Let X, Y be two independent random set variables which mass functions are $m_X, m_Y : 2^{\Theta} \to [0,1]$, respectively. Then,
(1) $bel(x = y | x \in X, y \in Y) = \sum_{U,V \subseteq \Theta; U \cap V \neq \emptyset} \frac{|U \cap V|}{|U| \times |V|} m_X(U) m_Y(V)$.
(2) $pls(x = y | x \in X, y \in Y) = bel(x = y | x \in X, y \in Y)$.

The definition of $bel(x = y | x \in X, y \in Y)$ is also quite intuitive: $\{(u, u) | u \in \Theta\}$ is used for the event $(U = V)$, and we have $k' = |U \cap V|$ pairs of (u, u)s in $k = |U| \times |V|$ pairs of (u, v)s $(u \in U, v \in V)$. So $bel(x = y | x \in X, y \in Y)$ is the sum of the products $\frac{|U \cap V|}{|U| \times |V|} m_X(U) m_Y(V)$ of the degrees of support from two different sources represented by random set variables X and Y for every element u in a domain Θ.

The definition of $bel(x = y|x \in X, y \in Y)$ is entirely different from the definition of $bel(X = Y)$. The former is on random element variables x, y in which $x, y \in \Theta$. The latter is on random set variables X, Y in which $X, Y \in 2^{\Theta}$.

And in the element case, we ought to define that the plausibility is equal to the belief.

Similar to the definition above, we can define evidential functions for other comparisons.

Definition 2.3 Let X, Y be two independent random set variables which mass functions are $m_X, m_Y : 2^{\Theta} \to [0, 1]$, respectively. Then

(1) $bel(x \neq y|x \in X, y \in Y) = 1 - pls(x = y|x \in X, y \in Y)$
$= 1 - bel(x = y|x \in X, y \in Y)$.

(2) $pls(x \neq y|x \in X, y \in Y) = 1 - bel(x = y|x \in X, y \in Y)$
$= bel(x \neq y|x \in X, y \in Y)$.

Example 2.1 Suppose we have some information about the blood types for Jim and Kim:

$BloodType_{Jim} = [\{A, B\}^{1.0}]$, $BloodType_{Kim} = [\{A, B\}^{1.0}]$.

Let X and Y denote the random set variables for the blood types for Jim and Kim, respectively. Then we obtain,

$bel(X = Y) = 1$,
$pls(X = Y) = 1.0 \times 1.0 = 1$;
$bel(x = y|x \in X, y \in Y) = \frac{|\{A,B\} \cap \{A,B\}|}{|\{A,B\}| \times |\{A,B\}|} = 1/2$,
$pls(x = y|x \in X, y \in Y) = 1/2$;
$bel(x \neq y|x \in X, y \in Y) = 1 - pls(x = y|x \in X, y \in Y) = 1 - 1/2 = 1/2$,
$pls(x \neq y|x \in X, y \in Y) = 1/2$.

Note that in this example we have $bel(X = Y) = 1$ instead of 0 (Lee 1992), whereas $bel(x = y|x \in X, y \in Y) = 1/2$ means the equal possibility of blood types for Jim and Kim to be 1/2.

For the non-equality comparisons, we have similar definitions. Suppose that Θ is an ordered set. First of all, we give the definitions for $bel(x < y|x \in X, y \in Y)$ and $pls(x < y|x \in X, y \in Y)$ as follows.

Definition 2.4 Let X, Y be two independent random set variables which mass functions are $m_X, m_Y : 2^{\Theta} \to [0, 1]$, respectively. Then,

(1) $bel(x < y|x \in X, y \in Y) = \sum_{U \subseteq \Theta}(m_X(U) \sum_{V \subseteq \Theta, U <^{\forall} V} m_Y(V))$,
where $U <^{\forall} V$ means that $u < v$ for all $u \in U, v \in V$;

(2) $pls(x < y|x \in X, y \in Y) = \sum_{U \subseteq \Theta}(m_X(U) \sum_{V \subseteq \Theta, U <^{\exists} V} m_Y(V))$,
where $U <^{\exists} V$ means that there exist $u \in U$ and $v \in V$ such that $u < v$.

The computation of $bel(x < y|x \in X, y \in Y)$ and $pls(x < y|x \in X, y \in Y)$ is based on the comparison between a given focal element of m_X and a given focal element of m_Y. When every member of a focal element of m_X is less than every member of a focal element of m_Y, the product of the corresponding probabilities is included in the computation of $bel(x < y|x \in X, y \in Y)$.

Similarly, when some member of a focal element of m_X is less than some member of a focal element of m_Y, the product of the corresponding probabilities is included in the computation of $pls(x < y|x \in X, y \in Y)$.

Extending our notation in a obvious way we can get another definition.

Definition 2.5 Let X, Y be two independent random set variables which mass functions are $m_X, m_Y : 2^\Theta \to [0,1]$, respectively. Then,
(1) $bel(x \leq y|x \in X, y \in Y) = \sum_{U \subseteq \Theta}(m_X(U) \sum_{V \subseteq \Theta, U \leq^\forall V} m_Y(V))$,
where $U \leq^\forall V$ means that $u \leq v$ for all $u \in U, v \in V$;
(2) $pls(x \leq y|x \in X, y \in Y) = \sum_{U \subseteq \Theta}(m_X(U) \sum_{V \subseteq \Theta, U \leq^\exists V} m_Y(V))$,
where $U \leq^\exists V$ means that there exist $u \in U$ and $v \in V$ such that $u \leq v$.

The definitions of $bel(x\&y|x \in X, y \in Y)$, and $pls(x\&y|x \in X, y \in Y)$, where $\& \in \{=, \neq, <, \leq; >, \geq\}$, satisfy the following properties:
(a) $0 \leq bel(x\&y|x \in X, y \in Y) \leq pls(x\&y|x \in X, y \in Y) \leq 1$.
(b) $bel(x\&y|x \in X, y \in Y) = 1 - pls(\neg(x\&y)|x \in X, y \in Y)$.
(c) $bel(x\&y|x \in X, y \in Y) + bel(\neg(x\&y)|x \in X, y \in Y) \leq 1$.
(d) $pls(x\&y|x \in X, y \in Y) + pls(\neg(x\&y)|x \in X, y \in Y) \geq 1$.
(e) If all the focal elements of m_X, m_Y are singletons, then $bel(x\&y|x \in X, y \in Y)$ is Bayesian: $bel(x\&y) = pls(x\&y) = P(x\&y)$.

For example, let us show (b) for & being "<", i.e.,

$$bel(x < y|x \in X, y \in Y) = 1 - pls(x \geq y|x \in X, y \in Y).$$

Indeed, we have

$$\begin{aligned}
&bel(x < y|x \in X, y \in Y) + pls(x \geq y|x \in X, y \in Y)\\
&= \textstyle\sum_{U \subseteq \Theta}(m_X(U) \sum_{V \subseteq \Theta, U <^\forall V} m_Y(V))\\
&\quad + \textstyle\sum_{U \subseteq \Theta}(m_X(U) \sum_{V \subseteq \Theta, U \geq^\exists V} m_Y(V))\\
&= \textstyle\sum_{U \subseteq \Theta}(m_X(U)[\sum_{V \subseteq \Theta, U <^\forall V} m_Y(V) + \sum_{V \subseteq \Theta, U \geq^\exists V} m_Y(V)])\\
&= \textstyle\sum_{U \subseteq \Theta}(m_X(U) \sum_{\{V \subseteq \Theta | U <^\forall V\} \cup \{V \subseteq \Theta | U \geq^\exists V\}} m_Y(V))\\
&= \textstyle\sum_{U \subseteq \Theta}(m_X(U) \sum_{V \subseteq \Theta} m_Y(V)) = \sum_{U \subseteq \Theta}(m_X(U)) = 1.
\end{aligned}$$

2.2 Linguistic Summaries on Equal Attributes Values

Having discussed extension of belief functions and plausibility functions for solving issues of comparison using evidence theory, we are now in the position to show how we can apply this extension to knowledge extraction.

The belief and plausibility functions in evidence theory may be regarded as pessimistic and optimistic measures of how strong a rule might be. They are used to determine the amount of support for a proposition and thus to induce rules based on the mass allocations for various propositions. The interval between these two functions can be used for expressing an extended *association rule* and so can be used for extracting knowledge by providing a taxonomy for rule type, expressed as a natural language summary which is then used as the basis of a user-friendly interface.

We attempt to extend linguistic summary to the issues of comparison, and then continue to extend knowledge extraction to the same issues.

For Example 2.1, we have

$BloodType_{Jim} = [\{A, B\}^{1.0}]$, $BloodType_{Kim} = [\{A, B\}^{1.0}]$.
That is, let $S = \{A, B\}$, we have

$m_J(S) = 1, m_J(\bar{S}) = 0, m_J(\Theta) = 0; m_K(S) = 1, m_K(\bar{S}) = 0, m_K(\Theta) = 0.$
Then we have the following rules.

Rule-1. *Always the blood type for Jim is A or B with truth T=1.*

Rule-2. *Always the blood type for Kim is A or B with truth T=1.*

Now the interest we concern with is how much the possibility of the blood types for Jim and Kim to be equal.

Let us answer this question in a linguistic summary. First of all, we

deduce a series of theorems for calculating the belief of existing equal attribute values for two summaries (S, Q_1, T_1), (S, Q_2, T_2).

We know that a summary (S, Q, T) for a frame of discernment Θ is associated with a mass function:

- $m(S) = \alpha$, $m(\bar{S}) = \beta$, $m(\Theta) = 1 - \alpha - \beta$, $(\alpha \geq \beta)$,
  when $Q \in \{always, typically, frequently\}$;
- $m(S) = \beta$, $m(\bar{S}) = \alpha$, $m(\Theta) = 1 - \alpha - \beta$, $(\alpha > \beta)$,
  when $Q \in \{never, atypically, infrequently\}$.

The quantifier Q has the corresponding belief interval $BI = [bel(S), pls(S)]$ with $bel(S) = \sum_{X \subseteq S} m(X)$ and $pls(S) = \sum_{X \cap S \neq \emptyset} m(X)$ as shown in the following table:

Q	*always*	*typically*	*frequently*	*never*	*atypically*	*infrequently*
BI	$[1, 1]$	$[\alpha, 1]$	$[\alpha, 1-\beta]$	$[0, 0]$	$[0, 1-\alpha]$	$[\beta, 1-\alpha]$

The quantifier *always, typically, frequently, never, atypically, infrequently* are also denoted by $Q_{a:[1,1]}, Q_{t:[\alpha,1]}, Q_{f:[\alpha,1-\beta]}, Q_{n:[0,0]}, Q_{at:[0,1-\alpha]}, Q_{inf:[\beta,1-\alpha]}$, respectively.

Theorem 2.1 For data D, if we have summaries (S, Q_1, T_1) and (S, Q_2, T_2), then we have summary

The belief of equal attribute values for S is E,

where $Q_1, Q_2 \in \{always, typically, frequently\}$ and the belief E depends on Q_1 and Q_2 as follows:

Q_1	Q_2	E
$Q_{f:[\alpha_1,1-\beta_1]}$	$Q_{f:[\alpha_2,1-\beta_2]}$	$\frac{\alpha_1\alpha_2}{\lvert S\rvert} + \frac{\beta_1\beta_2}{\lvert S\rvert} + \frac{C_1B_2+C_2B_1+B_1B_2}{\lvert\Theta\rvert}$
$Q_{f:[\alpha_1,1-\beta_1]}$	$Q_{t:[\alpha_2,1]}$	$\frac{\alpha_1\alpha_2}{\lvert S\rvert} + \frac{C_1A_2+\alpha_2B_1+B_1A_2}{\lvert\Theta\rvert}$
$Q_{f:[\alpha_1,1-\beta_1]}$	$Q_{a:[1,1]}$	$\lvert S\rvert^{-1}\alpha_1 + \lvert\Theta\rvert^{-1}B_1$
$Q_{t:[\alpha_1,1]}$	$Q_{t:[\alpha_2,1]}$	$\frac{\alpha_1\alpha_2}{\lvert S\rvert} + \frac{\alpha_1A_2+\alpha_2A_1+A_1A_2}{\lvert\Theta\rvert}$
$Q_{t:[\alpha_1,1]}$	$Q_{a:[1,1]}$	$\lvert S\rvert^{-1}\alpha_1 + \lvert\Theta\rvert^{-1}A_1$
$Q_{a:[1,1]}$	$Q_{a:[1,1]}$	$\lvert S\rvert^{-1}$

where $A_i = 1-\alpha_i$, $B_i = 1-\alpha_i-\beta_i$, $C_i = \alpha_i+\beta_i$, $i=1,2$.

Proof: Consider the case when $Q_1 = frequently$, $Q_2 = frequently$. The other cases can be obtained as special case by setting appropriately α_i, β_i ($\alpha_i > \beta_i$, $i=1,2$), respectively.

Starting with

$m_1(S)=\alpha_1$, $m_1(\bar{S})=\beta_1$, $m_1(\Theta)=1-\alpha_1-\beta_1$;

$m_2(S)=\alpha_2$, $m_2(\bar{S})=\beta_2$, $m_2(\Theta)=1-\alpha_2-\beta_2$,

according to definition 1.2, we have

$$bel(x=y|x\in X, y\in Y)$$
$$= \frac{|S|}{|S|\times|S|}\alpha_1\alpha_2 + \frac{|S|}{|S|\times|\Theta|}\alpha_1(1-\alpha_2-\beta_2) + \frac{|\bar{S}|}{|\bar{S}|\times|\bar{S}|}\beta_1\beta_2 + \frac{|\bar{S}|}{|\bar{S}|\times|\Theta|}\beta_1(1-\alpha_2-\beta_2)$$
$$+\frac{|S|}{|\Theta|\times|S|}(1-\alpha_1-\beta_1)\alpha_2 + \frac{|\bar{S}|}{|\Theta|\times|\bar{S}|}(1-\alpha_1-\beta_1)\beta_2$$
$$+\frac{|\Theta|}{|\Theta|\times|\Theta|}(1-\alpha_1-\beta_1)(1-\alpha_2-\beta_2)$$
$$= \frac{1}{|S|}\alpha_1\alpha_2 + \frac{1}{|S|}\beta_1\beta_2 + \frac{1}{|\Theta|}[(\alpha_1+\beta_1)(1-\alpha_2-\beta_2)+(\alpha_2+\beta_2)(1-\alpha_1-\beta_1)$$
$$+(1-\alpha_1-\beta_1)(1-\alpha_2-\beta_2)].$$

(1) When $Q_1 = frequently$, $Q_2 = typically$, we have

$m_1(S)=\alpha_1$, $m_1(\bar{S})=\beta_1$, $m_1(\Theta)=1-\alpha_1-\beta_1$;

$m_2(S)=\alpha_2$, $m_2(\bar{S})=0$, $m_2(\Theta)=1-\alpha_2$.

So $bel(x=y|x\in X, y\in Y) = \frac{1}{|S|}\alpha_1\alpha_2 + \frac{1}{|\Theta|}[(\alpha_1+\beta_1)(1-\alpha_2)+\alpha_2(1-\alpha_1-\beta_1)+(1-\alpha_1-\beta_1)(1-\alpha_2)]$.

(2) When $Q_1 = frequently$, $Q_2 = always$, we have

$m_1(S)=\alpha_1$, $m_1(\bar{S})=\beta_1$, $m_1(\Theta)=1-\alpha_1-\beta_1$;

$m_2(S)=1$, $m_2(\bar{S})=0$, $m_2(\Theta)=0$.

So $bel(x=y|x\in X, y\in Y) = \frac{1}{|S|}\alpha_1 + \frac{1}{|\Theta|}(1-\alpha_1-\beta_1)$.

(3) When $Q_1 = typically$, $Q_2 = typically$, we have

$m_1(S)=\alpha_1$, $m_1(\bar{S})=0$, $m_1(\Theta)=1-\alpha_1$;

$m_2(S)=\alpha_2$, $m_2(\bar{S})=0$, $m_2(\Theta)=1-\alpha_2$.

So $bel(x=y|x\in X, y\in Y) = \frac{1}{|S|}\alpha_1\alpha_2 + \frac{1}{|\Theta|}[\alpha_1(1-\alpha_2)+\alpha_2(1-\alpha_1)+(1-\alpha_1)(1-\alpha_2)]$.

(4) When $Q_1 = typically$, $Q_2 = always$, we have

$m_1(S)=\alpha_1$, $m_1(\bar{S})=0$, $m_1(\Theta)=1-\alpha_1$;

$m_2(S)=1$, $m_2(\bar{S})=0$, $m_2(\Theta)=0$.

So $bel(x = y|x \in X, y \in Y) = \frac{1}{|\bar{S}|}\alpha_1 + \frac{1}{|\Theta|}(1 - \alpha_1)$.

(5) When $Q_1 = always$, $Q_2 = always$, we have

$$m_1(S) = 1,\ m_1(\bar{S}) = 0,\ m_1(\Theta) = 0;$$

$$m_2(S) = 1,\ m_2(\bar{S}) = 0,\ m_2(\Theta) = 0.$$

So $bel(x = y|x \in X, y \in Y) = \frac{1}{|S|}$.

This theorem tells us how we can obtain directly the result about belief of being equal attribute values from different summaries, and get a new rule about equal value information. For example, for the blood type of Jim and Kim, from the example above, we get a rule as follows.

Rule-3. *The belief of equal blood types for Jim and Kim is E=1/2.*

Similiarly, we have the following theorems.

Theorem 2.2 For data D, if we have summaries (S, Q_1, T_1) and (S, Q_2, T_2), then we have summary

The belief of equal attribute values for S is N,

where $Q_1, Q_2 \in \{never, atypically, infrequently\}$ and the belief N depends on Q_1 and Q_2 as follows:

Q_1	Q_2	N
$Q_{inf:[\beta_1,1-\alpha_1]}$	$Q_{inf:[\beta_2,1-\alpha_2]}$	$\frac{\beta_1\beta_2}{\lvert S\rvert} + \frac{\alpha_1\alpha_2}{\lvert\bar{S}\rvert} + \frac{C_1B_2+C_2B_1+B_1B_2}{\lvert\Theta\rvert}$
$Q_{inf:[\beta_1,1-\alpha_1]}$	$Q_{at:[0,1-\alpha_2]}$	$\frac{\alpha_1\alpha_2}{\lvert\bar{S}\rvert} + \frac{C_1A_2+\alpha_2B_1+B_1A_2}{\lvert\Theta\rvert}$
$Q_{inf:[\beta_1,1-\alpha_1]}$	$Q_{n:[0,0]}$	$\lvert\bar{S}\rvert^{-1}\alpha_1 + \lvert\Theta\rvert^{-1}B_1$
$Q_{at:[0,1-\alpha_1]}$	$Q_{at:[0,1-\alpha_2]}$	$\frac{\alpha_1\alpha_2}{\lvert\bar{S}\rvert} + \frac{\alpha_1A_2+\alpha_2A_1+A_1A_2}{\lvert\Theta\rvert}$
$Q_{at:[0,1-\alpha_1]}$	$Q_{n:[0,0]}$	$\lvert\bar{S}\rvert^{-1}\alpha_1 + \lvert\Theta\rvert^{-1}A_1$
$Q_{n:[0,0]}$	$Q_{n:[0,0]}$	$\lvert\bar{S}\rvert^{-1}$

where $A_i = 1 - \alpha_i$, $B_i = 1 - \alpha_i - \beta_i$, $C_i = \alpha_i + \beta_i$, $i = 1, 2$.

Proof: The manner is similar to the proof of theorem 1.1.

Theorem 2.3 For data D, if we have summaries (S, Q_1, T_1) and (S, Q_2, T_2), then we have summary

The belief of equal attribute values S is C,

where $Q_1 \in \{always, typically, frequently\}$; $Q_2 \in \{never, atypically, infrequently\}$; and the belief C depends on Q_1 and Q_2 as follows:

Q_1	Q_2	C
$Q_{f:[\alpha_1,1-\beta_1]}$	$Q_{inf:[\beta_2,1-\alpha_2]}$	$\frac{\alpha_1\beta_2}{\lvert S\rvert} + \frac{\alpha_2\beta_1}{\lvert S\rvert} + \frac{C_1B_2+C_2B_1+B_1B_2}{\lvert\Theta\rvert}$
$Q_{f:[\alpha_1,1-\beta_1]}$	$Q_{at:[0,1-\alpha_2]}$	$\frac{\alpha_2\beta_1}{\lvert S\rvert} + \frac{C_1A_2+\alpha_2B_1+B_1A_2}{\lvert\Theta\rvert}$
$Q_{f:[\alpha_1,1-\beta_1]}$	$Q_{n:[0,0]}$	$\lvert S\rvert^{-1}\beta_1 + \lvert\Theta\rvert^{-1}B_1$
$Q_{t:[\alpha_1,1]}$	$Q_{inf:[\beta_2,1-\alpha_2]}$	$\frac{\alpha_1\beta_2}{\lvert S\rvert} + \frac{\alpha_1B_2+C_2A_1+A_1B_2}{\lvert\Theta\rvert}$
$Q_{t:[\alpha_1,1]}$	$Q_{at:[0,1-\alpha_2]}$	$\lvert\Theta\rvert^{-1}(\alpha_1A_2+\alpha_2A_1+A_1A_2)$
$Q_{t:[\alpha_1,1]}$	$Q_{n:[0,0]}$	$\lvert\Theta\rvert^{-1}A_1$
$Q_{a:[1,1]}$	$Q_{inf:[\beta_2,1-\alpha_2]}$	$\lvert S\rvert^{-1}\beta_2 + \lvert\Theta\rvert^{-1}B_2$
$Q_{a:[1,1]}$	$Q_{at:[0,1-\alpha_2]}$	$\lvert\Theta\rvert^{-1}A_2$
$Q_{a:[1,1]}$	$Q_{n:[0,0]}$	0

where $A_i = 1-\alpha_i$, $B_i = 1-\alpha_i-\beta_i$, $C_i = \alpha_i+\beta_i$, $i = 1,2$.

Proof: The manner is similar to the proof of theorem 1.1.

Example 2.2 Suppose we have some information about the blood types for Jim and Kim on the frame of discernment $\Theta = \{O, A, B, AB\}$:

$BloodType_{Jim} = [\{A,B\}^{0.8}, \Theta^{0.2}]$,

$BloodType_{Kim} = [\{O,AB\}^{0.5}, \{A,B\}^{0.4}, \Theta^{0.1}]$.

Let $S = \{A, B\}$. We have

$m_J(S) = 0.8$, $m_J(\bar{S}) = 0$, $m_J(\Theta) = 0.2$;

$m_K(S) = 0.4$, $m_K(\bar{S}) = 0.5$, $m_K(\Theta) = 0.1$.

Then from $[bel_J(S), pls_J(S)] = [0.8, 1] = [\alpha_1, 1]$ we can extract knowledge to obtain the following rule:

Rule-4. *Typically the blood type for Jim is A or B with truth T=0.8.*

Also, from $[bel_K(S), pls_K(S)] = [0.4, 0.5] = [0.4, 1-0.5] = [\beta_2, 1-\alpha_2]$ $(\alpha_2 > \beta_2)$ we can extract knowledge to obtain the following rule:

Rule-5. *Infrequently the blood type for Kim is A or B with truth T=0.9.*

Taking $\alpha_1 = 0.8$, $\beta_1 = 0$, $\alpha_2 = 0.5$, $\beta_2 = 0.4$, we find that $A_1 = 1-\alpha_1 = 0.2, A_2 = 1-\alpha_2 = 0.5$; $B_1 = 1-\alpha_1-\beta_1 = 0.2, B_2 = 1-\alpha_2-\beta_2 = 0.1$; $C_1 = \alpha_1+\beta_1 = 0.8, C_2 = \alpha_2+\beta_2 = 0.9$. By theorem 1.3 we obtain $C = \frac{\alpha_1\beta_2}{\lvert S\rvert} + \frac{\alpha_1B_2+C_2A_1+A_1B_2}{\lvert\Theta\rvert} = \frac{0.8*0.4}{2} + \frac{0.8*0.1+0.9*0.2+0.2*0.1}{4} = 0.23$. So we find a new rule as follows.

Rule-6. *The belief of equal blood types for Jim and Kim is 0.23.*

The procedure above can be easily generalized to consider the issue of the case that is not equal for attribute values. For example, for the boold type for Jim and Kim in Example 2.2, we have

$bel(x \neq y | x \in X, y \in Y) = 1 - bel(x = y | x \in X, y \in Y) = 1 - 0.23 = 0.77.$
So we obtain another new rule as follows.

Rule-7. *The belief of inequal blood types for Jim and Kim is 0.77.*

3 The Combination Operation in Evidential Reasoning

The *orthogonal sum* operator $\oplus$ in the Dempster-Shafer theory, known as the *Dempster-Shafer Rule of Combination* for mass functions m_1, m_2, is the following

$$(m_1 \oplus m_2)(A) = \frac{\sum_{X \cap Y = A} m_1(X) m_2(Y)}{\sum_{X \cap Y \neq \emptyset} m_1(X) m_2(Y)}$$

where $N = \sum_{X_i \cap Y_j \neq \emptyset} m_1(X_i) m_2(Y_j)$ is the *normalization constant*; and $E = 1 - N$ is called the *conflict factor.*

The *orthogonal sum* operator $\oplus$ in the Dempster-Shafer theory is remarkable.

On one hand, so many applications show this operator is the real strength of the theory since it can be used for pooling evidence from a variety of independent sources. This operator plays a main role in evidential reasoning. In D-S theory, masses (*basic probability assignments*) are assigned to only those propositions that are supported by evidence. As new evidence becomes available, masses are assigned to another set of propositions and these two sets of conclusions are *combined by the operator* to give a new set of propositions that are supported by the *combined* evidence. After masses have been assigned to the propositions supported by the evidence, the surplus belief is associated with *ignorance.* The evidential reasoning process is the following: *as more evidence being received the belief is spread out onto different propositions, reducing the belief in ignorance.* It can be called the *monotonicity in knowledge accumulation.*

Example 3.1 Consider a case where two doctors, equally competent, given two equal diagnoses for a patient. Let m denote the masses diagnosed by the two doctors, given by

$m(\{Brian\text{-}Haemorrhage\}) = 0.7$, $m(Do\text{-}Not\text{-}Know) = 0.3$.
Then from the intersection table

$m \backslash m$	$\{BH\}_{0.7}$	$\Theta_{0.3}$
$\{BH\}_{0.7}$	$\{BH\}_{0.7\times0.7}$	$\{BH\}_{0.7\times0.3}$
$\Theta_{0.3}$	$\{BH\}_{0.3\times0.7}$	$\Theta_{0.3\times0.3}$

we have

$(m \oplus m)(\{Brian\text{-}Haemorrhage\}) = 0.7 \times 0.7 + 0.7 \times 0.3 + 0.3 \times 0.7 = 0.91,$
$(m \oplus m)(Do\text{-}Not\text{-}Know) = 0.3 \times 0.3 = 0.09.$

This means the proposition that the patient has a Brain-Haemorrhage has been strengthened from 70% with ignorance 0.3 to 91% with the reduced ignorance $0.3^2 = 0.09$.

If the third doctor again makes the same diagnosis, then we have

$$(m \oplus m \oplus m)(\{Brian\text{-}Haemorrhage\}) = 0.91 \times 0.7 + 0.91 \times 0.3 + 0.09 \times 0.7 = 0.973,$$

$$(m \oplus m \oplus m)(Do\text{-}Not\text{-}Know) = 0.09 \times 0.3 = 0.027.$$

This means the proposition that the patient has a Brain-Haemorrhage has been strengthened further from 91% with ignorance 0.09 to 97.3% with the reduced ignorance $0.3^3 = 0.027$. Generally, the same diagnosis given by n doctors will reduce ignorance from 0.3 to 0.3^n.

The belief intervals at Brain-Haemorrhage for $m, m \oplus m, m \oplus m \oplus m, \ldots$ show the monotonicity as the belief intervals are nestedly reducing: $[1 - 0.3, 1] \supset [1 - 0.3^2, 1] \supset [1 - 0.3^3, 1] \supset \ldots$, and the ignorances are reducing: $0.3 > 0.3^2 > 0.3^3 > \ldots$.

On the other hand, the D-S orthogonal sum has been criticised when it is used for combining two largely conflicting pieces of evidence as the result obtained is controversial. It is called the *overwhelming trace problem*.

Example 3.2 Consider a case where two doctors, equally competent, given two different independent diagnoses for a patient. Let m_1 and m_2 denote the masses diagnosed by the two doctors, given by

$m_1(\{Brian\text{-}Haemorrhage\}) = 0.80$, $m_1(\{Meningitis\}) = 0.20$.
$m_2(\{Migraine\}) = 0.95$, $m_2(\{Meningitis\}) = 0.05$.

Then from the intersection table

$m_1 \backslash m_2$	$\{Men\}_{0.05}$	$\{Mig\}_{0.95}$
$\{BH\}_{0.8}$	$\emptyset_{0.8\times0.05=0.04}$	$\emptyset_{0.8\times0.95=0.76}$
$\{Men\}_{0.2}$	$\{Men\}_{0.2\times0.05=0.01}$	$\emptyset_{0.2\times0.95=0.19}$

we have the conflict factor to be $1 - 0.01 = 0.99$. These two pieces of evidence m_1 and m_2 are largely conflicting, and we have

$(m_1 \oplus m_2)(\{Brian\text{-}Haemorrhage\}) = 0,$
$(m_1 \oplus m_2)(\{Meningitis\}) = 0.01/(1 - 0.99) = 1,$
$(m_1 \oplus m_2)(\{Migraine\}) = 0.$

This means the combined support of two pieces of evidence to be focused on *meningitis* which is only weakly supported by the respective original pieces.

It is suggested that an acceptable result would be the *average* (let us denote the average operator by $\uplus$):

$(m_1 \uplus m_2)(\{Brian\text{-}Haemorrhage\}) = 0.8/2 = 0.4,$
$(m_1 \uplus m_2)(\{Meningitis\}) = (0.2 + 0.05)/2 = 0.125,$
$(m_1 \uplus m_2)(\{Migraine\}) = 0.95/2 = 0.475.$

The overwhelming trace problem initiates some proposals to modify the Dempster-Shafer combination rule.

In the next subsections we will analyse some proposals and propose a modification of the orthogonal sum to reach the average result.

3.1 Some Modifications of the Dempster-Shafer Combination

On the Dempster-Shafer framework, Yager (1987) presented an alternative rule of combination (denoted by $\perp$) as follows.

Assume m_3, m_4 are two mass functions on the frame of discernment Θ. Let $\{A_i\}, \{B_j\}$ be their sets of focal elements. Then
(1) $(m_3 \perp m_4)(\emptyset) = 0$;
(2) $(m_3 \perp m_4)(X) = \sum_{A_i \cap B_j = X} m_3(A_i)m_4(B_j)$ for $\emptyset \subset X \subset \Theta$;
(3) $(m_3 \perp m_4)(\Theta) = \sum_{A_i \cap B_j = \Theta} m_3(A_i)m_4(B_j) + E$,
where $E = \sum_{A_i \cap B_j = \emptyset} m_3(A_i)m_4(B_j)$ (the total conflict).

The fundamental distinction between the Dempster-Shafer combination rule and the alternative combination rule is that the former proportionally allocates the normalization factor $N = 1-E$ to the focal elements of $m_3 \oplus m_4$. With the use of this alternative rule the total conflict E is put back into the set Θ.

Thus, by using Yager's rule to Example 3.2, for those two different diagnoses m_1, m_2 above we have
$(m_1 \perp m_2)(\{Meningitis\}) = 0.01, \quad (m_1 \perp m_2)(\Theta) = 0.99.$
This is not an acceptable result since it is far different from the average.

Also, this means knowledge to be not accumulative since the combination $m_1 \perp m_2$ has ignorance 0.99 at Meningitis while m_1, m_2 both have ignorance 0.

Consider another example below.

Example 3.3 Consider a case where two doctors, equally competent, given two equal diagnoses for a patient. Let m_5 denote the masses diagnosed by the two doctors, given by
$m_5(\{Brian\text{-}Haemorrhage\}) = 0.80, \quad m_5(\{Meningitis\}) = 0.20.$
Then from the intersection table

$m_5 \backslash m_5$	$\{BH\}_{0.8}$	$\{Men\}_{0.2}$
$\{BH\}_{0.8}$	$\{BH\}_{0.8\times0.8=0.64}$	$\emptyset_{0.8\times0.2=0.16}$
$\{Men\}_{0.2}$	$\emptyset_{0.2\times0.8=0.16}$	$\{Men\}_{0.2\times0.2=0.04}$

we have the conflict factor to be $1-0.68 = 0.32$. These two pieces of evidence m_5 and m_5 are conflicting, and we have
$(m_5 \oplus m_5)(\{Brian\text{-}Haemorrhage\}) = \frac{0.64}{1-0.32} = \frac{0.64}{0.68} = 0.94,$
$(m_5 \oplus m_5)(\{Meningitis\}) = \frac{0.04}{1-0.32} = \frac{0.04}{0.68} = 0.06.$
This means the combined support of two pieces of evidence to be focused

on *Brain-Haemorrhage* (to degree 0.94) which is strongly supported by (to degree 0.8) the respective original pieces. This is a reasonable result.

On the other hand, using the Yager rule above, we find that

$(m_5 \perp m_5)(\{Brian\text{-}Haemorrhage\}) = 0.64.$
$(m_5 \perp m_5)(\{Meningitis\}) = 0.04.$
$(m_5 \perp m_5)(\Theta) = 0.16 + 0.16 = 0.32.$

Also, this means knowledge to be not accumulative since the Yager combination $m_5 \perp m_5$ has ignorance 0.32 while m_5 has ignorance 0.

On the Dempster-Shafer framework, Hau and Kashyap (1990) also presented an alternative rule of combination (denoted by $\odot$) as follows.

Assume m_3, m_4 are two mass functions on the frame of discernment Θ. Let $\{A_i\}, \{B_j\}$ be their sets of focal elements. Then

(1) $(m_3 \odot m_4)(\emptyset) = 0$;
(2) if $A_i \cap B_j \neq \emptyset$ then let $X = A_i \cap B_j$ and
$(m_3 \odot m_4)(X) = \sum_{A_i \cap B_j = X} m_3(A_i)m_4(B_j)$;
(3) if $A_i \cap B_j = \emptyset$ then let $X = A_i \cup B_j$ and
$(m_3 \odot m_4)(X) = \sum_{A_i \cup B_j = X} m_3(A_i)m_4(B_j)$.

The fundamental distinction between the Dempster-Shafer combination rule and the alternative combination rule is that with the use of this alternative rule the conflict $m_3(A_i)m_4(B_j)$ (for $A_i \cap B_j = \emptyset$) is put into the union $A_i \cup B_j$.

Thus, by using the Hau-Kashyap rule to Example 2.3, we find the intersection (and union) table as follows.

$m_5 \backslash m_5$	$\{BH\}_{0.8}$	$\{Men\}_{0.2}$
$\{BH\}_{0.8}$	$\{BH\}_{0.8\times0.8=0.64}$	$\Theta_{0.8\times0.2=0.16}$
$\{Men\}_{0.2}$	$\Theta_{0.2\times0.8=0.16}$	$\{Men\}_{0.2\times0.2=0.04}$

we have

$(m_5 \odot m_5)(\{Brian\text{-}Haemorrhage\}) = 0.64,$
$(m_5 \odot m_5)(\{Meningitis\}) = 0.04,$
$(m_5 \odot m_5)(DO\text{-}Not\text{-}Know) = 0.16 + 0.16 = 0.32.$

Again, this means knowledge to be not accumulative since the Hau-Kashyap combination $m_5 \odot m_5$ has ignorance 0.32 while m_5 has ignorance 0.

Also, for those two different diagnoses m_1, m_2 in Example 2.2, the intersection (and union) table now is the following.

$m_1 \backslash m_2$	$\{Men\}_{0.05}$	$\{Mig\}_{0.95}$
$\{BH\}_{0.8}$	$\{BH, Men\}_{0.8\times0.05=0.04}$	$\{BH, Mig\}_{0.8\times0.95=0.76}$
$\{Men\}_{0.2}$	$\{Men\}_{0.2\times0.05=0.01}$	$\{Men, Mig\}_{0.2\times0.95=0.19}$

So we have

$(m_1 \odot m_2)(\{Meningitis\}) = 0.01,$
$(m_1 \odot m_2)(\{Meningitis, Migraine\}) = 0.19,$
$(m_1 \odot m_2)(\{Brain\text{-}Haemorrhage, Migraine\}) = 0.76,$

$(m_1 \odot m_2)(\{Brain\text{-}Haemorrhage, Meningitis\}) = 0.04,$
Again, this is not an acceptable result since it is far different from the average.

Also, this means knowledge to be not accumulative since the Hau-Kashyap combination $m_1 \odot m_2$ has the belief interval $[0, 0.76 + 0.19]$ at Migraine and ignorance 0.95 while m_1, m_2 both have ignorance 0.

3.2 The Modified Combination Algorithm

Now, we propose an algorithm for modifying the Dempster-Shafer combination. In the case where conflict between two mass functions is high, we can use the algorithm to compute their modified combination.

Now the normalization constant $N \leq T$, i.e., the conflict factor $E = \sum_{X_i \cap Y_j = \emptyset} m_1(X_i) m_2(Y_j) = 1 - N \geq 1 - T$, where T is a *threshold* which can be arbitrary if there is no way of telling what value will be the best. When one thinks that the conflict between m_1 and m_2 is too high (i.e., $E >>$) this means that too many masses $m_1(X_i)m_2(Y_j)$ have been assigned to empty intersections $X_i \cap Yj = \emptyset$. So we need reassign these masses to non-empty subsets.

The Modified Combination Algorithm.
Given two mass functions m_1, m_2, this algorithm modifies the Dempster-Shafer orthogonal sum $m_1 \oplus m_2$ to our modified mass function $m = m_1 \ominus m_2$ as follows.
Suppose m_1 has focuses $X_1, X_2, ..., X_i, ..., X_I$, and m_2 has focuses $Y_1, Y_2, ..., Y_j$, $..., Y_J$. Make the conventional ***intersection table*** for the Dempster-Shafer orthogonal sum $m_1 \oplus m_2$.

Step 1. [Row by row in the table, delete all empty intersections and reassign masses to non-empty intersections to obtain a new mass function m_r.] For $i = 1, 2, ..., I$, delete all empty intersections and reassign masses to non-empty intersections in the ith row

$$X_{i, m_1(X_i)} \| \; X_i \cap Y_{1, m_1(X_i) m_2(Y_1)} \quad \cdots \quad X_i \cap Y_{j, m_1(X_i) m_2(Y_j)} \quad \cdots \quad X_i \cap Y_{J, m_1(X_i) m_2(Y_J)}$$

to sum to $m_1(X_i)$:

(1) If there is no non-empty intersection in the ith row, i.e., all intersections $X_i \cap Y_1, ..., X_i \cap Y_j, ..., X_i \cap Y_J = \emptyset$, then the first intersection

$$X_i \cap Y_{1, m_1(X_i) m_2(Y_1)}$$

is replaced by $X_{i, m_1(X_i)}$ and all the other (empty) intersections are deleted. (Masses on this row then sum to $m_1(X_i)$.)

(2) If there are non-empty intersections in the ith row, and they are $X_i \cap Y_{j_1}$, $X_i \cap Y_{j_2}, ..., X_i \cap Y_{j_s} \neq \emptyset$; $(s > 0)$, then for these intersections their masses $X_i \cap Y_{j_1, m_1(X_i) m_2(Y_{j_1})}$; $X_i \cap Y_{j_2, m_1(X_i) m_2(Y_{j_2})}$; $\cdots \ldots$; $X_i \cap Y_{j_s, m_1(X_i) m_2(Y_{j_s})}$

are replaced by the following masses

$$X_i \cap Y_{j1}, {}_{m_1(X_i)\frac{m_2(Y_{j1})}{m_2(Y_{j1})+m_2(Y_{j2})+...+m_2(Y_{js})}};$$

$$X_i \cap Y_{j2}, {}_{m_1(X_i)\frac{m_2(Y_{j2})}{m_2(Y_{j1})+m_2(Y_{j2})+...+m_2(Y_{js})}};$$

......

$$X_i \cap Y_{js}, {}_{m_1(X_i)\frac{m_2(Y_{js})}{m_2(Y_{j1})+m_2(Y_{j2})+...+m_2(Y_{js})}}$$

respectively and all the other (empty) intersections are deleted. (Masses on this row then sum to $m_1(X_i)$.)
Notice that masses on row X_i sum to $m_1(X_i)$ and $\sum_{i=1}^{I} m_1(X_i) = 1$. So we can find a combined mass function m_r in the modified intersection table. For a given subset A in $\emptyset \subset A \subseteq \Theta$, we collect all "$A$-intersections" in the table to obtain $m_r(A)$.

For example, our intersection table in Example 3.2 is modified to the following:

$(m_3 \backslash m_4)_r$	$\{Men\}_{0.05}$	$\{Mig\}_{0.95}$
$\{BH\}_{0.8}$	$\{BH\}_{0.8}$	
$\{Men\}_{0.2}$	$\{Men\}_{0.2}$	

So we can find a combined mass function m_r as follows.
$m_r(\{Brain\text{-}Haemorrhage\}) = 0.8$,
$m_r(\{Meningitis\}) = 0.2$.

Step 2. For each column j ($j = 1, 2, ..., J$) do the similar to obtain a mass function m_c.

For example, our intersection table in Example 3.2 is modified to the following.

$(m_3 \backslash m_4)_c$	$\{Men\}_{0.05}$	$\{Mig\}_{0.95}$
$\{BH\}_{0.8}$		$\{Mig\}_{0.95}$
$\{Men\}_{0.2}$	$\{Men\}_{0.05}$	

So we can find a combined mass function m_c as follows.
$m_c(\{Meningitis\}) = 0.05$,
$m_c(\{Migraine\}) = 0.95$.

Step 3. Obtain $m = (m_r + m_c)/2$.

For our example, we have
$m(\{Brain\text{-}Haemorrhage\}) = 0.8/2 = 0.4$,
$m(\{Meningitis\}) = (0.2 + 0.05)/2 = 0.125$,
$m(\{Migraine\}) = 0.95/2 = 0.475$.

Let us give one more example to illustrate our modified combination.

Example 3.4 This example is involved in knowledge discovery in heterogeneous databases. In particular, data stored at different levels of granularity can be incorporated into the discovery process.

Consider a situation where a company has data collected in two separate databases about the population of Scotland (SC-Database) and Northern Ireland (NI-Database). The data collected in the two databases is identical except that the AGE field in the two databases has different granularities. Now suppose we get the following sets of data about smokers from the two databases (Anand, Bell, & Hughes 1996):

AGE	[25,30]	[31,40]	[41,50]	[51,60]	[61,70]	[71,80]	[81,85]
SC: m_1	0.17	0.20	0.13	0.22	0.14	0.08	0.06

AGE	[25,40]	[41,55]	[56,70]	[71,85]
NI: m_2	0.10	0.35	0.40	0.15

where $[I, J]$ represents the interval of integers between I and J. For example, $[25, 30] = \{25, 26, 27, 28, 29, 30\}$.

Given these two sets of data, the important issue arising is as to what conclusion we can draw as a whole about the populations of Scotland and Northern Ireland; i.e., how these two sets of data can be combined to end up a sensible conclusion.

Evidence theory provides combination operators that allow the combination of evidence.

First of all, let us compute the Dempster-Shafer combination $m_1 \oplus m_2$ on the frame of discernment $\Theta = [25, 85]$. Make the intersection table as follows.

$m_1 \backslash m_2$	$[25,40]_{0.10}$	$[41,55]_{0.35}$	$[56,70]_{0.40}$	$[71,85]_{0.15}$
$[25,30]_{0.17}$	$[25,30]_{.17*.10}$	$\emptyset$	$\emptyset$	$\emptyset$
$[31,40]_{0.20}$	$[31,40]_{.20*.10}$	$\emptyset$	$\emptyset$	$\emptyset$
$[41,50]_{0.13}$	$\emptyset$	$[41,50]_{.13*.35}$	$\emptyset$	$\emptyset$
$[51,60]_{0.22}$	$\emptyset$	$[51,55]_{.22*.35}$	$[56,60]_{.22*.40}$	$\emptyset$
$[61,70]_{0.14}$	$\emptyset$	$\emptyset$	$[61,70]_{.14*.40}$	$\emptyset$
$[71,80]_{0.08}$	$\emptyset$	$\emptyset$	$\emptyset$	$[71,80]_{.08*.15}$
$[81,85]_{0.06}$	$\emptyset$	$\emptyset$	$\emptyset$	$[81,85]_{.06*.15}$

So the normalization constant is

$$
\begin{aligned}
N &= (0.17+0.20)*0.10+(0.13+0.22)*0.35+(0.22+0.14)*0.40+(0.08+\\
&\quad 0.06)*0.15\\
&= 0.37*0.10+0.35*0.35+0.36*0.40+0.14*0.15\\
&= 0.0370+0.1225+0.1440+0.0210 = 0.3245,
\end{aligned}
$$

and the Dempster-Shafer combination $m_1 \oplus m_2$ is the following:

$$\begin{aligned}
(m_1 \oplus m_2)[25,30] &= 0.0170/0.3245 \approx 0.056,\\
(m_1 \oplus m_2)[31,40] &= 0.0200/0.3245 \approx 0.066,\\
(m_1 \oplus m_2)[41,50] &= 0.0455/0.3245 \approx 0.109,\\
(m_1 \oplus m_2)[51,55] &= 0.0770/0.3245 \approx 0.242,\\
(m_1 \oplus m_2)[56,60] &= 0.0880/0.3245 \approx 0.276,\\
(m_1 \oplus m_2)[61,70] &= 0.0560/0.3245 \approx 0.178,\\
(m_1 \oplus m_2)[71,80] &= 0.0120/0.3245 \approx 0.041,\\
(m_1 \oplus m_2)[81,85] &= 0.0090/0.3245 \approx 0.032.
\end{aligned}$$

Here the normalization constant $N = 0.3245$ is not too small. Nevertheless, if one still thinks that the conflict factor $E = 1 - N = 0.6755$ is too high, then we can compute the modified combination $m_1 \ominus m_2$ as follows.

Step 1. Compute m_r

m_r	$[25,40]_{0.10}$	$[41,55]_{0.35}$	$[56,70]_{0.40}$	$[71,85]_{0.15}$
$[25,30]_{0.17}$	$[25,30]_{.17}$	$\emptyset$	$\emptyset$	$\emptyset$
$[31,40]_{0.20}$	$[31,40]_{.20}$	$\emptyset$	$\emptyset$	$\emptyset$
$[41,50]_{0.13}$	$\emptyset$	$[41,50]_{.13}$	$\emptyset$	$\emptyset$
$[51,60]_{0.22}$	$\emptyset$	$[51,55]_{.22*\frac{.35}{.35+.40}}$	$[56,60]_{.22*\frac{.40}{.35+.40}}$	$\emptyset$
$[61,70]_{0.14}$	$\emptyset$	$\emptyset$	$[61,70]_{.14}$	$\emptyset$
$[71,80]_{0.08}$	$\emptyset$	$\emptyset$	$\emptyset$	$[71,80]_{.08}$
$[81,85]_{0.06}$	$\emptyset$	$\emptyset$	$\emptyset$	$[81,85]_{.06}$

Step 2. Compute m_c

m_c	$[25,40]_{0.10}$	$[41,55]_{0.35}$	$[56,70]_{0.40}$	$[71,85]_{0.15}$
$[25,30]_{0.17}$	$[25,30]_{\frac{.17}{.17+.20}*.10}$	$\emptyset$	$\emptyset$	$\emptyset$
$[31,40]_{0.20}$	$[31,4]_{\frac{.20}{.17+.20}*.10}$	$\emptyset$	$\emptyset$	$\emptyset$
$[41,50]_{0.13}$	$\emptyset$	$[41,50]_{\frac{.13}{.13+.22}*.35}$	$\emptyset$	$\emptyset$
$[51,60]_{0.22}$	$\emptyset$	$[51,55]_{\frac{.22}{.13+.22}*.35}$	$[56,60]_{\frac{.22}{.22+.14}*.40}$	$\emptyset$
$[61,70]_{0.14}$	$\emptyset$	$\emptyset$	$[61,70]_{\frac{.14}{.22+.14}*.40}$	$\emptyset$
$[71,80]_{0.08}$	$\emptyset$	$\emptyset$	$\emptyset$	$[71,80]_{\frac{.08}{.08+.06}*.15}$
$[81,85]_{0.06}$	$\emptyset$	$\emptyset$	$\emptyset$	$[81,85]_{\frac{.06}{.08+.06}*.15}$

Step 3. Compute $m_1 \ominus m_2 = (m_r + m_c)/2$:

$$\begin{aligned}
(m_1 \ominus m_2)[25,30] &= \tfrac{1}{2}(0.17 + \tfrac{0.17}{0.17+0.20} * 0.10) \approx 0.108,\\
(m_1 \ominus m_2)[31,40] &= \tfrac{1}{2}(0.20 + \tfrac{0.20}{0.17+0.20} * 0.10) \approx 0.127,\\
(m_1 \ominus m_2)[41,50] &= \tfrac{1}{2}(0.13 + \tfrac{0.13}{0.13+0.22} * 0.35) \approx 0.131,
\end{aligned}$$

$$(m_1 \ominus m_2)[51,55] = \frac{1}{2}(0.22 * \frac{0.35}{0.35+0.40} + \frac{0.22}{0.13+0.22} * 0.35) \approx 0.162,$$
$$(m_1 \ominus m_2)[56,60] = \frac{1}{2}(0.22 * \frac{0.40}{0.35+0.40} + \frac{0.22}{0.22+0.14} * 0.40) \approx 0.181,$$
$$(m_1 \ominus m_2)[61,70] = \frac{1}{2}(0.14 + \frac{0.14}{0.22+0.14} * 0.40) \approx 0.147,$$
$$(m_1 \ominus m_2)[71,80] = \frac{1}{2}(0.08 + \frac{0.08}{0.08+0.06} * 0.15) \approx 0.082,$$
$$(m_1 \ominus m_2)[81,85] = \frac{1}{2}(0.06 + \frac{0.06}{0.08+0.06} * 0.15) \approx 0.062.$$

4 The Orthogonal Product and Multiple Attributes

So far we have discussed various aspects of data integration and linguistic summaries that involve only one attribute from distributed databases. In the cases of multiple attributes from distributed databases, we should extend a new fundamental operation, the *orthogonal product*, to provide means for combining the corresponding pieces of evidence and summaries from different sources.

Two evidential functions, each has its own attribute and own frame of discernment, can be combined on Cartesian products of the two frames of discernment.

Consider summaries of a form exemplified by "most people are tall and young". Assume x_1 and x_2 are two attributes appearing in the databases D_1 and D_2. Let S_1 and S_2 are summarizers associated with each of thesee attributes, respectively. The general form of the linguistic summary above is as follows.

Q objects in the database have x_1 is S_1 and x_2 is S_2 with truth T.

In this case, we need use the orthogonal product operation.

The orthogonal product combination. Let m_1 and m_2 be mass functions on the respective frames Θ_1 and Θ_2. Then the function m on $\{(A,B)|\emptyset \subseteq A \subseteq \Theta_1, \emptyset \subseteq B \subseteq \Theta_2\}$ defined by

$$m((A,B)) = m_1(A)m_2(B)$$

for all $\emptyset \subseteq A \subseteq \Theta_1, \emptyset \subseteq B \subseteq \Theta_2$ is a mass function since

$$m((\emptyset,\emptyset)) = m_1(\emptyset)m_2(\emptyset) = 0 \times 0 = 0;$$
$$\sum_{(A,B)\subseteq(\Theta_1,\Theta_2)} m((A,B)) = \sum_{(A,B)\subseteq(\Theta_1,\Theta_2)} m_1(A)m_2(B)$$
$$= (\sum_{A\subseteq\Theta_1} m_1(A))(\sum_{B\subseteq\Theta_2} m_2(B)) = 1 \times 1 = 1.$$

This mass function m is called the *orthogonal product* of m_1 and m_2, and denoted by $m_1 \otimes m_2$.

A useful device to compute the orthogonal products is the coordinate table.

A *coordinate table* with values of m_1 and m_2 along the rows and columns, respectively, is a helpful practical computational aid. Only non-zero values assigned by m_1 and m_2 need be considered since if $m_1(X)$ and/or $m_2(Y)$ is zero, then the product $m_1(X)m_2(Y)$ contributes 0 to $(m_1 \otimes m_2)((X,Y))$. Entry (i,j) in the table corresponds to the Cartesian product (X,Y) of the subsets X in row i and Y in column j. The product $m_1(X)m_2(Y)$ is calculated and inserted next to the product subset (X,Y) in the table.

The belief function, and plausibility function given by $m_1 \otimes m_2$ are called the *orthogonal product* of bel_1 and bel_2, pls_1 and pls_2, and are denoted by $bel_{m_1 \otimes m_2} = bel_1 \otimes bel_2$, $pls_{m_1 \otimes m_2} = pls_1 \otimes pls_2$, respectively. That is, the orthogonal belief and plausibility products $bel_1 \otimes bel_2$ and $pls_1 \otimes pls_2$ of mass functions m_1 and m_2 are the following, respectively:

(1) $(bel_1 \otimes bel_2)((A,B)) = \sum_{X \subseteq \Theta_1, Y \subseteq \Theta_2; (X,Y) \subseteq (A,B)} (m_1 \otimes m_2)((X,Y))$,
(2) $(pls_1 \otimes pls_2)((A,B)) = \sum_{X \subseteq \Theta_1, Y \subseteq \Theta_1; (X,Y) \cap (A,B) \neq (\emptyset,\emptyset)} (m_1 \otimes m_2)((X,Y))$.

Theorem 4.1 For all $(A,B) \subseteq (\Theta_1, \Theta_2)$, we have

$$(bel_1 \otimes bel_2)((A,B)) = bel_1(A) bel_2(B).$$

Proof. $(bel_1 \otimes bel_2)((A,B))$
$= \sum_{X \subseteq \Theta_1, Y \subseteq \Theta_2; (X,Y) \subseteq (A,B)} (m_1 \otimes m_2)((X,Y))$
$= \sum_{X \subseteq \Theta_1, Y \subseteq \Theta_2; X \subseteq A, Y \subseteq B} m_1(X) m_2(Y)$
$= (\sum_{X \subseteq \Theta_1, X \subseteq A} m_1(X))(\sum_{Y \subseteq \Theta_2, Y \subseteq B} m_2(Y))$
$= bel_1(A) bel_2(B)$.

Theorem 4.2 For all $(A,B) \subseteq (\Theta_1, \Theta_2)$, we have

$$pls_1 \otimes pls_2)((A,B)) = pls_1(A) pls_2(B).$$

Proof. $(pls_1 \otimes pls_2)((A,B))$
$= \sum_{X \subseteq \Theta_1, Y \subseteq \Theta_2; (X,Y) \cap (A,B) \neq (\emptyset,\emptyset)} (m_1 \otimes m_2)((X,Y))$
$= \sum_{X \subseteq \Theta_1, Y \subseteq \Theta_2; (X \cap A, Y \cap B) \neq (\emptyset,\emptyset)} (m_1 \otimes m_2)((X,Y))$
$= \sum_{X \subseteq \Theta_1, Y \subseteq \Theta_2; X \cap A \neq \emptyset, Y \cap B \neq \emptyset} m_1(X) m_2(Y)$
$= (\sum_{X \subseteq \Theta_1, X \cap A \neq \emptyset} m_1(X))(\sum_{Y \subseteq \Theta_2, Y \cap B \neq \emptyset} m_2(Y))$
$= pls_1(A) pls_2(B)$.

Example 4.1 Suppose we distributed databases DataSet-H and DataSet-A. Attribute x in DataSet-H is the *height* (of a person) with domain $\Theta_1 = [100, 250]$, attribute y in DataSet-A is the *age* (of a person) with domain $\Theta_2 = [10, 180]$, where $[I, J]$ represents the interval of integers between I and J. Now, suppose

DataSet-H = $\{[165,166], 168, 172, [174,176], [175,177], 175, 175, [176,177], 178, 181\}$;

DataSet-A = $\{18, [20,21], 23, 23, [24,26], 29, 30, 36, 47, [50,51]\}$.

That is, we have two data sets for ten persons in two distributed databases. Then, we get mass functions m_1, m_2 on $2^{\Theta_1}, 2^{\Theta_2}$ as follows.

$m_1(\{165,166\}) = m_1(\{168\}) = m_1(\{172\}) = m_1(\{174,176\}) = m_1(\{175, 177\}) = m_1(\{176,177\}) = m_1(\{178\}) = m_1(\{181\}) = .1, m_1(\{175\}) = .2.$
$m_2(\{18\}) = m_2(\{20,21\}) = m_2(\{24,26\}) = m_2(\{29\}) = m_2(\{30\}) = m_2 (\{36\}) = m_2(\{47\}) = m_2(\{50,51\}) = .1, m_2(\{23\}) = .2.$

So we find that

$bel_1(tall) = bel_1(\{heigth > 175\})$
$= m_1[176,177] + m_1\{178\} + m_1\{181\} = 0.3,$
$pls_1(tall) = pls_1(\{heigth > 175\})$
$= m_1[176,177] + m_1\{178\} + m_1\{181\} + m_1[174,176] + m_1[175,177] = 0.5.$
$bel_2(young) = bel_2(\{age < 25\})$
$= m_2\{18\} + m_2[20,21] + m_2\{23\} = 0.4,$
$pls_2(young) = pls_2(\{age < 25\})$
$= m_2\{18\} + m_2[20,21] + m_2\{23\} + m_2[24,26] = 0.5.$

According to theorems 3.1 and 3.2, we obtain the following results:

$(bel_1 \otimes bel_2)(tall\&young) = (bel_1 \otimes bel_2)((tall, young))$
$= (bel_1(tall)) \times (bel_2(young)) = 0.3 \times 0.4 = 0.12.$
$(pls_1 \otimes pls_2)(tall\&young) = (pls_1 \otimes pls_2)((tall, young))$
$= (pls_1(tall)) \times (pls_2(young)) = 0.5 \times 0.5 = 0.25.$

Finally, from $T = 1 - ign(tall\&young) = 1 - (0.25 - 0.12) = 0.87$ we reach a summary as follows.

Rule-8 *12% to 25% people are tall and young with truth T=0.87.*

The above procedure can be easily generalized to consider any number of attributes.

5 Summary

This work discusses some extensions of the Dempster-Shafer theory of evidence. We extend the belief functions to handle the issues of comparison; present a modified combination rule, the modified orthogonal sum; and introduce an operation, orthogonal product operation, to combine the evidence comes from the different frames of discernment. Also, we investigate their applications to knowledge discovery and linguistic summaries.

Acknowledgment

This research was supported in part by EPSRC.

References

1. Anand, S. S.; Bell, D. A.; Hughes, J. G. 1996, "EDM: a general framework for data mining based on evidence theory", *Data & Knowledge Engineering*, 18 (1996), 189-223.

2. Anand, S. S.; Scotney, B. W.; Tan, M. G.; McClean, S. I.; Bell, D. A.; Hughes, J. G.; Magill, I. C. 1997, "Designing a kernel for data mining", *IEEE Expert*, Vol.12(1997), No.2, 65-74.
3. Bell, D. 1993, "From data properties to evidence", *IEEE Transactions of Knowledge & Data Engineering*, Vol.5, No.6 (December 1993), 965-969.
4. Bell, D.; Guan, J.; Lee, S. K. 1996, "Generalized union and project operations for pooling uncertain and imprecise information", *Data & Knowledge Engineering*, Vol.18, 89-117.
5. Dubois, D.; Prade, H. 1986, Possibility theory: An Approach to Computerized Processing of Uncertainty, (1986) Plenum Press, New York.
6. Dubois, D.; Prade, H. 1987, "An Approach to Approximate Reasoning Based on the Dempster Rule of Combination", International Journal of Expert Systems, Vol.1, No.1 (1987) 67-85.
7. Dubois, D.; Prade, H. 1990, "Modeling Uncertain and vague knowledge in Possibility and Evidence Theories", Uncertainty in Artificial Intelligence 4 (1990) 303-318.
8. Guan, J.; Bell, D. 1991-2, Evidence Theory and its Applications, Vol.1-2, (1991-2) North-Holland.
9. Hau, H. Y.; Kashyap, R. L. 1990, Belief combination and propagation in a lattice-structured inference network, *IEEE Trans. on Systems, Man, and Cybernatics*, Vol.20(1990), No.1, 45-57.
10. Lee, S.K. 1992a, "Imprecise and Uncertain Information in Databases: An Evidential Approach", In Proc. 8th Int. Conf. Data Engineering (1992) 614-621.
11. Lee, S. K. 1992b, "A extended relational database model for uncertain and imprecise information", *Proceedings of the 18th VLDB Conference, Vancouver, British Columbia, Canada, 1992*, 211-220.
12. Lim, E. P.; Srivastava, J.; Shekhar, S. 1996, "An evidentail reasoning approach to attribute value conflict resolution in database integration", *IEEE Transactions of Knowledge & Data Engineering*, Vol.8, No.6 (October 1996), 707-722.
13. Shafer, G. 1976, A Mathematical theory of evidence. Princeton, NJ: Princeton Univ. Press, (1976).
14. Yager, R. R. 1982, "A new approach to the summarization of data", in *Information Sciences*, 28, 69-86.
15. Yager, R. R. 1987, "On the Dempster-Shafer framework and new combination rules", in *Information Sciences*, 41, 93-137.
16. Yager, R. R. 1991, "On linguistic summaries of data", in *Knowledge Discovery in Databases*, G. Piatetsky-Shapiro and B. Frawley, Eds., MIT Press, Cambridge, MA, 347-363.
17. Yager, R. R. 1996, "Database discovery using fuzzy sets", in *International Journal of Intelligent Systems*, Vol.11, 691-712.

2.

ANALYSIS OF EVIDENCE AND DATA

Data Mining Based on Evidence Theory

Di Cai

Department of Computing Science
University of Glasgow
Glasgow G12 8RZ
Scotland, UK
caid@dcs.gla.ac.uk

Data mining is currently one of the most exciting and challenging areas. The concept of linguistic summaries is a user friendly way to express information contained in a database. Commonsense knowledge is a collection of linguistic propositions, that is, propositions with implied imprecise and uncertain quantifiers. The Dempster-Shafer (D-S) theory of evidence fits in handling both imprecision and uncertainty very well. This work uses the D-S theory to establish a framework for dealing with integration of data for distributd databases. Using evidence theory, this work also introduces concept of linguistic summaries and studies their applications to knowledge discovery in distributed databases. We illustrate the use of linguistic summaries by means of running examples using data of risk status conditioned on savings accounts from banks.

1 Introduction

Knowledge discovery, or data mining, in distributed databases is currently one of the most exciting and challenging areas where database techniques are coupled with techniques of artificial intelligence and various mathematical sub-disciplines.

Databases are frequently used for purposes of storing and providing access to statistical data typically concerning government, or collected for research purposes. Such data are frequently held in a number of databases which may be distributed among computing facilities at different sites. Knowledge discovery has been defined as the non-trivial extraction of implicit, previously unknown and potentially useful information from data (Frawley, Piatetsky-Shapiro, Matheus 1991; Matheus, Chan, Piatetsky-Shapiro 1993). It is becoming increasingly common for the distributed data to be used for purposes of data mining as part of a distributed database management system. Dao and Perry (1995) have discussed applying a data miner to heterogeneous schema integration. Ribeiro, Kaufman, and Kerschberg (1995) have discussed knowledge discovery from multiple databases. By integrating such data it may be able to discover new knowledge which is nontrivial and was previously unknown. In order to provide a suitable framework for knowledge discovery, a means must be established of translating rules or interesting statements into a representation which is easily understood by the computer user. This typically involves using natural language to provide linguistic summaries of the data such as "In the companies with share option schemes, 85% of persons between 35 and 40 years old and in management positions buy the maximum number of shares allowed". Such a rule would obviously constitute valuable knowledge for both the companies involved and the financial institutes wishing to target their advertising campaigns.

An ability to handle commonsense knowledge is a crucial need in the development of artificial intelligence. A number of variants of so-called non-monotonic logics have been introduced as aids in developing these commonsense reasoning mechanisms. Commonsense knowledge is a collection of linguistic propositions, that is, propositions with implied imprecise and uncertain quantifiers. A computational framework for dealing with uncertainty can achieve a much higher level of expressive power and provide a basis for representing the meaning of a much wider variety of linguistic propositions.

Zadeh (1978, 1984) presented a theory of *commonsense knowledge*. The idea is that commonsense knowledge may be viewed as a collection of *dispositions*. A disposition is a proposition with implied fuzzy quantifiers. Upon restoration of fuzzy quantifiers, a disposition can be converted into a proposition with explicit fuzzy quantifiers. For example, the disposition "Tall men are not very agile" is converted into the proposition "*Most* tall men are not very agile" with a fuzzy quantifier "most" which provides an approximate characterization of the proportion of "men who are not very agile" among "men who are tall". Since traditional logical systems provide no methods for representing the meaning of propositions containing fuzzy quantifiers, such systems are not suitable for dealing with commonsense knowledge. It is suggested that an appropriate computational framework is provided by fuzzy logic. The inapplicability of traditional logical systems to the representation of commonsense knowledge reflects the fact that such systems make no provision for dealing with uncertainty. Thus, in predicate logic, for example, a proposition is either true or false and no gradations of truth or membership are allowed. By contrast, in the case of commonsense knowledge, a typical proposition contains a multiplicity of sources of uncertainty. A proposition is a collection of fuzzy constraints. For example, the proposition "*Debbie is a few years older than Dana*" involves the fuzzy constraint few. The constrained variable is the attribute Age, or the difference in the ages of Debbie and Dana, $Age(Debbiee) - Age(Dana)$. For a suitable data base $DB = Population[Name, Age] + Few[Number, \mu]$, we test constraint few on DB to find its *truth* T, where $Population[Name, Age]$ is a non-fuzzy relation which lists names of individuals and their age; $Few[Number, \mu]$ is a fuzzy relation which associates with each value of $Number$ the degree μ (to which $Number$ is compatible with the intended meaning of few). That is,

$\mu_{Few}(Number)$ is a membership function of *Few* on the set of numbers.

The ability to summarize data, or observations, provides an important method for getting a grasp of the meaning of a large collection of data, and allows us to communicate our observations about the world in a useful and understandable manner. It also provides a starting point for making useful inferences from large collections of data. In a broad sense, summarization plays an essential role in the manner of how we learn about our environment.

The concept of linguistic summaries is a user friendly way to express information contained in a database. The key idea behind "linguistic summaries for distributed databases" is to use linguistic, natural language, terms in expressing information and knowledge which are hidden in distributed databases. Yager (1982, 1991, 1995, 1996) suggested the use of linguistic summaries as a tool for representing various kinds of knowledge discovered in a database. Kacprzyk and Yager (1984) introduced fuzzy linguistic quantifiers exemplified by most, almost all, etc. Yager introduced the summarization of data based upon the theory of fuzzy subsets and defined a summary of a data set D by means of the tuple (Q, S, T), where S was a summarizer, e.g. "age between 30 and 40", Q was a quantifier, e.g. *most*, and T was a measure of validity, in Yager's case a *truth* function. So, for example, a summary might say that "most engineers in the firm are aged between 30 and 40" with truth function value 0.8. Yager described S and Q in terms of fuzzy membership functions and T as a function of S and Q which took values in the interval $[0, 1]$: $T = T(S, Q)$. Assume $D = [a_1, a_2, ..., a_n]$ is the collection of values that appear in the database for the attribute x. The Yager methodology used to calculate the truth T of a linguistic summary is the following.

Step 1. For each $a_i \in D$, calculate $S(a_i)$, the degree to which a_i satisfies the summarizer S.

Step 2. Let $r = \frac{1}{n}\sum_{i=1}^{n} S(a_i)$.

Step 3. $T = Q(r)$, the grade of membership of r in the quantifier.

Summarizing, the truth is $T(S, Q) = Q(\frac{1}{n}\sum_{i=1}^{n} S(a_i))$.

For example, let x be the attribute *Age* and let data set be D=[19,21,25,28, 29,30,33,47,50,52]. Yager(1996) gave the following summaries:

- *Most people in database D are at least 25 with truth T=0.77.*
- *Most people in database D are young with truth T=0.0*
- *About half the ages in database D are near 30 with truth T=0.95.*

Distributed databases allow us to integrate data from different sources which have not previously been combined. The Dempster-Shafer theory of evidence and evidential reasoning are particularly suited to the integration of distributed databases. A number of approaches have been explored in the literature in an attempt to produce a suitable representation for using natural language to provide linguistic summaries of the data contained in the distributed databases. The approach presented in this chapter draws on these sources to provide a methodology which allows the Dempster-Shafer representation of the data to be mapped into a natural language summary, that is, our ability to meaningfully give these kinds of summaries is based on the use of evidence theory to provide a formal semantics for the terms used in the linguistic summary. In the procedure to be subsequently described, we shall use this ability to formalize the quantifiers by a belief function over the domain of attribute to represent the meaning of a linguistic concept or value. For a given attribute we should be able to derive the mass distribution associated with this attribute from some information that is being in the databases. Then with the aid of the mass distribution we can evaluate truth to determine which are the valid summaries. We will also investigate the properties of such a representation scheme.

This chapter is organised as follows. Section 1 initially gives a brief introduction to Dempster-Shafer's theory of evidence which is used as the basis of our methodology. Section 2 presents some concepts of linguistic summaries based on D-S theory, and discusses its applications to knowledge discovery in distributed databases by giving some examples for risk analysis using data of credit risk status conditioned on savings accounts from banks. Section 3 is to treat samples with different weights. Section 4 discusses how the summaries derived from distributed database can be combined by using D-S rule of combination – orthogonal sum. Section 5 is to integrate distributed databases with different granularities for the summaries. Section 6 introducts the concepts of antonym and negation of summaries. Section 7 investigates the properties of summaries. And finally section 8 discusses the further issue of knowledge extraction.

2 The Dempster-Shafer Theory of Evidence

Most database systems have been designed to handle only accurate data, even though in the real world, data we encounter everyday is often *imprecise* and *uncertain.* Dubois and Prade (1986, 1987, 1990) distinguished between these two complementary aspects of imperfect information by stating that the concept of *imprecision* is relevant to the content of an attribute value of a data object, while the concept of *uncertainty* is relevant to the degree of truth of its attribute value.

An attribute value of data object is called *precise* when the set corresponding to its value component is a singleton; otherwise, it is called *imprecise.* For example, if you do not know what is the precise value of blood type for Jim then the imprecise value can be stored in a database as $\{O, A, B, AB\}$.

When the fullest amount of belief (confidence) cannot be given to a given data, the data is said to be *uncertain.* During the last two decades, several ways of representing and maintaining uncertainty in artificial intelligence have been proposed and used: certainty factors (used in MYCIN), probability theory, and possibility theory (or fuzzy set theory).

The Dempster-Shafer theory of evidence can deal with both uncertainty and imprecision very effectively. In this theory, a basic probability assignment (mass function) is used to represent imprecise and uncertain evidence. The basic probability assignment is a mapping from the power set of a domain to the real number interval between 0 and 1. That is, a subset of the domain is used to represent the aspect of imprecision, while a probability is used to handle the aspect of uncertainty.

In evidence theory, evidence is represented in terms of evidential functions (Shafer 1976, Guan & Bell 1991-2). These functions are in one-to-one correspondence and each representation conveys the same information as any of the others. The essential ingredient is the mass function, and masses are assigned to only those propositions that are supported by evidence. In the remainder of this chapter we indicate the space, the frame of discernment, by the symbol Θ, from which choices are to be made, and the contenders for choice are required to be mutually exclusive and exhaustive.

A function $m: 2^{\Theta} \rightarrow [0,1]$ is a *mass* function if it satisfies

$m(\emptyset) = 0, \quad \sum_{X \subseteq \Theta} m(X) = 1.$

Closely related to the mass function is the belief function and plausibility function.

For a mass function m, the corresponding *belief* function *bel* and *plausibility* function *pls* are defined as follows:

(1) $bel(A) = \sum_{X \subseteq A} m(X) = 1 - pls(\bar{A})$,

(2) $pls(A) = \sum_{X \subseteq \Theta, X \cap A \neq \emptyset} m(X) = 1 - bel(\bar{A})$

for all $A \subseteq \Theta$, where $\bar{A} = \Theta - A$.

Since $bel(A) \leq pls(A)$ for all $A \subseteq \Theta$, we can introduce the *belief interval* $[bel(A), pls(A)]$ for subset A. Here $bel(A)$ gives the degree to which the current evidence supports A, and $pls(A)$ gives the degree to which A remains plausible. We also call $bel(A)$ the *lower probability* of A, $pls(A)$ the *upper probability* of A. The difference of them represents the residual ignorance, and the corresponding *ignorance* function is defined as follows:

(3) $ign(A) = pls(A) - bel(A)$

for all $A \subseteq \Theta$.

3 The Concepts of Linguistic Summaries

Let (U, A) be a database, where $U = \{u_1, u_2, ..., u_{|U|}\}$ is a universe and A is the set of attributes. If U is a group of people, then examples of attribute could be age, nationality, or occupation. In our definition below, an attribute $x \in A$ has a frame of discernment Θ, and a set $X = 2^{\Theta}$ as its *domain*. Associated with attribute x is a multiset $D_x = [x(u_1), x(u_2), ..., x(u_{|U|})]$ called the *data set* of attribute x, where $x(u_i) \in X$ for $i = 1, 2, ..., |U|$, is called the value of attribute x at object u_i. It is important to note here that each $x(u_i)$ is a subset on Θ, $i = 1, 2, ..., |U|$.

A linguistic summary associated with attribute x is a proposition that contains some information about objects in the databases to which x belongs. Now we formally present our definition of *summary* as follows:

(S, Q, T): Q objects in the database have that the attribute x is concept S with truth T.

In the formulation above S is called the *summarizer*, represented by an ordinary subset in domain X of attribute x. For example, if we are considering the attribute x as *age*, then the concepts *young* and *near 30* would be indicated with summarizers as S_1=*young*=$\{$*natural number* $n \mid n < 25\}$, and S_2=*near 30*=$\{25, 26, 27, 28, 29, 30, 31, 32, 33, 34, 35\}$, respectively.

The second component in our linguistic summary is the *quantifier* Q, some objects belong to a class of concepts, represented by a belief interval $[bel(S), pls(S)]$ from evidence theory which is a sub-interval of the unit interval $[0, 1]$, where *bel* and *pls* are belief and plausibility functions on attribute domain $X = 2^{\Theta}$. Alternatively, we can take Q as "$bel(S) \times 100\%$ to $pls(S) \times 100\%$". For example, we can take $Q_1 = [0.4, 0.6]$ as "40% to 60%" (about half), and $Q_2 = [0.95, 1.0]$ as "95% to 100%" (nearly all).

Finally, in the formulation above T is the *truth*: $T(S) = 1 - ign(S) = bel(S) + bel(\bar{S})$. The truth T associated with each linguistic summary is called the measure of validity of the summary. Given the data set D associatd with attribute x the value T is used to indicate the truth of the statement that Q objects have the property that attribute x is concept S.

It should be noted that in evidence theory we have the property $bel(S) + bel(\bar{S}) \leq 1$. This means that we are now allowed to reserve some of our belief for something rather than either S or not S. As mentioned above, here $bel(S)$ gives the degree to which the current evidence supports S, and $pls(S) = 1 - bel(\bar{S})$ gives the degree to which S remains plausible. The difference $ign(S) = pls(S) - bel(S)$ represents the residual unknown, namely *ignorance*. So it is natural and reasonable to define truth as "not unknown", i.e., $T(S) = 1 - ign(S)$.

A fundamendtal characteristic of the formulation above is that the summarizer and quantifier are expressed in linguistic terms. The benefit of using these linguistic summaries is that we can provide some information about the databases in terms or statements that are very natural for people to comprehend. Now we indicate the advantage of our definition. The quantifier Q is represented by a belief interval $Q(S) = [bel(S), pls(S)]$, and the summary "80% to 86% customers are credit worthy" (here $[bel(S), pls(S)] = [80\%, 86\%]$) is more informative and valuable than the summary "many customers are credit worthy" for inference about the viability of opening a new kind of account for a bank.

If x is the attribute *Credit Risk*, then the application of the definition of summarization above is illustrated by the example below.

Example 1. Consider the following 2 samples (McClean & Scotney 1996).

DataSet-1: Sample of credit risk status conditioned on savings accounts from bank 1 (weight w_1=100)

Status-of-savings-account	*Total*	m_1
1. *less-than* 100 *DM*	60	0.60
2. 100-499 *DM*	9	0.09
3. 500-999 *DM*	6	0.06
4. 1000 *DM or more*	6	0.06
Θ. *unknown*	19	0.19

DataSet-2: Sample of credit risk status conditioned on savings accounts from bank 2 (weight w_2=200)

Status-of-savings-account	*Total*	m_2
1. *less-than* 100 *DM*	114	0.570
2. 100-499 *DM*	23	0.115
3. 500-999 *DM*	14	0.070
4. 1000 *DM or more*	13	0.065
Θ. *unknown*	36	0.180

The masses m_1 and m_2 in the tables above are obtained via the proportions of the total number of some class of customers to the number of samples. From m_1 and m_2 we can readly obtain their belief and plausibility, and then have the following summaries for each sample:

- ***60% to 79% customers with savings accounts being less than 100 DM in bank 1 are credit risk with truth T=0.81.***
- ***57% to 75% customers with savings accounts belng less than 100 DM in bank 2 are credit risk with truth T=0.82.***

4 Treating Different Weights

Discounting factors may be needed if there are some reasons to believe that some of the samples are in some way suspect. In evidence theory, the discount operation on evidential functions can be used to provide a weighted combination.

Let m be a mass function on 2^Θ. Given a real number α such that $0 < \alpha < 1$, the function m^α defined by
(1) $m^\alpha(\Theta) = (1-\alpha)m(\Theta) + \alpha$,
(2) $m^\alpha(A) = (1-\alpha)m(A)$
for all $A \subset \Theta$ is said to be the *discounted function with rate* α of mass function m.

In accordance with the usual pattern, the ***discounted belief*** function and the ***discounted plausibility*** function of the ***discounted mass*** function m^α are denoted by bel^α, pls^α and defined by

(3) $bel^\alpha(A) = \sum_{X\subseteq A} m^\alpha(X)$,
(4) $pls^\alpha(A) = \sum_{X\cap A\neq\emptyset} m^\alpha(X)$
for all $A \subseteq \Theta$, respectively.

The discount operation can be widely used to a weighted combination. Here and next section we will give some examples to show how the discount operation can be used to integrate distributed databases with different weights.

Example 2. Discounting mass functions m_1, m_2 in example 1 by respective rates $\alpha_1 = 200/300$, $\alpha_2 = 100/300$ since DataSet-1 and DataSet-2 have respective wights $w_1 = 100$, $w_2 = 200$, we obtain

$m_1^{\alpha_1}(\{1\}) = 0.60(1/3) = 0.20$, $m_1^{\alpha_1}(\{2\}) = 0.09(1/3) = 0.03$,
$m_1^{\alpha_1}(\{3\}) = 0.06(1/3) = 0.02$, $m_1^{\alpha_1}(\{4\}) = 0.06(1/3) = 0.02$,
$m_1^{\alpha_1}(\Theta) = 1 - \sum_{A\subset\Theta} m_1^{\alpha_1}(A) = 0.71.$
$m_2^{\alpha_2}(\{1\}) = 0.57(2/3) = 0.373$, $m_2^{\alpha_2}(\{2\}) = 0.115(2/3) = 0.077$,
$m_2^{\alpha_2}(\{3\}) = 0.07(2/3) = 0.047$, $m_2^{\alpha_2}(\{4\}) = 0.065(2/3) = 0.043$,
$m_2^{\alpha_2}(\Theta) = 1 - \sum_{A\subset\Theta} m_2^{\alpha_2}(A) = 0.48.$

So we have the following summaries for each sample:

- *20% to 91% customers with savings accounts being less than 100 DM in bank 1 are credit risk with truth T=0.29.*
- *37.3% to 85.3% customers with savings accounts being less than 100 DM in bank 2 are credit risk with truth T=0.52.*

5 Combination Operation in Evidential Reasoning

An importment problem in any knowledge discovery type problem is the issue of combination. In this section we discuss how such summaries, derived from distributed databases, may be combined by using the Orthogonal Sum Operator.

The *orthogonal sum* operator $\oplus$ in the Dempster-Shafer theory, known as the *Dempster-Shafer Rule of Combination*, is the following:

$$(m_1 \oplus m_2)(A) = \frac{\sum_{X\cap Y=A} m_1(X)m_2(Y)}{\sum_{X\cap Y\neq\emptyset} m_1(X)m_2(Y)}$$

for mass functions m_1, m_2.

The orthogonal sum operator $\oplus$ plays a main role in evidential reasoning. So many applications show this operator is the real strength of the evidence theory since it can be used for pooling evidence from a variety of independent sources. We now give the example below to illustrate its application.

Example 3. Combining mass functions in example 2, the process we find the orthogonal sum $m_1^{2/3} \oplus m_2^{1/3}$ is the following.

$m_1^{2/3} \backslash m_2^{1/3}$	$\{1\}_{.373}$	$\{2\}_{.077}$	$\{3\}_{.047}$	$\{4\}_{.043}$	$\Theta_{.480}$
$\{1\}_{.20}$	$\{1\}_{.0746}$	$\emptyset$	$\emptyset$	$\emptyset$	$\{1\}_{.0960}$
$\{2\}_{.03}$	$\emptyset$	$\{2\}_{.0023}$	$\emptyset$	$\emptyset$	$\{2\}_{.0144}$
$\{3\}_{.02}$	$\emptyset$	$\emptyset$	$\{3\}_{.0009}$	$\emptyset$	$\{3\}_{.0096}$
$\{4\}_{.02}$	$\emptyset$	$\emptyset$	$\emptyset$	$\{4\}_{.0009}$	$\{4\}_{.0096}$
$\Theta_{.71}$	$\{1\}_{.2648}$	$\{2\}_{.0547}$	$\{3\}_{.0334}$	$\{4\}_{.0305}$	$\Theta_{.3408}$

The normalization constant is $N = \sum_{X \cap Y \neq \emptyset} m_1(X) m_2(Y) = 0.9325$. So we have

$(m_1^{2/3} \oplus m_2^{1/3})(\{1\}) = (.0746 + .0960 + .2648)/N = .4669,$
$(m_1^{2/3} \oplus m_2^{1/3})(\{2\}) = (.0023 + .0144 + .0547)/N = .0766,$
$(m_1^{2/3} \oplus m_2^{1/3})(\{3\}) = (.0009 + .0096 + .0334)/N = .0471,$
$(m_1^{2/3} \oplus m_2^{1/3})(\{4\}) = (.0009 + .0096 + .0305)/N = .0439,$
$(m_1^{2/3} \oplus m_2^{1/3})(\Theta) = .3408/N = .3655.$

We obtain combined data for credit risk status conditioned on savings accounts from bank 1 and bank 2 as follows:

Status-of-Savings-Account	$m_1^{2/3} \oplus m_2^{1/3}$
1. *less-than* 100 *DM*	0.4669
2. 100-499 *DM*	0.0766
3. 500-999 *DM*	0.0471
4. 1000 *DM or more*	0.0439
Θ. *unknown*	0.3655

Then, we can compute the combined belief function and plausibility function and ignorance function $bel_1^{2/3} \oplus bel_2^{1/3}$, $pls_1^{2/3} \oplus pls_2^{1/3}$, $ign_1^{2/3} \oplus ign_2^{1/3}$ from $m_1^{2/3} \oplus m_2^{1/3}$. For example,

$(bel_1^{2/3} \oplus bel_2^{1/3})(\{1\}) = \sum_{X \subseteq \{1\}} (m_1^{2/3} \oplus m_2^{1/3})(X) = 0.4669$
$(pls_1^{2/3} \oplus pls_2^{1/3})(\{1\}) = \sum_{X \cap \{1\} \neq \emptyset} (m_1^{2/3} \oplus m_2^{1/3})(X) = 0.8324.$
$(ign_1^{2/3} \oplus ign_2^{1/3})(\{1\}) = 0.8324 - 0.4669 = 0.3655.$

So we have the following summary for the combined data:

46.69% to 83.24% customers with savings accounts being less than 100 DM in bank 1 and bank 2 are credit risk with truth T=0.6345.

6 Multiple Granularities in Frames of Discernment

As we know that the evidence from different databases can be combined only if they have the same granularity in frame of discernment. In the case of different granularity of frames some operations are needed to unify these granularities. Refining and coarsening operations are ones which can be widely used to move up and down a concept hierarchy to unify the granularities, and thereby unify the different frames in the distributed databases.

6.1 Refining and Coarsening Mapping

Refining and coarsening operations are based on refining and coarsening mappings between frames of discernment. So we introduce these mappings firstly.

Let $\Xi = \{x_1, x_2, ..., x_n\}$ and Θ be two frames of discernment. A mapping $\sigma : 2^\Xi \to 2^\Theta$ is said to be a *refining mapping* from 2^Ξ to 2^Θ if for all singletons $\{x_i\}$ $(i = 1, 2, ..., n)$ in 2^Ξ the corresponding subsets $\sigma(\{x_i\}) = R_i$ $(i = 1, 2, ..., n)$ in 2^Θ form a partition of Θ, and
(1) $\sigma : \emptyset \to \emptyset$,
(2) $\sigma : A = \cup_{x_i \in \Xi}\{x_i\} \to \cup_{x_i \in \Xi} R_i$.

The *inner* and *outer* coarsening mappings $\sigma^{-1^-} : 2^\Theta \to 2^\Xi$ and $\sigma^{-1^+} : 2^\Theta \to 2^\Xi$ of σ are defined as follows.
(3) $\sigma^{-1^-}(R) = \{x \mid x \in \Xi \text{ and } \sigma(\{x\}) \subseteq R\}$,
(4) $\sigma^{-1^+}(R) = \{x \mid x \in \Xi \text{ and } \sigma(\{x\}) \cap R \neq \emptyset\}$,
for all $R \subseteq \Theta$.

For example, let $\Xi = \{w, n\}, \Theta = \{a, b, c\}$, where: w — Credit Worthy, n — Not Credit Worthy; a — High Credit Worthy, b — Average Credit Worthy, c — Low Credit Worthy. Here we have two frames of discernment Ξ and Θ, associated with different granularities for the attribute *Credit Worthy*. In order to unify these frames we can, for example, let σ: $2^\Xi \to 2^\Theta$, i.e., σ: $\emptyset \to \emptyset$, $\{w\} \to \{a, b\}$, $\{n\} \to \{c\}$, $\{w, n\} \to \{a, b, c\}$. Then σ is a refining mapping from 2^Ξ to 2^Θ. And using the inner and outer coarsening mapping formulae above, we have some results in the following:

$\sigma^{-1^-}(\{a\}) = \{x \mid x \in \Xi \text{ and } \sigma(\{x\}) \subseteq \{a\}\} = \emptyset$,
$\sigma^{-1^-}(\{a, b\}) = \{x \mid x \in \Xi \text{ and } \sigma(\{x\}) \subseteq \{a, b\}\} = \{w\}$;
$\sigma^{-1^+}(\{a\}) = \{x \mid x \in \Xi \text{ and } \sigma(\{x\}) \cap \{a\} \neq \emptyset\} = \{w\}$,
$\sigma^{-1^+}(\{a, b\}) = \{x \mid x \in \Xi \text{ and } \sigma(\{x\}) \cap \{a, b\} \neq \emptyset\} = \{w\}$.

6.2 Refining and Coarsening Operations

Having refining and coarsening mappings between frames of discernment we now proceed to introduce the corresponding operations on the evidential functions. Let σ be a refining mapping from 2^Ξ to 2^Θ and m_Θ be a mass function on 2^Θ. The *mass* function $m_\Xi = m_\Theta^{\sigma^{-1}}$ on 2^Ξ *coarsened* from m_Θ and σ^{-1} is represented as follows:
(1) $m_\Xi(X) = m_\Theta^{\sigma^{-1}}(X) = \sum_{R \subseteq \Theta, \sigma^{-1^+}(R) = X} m_\Theta(R)$
for all non-empty subsets $X \subseteq \Xi$.

And we have
(2) $bel_\Xi(X) = bel_\Theta^{\sigma^{-1}}(X) = bel_\Theta(\sigma(X)) = bel_\Theta(R)$,
(3) $pls_\Xi(X) = pls_\Theta^{\sigma^{-1}}(X) = pls_\Theta(\sigma(X)) = pls_\Theta(R)$,
for $X \subseteq \Xi$.

Conversely, let σ be a refining mapping from 2^Ξ to 2^Θ and m_Ξ be a mass function on 2^Ξ. The *mass* function $m_\Theta = m_\Xi^\sigma$ on 2^Θ *refined* from m_Ξ and σ is represented as follows:

(1) $m_\Theta(R) = m^\sigma_\Xi(R) = m_\Xi(X)$ if $R = \sigma(X)$ for a (unique) $X \subseteq \Xi$.
(2) $m_\Theta(R) = m^\sigma_\Xi(R) = 0$ if $R \neq \sigma(X)$ for any $X \subseteq \Xi$.
And we have
(3) $bel_\Theta(R) = bel^\sigma_\Xi(R) = bel_\Xi(\sigma^{-1^-}(R))$,
(4) $pls_\Theta(R) = pls^\sigma_\Xi(R) = pls_\Xi(\sigma^{-1^+}(R))$,
for $R \subseteq \Theta$.

Notice that, in general, for refining mapping $\sigma : 2^\Xi \to 2^\Theta$, and mass function m_Θ on 2^Θ, we have $(m_\Theta^{\sigma^{-1}})^\sigma = m^\sigma_\Xi = m'_\Theta \neq m_\Theta$.

We now give an example to show how the refining and coarsening operations can be used to integrate distributed databases with different frames of discernment associated with the attribute *Credit Worthy*, and to conduct knowledge discovery.

Example 4. Suppose that we have two data sets for credit worthiness of bank customers as follows.

Credit-Worthiness\Customer-Number	*DataSet-1*
Have-Credit-Worthy	80
No-Credit-Worthy	14
Do-not-know	6

Worthiness\Customer-Number	*DataSet-2*
High-Credit	100
Average-Credit	40
Low-Credit	0
High-Credit, Average-Credit	0
Average-Credit, Low-Credit	50
Do-not-know	10

Notice that the Worthiness fields in the two data sets have different *granularities*.
For DataSet-1, we have frame of discernment $\Xi = \{w, n\}$, where: w – Have Credit Worthy, n – No Credit Worthy. Its mass function m_3 is: $m_3(\{w\}) = 0.80$, $m_3(\{n\}) = 0.14$, $m_3(\Xi) = 0.06$.
For DataSet-2, we have frame of discernment $\Theta = \{a, b, c\}$, where: a – High Credit Worthy, b – Average Credit Worthy, c – Low Credit Worthy. Its mass function m_{04} is: $m_{04}(\{a\}) = 0.50$, $m_{04}(\{b\}) = 0.20$, $m_{04}(\{b, c\}) = 0.25$, $m_{04}(\Theta) = 0.05$.
In order to combine m_3 on 2^Ξ and m_{04} on 2^Θ, we needed unify their frames of discernment. We can coarsen m_{04} to mass function $m_{04}^{\sigma^{-1}} = m_4$ on 2^Ξ as follows:

$m_4(\{w\}) = m_{04}^{\sigma^{-1}}(\{w\}) = \sum_{R \subseteq \Theta, \sigma^{-1^+}(R)=\{w\}} m_{04}(R) = 0.7$,
$m_4(\{n\}) = m_{04}^{\sigma^{-1}}(\{n\}) = \sum_{R \subseteq \Theta, \sigma^{-1^+}(R)=\{n\}} m_{04}(R) = 0.0$,
$m_4(\{w, n\}) = m_{04}^{\sigma^{-1}}(\{w, n\}) = \sum_{R \subseteq \Theta, \sigma^{-1^+}(R)=\{w,n\}} m_{04}(R) = 0.0$,

$m_4(\Xi) = m_{04}^{\sigma^{-1}}(\Xi) = \sum_{R \subseteq \Theta, \sigma^{-1+}(R)=\{\Xi\}} m_{04}(R) = 0.3.$

Then we can combine m_3 and m_4 on the unified frame of discernment $\Xi = \{w, n\}$ to get $m_3 \oplus m_4$ as follows.

2^{Ξ}	$m_3 \oplus m_4$	$bel_3 \oplus bel_4$	$pls_3 \oplus pls_4$	$ign_3 \oplus ign_4$
$\{w\}$	0.934	0.934	0.953	0.019
$\{n\}$	0.047	0.047	0.066	0.019
Ξ	0.019	1.000	1.000	0.000

By using the coarsening operation and combining operation we pool the information which are from two different data set, and discover new knowledge represented by the summary:

93.4% to 95.3% customers in DataSet-1 and DataSet-2 are Have Credit Worthy with truth T=0.981

7 The Antonym and Negation of Summaries

An important idea associated with a linguistic quantifier is the concept of antonym and negation. If Q is a linguistic quantifier then its antonym and negation are also a linguistic quantifier. Here are their formal definitions.

Definition 1. For some data set D, let S be a summarizer. A quantifier is said to be the antonym of quantifier $Q(S) = [q_1, q_2]$, and denoted by $\widetilde{Q}(S)$, if $\widetilde{Q}(S) = [1 - q_2, 1 - q_1]$.

According to this definition, from $Q(S) = [bel(S), pls(S)]$, we find the following.

- $\widetilde{Q}(S) = [1 - pls(S), 1 - bel(S)] = [bel(\bar{S}), pls(\bar{S})] = Q(\bar{S})$.
- $\widetilde{Q}(\bar{S}) = Q(\bar{\bar{S}}) = Q(S)$.

The operation of taking an antonym is involutionary, that is,

- $\widetilde{\widetilde{Q}}(S) = \widetilde{Q}(\bar{S}) = Q(\bar{\bar{S}}) = Q(S)$.

From this we see that antonyms come in pairs. Prototypicaly examples of antonym pairs are *all* - *none*, *most* — *few*, and *at least 20%* — *at most 80%*.

Care must be taken to distinguish between the antonym of a quantifier and its negation that is now definited as follows.

Definition 2. For some data D, let S be a summarizer. A quantifier is said to be the *negation* of quantifier $Q(S) = [q_1, q_2]$, and denoted by $\neg Q(S)$, if $\neg Q(S) = \neg[q_1, q_2] = [0, 1] - [q_1, q_2]$ and
(1) $\neg Q = [0, q_1) \cup (q_2, 1]$ when $q_1 \neq 0$ and $q_2 \neq 1$,
(2) $\neg Q = (q_2, 1]$ when $q_1 = 0$ and $q_2 \neq 1$,
(3) $\neg Q = [0, q_1)$ when $q_1 \neq 0$ and $q_2 = 1$,
(4) $\neg Q$ undefined, when $q_1 = 0$ and $q_2 = 1$.

According to this definition, from $Q(S) = [bel(S), pls(S)]$, we can find, for example, when $q_1 \neq 0$, $q_2 \neq 1$, the following.

- $\neg Q(S) = \neg[bel(S), pls(S)] = [0, bel(S)) \cup (pls(S), 1]$.
- $\neg Q(\bar{S}) = \neg[bel(\bar{S}), pls(\bar{S})] = [0, bel(\bar{S})) \cup (pls(\bar{S}), 1]$.

It is the same as the antonym that the operation of taking a negation is also involutionary. That is,

- $\neg\neg Q(S) = \neg([0, bel(S)) \cup (pls(S), 1]) = \neg[0, bel(S)) \cap \neg(pls(S), 1]$
  $= [bel(S), 1] \cap [0, pls(S)] = [bel(S), pls(S)] = Q(S)$.

From this we see that negations come in pairs, too. Prototypicaly examples of negation pairs are *all — not all*, *most — not most*, and *at least 20 % — less than 20%*.

For the negation of the quantifier Q we can explain their meanings, say, $[0, q_1)$ as *less than* $q_1 \times 100\%$, $(q_2, 1]$ as *greater than* $q_2 \times 100\%$, and $[0, q_1) \cap (q_2, 1]$ as *less than* $q_1 \times 100\%$ *or greater than* $q_2 \times 100\%$, etc.

For concept S with the quantifier $Q(S) = [bel(S), pls(S)]$ the truth

$$T(S, Q) = 1 - ign(S) = 1 - (pls(S) - bel(S))$$

indicates that we are now already aware of the degree of validity about the summary (S, Q, T). By the same way we can definite the truth

$$T(S, \neg Q) = pls(S) - bel(S) = ign(S)$$

for the concept S with the quantifier $\neg Q(S) = [0, bel(S)) \cup (pls(S), 1]$. The truth $T(S, \neg Q)$ indicates that we are now already aware of the degree of validity about the summary (S, $\neg Q$, T).

Obviously, for the quantifier $\neg Q = [0, q_1) \cup (q_2, 1]$ with $q_1 = 0$ and $q_2 \neq 1$, we have $T(S, \neg Q) = pls(S)$; and for $\neg Q = [0, q_1) \cup (q_2, 1]$ with $q_1 \neq 0$ and $q_2 = 1$, we have $T(S, \neg Q) = 1 - bel(S)$.

We now give an example to illustrate representations and their meanings of antonyms and negations of some quantifiers as follows.

Example 5. Consider some quantifiers below:

Q_1="*all*", i.e., $Q_1 = [1.0, 1.0]$,
Q_2="*most*", i.e., $Q_2 = [0.95, 1.0]$.
Q_3="*at least 20%*", i.e., $Q_3 = [0.2, 1.0]$.
Q_4="*at most a half*", i.e., $Q_4 = [0.0, 0.5]$.

We have their antonyms and negations as follows:

$\tilde{Q}_1 = [1 - 1.0, 1 - 1.0] = [0.0, 0.0]$, that means "*none*";
$\neg Q_1 = [0.0, 1)$, that means "*not all*".
$\tilde{Q}_2 = [1 - 1.0, 1 - 0.95] = [0.0, 0.05]$, that means "*few*";
$\neg Q_2 = [0.0, 0.95)$, that means "*not most*".
$\tilde{Q}_3 = [1 - 1.0, 1 - 0.2] = [0.0, 0.8]$, that means "*at most 80%*";
$\neg Q_3 = [0.0, 0.2)$, that means "*less then 20%*".
$\tilde{Q}_4 = [1 - 0.5, 1 - 0.0] = [0.5, 1.0]$, that means "*at laest a half*";
$\neg Q_4 = [0.5, 1.0)$, that means "*greater than a half*".

8 The Properties of Summaries

A number of interesting properties can be associated with these linguistic summaries.

8.1 The Measure of Validity of Summaries

Let $\tilde{Q}$ and $\neg Q$ represent the antonym and negation of Q, respectively. Consider a proposed summary (S, Q, T) with a quantifier $Q = Q(S) = [q_1, q_2] = [bel(S), pls(S)]$. Summarizing the results in the last section, the following theorems can be obtained.

Theorem 1. For data D, if we have a summarizer S with $Q(S) = [q_1, q_2]$ and $T = T(S, Q(S))$ then we have $T(S, \tilde{Q}(S)) = T(\bar{S}, Q(\bar{S})) = T(S, Q(S))$.

Proof: From the discussion in the last section we have

$$\tilde{Q}(S) = Q(\bar{S}) = [bel(\bar{S}), pls(\bar{S})].$$

Then we have

$$T(S, \tilde{Q}(S)) = 1 - (pls(\bar{S}) - bel(\bar{S})) = 1 - ign(\bar{S}) = T(\bar{S}, Q(\bar{S})).$$

And also we have

$$\begin{aligned} T(S, \tilde{Q}(S)) &= 1 - [(1 - bel(S)) - (1 - pls(s))] = 1 - (pls(S) - bel(S)) \\ &= 1 - ign(S) = T(S, Q). \end{aligned}$$

The prototypical manifestation of this is that the summaries "*Most people in this firm are interested in buying stocks*" will have the same measure of validity as "*Few people in this firm are not interested in buying stocks*".

Theorem 2. For data D, if we have a summarizer S with $Q(S) = [q_1, q_2]$ and truth $T = T(S, Q)$ then we have $T(S, \neg Q(S)) = 1 - T(S, Q(S))$.

Proof: Again from the discussion in the last section, we have

(1) when $q_1 \neq 0$ and $q_2 \neq 1$, $T(S, Q(S)) = 1 - (pls(S) - bel(S))$ and

$$\begin{aligned} T(S, \neg Q(S)) &= pls(S) - bel(S) = 1 - [1 - (pls(S) - bel(S))] \\ &= 1 - T(S, Q(S)). \end{aligned}$$

(2) when $q_1 = 0$ and $Q_2 \neq 1$, $T(S, Q(S)) = 1 - pls(S)$ and

$$T(S, \neg Q(S)) = pls(S) = 1 - [1 - (pls(S))] = 1 - T(S, Q).$$

(3) when $q_1 \neq 0$ and $Q_2 = 1$, $T(S, Q(S0) = bel(S)$ and

$$T(S, \neg Q(S)) = 1 - bel(S) = 1 - T(S, Q).$$

From the theorem above we can see that $T(S, Q) + T(S, \tilde{Q}) = 1$. Thus the statement $(S, \tilde{Q}, T)$ has a measure of validity complement to our original proposition (S, Q, T). The prototypical manifestation of this is that the summaries "*Most people in this firm are interested in buying stocks*" will have a measure of validity complement to our original proposition as "*Not most people in this firm are interested in buying stocks*".

Summarizing, for summarizer S with quantifier $Q(S)$ and the measure T of validity in a database, we have the following results:

$Q(S)$	$Q(\bar{S})=\bar{Q}(S)$	$\bar{Q}(\bar{S})=Q(S)$	$\bar{\bar{Q}}(S)=Q(S)$
$T(S,Q)$	$T(\bar{S},Q)=T(S,\bar{Q})=T(S,Q)$	$T(\bar{S},\bar{Q})=T(S,Q)$	$T(S,\bar{\bar{Q}})=T(S,Q)$

$\neg Q(S)$	$\neg Q(\bar{S})$	$\neg\neg Q(S)=Q(S)$
$T(S,\neg Q)=1-T(S,Q)$	$T(\bar{S},\neg Q)=1-T(\bar{S},Q)=1-T(S,Q)$	$T(S,\neg\neg Q)=T(S,Q)$

8.2 The Strength of Summaries

Another important idea associated with a linguistic quantifier is the concept of its strength. Here is its definition.

Definition 3. Two quantifiers $Q_1 = [q_{11}, q_{12}]$ and $Q_2 = [q_{21}, q_{22}]$ are said to have *the same strength*, and denoted by $Q_1 \equiv Q_2$, if $q_{22} - q_{21} = q_{12} - q_{11}$.

According to this definition, we also have

Theorem 3. Two quantifiers Q_1, Q_2 have the same strength, i.e., $Q_1 \equiv Q_2$ if and only if they have the same truth $T(S, Q_1) = T(S, Q_2)$.

Proof: The result follows immediately from the concept of summary and definition 3.

The theorem above shows that we can measure quantifiers with the same degree of strength by calculating their truths, and any quantifiers with the same truth are viewed as having the same strength. As we shall see later, however, it does not mean when two quantifiers with different truths, respectively, the larger the truth is, the stronger the quantifier is. That is, the strength logic associated with the quantifier $Q(S)$ is non-monotonic. We will discuss this issue in detail in the next section.

9 Knowledge Extraction

In the preceding we have discussed the basic ideas and concepts of summaries based on the use of the evidence theory to represent the quantifiers and the measure of validity. We are now in a position to discuss knowledge extraction, and attempt to propose a further extension about language summary representation. In order to more formally discuss the idea of knowledge extraction associated with a linguistic summary, first of all, we suggest that, for quantifiers there are several kinds of representations. Then some further properties about these representations have been introduced. Finally, we show an example to illustrate how it is possible to combine data summaries from different samples directly without having to return to the raw data or mass assignments.

9.1 Representing Summaries

Let (U, A) be a database, and assume that $x \in A$ is an attribute with respect to the database. We can establish a summary (S, Q, T), where $S \in 2^{\Theta}$ is a summarizer and Θ is the frame of discernment, $Q = Q(S)$ is the quantifier represented as an interval $[q_1, q_2]$ in which the quantity lies, and $T = (S, Q)$ is a measure of validity with respect to the summary, in our case it is a truth function.

We suggest that for quantifiers Q there are several kinds of representations: *always*, *typically*, *frequently*, and they are represented, respectively, as follows.

- *always* is denoted by $Q_{a:[1,1]}$ and represented by a mass function
  $m(S) = 1, \quad m(\bar{S}) = 0, \quad m(\Theta) = 0;$ i.e.,
  $[bel(S), pls(S)] = [1, 1]$, and summary for S with truth $T(S) = 1.0$.
- *typically* is denoted by $Q_{t:[\alpha,1]}$ and represented by a mass function
  $m(S) = \alpha, \quad m(\bar{S}) = 0, \quad m(\Theta) = 1 - \alpha;$ i.e.,
  $[bel(S), pls(S)] = [\alpha, 1]$, and summary for S with truth $T(S) = \alpha$.
- *frequently* is denoted by $Q_{f:[\alpha,1-\beta]}$ and represented by a mass function
  $m(S) = \alpha, \quad m(\bar{S}) = \beta, \quad m(\Theta) = 1 - \alpha - \beta, \quad \alpha \geq \beta \; (m(S) \geq m(\bar{S}))$; i.e.,
  $[bel(S), pls(S)] = [\alpha, 1 - \beta]$ and summary for S with truth $T = \alpha + \beta$.

To these three quantifier definitions, their antonyms *never*, *atypically*, *infrequently* are represented as follows.

- *never* is denoted by $Q_{n:[0,0]}$ and represented by a mass function
  $m(S) = 0, \quad m(\bar{S}) = 1, \quad m(\Theta) = 0;$ i.e.,
  $[bel(S), pls(S)] = [0, 0]$, and summary for S with truth $T(S) = 1.0$.
- *atypically* is denoted by $Q_{at:[0,1-\alpha]}$ a and represented by a mass functionn
  $m(S) = 0, \quad m(\bar{S}) = \alpha, \quad m(\Theta) = 1 - \alpha;$ i.e.,
  $[bel(S), pls(S)] = [0, 1 - \alpha]$, and summary for S with truth $T(S) = \alpha$.
- *infrequently* is denoted by $Q_{inf:[\beta,1-\alpha]}$ and represented by a mass function
  $m(S) = \beta, \quad m(\bar{S}) = \alpha, \quad m(\Theta) = 1 - \alpha - \beta; \alpha > \beta \; (m(S) < m(\bar{S}))$; i.e.,
  $[bel(S), pls(S)] = [\beta, 1 - \alpha]$, and summary for S with truth $T(S) = \alpha + \beta$.

Note that for the antonym definition, the validity value T is for S not for $\bar{S}$.

From the discuss above and the definition 1 we have $\tilde{Q}_{a:[1,1]} = Q_{n:[0,0]}$, $\tilde{Q}_{t:[\alpha,1]} = Q_{at:[0,1-\alpha]}$, $\tilde{Q}_{f:[\alpha,1-\beta]} = Q_{inf:[\beta,1-\alpha]}$.

And from definition 3, we have $Q_{a:[1,1]} \equiv Q_{n:[0,0]}$, $Q_{t:[\alpha,1]} \equiv Q_{at:[0,1-\alpha]}$, $Q_{f:[\alpha,1-\beta]} \equiv Q_{inf:[\beta,1-\alpha]}$.

Also we present a useful definition in the following.

Definition 4. A quantifier $Q_1 = [q_{11}, q_{12}]$ is said to be *stronger* than a quantifier $Q_2 = [q_{21}, q_{22}]$, and denoted $Q_1 \models Q_2$, if the tuple (q_{11}, q_{12}) is a special case of the tuple (q_{21}, q_{22}).

Obviously, a stronger quantifier may be derived from a weaker quantifier by appropriate choices of α and β.

Theorem 4. We have the following *always* $\models$ *typically* $\models$ *frequently, never* $\models$ *atypically* $\models$ *infrequently.*

Proof. First of all, for $Q \in$ {*always, typically, frequently*} we can easily verify the following:

(1) $Q_{a:[1,1]} \models Q_{t:[\alpha,1]}$ since the tuple $(1,1)$ in $Q_{a:[1,1]}$ is a special case of the tuple $(\alpha, 1)$ in $Q_{t:[\alpha,1]}$ where $\alpha = 1$.

(2) $Q_{t:[\alpha,1]} \models Q_{f:[\alpha,1-\beta]}$ since the tuple $(\alpha, 1)$ in $Q_{t:[\alpha,1]}$ is a special case of the tuple $(\alpha, 1-\beta)$ in $Q_{f:[\alpha,1-\beta]}$ where $\beta = 0$.

Similarly, for $Q \in$ {*never, atypically, infrequently*} we have

(3) $Q_{n:[0,0]} \models Q_{at:[0,1-\beta]}$ since the tuple $(0,0)$ in $Q_{n:[0,0]}$ is a special case of the tuple $(0, 1-\beta)$ in $Q_{at:[0,1-\beta]}$ where $\beta = 1$.

(4) $Q_{at:[0,1-\alpha]} \models Q_{inf:[\beta,1-\alpha]}$ since the tuple $(0, 1-\alpha)$ in $Q_{at:[0,1-\alpha]}$ is a special case of the tuple $(\beta, 1-\alpha)$ in $Q_{inf:[\beta,1-\alpha]}$ where $\beta = 0$.

As we mentioned in the last section, when two quantifiers have different truths, it did not mean the larger the truth was, the stronger the quantifier was. That is, the strength logic is non-monotonic. For example, the quantifier $Q_1 = typically = [\alpha, 1]$ with truth $T_1(S) = \alpha$, and the quantifier $Q_2 = frequently = [\alpha, 1-\beta]$ with truth $T_2(S) = \alpha + \beta$ $(\beta > 0)$, $T_1(S) < T_2(S)$, but $Q_1 \models Q_2$.

When inducing a rule concerning a summary, however, it is usual to induce the strongest rule firstly. After obtaining the belief interval for S: $Q(S)=[q_1, q_2]$, the values of belief and plausibility functions for S, we proceed as follows:

(1) if $Q(S)=[1,1]$ we infer that $Q(S)=$*always*; and
if $Q(S)=[0,0]$ we infer that $Q(S)=$*never*;

(2) if $Q(S)=[q_1, 1]$ with $0 < q_1 < 1$ we infer that $Q(S)=$*typically*; and
if $Q(S)=[0, q_2]$ with $0 < q_2 < 1$ we infer that $Q(S)=$*atypically*;

(3) if $Q(S)=[q_1, q_2]=[q_1, 1-(1-q_2)]$ with $0 < 1-q_2 < q_1 < 1$ we infer that $Q(S)=$*frequently*; and
if $Q(S)=[q_1, q_2]=[q_1, 1-(1-q_2)]$ with $0 < q_1 < 1-q_2 < 1$ we infer that $Q(S)=$*infrequently*.

9.2 Combining Summaries

Having given some representations of summaries it is now in a position to show how it is possible to combine data summaries from different samples directly without having to return to the raw data or mass assignments. The following theorems establish a set of rules which allow such high level integration to be carried out. By using the Orthogonal Sum these results are, for example, if a *typically* summary is combined with an *always* summary then the result is an *always* summary, etc.

In order to discuss the central issue of knowledge discovery, i.e., the issue of combination, we introduce the following definition.

Definition 5. Consider data summaries on two samples, given by (S, Q_1, T_1) and (S, Q_2, T_2), where $Q_1 = [bel_1(S), pls_1(S)]$ and $Q_2 = [bel_2(S), pls_2(S)]$. Combining their mass functions $m = m_1 \oplus m_2$ by the Dempster-Shafer orthogonal sum, we can obtain their combined summary (S, Q, T) with $Q = [bel(S), pls(S)]$. Then the quantifier Q is called the *combination* of Q_1 and Q_2, and denoted by $Q = Q_1 \oplus Q_2$.

We now present a series of theorems that are effective mechanisms for knowledge discovery in the distributed databases.

First of all, consider the summaries with their respective quantifiers $Q_1, Q_2 \in$ {*always, typically, frequently*}. Without loss of generality, we may only discuss the case where Q_1, Q_2=*frequently*. The other cases of Q_1, Q_2 are special ones which may be obtained by appropriately setting α_i, β_i, $i = 1, 2$, respectively.

Lemma 1. *frequently* $\oplus$ *frequently* = *frequently*.

Proof. With orthogonal sum of m_1 and m_2 we have the intersection table as follows:

$m_1 \backslash m_2$	S_{α_2}	$\bar{S}_{\beta_2}$	$\Theta_{1-\alpha_2-\beta_2}$
S_{α_1}	$S_{\alpha_1\alpha_2}$	$\emptyset_{\alpha_1\beta_2}$	$S_{\alpha_1(1-\alpha_2-\beta_2)}$
$\bar{S}_{\beta_1}$	$\emptyset_{\beta_1\alpha_2}$	$\bar{S}_{\beta_1\beta_2}$	$\bar{S}_{\beta_1(1-\alpha_2-\beta_2)}$
$\Theta_{1-\alpha_1-\beta_1}$	$S_{(1-\alpha_1-\beta_1)\alpha_2}$	$\bar{S}_{(1-\alpha_1-\beta_1)\beta_2}$	$\Theta_{(1-\alpha_1-\beta_1)(1-\alpha_2-\beta_2)}$

From the table above we obtain

$$\begin{aligned}(m_1 \oplus m_2)(S) &= \frac{\alpha_1\alpha_2+\alpha_1(1-\alpha_2-\beta_2)+\alpha_2(1-\alpha_1-\beta_1)}{1-\alpha_1\beta_2-\alpha_2\beta_1} \\ &= \frac{(\alpha_1+\alpha_2-\alpha_1\alpha_2)-(\alpha_1\beta_2+\alpha_2\beta_1)}{1-\alpha_1\beta_2-\alpha_2\beta_1} = \alpha, \\ (m_1 \oplus m_2)(\bar{S}) &= \frac{\beta_1\beta_2+\beta_1(1-\alpha_2-\beta_2)+\beta_2(1-\alpha_1-\beta_1)}{1-\alpha_1\beta_2-\alpha_2\beta_1} \\ &= \frac{(\beta_1+\beta_2-\beta_1\beta_2)-(\alpha_1\beta_2+\alpha_2\beta_1)}{1-\alpha_1\beta_2-\alpha_2\beta_1} = \beta,\end{aligned}$$

$$(m_1 \oplus m_2)(\Theta) = \frac{(1-\alpha_1-\beta_1)(1-\alpha_2-\beta_2)}{1-\alpha_1\beta_2-\alpha_2\beta_1} = 1-\alpha-\beta.$$

From the condition $0 \le \beta_i \le \alpha_i \le 1$ and $0 \le \beta_i + \alpha_i \le 1$, $i = 1, 2$, we show that $0 \le \beta \le \alpha \le 1$ as follows.

From $\beta_1 \le \alpha_1$, $\beta_2 \le \alpha_2$ we have $-\beta_1 \ge -\alpha_1$, $-\beta_2 \ge -\alpha_2$, and
$1-\beta_1 \ge 1-\alpha_1 \ge 0$, $1-\beta_2 \ge 1-\alpha_2 \ge 0$, and
$(1-\beta_1)(1-\beta_2) \ge (1-\alpha_1)(1-\alpha_2)$, and
$-(1-\beta_1)(1-\beta_2) \le -(1-\alpha_1)(1-\alpha_2)$, and
$1-(1-\beta_1)(1-\beta_2) \le 1-(1-\alpha_1)(1-\alpha_2)$, and
$\beta_1+\beta_2-\beta_1\beta_2 \le \alpha_1+\alpha_2-\alpha_1\alpha_2$, and $\beta \le \alpha$.

That is, the combined quantifier Q is *frequently*, and $Q(S) = [\alpha, 1-\beta]$ with $\alpha \ge \beta$.

Now we consider the summaries with their respective quantifiers $Q_1 \in \{always, typically, frequently\}$ and $Q_2 \in \{never, atypically, infrequently\}$. Again, without loss of generality, we may only discuss the case where Q_1=*frequently* and $Q_2 = infrequently$. The other cases of Q_1 and Q_2 are special ones which may be obtained by appropriately setting α_i and β_i, $i = 1, 2$, respectively.

Lemma 2. *frequently* $\oplus$ *infrequently* = *frequently* or *infrequently*, but *always* $\oplus$ *never* does not exist.

Proof. With orthogonal sum we have the intersection table as follows.

$m_1 \backslash m_2$	S_{β_2}	$\bar{S}_{\alpha_2}$	$\Theta_{1-\beta_2-\alpha_2}$
S_{α_1}	$S_{\alpha_1\beta_2}$	$\emptyset_{\alpha_1\alpha_2}$	$S_{\alpha_1(1-\beta_2-\alpha_2)}$
$\bar{S}_{\beta_1}$	$\emptyset_{\beta_1\beta_2}$	$\bar{S}_{\beta_1\alpha_2}$	$\bar{S}_{\beta_1(1-\beta_2-\alpha_2)}$
$\Theta_{1-\alpha_1-\beta_1}$	$S_{(1-\alpha_1-\beta_1)\beta_2}$	$\bar{S}_{(1-\alpha_1-\beta_1)\alpha_2}$	$\Theta_{(1-\alpha_1-\beta_1)(1-\beta_2-\alpha_2)}$

From the table above we obtain

$$\begin{aligned}
(m_1 \oplus m_2)(S) &= \frac{\alpha_1\beta_2+\alpha_1(1-\alpha_2-\beta_2)+\beta_2(1-\alpha_1-\beta_1)}{1-\alpha_1\alpha_2-\beta_1\beta_2} \\
&= \frac{(\alpha_1+\beta_2-\alpha_1\beta_2)-(\alpha_1\alpha_2+\beta_1\beta_2)}{1-\alpha_1\alpha_2-\beta_1\beta_2} = \alpha \text{ (or } \beta), \\
(m_1 \oplus m_2)(\bar{S}) &= \frac{\alpha_2\beta_1+\alpha_2(1-\alpha_1-\beta_1)+\beta_1(1-\alpha_2-\beta_2)}{1-\alpha_1\alpha_2-\beta_1\beta_2} \\
&= \frac{(\alpha_2+\beta_1-\alpha_2\beta_1)-(\alpha_1\alpha_2+\beta_1\beta_2)}{1-\alpha_1\alpha_2-\beta_1\beta_2} = \beta \text{ (or } \alpha), \\
(m_1 \oplus m_2)(\Theta) &= \frac{(1-\alpha_1-\beta_1)(1-\alpha_2-\beta_2)}{1-\alpha_1\alpha_2-\beta_1\beta_2} = 1-\alpha-\beta, \text{ where } \alpha \ge \beta (\text{or } \alpha > \beta).
\end{aligned}$$

So Q is *frequently/infrequently*. That is, Q is *frequently* and $Q(S) = [\alpha, 1-\beta]$ with $\alpha \ge \beta$ $(m(S) \ge m(\bar{S}))$, or Q is *infrequently* and $Q(S) = [\beta, 1-\alpha]$ with $\alpha > \beta$ $(m(S) < m(\bar{S}))$, depending on the values α_i, β_i, $(i = 1, 2)$.

It should be noted that, summaries *always* and *never* cannot be combined each other. In other words, for The Orthogonal Sum Operator $\oplus$, their combination does not exist. In this case, we can use The Modified Combination Operator $\ominus$ to combine them (see another chapter "*Extensions and Applications of Evidence Theory*" in this book).

Consider data summaries on two samples, given by (S, *always*, T_1), and (S, *never*, T_2), where
always $= Q_{a:[1,1]}(S)$ with the belief interval $[bel(S), pls(S)] = [1,1]$, and its mass fumction m_1 is

$m_1(S) = 1, m_1(\bar{S}) = 0, m_1(\Theta) = 0;$

never $= Q_{n:[0,0]}(S)$ with the belief interval $[bel(S), pls(S)] = [0,0]$, and its mass function m_2 is

$m_2(S) = 0, m_2(\bar{S}) = 1, m_2(\Theta) = 0.$

Using The Modified Combination Operator $\ominus$, we have

$(m_1 \ominus m_2)(S) = \frac{1}{2}, (m_1 \ominus m_2)(\bar{S}) = \frac{1}{2}.$

So the combined belief interval is $[bel(S), pls(S)] = [0.5, 0.5]$, and the modified combination of *always* $Q_{a:[1,1]}$ and *never* $Q_{n:[0,0]}$ is *frequently* $Q_{f:[0.5,0.5]}$. That is,

Lemma 3. *always* $\ominus$ *never* = *frequently.*

Finally, we consider the summaries with their respective quantifiers $Q_1 \in \{never, atypically, infrequently\}$ and $Q_2 \in \{never, atypically, infrequently\}$. Again, without loss of generality, we may only discuss the case where $Q_1, Q_2 =$ *infrequently*. The other cases of Q_1 and Q_2 are special ones which may be obtained by appropriately setting α_i and β_i, $i = 1, 2$, respectively.

Lemma 4. $infrequently \oplus infrequently = infrequently.$

Proof. With orthogonal sum we have the intersection table as follows.

$m_1 \backslash m_2$	S_{β_2}	$\bar{S}_{\alpha_2}$	$\Theta_{1-\alpha_2-\beta_2}$
S_{β_1}	$S_{\beta_1\beta_2}$	$\emptyset_{\beta_1\alpha_2}$	$S_{\beta_1(1-\alpha_2-\beta_2)}$
$\bar{S}_{\alpha_1}$	$\emptyset_{\alpha_1\beta_2}$	$\bar{S}_{\alpha_1\alpha_2}$	$\bar{S}_{\alpha_1(1-\alpha_2-\beta_2)}$
$\Theta_{1-\alpha_1-\beta_1}$	$S_{(1-\alpha_1-\beta_1)\beta_2}$	$\bar{S}_{(1-\alpha_1-\beta_1)\alpha_2}$	$\Theta_{(1-\alpha_1-\beta_1)(1-\alpha_2-\beta_2)}$

From the table above we obtain

$$\begin{aligned}
(m_1 \oplus m_2)(S) &= \frac{\beta_1\beta_2+\beta_1(1-\alpha_2-\beta_2)+\beta_2(1-\alpha_1-\beta_1)}{1-\beta_1\alpha_2-\alpha_1\beta_2} \\
&= \frac{(\beta_1+\beta_2-\beta_1\beta_2)-(\beta_1\alpha_2+\alpha_1\beta_2)}{1-\beta_1\alpha_2-\alpha_1\beta_2} = \beta, \\
(m_1 \oplus m_2)(\bar{S}) &= \frac{\alpha_2\alpha_1+\alpha_2(1-\alpha_1-\beta_1)+\alpha_1(1-\alpha_2-\beta_2)}{1-\beta_1\alpha_2-\alpha_1\beta_2} \\
&= \frac{(\alpha_2+\alpha_1-\alpha_2\alpha_1)-(\beta_1\alpha_2+\alpha_1\beta_2)}{1-\beta_1\alpha_2-\alpha_1\beta_2} = \alpha, \\
(m_1 \oplus m_2)(\Theta) &= \frac{(1-\alpha_1-\beta_1)(1-\alpha_2-\beta_2)}{1-\beta_1\alpha_2-\alpha_1\beta_2} = 1-\alpha-\beta.
\end{aligned}$$

From conditions $0 \leq \beta_i < \alpha_i \leq 1$ and $0 \leq \alpha_i + \beta_i \leq 1; i = 1, 2$, we can show that $\beta_1 + \beta_2 - \beta_1\beta_2 < \alpha_1 + \alpha_2 - \alpha_1\alpha_2; i.e., \beta < \alpha$.

That is, the combined quantifier Q is *infrequently*, and $Q(S) = [\beta, 1-\alpha]$ with $\alpha > \beta$.

From the lemmas above, we can obtain a series of theorems as follows immediately.

Theorem 5. If Q_1=*frequently* or *infrequently* and Q_2 is an arbitrary quantifier then $Q_1 \oplus Q_2 = frequently$ or $infrequently$, but $always \oplus never$ and $never \oplus always$ do not exist.

Theorem 6. If Q_1=*typically* or *atypically* and Q_2 is an arbitrary quantifier then $Q_1 \oplus Q_2 = Q_2$ or $\bar{Q}_2$, but *typically* $\oplus$ *atypically*, *atypically* $\oplus$ *typically*= *frequently* or *infrequently*.

Theorem 7. Let Q_2 be an arbitrary quantifier, then $always \oplus Q_2 = always$, $never \oplus Q_2 = never$ but $always \oplus never$, $never \oplus always$ do not exist.

Summarizing the computation above, we obtain the following table.

Q_1	Q_2	$Q_1 \oplus Q_2$
$Q_{a:[1,1]}$	$Q_{a:[1,1]}$	$Q_{a:[1,,1]}$
$Q_{a:[1,1]}$	$Q_{t:[\alpha_2,1]}$	$Q_{a:[1,,1]}$
$Q_{a:[1,1]}$	$Q_{f:[\alpha_2,1-\beta]}$	$Q_{a:[1,1]}$
$Q_{t:[\alpha_1,1]}$	$Q_{t:[\alpha_2,1]}$	$Q_{t:[\alpha,]}$
$Q_{t:[\alpha_1,1]}$	$Q_{f:[\alpha_2,1-\beta_2]}$	$Q_{f:[\alpha,,1-\beta]}$
$Q_{f:[\alpha_1,1-\beta_1]}$	$Q_{f:[\alpha_2,1-\beta_2]}$	$Q_{f:[\alpha,,1-\beta]}$
$Q_{n:[0,0]}$	$Q_{a:[1,1]}$	$Q_{f:[0.5,0.5]}$
$Q_{n:[0,0]}$	$Q_{t:[\alpha_2,1]}$	$Q_{n:[0,,0]}$
$Q_{n:[0,0]}$	$Q_{f:[\alpha_2,1-\beta]}$	$Q_{n:[0,,0]}$
$Q_{at:[0,1-\alpha_1]}$	$Q_{a:[1,1]}$	$Q_{a:[1,1]}$
$Q_{at:[0,1-\alpha_1]}$	$Q_{t:[\alpha_2,1]}$	$Q_{f:[\alpha,1-\beta]}$ with $\alpha \geq \beta$, or $Q_{inf:[\beta,1-\alpha]}$ with $\alpha > \beta$
$Q_{at:[0,1-\alpha_1]}$	$Q_{f:[\alpha_2,1-\beta_2]}$	$Q_{f:[\alpha,1-\beta]}$ with $\alpha \geq \beta$, or $Q_{inf:[\beta,1-\alpha]}$ with $\alpha > \beta$
$Q_{inf:[\beta_1,1-\alpha_1]}$	$Q_{a:[1,1]}$	$Q_{a:[1,1]}$
$Q_{inf:[\beta_1,1-\alpha_1]}$	$Q_{t:[\alpha_2,1]}$	$Q_{f:[\alpha,1-\beta]}$ with $\alpha \geq \beta$, or $Q_{inf:[\beta,1-\alpha]}$ with $\alpha > \beta$
$Q_{inf:[\beta_1,1-\alpha_1]}$	$Q_{f:[\alpha_2,1-\beta_2]}$	$Q_{f:[\alpha,1-\beta]}$ with $\alpha \geq \beta$, or $Q_{inf:[\beta,1-\alpha]}$ with $\alpha > \beta$

The theorems above show that the rules with the quantifier *always* or *never* are the strongest rules: any rule with different quantifier is combined with them, the resulting rule must be *always/never*. In our case, however, it does not mean when two rules with different quantifiers, respectively, to be combined, the resulting rule to take the quantifier of the stronger one of the two original rules. That is, the resulting logic is non-monotonic. Thus, for example, if a rule which is Q_1=*typically* true is combined with a rule derived from a different sample which is Q_2=*frequently* true, the result is *frequently* true. The original *typically*-true rule has been revised downwards in the light of new data.

An implicit consequence of the combination rule is, of course, that when two rules with the same quantifier are combined, then the rule obtained also has that quantifier.

Finally we give a simple example to conclude this section.

Example 6. Suppose that we have some information about the credit worthiness for two different DataSets in a bank as follows:

DataSet-1 $= [\{a, b\}^{0.80}, \{c\}^{0.10}, \Theta^{0.10}]$;

DataSet-2 $= [\{a, b\}^{0.70}, \Theta^{0.30}]$;

where $\Theta = \{a, b, c\}$ and a - High Credit Worthy, b - Average Credit Worthy and c - Low Credit Worthy.

Let $S = \{a, b\}$, then we have

$m_1(S) = 0.80$, $m_1(\bar{S}) = 0.10$, $m_1(\Theta) = 0.10$;

$m_2(S) = 0.70$, $m_2(\Theta) = 0.30$.

We can extract knowledge from the information above to obtain the following rules:

Rule-1. *Frequently* $[0.8, 0.9]$ *the credit worthiness for DataSet-1 is High Credit Worthy or Average Credit Worthy with truth T=0.90.*

Rule-2. *Typically* $[0.7, 1.0]$ *the credit worthiness for DataSet-2 is High Credit Worthy or Average Credit Worthy with truth T=0.70.*

Combining these two rules by using the formulae above we obtain a new rule:

Rule-3. *Frequently* $[0.936, 0.968]$ *the credit worthiness for DataSet-1 and DataSet-2 is High Credit Worthy or Average Credit Worthy with truth T=0.968.*

10 Conclusion

In this chapter we discuss how evidential reasoning can be used to integrate distributed databases. We establish a general framework and propose some approaches about knowledge discovery techniques by using Dempster-Shafer evidence theory. We introduce the concepts of linguistic summaries based upon D-S evidence theory and describe the methodology used to represent information contained in the distributed databases. The method for determining the quantifiers and measuring the validity of the rules provided by the linguistic summaries is discussed. The essential and central issues of knowledge discovery, that is, the issues of combination for multiple mass distributions associated with a certain attribute on both the same frame of discernment and the different frames of discernment are investigated. Finally, a further extension of linguistic summary representation for knowledge extraction is presented.

Acknowledgment

This research was supported in part by EPSRC.

References

1. Dao, S.; Perry, B. 1995, Applying a data miner to heterogeneous schema integration, *Proceedings of the First international Conference on Knowledge Discovery and Data Mining*, U.M.Fryyad and R.Uthurusamy, Eds., montreal, 1995, 63-68.
2. Dubois, D.; Prade, H. 1986, Possibility theory: An Approach to Computerized Processing of Uncertainty, (1986) Plenum Press, New York.
3. Dubois, D.; Prade, H. 1987, "An Approach to Approximate Reasoning Based on the Dempster Rule of Combination", International Journal of Expert Systems, Vol.1, No.1 (1987) 67-85.
4. Dubois, D.; Prade, H. 1990, "Modeling Uncertain and vague knowledge in Possibility and Evidence Theories", Uncertainty in Artificial Intelligence 4 (1990) 303-318.
5. Frawley, W. J.; Piatettsky-Shapiro, G.; Matheus, C. J. 1991, Knowledge discovery in databases: a overview. In *Knowledge Discovery in Databases*, eds. G. Piatetsky-Shapiro and W. J. Frawley, AAAI Press/The MIT Press, 1-27.
6. Guan, J.; Bell, D. 1991, Evidence Theory and its Applications, Vol.1, (1991) North-Holland.
7. Guan, J.; Bell, D. 1992, Evidence Theory and its Applications, Vol.2, (1991) North-Holland.
8. Kacprzyk, J.; Yager, R. R. 1984, Linguistic quantifiers and belief qualification in fuzzy multicriteria and multistage decision making, *Control and Cybernetics*, Vol.13(1984), No.3, 153-172.
9. Matheus, C. J.; Chan, P. K.; Piatetsky-Shapiro G. 1993, *Systems for knowledge discovery in databases*, IEEE Trans. Knowledge and Data Engineering, 5, 903-913.
10. McClean, S.; Scotney B. 1996, "Using evidence theory for knowledge discovery and extraction in distributed databases", *Informatics Research Reports, University of Ulster*, No.11 (December 1996), 47-54.
11. McClean, S.; Scotney B. 1997, "Using evidence theory for the integration of distributed databases", *International Journal of Intelligent Systems*, Vol.12(1997), 763-776.
12. Ribeiro, J. S.; Kaufman, K. A.; Kerschberg, L. 1995, Knowledge discovery from multiple database, *Proceedings of the First international Conference on Knowledge Discovery and Data Mining*, U.M.Fryyad and R.Uthurusamy, Eds., montreal, 1995, 240-245.
13. Shafer, G. 1976, A Mathematical theory of evidence. Princeton, NJ: Princeton Univ. Press, (1976).
14. Yager, R. R. 1982, "A new approach to the summarization of data", in *Information Sciences*, 28, 69-86.
15. Yager, R. R. 1991, "On linguistic summaries of data", in *Knowledge Discovery in Databases*, G. Piatetsky-Shapiro and B. Frawley, Eds., MIT Press, Cambridge, MA, 347-363.

16. Yager, R. R. 1995, "Linguistic summaries as a tool for database discovery", in *Workshop on Fuzzy Database Systems and Information Retrieval at FUZZ-IEEE/IFES*, Yokohama, 79-82.
17. Yager, R. R. 1996, "Database discovery using fuzzy sets", in *International Journal of Intelligent Systems*, Vol.11, 691-712.
18. Zadeh, L. A. 1978, "Fuzzy sets as a basis for a theory of possibility", Fuzzy Sets Syst. Vol.1, No.1 (1978) 3-28.
19. Zadeh, L. A. 1984, "A theory of commonsense knowledge", in *Aspects of Vagueness*, H. J. Skala, S. Termini, E. Trillas, Eds., D. Reidel Publishing Company, 257-295.

Interesting Rough Lattice-based Implication Rules Discovery

Yi Zhao[1], Jianliang Shi[2], Da Ruan[3], and Pengfei Shi[1]

[1]Institute of Image Processing & Pattern Recognition
Shanghai Jiaotong University, Shanghai 200030, China
[2]Guoxin Lucent Technologies Network Technologies Co., Ltd.
Pudong, Shanghai 200122, China
[3]Reactor Physics & Myrrha Department
Belgian Nuclear Research Centre (SCK•CEN), 2400 Mol, Boeretang 200, Belgium

As an important data mining technique, implication rules help to explore the dependencies among attributes of a database. In this paper, we aim at finding an efficient algorithm to discover the implication rules in a data set. This algorithm first extends the concept lattice theory by building the lattice according to the data set with the interesting attributes to improve user interaction and mining efficiency. The interesting concept lattice, together with the rough set theory, is then incorporated into our method to implement a new interesting rough lattice-based implication rules discovery (IRLIRD) approach to interactively acquire the rules with the specific rough upper and lower approximation. It generates implication rules without multiple passes over the data set and computing frequent itemsets. For the application of the proposed method to the transaction data set of the large-scale supermarkets, a novel data structure, which is based on the numeric attribute and termed as linked list of transaction set, is introduced here to reduce the needed memory. A simulation is implemented to illustrate the whole mining process, which demonstrates that the approach reduces the computational time greatly comparing with that of the classical Apriori algorithm. The algorithm can also be extended to many other application areas such as stock analysis, credit card distribution and agricultural application, etc.

1 Introduction

In today's world, we are inundated with volumes of data. It is estimated that the vast amount of information in the world doubles every 20 months. Though the massive amounts of data contain potential gold of mine of valuable information, they are beyond the expert's ability to be analyzed, which has spurred the interest of researches from databases to the work of Knowledge Discovery & Data Mining (KDD). Within the area of data mining, the problem of deriving dependencies, i.e., implication rules, from data has recently received a great deal of attention. An implication rule typically strives for discovering a dependency among attributes with respect to the externally defined parameters like support threshold and confidence threshold. It is an important field of data mining to represent the support and confidence of items that are purchased together in supermarket domain. The solution of market basket problem can help a retailer to learn about customers' purchasing trends to make useful decision, i.e., strolling route and layout design. Moreover, the retailer can decide how to implement sales promotion periodically.

Rough set theory (RST) [1,8,10] has made important contributions to the knowledge discovery process, and its extension to solve problems of KDD is a challenging research topic and has the promise of high payoffs in many business and scientific domains. The rough set theory approximates given concepts using upper and lower sets of concepts, which has been successfully used in data classification and knowledge discovery.

One of the main and traditional problems of rough set is how to find the reduct of attributes. Using the reduct of attributes in data modeling, the computational burden on data mining algorithm can be reduced by incorporating only the important attributes in the rules.

Finding reduct of attributes can be viewed as the attribute selection. Many researchers have proposed all kinds of methods and techniques. Whereas in the transaction basket analysis, reduct is of no use, for the attributes of a certain basket indeed implicate a specific consumption pattern. We can obtain different kinds of significant purchasing preferences directly from the transaction baskets themselves.

Concept lattice [9], also called Galois lattice, is a mathematical framework that allows building embedded classes from a set of objects. As a kind of machine learning techniques, concept lattice theory (CLT) can be used to knowledge representation and knowledge discovery. Also, it is a form of concept hierarchy in which each node represents a subset of objects (extent) with their common attributes (intent). The most important characteristic of CLT lies in reasoning the possible attributes of data sets. The Hasse diagram of the lattice represents a generalization/specification relationship between the concepts (nodes). Therefore, building the lattice and Hasse diagram corresponding to a set of objects described by some attributes, can be used as an effective tool for symbolic data analysis and knowledge acquisition.

According to the recent data mining research, it is evident that concept lattice based knowledge discovery approach [6] can improve the mining efficiency greatly. The increasing development of machine learning lattice based method faces a hard problem due to the time consuming of lattice construction. This part has an exponential time and space complexity that overlaps the rest in the whole complexity. Many approaches have been presented to reduce the construction time. For example, the pruned lattice [11] is developed by Oosthuizen to remove nodes selectively from the lattice based on statistic and heuristic measures, but it only works after the complete lattice is built. The join-semi-lattice is constructed by Nguifo in LEGAL (Learning with Galois Lattice) system [3] with some learning parameters, whereas it has the exponential complexity of the learning process. Therefore, the time cost cannot decrease that much.

In this paper, we will combine the lattice theory and rough set theory together to propose a new algorithm to accelerate the exploration of implication rules. The concept lattice theory is first extended by building the lattice according to the data set with the interesting attributes to improve user interaction and mining efficiency. The interesting concept lattice theory, together with the rough set theory, called interesting rough concept lattice theory (IRLT), is then incorporated into our method to implement a new discovery approach (*IRLIRD*) to interactively acquire the implication rules with the specific rough upper and lower approximation. It's unnecessary to compute frequent itemsets with multiple passes over the data set. The algorithm is applied to the transaction data set of the large-scale supermarkets and the simulation results show that the computational time can be reduced greatly.

The paper is organized as follows: in Section 2, data mining oriented concept lattice theory is described; Section 3 introduces the IRLT; the new algorithm *IRLIRD* is introduced in Section 4; Section 5 is the algorithm analysis and simulation results and Section 6 draws the conclusion.

2 Data Mining Oriented Concept Lattice Theory

Concept lattice is a machine learning technique to classify object set and to improve the efficiency of the learning process.

From the data mining context $C = (O, I, R')$ describing a set O of objects, a set I of itemsets and a binary relation R' between O and I, $R' \subseteq O \times I$, there is a unique ordered set which describes the inherent lattice structure defining natural groupings and relationship among the objects and their itemsets. This structure is known as data mining oriented concept lattice L (Fig. 1 is an example of the concept lattice). Each element of the lattice L derived from the context (O, I, R') is a couple, noted (X, X'), composed of an object set X of the power set $P(O)$ and itemset $X' \subseteq P(I)$. Each couple must be complete with respect to R', which means that the following two properties are satisfied:

(1) $X' = \{x' \in I \mid \forall x \in X, xR'x'\}$;

(2) $X = \{x \in O \mid \forall x' \in X', xR'x'\}$.

X is the largest set of objects described by the properties found in X', and symmetrically, X' is the largest itemset common to the objects in X.

The partial order in the lattice is used to generate the graph in the following way: there is an edge from H_1 to H_2, if $H_1 < H_2$ and there is no other element H_3 in the lattice such that $H_1 < H_3 < H_2$, then H_1 is ancestor of H_2, and H_2 is descendant of H_1. The graph is usually called a Hasse diagram, the couple in the diagram is called node. As a valid visualization tool to data analysis and knowledge discovery, the Hasse diagram reveals the generalization/specification relationship between the concept nodes. Therefore, building the lattice and Hasse diagram corresponding to a set of objects, each described by some attributes, can be used as an effective tool for symbolic data analysis and knowledge acquisition.

Clearly, the edge in the Hasse diagram is different from that in rough set, while the former is the real relationship between the nodes, the latter is the edge area that is hard to describe.

A Hasse diagram is a directed acyclic graph with an additional constraint [5]: every pair of nodes in the graph has a unique nearest common descendant – or *meet* – and a unique nearest common ancestor – their *join*. These are referred to as greatest lower bounds and least upper bounds respectively. Each pair of nodes in the concept lattice has a unique meet and join. From Fig.1, we can see that node {(1 2 3 4), b} is the *join* of node {(1 2 3), a b} and node {(3 4), b c}, node {(3), a b c} is the *meet* of node {(1 2 3), a b} and node {(3 4), b c}.

The *downward closure* of the node includes itself and its descendants, or nodes that can be covered. And the *upward closure* of the node just consists of its descendants. In Fig. 1, the downward closure of the node {(1 2 3 4), b} is {{(1 2 3 4), b}, {(1 2 3), a b}, {(3 4), b c}, {(3), a b c}}, its upward closure is {{(1 2 3), a b}, {(3 4), b c}, {(3), a b c}}.

R'	***a***	***b***	***c***
1	1	1	
2	1	1	
3	1	1	1
4		1	1

(a)

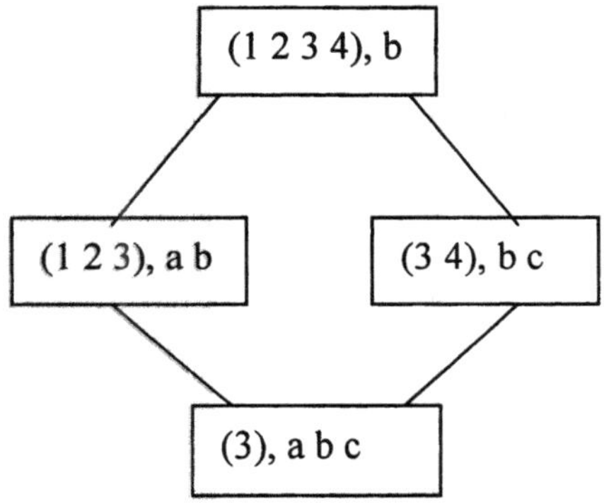

(b)

Fig. 1. A matrix data mining context $C = (O, I, R')$ and its concept lattice. O is a set of objects, I is a set of items and R describes a relation between O and I. A rectangle in Fig.1 (b) represents a concept node and edges between rectangles denote the top-down order relation

For our convenience, it is sufficient to concentrate just on the intentional part (items) of each node, and record the frequency of the itemset in the transaction relation. The concept lattice in Fig. 1 can be transformed into Fig. 2, where the left part of the node is the itemset and right part is the frequency of the itemset in the transaction.

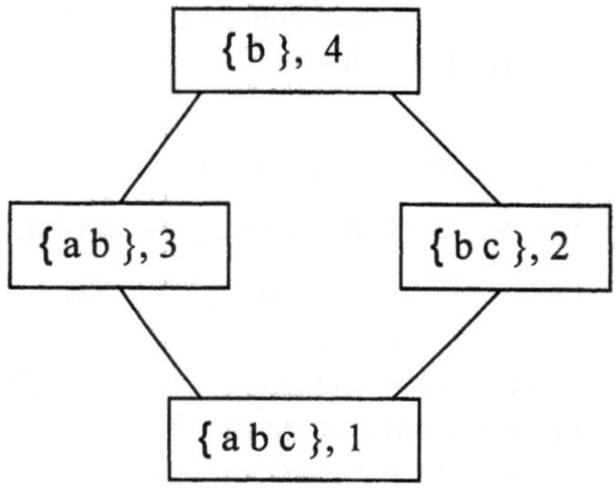

Fig. 2. The transformed concept lattice of Fig. 1

3 Interesting Rough Lattice Theory (IRLT)

3.1 Rough Set Fundamentals

The rough set theory, based on either algebraic or probabilistic approximation spaces, is used to reason about data that may contain uncertain information for a particular knowledge discovery task. The main virtue of RST is that it is unnecessary to know any information of the data set, such as the membership function in advance.

Let $A = (U, R)$ be an approximation space, where U is a finite set of objects in the universe and R is an equivalence relation on U. The equivalence classes of R and the empty set are called elementary sets. Let X be a subset of U, the least definable set containing X in A is called *closure set* (also known as *upper set*). Similarly, the greatest definable set in A that is contained in X is called *interior set* (also known as *lower set*). The set X is definable in A if the upper set of X is equivalent to the lower one; otherwise, X is a *rough* set or *non-definable*.

For a given $x \in U$, a decision algorithm, denoted by $D_A(X)$, yields one of the three answers: 1) x is in X, 2) x is not in X, 3) unknown. We now can get the corresponding set of X in A for each answer. Let $POS_A(X)$ be a set of objects in which each object is considered a member of the concept X by $D_A(X)$. Let $BND_A(X)$ be a set of objects in which $D_A(X)$ gives the answer of unknown. Finally, let $NEG_A(X)$ be a set of objects that are not regarded as members of X by $D_A(X)$.

In the algebraic space, RST approximates given concept(s) using lower and upper sets of the concept(s). If a concept is known, then the objects belonging to the lower approximation form the positive region. Those objects that certainly do not belong to the classification form the negative region. If the information is not sufficient to classify an object then such objects belong to the boundary regions. The upper approximation is theoretically the union set of the positive and boundary regions.

An attribute a is said to be *dispensable* if $POS_{A-\{a\}}(Y) = POS_A(Y)$; otherwise a is an *indispensable* attribute [1,2]. If attributes $A' \subseteq A$ satisfies: $POS_{A'}(X) = POS_A(X)$, and $\forall A'' \subset A'$, $POS_{A''}(X) \neq POS_A(X)$, A' is called *reduct* of A with respect to concept X (or simply called reduct). The set of attributes belonging to the intersection of all reducts is referred to as the *core* of conditional attributes A. One of the main problems in the rough set is how to find the reduct of attributes. The reduct can help to reduce computational burden on data mining algorithm to incorporate only the important attributes in the rules.

3.2 Interesting Rough Lattice

The interesting concept lattice can be built according to the user's need. With so many kinds of power sets of transaction data set, it is unnecessary to build the complete concept lattice, and usually implication rules that have been discovered are too much to the retailer. If the user wants to make sure whether sales of some certain categories is dependent on some other categories, he/she can select the interesting categories. In this way, with the basic database query operation, those that contain the chosen categories can be obtained to build the concept lattice. Data preprocessing is needed to eliminate the noise beforehand. For instance, from Fig. 3, with the expert knowledge, the row with "*" ahead can be excluded for it belongs to group sale and does not matter to the personal basket analysis. For the detail of data preprocessing, please refer to the work in [12].

POSID	Serialno	TranTime	BarCode	CatID	Category	Quantity	Price
B014	5	09:43:13	692024119705	28	2801	1	2.3
B014	5	09:43:25	691470622232	5	0530	4	1.2
B014	5	09:43:32	691470611129	5	0530	4	1.2
B016	1	09:43:34	692131499701	25	2507	2	2.1
B004	7	09:43:59	691470611129	5	0530	2	1.2
B004	7	09:44:04	690898278917	5	0508	1	5.4
*B024	5	09:44:21	288871780000	51	5101	324	24.0

Fig. 3. Transaction database structure

In the viewpoint of RST, finding reduct of attributes can be considered as the attribute selection in data modeling. Many methods and techniques have been proposed to acquiring the reduct. In the transaction basket analysis, the attributes in the transaction data set are the basket itemsets. All the attributes of a certain basket indeed implicate a specific consumption pattern, and so, reduct of attributes is of no use. Instead, selecting interesting categories can improve users' interaction to build the specific concept lattice and speed up the rules discovery process. In this paper, Njiwoua's parallel algorithm [4] is adopted to construct the interesting lattice. In the interesting rough lattice, the lower and upper approximation of a data set can be computed by extending the closure operation of RST. The lower approximation of a data set can be obtained by computing the downward closure of the *meet* of the attributes, and the upper approximation is the union of the downward closure of the attributes themselves.

4 Interesting Rough Lattice-based Implication Rules Discovery Algorithm (*IRLIRD*)

4.1 Principles in Implication Rules Mining

The problem of mining implication rules in transaction basket data set, a weakly correlated data, is defined as follows.

Given a set of item $I = \{i_1, i_2, \ldots, i_m\}$, ($m$ is the quantity of items). A set of items $X \subseteq I$ is called an itemset. Let D be a set of transaction, where each transaction T (identified by *TID*) is an itemset such that $T \subseteq I$. We say that a transaction T contains an itemset X, if $X \subseteq T$. An implication rule can be expressed as the form of $X \Rightarrow Y$, where $X \subset I$, $Y \subset I$ and $X \cap Y = \phi$. The rule $X \Rightarrow Y$ holds in the transaction set D with confidence c if c % of transaction set D that contain Y on the condition of X. The rule $X \Rightarrow Y$ has support s in the transaction set D if s % of transactions in D contain $X \cup Y$.

Given a transaction set D, the problem of mining implication rules is to generate all implication rules that have certain user-specified support threshold (called *minsup*) and confidence threshold (called *minconf*).

The mining problem can be separated into two sub-problems:

1. Find all frequent itemsets in the data set, i.e., itemsets with support greater or equal to *minsup*. Frequent itemsets are also called *large* itemsets.
2. For each frequent itemset I_1 found, generate all rules $I_2 \Rightarrow I_1 - I_2 \mid I_2 \subseteq I_1$, with confidence greater or equal to *minconf*. (I_1 and I_2 are two different itemsets)

The second sub-problem can be solved in the memory in a straightforward manner once all frequent itemsets and their support are known. Hence, the problem of mining implication rules is simplified to the problem of seeking frequent itemsets. Many algorithms have been proposed to find implication rules, though dissimilar, they are all based on the Apriori [7] mining method, that is to prune the subset of itemsets in order to find frequent itemsets.

The size of the data set has an essential role in frequent itemsets discovery. Large data sets are necessary for reliable results; unfortunately, the efficiency of mining algorithms depends heavily on the scale of the database. The classical algorithm requires multiple passes over the whole database to calculate candidate frequent itemsets and their corresponding support and confidence value, consequently, the database size is by far the most influential factor of the execution time for very large databases.

4.2 Linked List of Transaction Set based on Numeric Attribute

Interesting rough lattice is a particular rough concept lattice theory in the transaction basket analysis. In the transaction data set of the supermarket, a record is obtained when a product's barcode is scanned through POS. So, all the records are indexed by the transaction time over the POS, consisting of transaction time, POS identity number, barcode, CatID, etc. (Fig. 3). Obviously, each record cannot represent all the items in a basket.

The value of attributes CatID and Category are numeric values specified by the supermarket, for example, there is ninety-nine categories (attribute CatID) of products in the data set, each of which has different subcategories (attribute Category), such as clothing, cosmetics, etc.

Due to the numeric attribute feature of the transaction database, different items in a basket can be arranged in a set sorted in their lexicographic order. Then, all the sets of basket items are linked to be a list. The data structure is called linked list of transaction set. With the list structure, the main memory can be distributed by allocation. Moreover, the structure is suitable for the dynamically update transaction data. Traditional transaction data set is expressed with sparse matrix. Comparing with the quantity of product categories, the number of items in one basket is so limited. Therefore, though the matrix is sparse enough, it needs much memory. Also, the sparse matrix cannot change the size, it is not fit for the update of transaction data.

4.3 Description of the Algorithm *IRLIRD*

In this section, implication rules is discovered in a novel aspect by using concept lattice (or concept hierarchy) structure and Hasse diagram with the structure of linked list of transaction data set in data modeling. The *IRLIRD* algorithm is fundamentally different from existing algorithms with the combination of interesting concept lattice and rough set. It can generate implication rules without multiple passes over the data set and calculation of frequent itemsets.

The support value can be expressed with an integer during the computation. Given a concept lattice L, its support threshold σ and its confidence threshold λ, the *IRLIRD* algorithm description of a certain attribute set is as follows (Fig. 4).

Algorithm IRLIRD description

Input: An interesting concept lattice L, support threshold σ and confidence threshold λ. An attribute set H.
Output: A set ε of implication rules of H.

```
Method:
  Let ε= φ, UA []=φ, cn = 0, cn′ = 0, cn″ = 0;
  Count the number of sets in H's upper approximation
   set UA[H] and record the number of frequency ==> cn;
  while  ( ∃ H′≠H in L, the number of H′ ≥ σ)
  {
        Compute the upper approximation set UA[H′] and
           count the number of sets in UA[H′] with
           frequency ≥ σ ==> cn′;
        if (∃ edge between H′ and H ) // H′ is the
             descendant of H
         { if (||H′||>||H||)//H′ is the descendant of H
             { if (cn' / cn ≥ λ ) ε = ε∪(H ⇒H′-H ) }
           else if (||H′||<||H||) // H is the descendant
               of H′
             { if ( cn / cn′ ≥ λ) ε = ε∪(H' ⇒H-H′) }
        }
         else
         { if  (∃ meet M of H and H′ )  {
              Compute the lower approximation set LA[H]
                 and count the number of sets in LA[H]
                  with frequency ≥ σ ==> cn″;
              if  (cn" / cn ≥ λ) ε = ε∪(H ⇒H′)   }
         }
  }
```

Fig. 4. *IRLIRD* algorithm description

5 Simulation and Results

Given a market basket data of a large-scale supermarket, each row in the table represents an individual record (*TID*), each column represents items of the records, which consists of the value of attributes CatID and Category (Fig. 3). After the representation with linked list of the transaction set, the binary relation *R'* is presented by the matrix in Fig. 5. Since implication rule means to discover the dependencies among different attributes, there is no need to take the objects into account when building concept lattice except for the number of frequency of the itemsets. Fig. 6 is the corresponding simple Hasse diagram, where each node is identified by the number and "#", the middle part is the number of frequency of the node.

R'	*a*	*b*	*c*	*d*	*e*	*f*	*g*	*h*	*i*
1	1	0	1	0	0	1	0	1	0
2	1	0	1	0	0	0	1	0	1
3	1	0	0	1	0	0	1	0	1
4	0	1	1	0	0	1	0	1	0
5	0	1	0	0	1	0	1	0	0

Fig. 5. The matrix representation of the binary relation R'

Assuming the support threshold be 2 and confidence threshold be 0.2. Considering node 6# {a, c}, there are three kinds of the surrounding nodes (Fig. 6 shows that support values of the itemsets are all greater than or equal to 2).

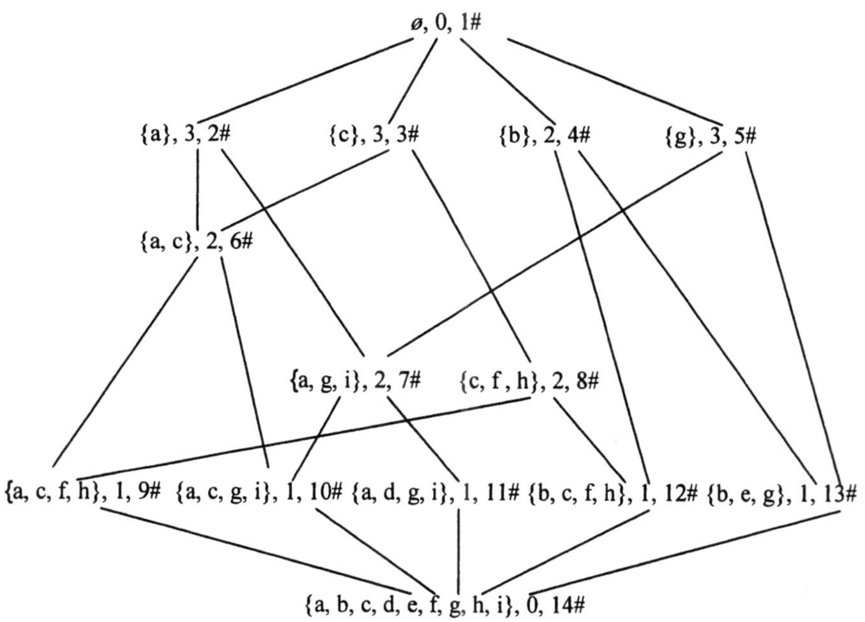

Fig. 6. Simplified Hasse diagram of the concept lattice in Fig. 5

(1) node 3# {c} is directly connected with node 6# by an edge, and node 3# is one of the ancestors of node 6#. Therefore, the confidence between nodes 3# and 6# is: (*fr* () is the calculation of number of frequency)

$$Conf(\{c\}, \{a, c\}) = fr\,(\mathrm{UA}\,(\{a, c\})) \,/\, fr\,(\mathrm{UA}\,(\{c\})) = 4/10 > 0.2.$$

The implication rule {c} $\Rightarrow$ {a} holds.

(2) node 10# {a, c, g, i} is also directly connected with node 6#, while node 10# is one of the descendants of node 6#. The confidence value between these two is:

$$Conf(\{a, c\}, \{a, c, g, i\}) = fr\,(UA\,(\{a, c, g, i\})) / fr\,(UA\,(\{a, c\})) = 1/4 > 0.2.$$

Therefore, there lies in the implication rules: $\{a, c\} \Rightarrow \{g, i\}$.

(3) node 8# {c, f, h}, which has the same meet as node 6# (the meet of node 6# and 8# is 9# {a, c, f, h}). Then the confidence value between node 6# and 8# is:

$$Conf(\{a, c\}, \{c, f, h\}) = fr\,(LA\,(\{c, f, h\})) / fr\,(UA\,(\{a, c\})) = fr\,(UA(\{a, c, f, h\})) / fr(UA(\{a, c\})) = 1/4 > 0.2.$$

Then rule $\{a, c\} \Rightarrow \{c, f, h\}$ holds.

From the simulation with 40K transaction records, computing implication rules in the worst case (discovering all the rules without interaction), the comparison between the classical Apriori algorithm and *IRLIRD* is shown in Fig.7.

The experiment results show that with the concept lattice, the algorithm *IRLIRD* can generate all implication rules whose support and confidence are greater than the relative threshold. The upper and lower approximation, which are characteristic of rough concept lattice, can improve the robustness to the noise in a certain extent. The whole mining process can be visualized with the Hasse diagram.

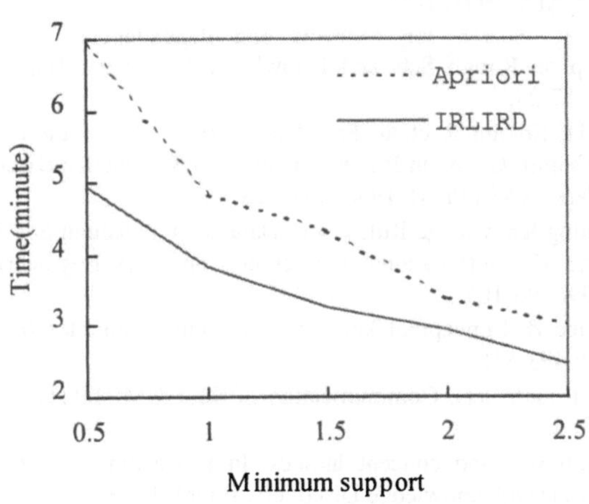

Fig. 7. Comparison between Apriori and *IRLIRD* algorithms

6 Conclusion

The CLT can be used to describe the dependency relation among the attributes, while RST can implement data modeling and approximate classification to the incomplete and imprecise knowledge. Based on the linked list of transaction set structure, the algorithm *IRLIRD* is presented with the interesting rough lattice (combined with rough set and interesting concept lattice) to select interesting data set and find the implication rules. *IRLIRD* improves the user's interaction during the mining process and accelerates the rule mining greatly. It also avoids the computation of the complex frequent itemsets without multiple passes over the data set.

The *IRLIRD* algorithm can also be applied to many other areas, such as stock market analysis, credit cards distribution, agriculture application, etc.

References

[1] Pawlak Z. Rough Sets. Kluwer Academic Publishers, Dordrecht, 1991.

[2] Ziarko W. Variable precision rough set model. Journal of Computer and System Scientifics 46, 1993, 39-59.

[3] Njiwoua P., Nguifo E.M. IGLUE: an instance-based learning system over lattice theory. In Proceedings of International Conference on Tools with Artificial Intelligence, Newport Beach, California, 1997: 75-76.

[4] Njiwoua P., Nguifo E.M. A parallel algorithm to build concept lattice. In Proceedings of 4th Groningen International Information Technique Conference for students, 1997: 103-107.

[5] Frawley W.J., Shapiro G.P., et al. Knowledge discovery in databases: an overview. In Shapiro G.P. and Frawley W.J., editors, Knowledge Discovery in Databases. AAAI/MIT, Cambridge, MA, 1991: 1-27.

[6] Missaoui R., Godin R. Search for concepts and dependencies in databases. International Workshop on Rough Sets and Knowledge Discovery, Banff, Alberta, Springer-Verlag. 1994: 17-23.

[7] Agrawal R, Mannila H, Srikant R et al. Fast Discovery of Association Rules. In: Fayyad M, Piatetsky-Shapiro G, Smyth P Eds. Advances in Knowledge Discovery and Data Mining, Menlo Park, AAAI Press, 1996: 307-328.

[8] Hu X, Cercone N. Mining Knowledge Rules from Databases: A Rough Set Approach. In: Proceedings of the 12th International Conference on Data Engineering, New Orleans, Louiseana, 1996: 96-105.

[9] Wille R. Concept lattice & Conceptual knowledge system. Computer Mathematics Applied, 23(6-9), 1992: 493-515.

[10] Pawlak Z., Ziarko W. Rough Sets. Communication of the ACM, 38(11), Nov. 1995: 36-41.

[11] Oosthuizen G.D. Rough sets and concept lattices. In Proceedings of International Conference on Rough Sets and Knowledge Discovery, Bannf, 1993.

[12] Wu X. Induction as pre-processing. In Proceedings of 3rd PAKDD-99, Beijing, China, Springer, 1999: 114-122.

Concept Lattice-based Approach for Incrementally Association Mining

Yi Zhao[1], Da Ruan[2], and Pengfei Shi[1]

[1]Institute of Image Processing & Pattern Recognition
Shanghai Jiaotong University, Shanghai 200030, China
[2]Reactor Physics & Myrrha Department
Belgian Nuclear Research Centre (SCK•CEN), 2400 Mol, Boeretang 200, Belgium

In this paper, a novel Concept Lattice-based Incrementally Large Itemset Generation Algorithm (CLLGA) is presented to discover association rules. As an important database discovery method, the kernel of association mining is the acquisition of large itemsets. According to Hu's [5] algorithm to generate market-basket association rules, an improved concept lattice based approach for incrementally acquiring large itemsets is introduced. The approach is especially efficient when the database is dynamically updated (insertion or deletion). By means of attaching frequency information to each itemset, i.e., each node in the lattice, the corresponding support and confidence measure can be obtained without constructing the complete lattice. Therefore, association rules can be derived from the concept lattice. Compare with Hu's approach, the time complexity of our algorithm can be reduced greatly.

1. Introduction

More and more attention is drawn to the application of concept (Galois) lattice [6], a mathematical theory, to machine learning. Concept lattice is a form of a concept hierarchy in which each node represents a subset of objects (extent) with their common properties (intent). Hasse diagram of the lattice represents a generalization/specialization relationship between the concepts. Therefore, the lattice and Hasse diagram corresponding to a set of objects described by some properties can be used as an effective tool for symbolic data analysis and knowledge acquisition.

Discovering association rules among market baskets in large databases is an important data mining issue [1]. Association rules are based on support and confidence. The support of an itemset, is the fraction of transactions that the itemset occurs in. An itemset is called large if its support exceeds a given threshold σ. Assuming itemset X and Y, an association rule is $X \Rightarrow Y$, where $X \neq \phi$, $Y \neq \phi$, $X \cap Y = \phi$. The confidence of the rule is the fraction of transactions containing X that also contain Y. For the association rule $X \Rightarrow Y$ to hold, $X \cup Y$ must be large and the confidence of the rule must exceed a given confidence threshold γ. In probability terms, we have $P(X \cup Y) > \sigma$ and

$P(Y \mid X) > \gamma$.

The process of extracting rules consist of two steps, that is, to generate large itemsets for a given σ and to construct rules with the confidence exceeding γ from the large itemsets. Mining efficiency is thus of the outmost importance for association rules, which depends greatly on the first step.

Concept lattice is applicable in information management and knowledge discovery [3]. In [4], Missaoui *et al.* introduced an algorithm to derive rules using concept lattice. According to [2], concept lattice and Hasse diagram can be constructed incrementally to find large itemsets. A valid algorithm is proposed by Hu *et al.* [5] accordingly to explore market-basket association rules more efficiently. But the algorithm neglects the large itemsets. Moreover, the cardinal of the itemset is employed in the algorithm to acquire the large itemset. In fact, the cardinal is of no use to the calculation of support and confidence measures. At last, the algorithm can only deal with the insertion of transaction.

A modified method CLLGA based on K. Hu's approach to incrementally find association rules using concept lattice and Hasse diagram is introduced in this paper to fit the data update, i.e., insertion or deletion. By recording frequency information of each lattice node in the database, large itemsets can be acquired without constructing the complete lattice. Comparatively, our approach can reduce time complexity greatly.

The paper is organized as follows: in Section 2, concept lattice theory and associations are described. A new algorithm CLLGA for incrementally deriving association rules is proposed in Section 3. Sections 4 and 5 are the algorithm analysis and conclusion respectively.

2. Concept Lattice and Associations

In this section, we describe the framework of concept lattice and its connection with association rules.

2.1 Basic Definitions

Given two finite sets O (sets of object) and D (sets of description), and a binary relation R between these two sets, there is a unique ordered set which describes the inherent lattice structure defining natural groupings and relationships among the elements of O and D. This structure is known as concept lattice.

Each node of the lattice L is a pair, called concept (X', X), composed of a set $X' \in P(O)$ (intent) and a set $X \in P(D)$ (extent). $P(O)$ is the power set of O and $P(D)$ is the power set of D. Each pair must be complete with respect to R if the two following properties are satisfied:

(1) $X' = \{x' \in O \mid \forall x \in X, xRx'\}$;

(2) $X = \{x \in D \mid \forall x' \in X', xRx'\}$.

The partial order is used to generate the lattice graph: given two nodes H_1 and H_2, if $H_1 < H_2$ and there is no other element H_3 in the lattice such that $H_1 < H_3 < H_2$. H_1 is called parent of H_2 and H_2 child of H_1. The graph is usually called a Hasse diagram. It is a valid visualization tool to data analysis and knowledge discovery, which revealing the generalization/specification relationship between the concept nodes.

2.2 Hu's Large Itemset Incremental Generation Method

Given a concept lattice L, let T be a new transaction and M be the cardinal of T. Each node in L is expressed with a modified one to the (extent, intent) pair, consisting of the cardinal of the itemset and the itemset itself. Let the cardinal threshold be t. There are three kinds of nodes in the new lattice after insertion, one remains invariable, one is modified, and the other is a new generated node. The following are some propositions presented by Hu *et al.*:

Proposition 1. A node $H = (M,T)$ is to be inserted in L. Let each node $N = (C,X)$. If $X \subseteq T$, then N is updated to be $N = (C+1,X)$.

Proposition 2. If $(C,X) = \inf\{(D,Y) \in L \mid X = Y \cap T\}$ and there is no node $(E,X) \in L$, then a new node $(C = D+1, X = Y \cap T)$ can be generated, where (D,Y) is the child of the new node.

Proposition 3. For each node (C,X) in the lattice, if the cardinal C is bigger than t, then X is the large itemset.

Besides, the edge in the new lattice must be changed, the generator of new node is always the child of new one, and original parent of the generator is updated. The key point lies in adding edges when search new nodes' parents, then the large itemsets can be obtained by judging whether the node's cardinal exceeds the threshold t.

3. Modified Concept Lattice-based Incrementally Large Itemset Generation Algorithm (CLLGA)

The following cases should be improved in Hu's algorithm:

(1) Part of large itemsets is neglected. For example, let $T_1 = \{a,c,d\}$, $T_2 = \{b,c,e\}$, $T_3 = \{a,b,c,e\}$, $T_4 = \{b,e\}$, and the threshold t be 2, then large itemsets obtained through Hasse diagram are: $\{b,c,e\}$ for large-3 one, $\{a,c\}$ and $\{b,e\}$ for large-2 one, $\{c\}$ for large-1 one. In fact, we can get different results with classical Apriori [7] algorithm, $\{b,c,e\}$ for large-3 one $\{a,c\}$, $\{b,e\}$, $\{b,c\}$, $\{c,e\}$ for large-2 one, $\{c\}$, $\{a\}$, $\{b\}$, $\{e\}$ for large-1 one.

(2) The cardinal value of the itemset is of little use to calculate support and confidence measuring. In other words, cardinal cannot reflect the two measures wholly.

(3) Frequency of itemsets is out of consideration. Once a certain itemset is

deleted, the relative node in the lattice will also be deleted, while in fact, the number of appearances of the node is more than once.

Current databases are always updated periodically or by random, which affects rule generation and rule's validity. Therefore, maintenance of association rules is important. Hu's algorithm works with transaction insertion, but not deletion.

The main idea of CLLGA is as follows. Given threshold t, the subsets of nodes whose cardinal large itemsets are less than or equal t can construct the large itemsets. In other words, large itemsets can be computed without constructing the complete lattice.

Assuming node in the lattice be (C,F,X), where C and X share the same meaning in this approach, F is used to record the frequency of X:

$$F(X)=\begin{cases}\text{frequency value}, & \text{X, node keep intact}\\ -1, & \text{X, new node}\\ 0, & \text{X, deleted node.}\end{cases}$$

where frequency measure of the first kind of nodes can help to compute the support and confidence measure.

The following sections describe the three cases in detail.

3.1 The Case of Transaction Incremental Insertion

Record the frequency (F) of each itemset when constructing Hasse diagram. Given threshold t, when $F(X) \neq -1$, large itemsets can be obtained with nodes where $C+F=t$, otherwise, with nodes where $C=t$.

Algorithm CLLGA (Concept lattice-based large itemsets generation algorithm with data insertion)

Input: the original lattice $L: N_0.item=(0,0,Sup(L))$,

$N_0.chd=\phi$, $N_0.par=\phi$, a new transaction X;

Output: modified lattice L'.

Method:

$\forall$ node $N_i \neq NULL$, $(N_i \in L)$ i is the quantity of the nodes;

if $(IsContained(X,N_i))$

$\{X.par=N_i, N_i.chd.par=X$, $C(N_i)++, F(N_i)++$;

if $(C(N_i)+F(N_i)\geq t)$ *then* $N_i \rightarrow AddMark$;

/* *AddMark* is the set of tagged large itemsets */

if $(N_i = X)$ *then* $\{F(N_i)++, break\}$;

};

else if $(\neg\exists X_k \subseteq AddMark, X_k = N_i \cap X)$

{ *new* node $(C(N_i)+1,\ 0,\ N_i \cap X)$;

$N_i.par = N_i \cap X$, $X.par = N_i \cap X$

$(N_i \cap X).par = N_i.par, \quad (N_i \cap X).par = X.par$

$if\ (C(N_i)+1 \geq t)\ then\ \ N_i \cap X \rightarrow AddMark\ ;\}$

}

Where *AddMark* consists of the itemsets whose $(C+F)$ value is greater than threshold t. Correspondingly, the subsets of these itemsets are large, too.

With the example in K. Hu's approach (t is 2), the results are listed in Fig. 1 and Fig. 2.

3.2 The Case of Transaction Deletion

First, locate the node X to be deleted, decrement its corresponding frequency information. If the frequency turns to zero (which indicates X is nonexistent), the common method is to delete all the parents and children of X. But there is still any probability for X to be re-inserted, the node must be kept for the moment by tagging its parents and children, putting its nodes and edges into the set *DeleteMark*. Once X is inserted again, return to its original status and delete the relative nodes and edges in the set *DeleteMark*.

In other words, deletion depends mostly on frequency information. As to the case where insertion and deletion occur simultaneously, the large itemsets can be obtained using algorithms in 2.1 and 2.2 at the same time, it is not necessary to give details here.

4. Algorithm Analysis

The time complexity of CLLGA depends on the number of new nodes (new transaction) to be added, which greatly improves the execution efficiency. Let each itemset consist of k items on average, and assuming the worst case be that there are no two same itemsets in the transaction (i.e., frequency of each itemset is 1). Given threshold t, the cardinal of the node is at most $(k+1)$, in other way, at most $t \cdot 2^k/(k+1)$ new nodes can be generated. In conclusion, the time complexity of algorithm CLLGA is not more than $O(t \cdot 2^k/(k+1) \cdot \|D\|)$, where $\|D\|$ is the size of database, k and t are constants. It is obvious from the up mentioned analysis, the efficiency of CLLGA is greatly improved comparative with the Hu method.

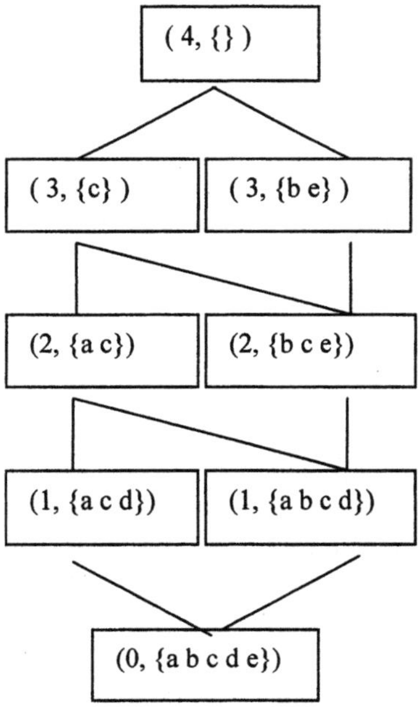

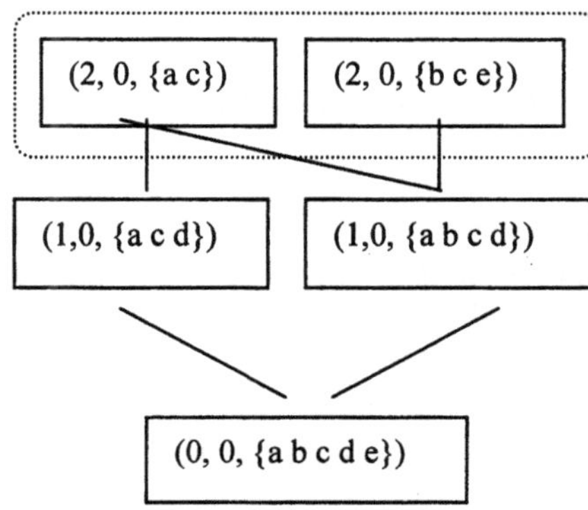

Fig. 1. The original Hasse diagram.

Fig. 2. The modified Hasse diagram

5. Conclusion

A valid approach for incrementally deriving large itemsets based on concept lattice is described. It is especially efficient when the database is dynamically updated, i.e., whether insertion or deletion. Frequency of each itemset is recorded to compute the corresponding support and confidence measure. Without constructing the complete concept lattice, association rules can be acquired. From the analysis above, our algorithm's time complexity can be reduced greatly.

References

[1] Agrawal R, Imielinski T, Swami A. Mining association rules between sets of items in large databases. ACM Press, Washington, DC., USA: Proceedings of ACM SIGMOD, 1993: 207-216

[2] Godin R, Missaoui R, Alaui H. Incremental concept formation algorithms based on Galois(Concept) Lattice. Computational Intelligence, 1995,11(2): 246-267

[3] Oosthuizen G D. Rough sets and concept lattices. Ed. Ziarko W P. Rough sets, and fuzzy sets and knowledge discovery(RSKD'93). London: Springer-Verlag, 1994: 24-31

[4] Missaoui R, Godin R. Search for concepts and dependencies in databases. Ed. Ziarko W P. Rough sets, and fuzzy sets and knowledge discovery (RSKD'93). London: Springer-Verlag, 1994: 16-23

[5] Hu K, Lu Y, Shi C. Incremental discovering association rules: a concept lattice approach. PAKDD'99, London: Springer-Verlag, 1999: 109-113

[6] Wille R. Restructuring lattice theory: an approach based on hierarchies of concepts. Rival I (Ed.). Dordrecht Reidel, 1982: 445-470

[7] R. Agrawal and R. Srikant. Fast algorithms for mining association rules in large databases. Proc. 20th International Conference Very Large Data Bases, Sept.1994: 478-499

Neuro-Fuzzy Modelling of Time Series

Jianwei Zhang and Alois Knoll

Faculty of Technology, University of Bielefeld, 33501 Bielefeld, Germany
E-mail: {zhang, knoll}@techfak.uni-bielefeld.de

Abstract

This paper proposes an approach for the multivariate modelling of time series with neuro-fuzzy systems. The fuzzy rule model is based on adaptive B-splines which can approximate any given input-output data series of low dimension. To efficiently describe high-dimensional input data, statistical indices are extracted to feed the fuzzy controller. The original input space is transformed into an eigenspace. If a sequence of training data are sampled in a local context, a small number of eigenvectors which possess larger eigenvalues provide a good summary of all the original variables. Fuzzy controllers can be trained for mapping the input projection in the eigenspace to the outputs. Experiments of time series prediction validate the concept.

1 Introduction

For efficiently modelling complex time series it is desirable that not only the extracted model approximate the training data as precisely as possible and without overfitting but also that the model be human-understandable, e.g., interpretable with fuzzy linguistic rules. However, it is well known that general fuzzy rule descriptions of systems with a large number of input variables suffer from the problem of the "curse of dimensionality." In many real applications it is difficult to identify the correct influential factors and reduce their number to the minimum. A general solution to build fuzzy models is not only interesting from a theoretical point of view but also practically meaningful because the application of fuzzy systems could then be extended to a wider range of complex modelling/prediction problems.

In the recent literature on neuro-fuzzy modelling and machine learning, some standard benchmark problems with medium-dimensional input spaces are frequently discussed and simulation results are presented. As training data set for nonlinear system identification, the Box-Jenkins gas furnace data [3] is often studied and compared. The furnace input $x(t)$ is the gas flow rate, the output $y(t)$ is the CO_2 concentration. At least 10 candidate inputs are considered: $x(t-6), x(t-5), \cdots, x(t-1), y(t-1), \cdots, y(t-4)$. If all of them are used, building a fuzzy prediction system means to solve a 10-input-1-output problem. If each input is defined by 5 linguistic terms, this would result in a fuzzy rule system of about 10 million rules. The modelling and prediction of financial markets are based on much more influential factors which are hard to differentiate since they are inter-related.

2 Existing Solutions

The two main methods addressing the multivariate problem are "input selection" [6, 4] and "hierarchy" [7].

Input Selection is in principle an experimental method to find the most important input variables among a large number of them. There are two obvious problems with such a procedure. First, all the other less influential inputs are discarded, which means an information loss for the controller. Secondly, the combinational number of lower-dimensional fuzzy controllers for a system with thousands or even millions of inputs is too large to enumerate and to evaluate.

Hierarchy A hierarchical structure assumes that the input information can be classified into groups. Within each group the inputs determine an intermediate variable; they can be decoupled from the inputs of the other groups. To realise such a grouping, there exists no general automatic approach but heuristics based on fusion of physical sensors.

If a fuzzy controller is only observed at each moment in time, the only input information to the controller comes from the whole set or a selected subset of the input variables. This kind of information is called *horizontal index* in this paper. Both "input selection" and hierarchical structure consider only the horizontal indices. They presume that the inputs are independent of each other and give no priority or importance of the selected input variables.

3 B-Splines for Approximating Low-Dimensional Data

In principle, the evaluation of a fuzzy rule base is an *interpolation process*. Therefore, if we consider an automatic method for designing a fuzzy controller, it is useful to review the interpolation methods that use analytical functions. Lagrange polynomials supply a set of functions that can be used for interpolating a given number of data points. Newton polynomials can realise the same task but they can be recursively computed so that a new polynomial does not need to be totally re-calculated for new data. Bernstein polynomial functions are based on a parameter set and can also interpolate data quite well. However, with B-splines, the order of basis functions is independent of the number of interpolation data. Moreover, basis functions of the B-spline model can be used as a convenient tool to specify linguistic terms.

3.1 B-Splines

Assume x is a general input variable of a control system that is defined on the universe of discourse $[x_0, x_s]$. Given a sequence of ordered parameters (knots): $(x_0, x_1, x_2, \ldots, x_s)$, the ith normalised B-splines $X_{i,k}$ of order k is recursively defined as:

$$X_{i,k}(x) = \begin{cases} \begin{cases} 1 & \text{for } x_i \le x < x_{i+1} \\ 0 & \text{otherwise} \end{cases} & \text{if } k = 1 \\ \frac{x - x_i}{x_{i+k-1} - x_i} X_{i,k-1}(x) + \frac{x_{i+k} - x}{x_{i+k} - x_{i+1}} X_{i+1,k-1}(x) & \text{if } k > 1 \end{cases} \tag{1}$$

with $i = 0, 1, \ldots, s - k$. Figure 1 shows B-splines from order $k = 1, 2, 3, 4$ which can be compared with fuzzy membership functions (MFs).

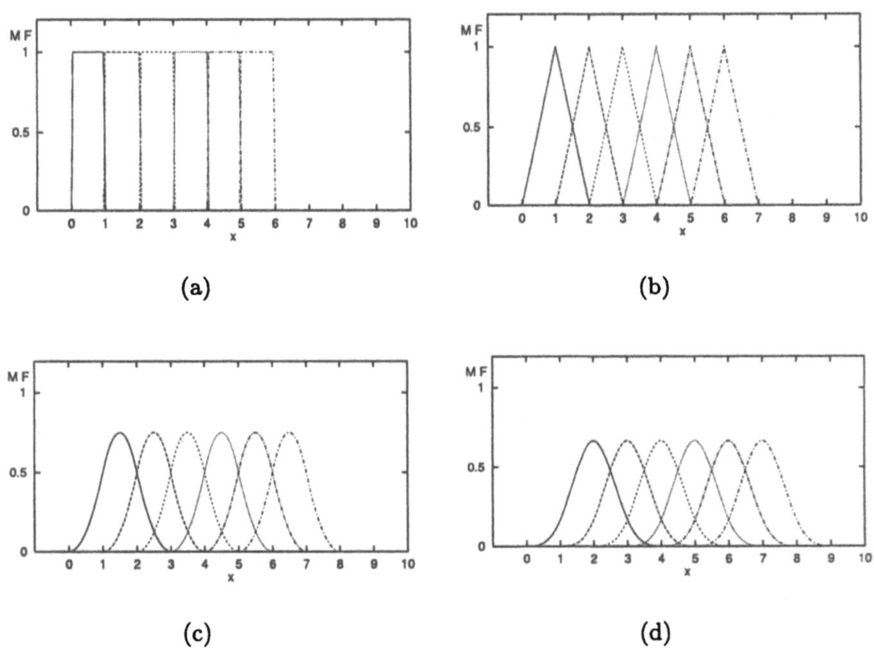

Fig. 1. Fuzzy sets defined by B-spline basis functions of different orders. (a) k=1. (b) k=2. (c) k=3. (d) k=4.

3.2 A B-Spline Interpolator

Since an MIMO (multiple-input–multiple-output) rule base is normally divided into several MISO (multiple-input–single-output) rule bases, we consider only the MISO case. In [13], we showed that under the following conditions the computation of the output of such a fuzzy controller is equivalent to that of a *general B-spline hypersurface*:

- The membership functions for the inputs are periodical B-spline basis functions;
- the membership functions for the outputs are fuzzy singletons;
- "Product" is used as fuzzy conjunctions;
- "Centroid" is the defuzzification method;
- "Virtual linguistic terms" are added at both ends of each input variable; and
- the rule base for the "virtual linguistic terms" is extended by copying the output values of the "nearest" neighbourhood.

Generally, we consider a MISO system with n inputs $x_1, x_2, \ldots, x_n$, and rules with the n conjunctive terms in the premise given in the following form:

IF (x_1 is $X^1_{i_1,k_1}$) and (x_2 is $X^2_{i_2,k_2}$) and ... and (x_n is $X^n_{i_n,k_n}$)
THEN y is $Y_{i_1 i_2 \ldots i_n}$

where

- x_j: the jth input ($j = 1, \ldots, n$),
- k_j: the order of the B-spline basis functions used for x_j,
- $X^j_{i_j,k_j}$: the ith linguistic term of x_j defined by B-spline basis functions,
- $i_j = 1, \ldots, s_j$, representing how fine the jth input is fuzzy partitioned,
- $Y_{i_1 i_2 \ldots i_n}$: the control vertex (de Boor points) [2] of the output[1] if *Rule*$(i_1, i_2, \ldots, i_n)$ fires 100%.

Then, the output y of a MISO fuzzy controller is [13]:

$$y = \frac{\sum_{i_1=1}^{m_1} \cdots \sum_{i_n=1}^{m_n} (Y_{i_1,\ldots,i_n} \prod_{j=1}^{n} X^j_{i_j,k_j}(x_j))}{\sum_{i_1=1}^{m_1} \cdots \sum_{i_n=1}^{m_n} \prod_{j=1}^{n} X^j_{i_j,k_j}(x_j)} = \sum_{i_1=1}^{m_1} \cdots \sum_{i_n=1}^{m_n} (Y_{i_1,\ldots,i_n} \prod_{j=1}^{n} X^j_{i_j,k_j}(x_j)) \tag{2}$$

This is a multivariate B-spline model, representing a general NUBS (nonuniform B-spline) *hypersurface*.

3.3 Properties of Parameter Learning

Computing the parameters of such a B-spline fuzzy system is divided into two steps: one for the IF-part and the other one for the THEN-part.

Properties of modelling the IF-part are:

[1] The THEN-part has the same form of a zero-order TSK model [11].

- Based on the granularity of the input space and the distribution of extrema in the control space (if known), the fuzzy sets can be generated using the recursive computation of B-spline basis functions. This approach provides an automatic approach to generate the information granularity as proposed by Zadeh [12].
- These fuzzy sets can be further adapted during the generation of the whole system by modifying the knots with neural networks [14] or genetic algorithms [15].

Properties for generating the THEN-part are:

- Fuzzy singletons represented by control vertices can be initialised with the values acquired from expert knowledge. These parameters will be fine-tuned by a learning algorithm.
- For supervised learning, we show in the following that the squared errors with respect to control vertices are convex functions. Therefore, rapid convergence for supervised learning is guaranteed. The control space changes locally due to the "local support" property of B-splines while the control vertices are modified. Based on this feature, the control vertices can be optimised gradually area-by-area.

3.4 Supervised Learning of Control Vertices

A fuzzy system constructed as outlined above can be optimised with the gradient descent approach. Assume $\{(\mathbf{x}, y_d)\}$ is a set of training data, where $\mathbf{x} = (x_1, x_2, \ldots, x_n)$ is a training data vector, and y_d is the desired output for $\mathbf{x}$.

If we apply the following squared error function:

$$E = \frac{1}{2}(y_r - y_d)^2 \tag{3}$$

where y_r is the current real output value during training computed with (2), the parameters to be found are $Y_{i_1,i_2,\ldots,i_n}$. To minimise the error in (3), each control vertex $Y_{i_1,\ldots,i_n}$ can be modified by using the gradient descent method:

$$\begin{aligned} \Delta Y_{i_1,\ldots,i_n} &= -\epsilon \frac{\partial E}{\partial Y_{i_1,\ldots,i_n}} \\ &= -\epsilon(y_r - y_d) \prod_{j=1}^{n} X^j_{i_j,k_j}(x_j) \end{aligned} \tag{4}$$

where $0 < \epsilon \leq 1$ represents the learning rate.

The gradient descent method guarantees that the learning algorithm converges to the global minimum of the error function because the second partial differentiation with respect to $Y_{i_1,i_2,\ldots,i_n}$ is constant:

$$\frac{\partial^2 E}{\partial^2 Y_{i_1,\ldots,i_n}} = -\epsilon(\prod_{j=1}^{n} X^j_{i_j,k_j}(x_j))^2 \geq 0. \tag{5}$$

This means that the error function (3) is convex in the space $Y_{i_1,i_2,\ldots,i_n}$ and therefore possesses only one minimum.

3.5 Identification of a Nonlinear System

The examples in [5] and [9] were implemented using our approach to demonstrate the learning ability for function approximation and system identification. In the following figure, the modelling of the linguistic terms using B-splines are not shown. Our observation mainly focuses on the learning process of the input–output relation, i.e. the control surface, the training error, and the checking error.

One interesting task is to identify a nonlinear component in a dynamic system: $y(t+1) = 0.3y(t)+0.6y(t-1)+f(u(t))$, where $y(t)$ is the output of moment t and $u(t)$ is the input, see [5].

In this simulation $f(\cdot)$ has the the form: $f(u) = 0.6\sin(\pi u) + 0.3\sin(3\pi u) + 0.1\sin(5\pi u)$. The input signal is:

$$u(t) = \begin{cases} \sin(2\pi t/250), & \text{for } t \leq 500, \\ \\ 0.5\sin(2\pi t/250) + 0.5\sin(2\pi t/25), & \text{for } t > 500. \end{cases}$$

B-spline fuzzy controllers of order three are trained with 250 data points from $t = 1$ to $t = 250$. The data points between $t = 251$ to $t = 750$ are used for checking. Figure 2 shows the results with 30 B-functions after one and ten training epochs. In Figure 2(a), (c) and (e), the lighter curves represent the desired data, while in Figure 2(b), (d) and (f), it is shown that these curves are approximated to such a degree where there approximation error is almost indescernible.

4 Concept of Using Statistical Indices

4.1 High Input Dimensionalities

If the dimension of the input space is small, the input variables can be directly covered by fuzzy sets, thus fuzzy rules are formed in a tabular form. Each item of the rule table can be interpreted as a general situation. Such a "situation" is still intuitively understandable for a small number of inputs. This is certainly no longer the case if the number of the input variables is no more 2, 3, 4, but 10, 10.000, or even millions.

If all inputs are normalised, e.g., mapped to the interval $[0, 1]$ and all inputs are arranged in matrix form, then the ensemble of all input varibles can be considered as a "brightness image" with the brightness of each matrix element varying between 0 and 1. Taking this point of view, the multivariate modelling problem can be likened to the supervised learning control problem as known from computer vision and the matrix corresponds to the image plane of a camera.

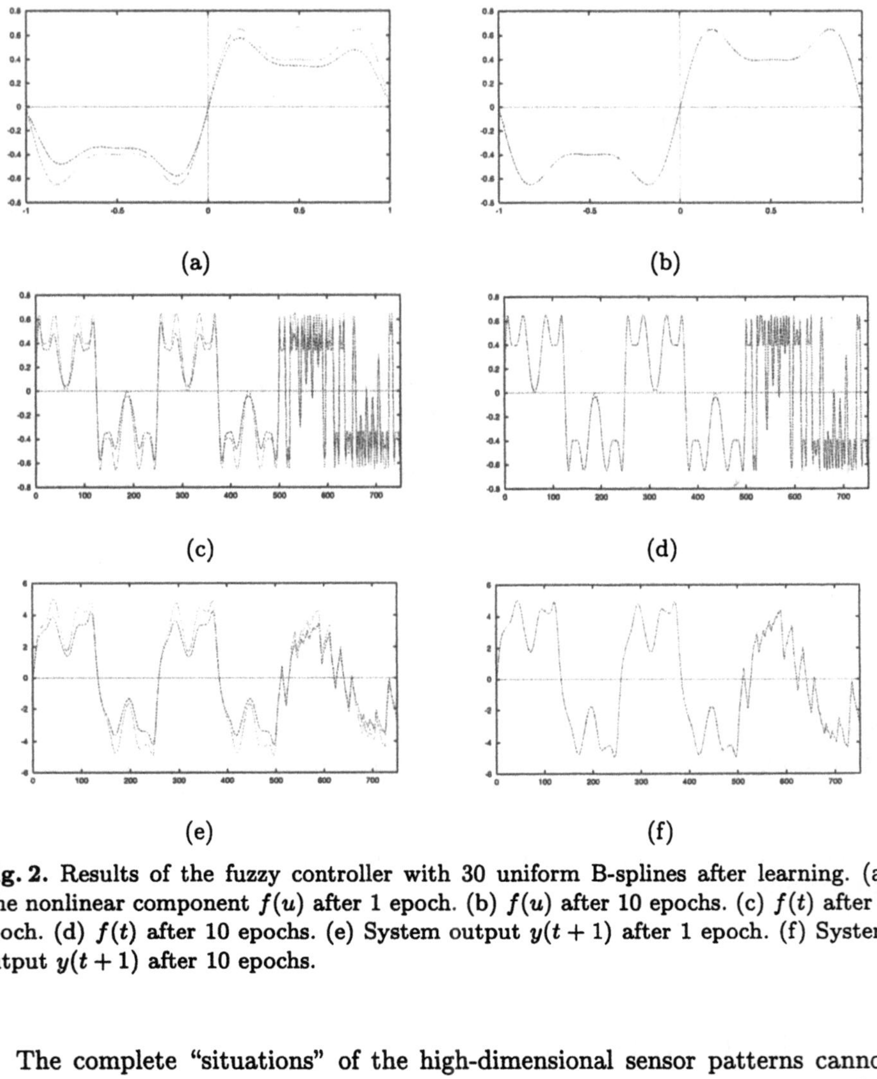

Fig. 2. Results of the fuzzy controller with 30 uniform B-splines after learning. (a) The nonlinear component $f(u)$ after 1 epoch. (b) $f(u)$ after 10 epochs. (c) $f(t)$ after 1 epoch. (d) $f(t)$ after 10 epochs. (e) System output $y(t+1)$ after 1 epoch. (f) System output $y(t+1)$ after 10 epochs.

The complete "situations" of the high-dimensional sensor patterns cannot be represented as a number of computer internal memorable and computable tabular items. However, if we observe the local context of the camera sensor patterns in a certain local context, it can often be found that they concentrate within certain parts of the whole input space, i.e., they are correlated.

4.2 Vertical versus Horizontal Indices

If the input vectors are considered to be a pile of training data originating from a continuously varying process *vertical indices* can be extracted by statistical analysis. These vertical indices describe data distribution, variances, and the most interesting features for control-projections on the eigenvectors (Fig. 3).

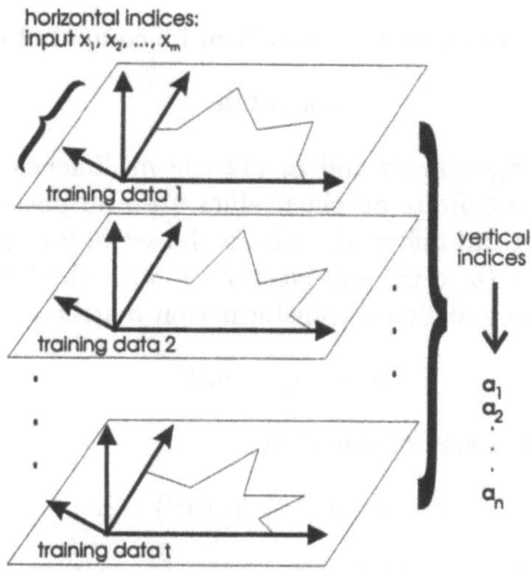

Fig. 3. Vertical information about the distribution of the data in the input space can be extracted by statistics on a sequence of the training data set.

For an input space $X_1 \times X_2 \times \cdots \times X_m$, if all the variables $x_1 \ldots x_m$ could vary in all their universes in the observed sampling procedure, the input data would scatter in the whole input space. Nevertheless, in a high-dimensional input space, if the observed process runs continuously, the input vector varies gradually. Under an appropriately separated observation, e.g., within a local scenario of the camera's environment, the input vectors possess a large grade of similarity in most cases. That means in the high-dimensional input space, the observed input data are correlated to a large degree.

4.3 Projection in Eigenspace

A well-known technique for dealing with multivariate problems in statistics is the *principal component analysis* (PCA). Until now, it has been mainly applied to data compression and pattern recognition, [10]. In our opinion, this technique is also suitable for reducing the dimension of the input space of a general control problem.

Let us assume l sample input vectors $\mathbf{x}^1, \ldots, \mathbf{x}^l$ with $dim(\mathbf{x}^i) = m$ originating from a pattern-generating process. The PCA can be applied to them as follows: First the (approximate) mean value $\mathbf{v}$ and the covariance matrix $\mathbf{Q}$ of these vectors are computed according to

$$\mathbf{v} = \frac{1}{l} \sum_{i=1}^{l} \mathbf{x}^i$$

$$\mathbf{Q} = \frac{1}{l} \sum_{i=1}^{l} (\mathbf{x}^i - \mathbf{v})(\mathbf{x}^i - \mathbf{v})^T$$

The eigenvectors and eigenvalues can then be computed by solving

$$\lambda_i \mathbf{a_i} = \mathbf{Q}\mathbf{a_i}$$

where λ_i are the m eigenvalues and $\mathbf{a_i}$ are the m-dimensional eigenvectors of $\mathbf{Q}$. Since $\mathbf{Q}$ is positive definite all eigenvalues are also positive. Extracting the most significant structural information from the set of input vectors $\mathbf{x^i}$ is equal to isolating the first n ($n < m$) eigenvectors $\mathbf{a_i}$ with the largest corresponding eigenvalues λ_i. If we now define a transformation matrix

$$\mathbf{A} = (\mathbf{a_1} \ldots \mathbf{a_n})^{\mathbf{T}}$$

we can reduce the dimension of the $\mathbf{x^i}$ by

$$\mathbf{p^i} = \mathbf{A} \cdot \mathbf{x^i}; \quad dim(\mathbf{p^i}) = n$$

The eigenvectors form an orthogonal basis for representing the original individual sensor patterns. The dimension n should be determined depending on the discrimination accuracy needed for further processing steps vs. the computational complexity that can be afforded. as exAssume that the eigenvectors $\mathbf{a_1}, \mathbf{a_2}, \cdots, \mathbf{a_n}$ are sorted according to their eigenvalues in a descending order. An eigenspace with a reduced dimension n can be formed with the first n eigenvectors. The projection of an input vector $[x_1, x_2, \cdots, x_m]$ on eigenvector $\mathbf{a_i}$, called the ith principal component, is $a_{1,i}x_1 + a_{2,i}x_2 + \cdots + a_{m,i}x_m$. All projections of the sample data sequence form a manifold in the eigenspace.

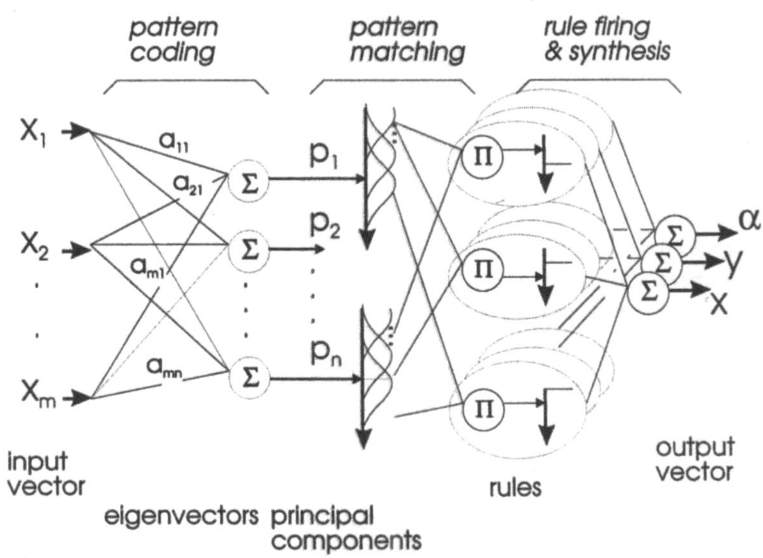

Fig. 4. The task based mapping can be interpreted as a neuro-fuzzy model.

4.4 The Neuro-Fuzzy Model

Depending on how "local" the measuring data are and, therefore, how similar the observed input patterns appear, a more or less small number of eigenvectors can provide a sufficient summary of the state of all input variables. Moreover, in the case of very high input dimensions, an effective dimension reduction can be achieved by PCA. This step is illustrated in the left part of Fig. 4.

Eigenvectors can be partitioned by covering them with linguistic terms (see the right part of Fig. 4). In our implementations, fuzzy controllers constructed according to the B-spline model are used (section 3). This model provides an ideal implementation of the CMAC proposed by Albus [1]. We define linguistic terms for input variables with B-spline basis functions and for output variables with singletons. Such a method requires fewer parameters than other set functions such as trapezoid, Gaussian function, etc. The output computation is very simple and the interpolation process is transparent. Through comparative studies, B-spline model generally achieves better approximation capabilities and rapid convergence than the other fuzzy models.

4.5 Off-line and On-line Phases

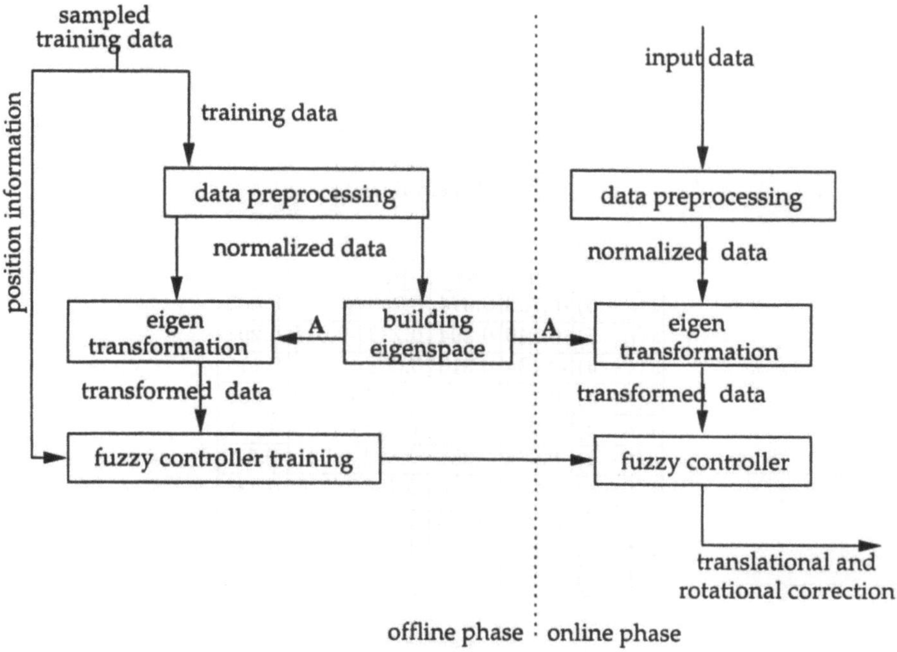

Fig. 5. The training and the application of the PCA neuro-fuzzy controller.

The working system implements two phases: off-line training and on-line evaluation. In the off-line phase, a sequence of training input patterns and their corresponding outputs are used. In the on-line phase the input pattern is transformed into the eigenspace and is then processed by the fuzzy controller. The controller output is the system prediction.

5 Implementations

5.1 Prediction with Box-Jenkins Data

296 data points $(x(t), y(t))$ are first transformed into the form $((x(t-6), x(t-5), \ldots, x(t), y(t-1), \ldots, y(t-4)), y(t))$. The computed eigenvectors are shown in Fig. 6(a), the eigenvalues of each eigenvector are depicted in Fig. 6(b). The projection of the data into the eigenspace constructed by the first two and first three eigenvectors are shown in Fig. 7(a) and (b), respectively.

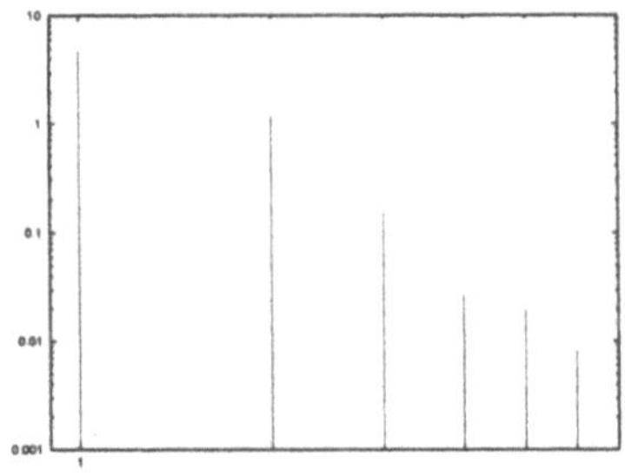

(a) Eigenvalues

	p_1	p_2	p_3	p_4	p_5
$x(t-6)$	-0.158	0.074	0.239	0.354	0.602
$x(t-5)$	-0.149	0.204	0.233	-0.015	0.291
$x(t-4)$	-0.134	0.310	0.081	-0.382	0.115
$x(t-3)$	-0.117	0.372	-0.164	-0.419	0.288
$x(t-2)$	-0.099	0.386	-0.413	-0.044	0.228
$x(t-1)$	-0.084	0.361	-0.582	0.531	-0.222
$y(t-1)$	0.477	-0.334	-0.295	0.136	0.523
$y(t-2)$	0.487	-0.015	-0.236	-0.263	0.141
$y(t-3)$	0.480	0.278	0.053	-0.225	-0.248
$y(t-4)$	0.457	0.501	0.448	0.362	-0.017

(b) Eigenvectors

Fig. 6. The important eigenvectors and eigenvalues of the Box-Jenkins data.

In the following implementations, linguistic terms of the eigenvectors are defined by B-spline basis function of order three. The output value of each rule is represented as a fuzzy singleton which is called a control vertex in the B-spline fuzzy controller. The control vertices are adaptively determined by minimising the RMS Error' (as in [6]) for the supervised learning.

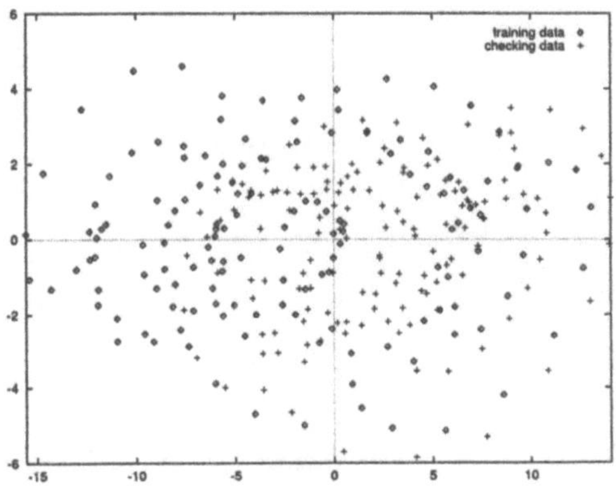

(a) 2D projection

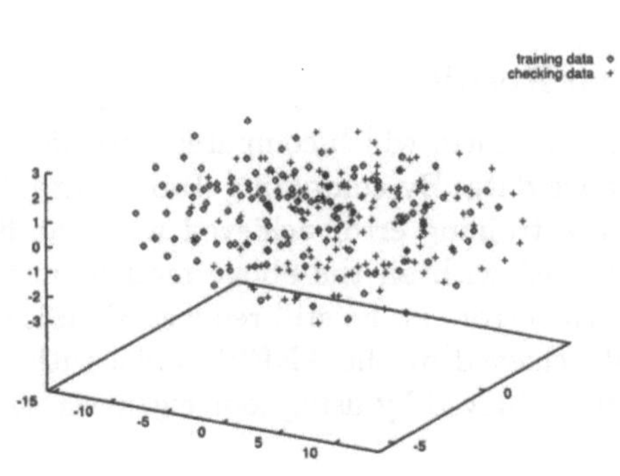

(b) 3D projection

Fig. 7. Distribution of the data in the 2D and 3D eigenspace.

The first, second, third and fourth eigenvector are defined by 10, 7, 7 and 5 linguistic terms respectively. The RMS (and MS) training errors achieved with the first three eigenvectors are listed in Tab. 1.

	RMS (MS) Error		
epochs	1	2	3
100	0.73 (0.533)	0.22 (0.048)	0.25 (0.063)
1000	0.71 (0.504)	0.19 (0.036)	0.20 (0.04)
10000	0.71 (0.504)	0.19 (0.036)	0.17 (0.029)

Table 1. RMS and MS (in parentheses) training error by using 1, 2, and 3 first eigenvectors if all 296 data are used for training.

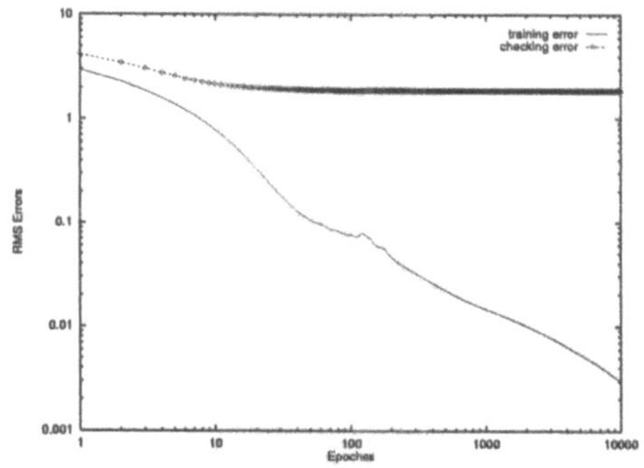

Fig. 8. RMS errors of modelling the Box-Jenkins data with four eigenvectors. 148 data are used for training and 148 data for checking.

5.2 Comparative Results

In [8], Table 2 was provided, which compared with the results for solving the Box-Jenkins gas furnace data. By comparing Tab. 1 with Tab. 2, it can be seen that –as expected– the training error achieved with the first two eigenvectors is less than that achieved with all the above models, and with the first three or four eigenvectors, the error can be still reduced significantly. The RMS error is also less than that achieved by the ANFIS model with input selection in [6]. Fig. 8 shows the result achieved by using four eigenvectors.

6 Discussion

The main advantage of the proposed approach to "input selection" is that less information is lost after the dimension reduction for problems with correlated input training data. Even if the training data are not correlated at all, the projection in the eigenspace provides information which input variables have larger variance. These variables can be good candidates for the inputs to be selected. Therefore, no "trial-comparison-select" procedure is necessary.

To generally deal with the high-dimensional input space, the solution based on the low-dimensional fuzzy controllers would need the partitioning of the com-

Model	Input	No. of rules	MS Error
Tong's	$x(t-1), y(t-4)$	19	0.469
Pedrycz's	$x(t-1), y(t-4)$	81	0.320
Xu/Lu's	$x(t-1), y(t-3)$	25	0.328
Chiu's TSK 2	$x(t-1), y(t-3)$	3	0.146
Chiu's TSK 3	$x(t-1), x(t-3), y(t-3)$	3	0.072
[8] GA-fuzzy	$x(t-1), y(t-4)$	25	0.257

Table 2. Comparison of different models derived using the Box-Jenkins gas furnace data, excerpted from [6].

plete high-dimensional input data set into clusters, within which the data are correlated to a large degree. Such a partition would be intrinsically fuzzy, since there are no crisp boundaries between two continuous "situations". A "behaviour arbiter" would coordinates multiple simultaneously active local controllers to achieve a high-level task and can be realised with a set of meta-rules like: "IF *Situation_Evaluation* IS for_C_i THEN Apply Controller C_i."

References

1. J. S. Albus. A new approach to manipulator control: The Cerebellar Model Articulation Controller (CMAC). *Transactions of ASME, Journal of Dynamic Systems Measurement and Control*, 97:220–227, 1975.
2. W. Böhm, G. Farin, and J. Kahmann. A survey of curve and surface methods in cagd. *Computer Aided Geometric Design*, 1:1–60, 1984.
3. G. E. P. Box and G. M. Jenkins. *Time series analysis*. Holden Day, San Francisco, 1970.
4. S. L. Chiu. Selecting input variables for fuzzy models. *Journal of Intelligent and Fuzzy Systems*, 4:243–256, 1996.
5. J.-S. R. Jang. ANFIS: Adaptive-network-based fuzzy inference system. *IEEE Transactions on System, Man and Cybernetics*, 23(3):665–685, 1993.
6. J.-S. R. Jang, C.-T. Sun, and E. Mizutani. *Neuro-Fuzzy and Soft Computing*. Prentice Hall, 1997.
7. V. Lacrose and A. Tilti. Fusion and hierarchy can help fuzzy logic controller designers. In *IEEE International Conference on Fuzzy Systems, Barcelona*, 1997.
8. S. Lotvonen, S. Kivikunnas, and E. Juuso. Tuning of a fuzzy system with genetic algorithms and linguistic equations. In *Proceedings of Fourth European Congress on Intelligent Techniques and Soft Computing, Aachen*, 1997.
9. S. Mitaim and B. Kosko. What is the best shape of a fuzzy set in function approximation. In *IEEE International Conference on Fuzzy Systems*, pages 1237–1243, 1996.
10. E. Oja. *Subspace methods of pattern recognition*. Research Studies Press, Hertfordshire, 1983.

11. T. Takagi and M. Sugeno. Fuzzy identification of systems and its application to modelling and control. *IEEE Transactions on System, Man and Cybernetics*, SMC-15(1):116–132, 1985.
12. L. A. Zadeh. Fuzzy logic = computing with words. *IEEE Trans. on Fuzzy Systems*, 4(2):103–111, 1996.
13. J. Zhang and A. Knoll. Constructing fuzzy controllers with B-spline models - principles and applications. *International Journal of Intelligent Systems*, 13(2/3):257–285, Feb./Mar. 1998.
14. J. Zhang and A. Knoll. Designing fuzzy controllers by rapid learning. *Fuzzy Sets and Systems*, 13(2), 1998.
15. J. Zhang, A. Knoll, and I. Renners. Efficient learning of non-uniform B-splines for modelling and control. In *International Concerence on Computational Inteligence for Modelling, Control and Automation, Viena*, pages 282–287, Viena, 1999.

A Fast Logically-Complete Preferential Reasoner for the Assessment of Critical Situations

Éric Grégoire

CRIL
Université d'Artois
rue de l'Université SP16
F-62307 Lens Cedex France
gregoire@cril.univ-artois.fr

We propose a new computational approach to logic-based systems that should reason in a fast but logically sound and complete manner about large-scale complex critical devices and systems that can exhibit unexpected faulty behaviors. The approach is original from at least two points of view. First, it makes use of local search techniques while preserving logical deductive completeness. Second, it proves experimentally efficient for very large knowledge bases thanks to new heuristics in the use of local search techniques

1 Introduction

In this chapter, we are concerned with monitoring-like systems that need to reason in a fast but logically sound and complete manner about large-scale complex critical devices and systems that can exhibit unexpected faulty behaviors. When the reasoning systems are based on a deep knowledge that is expressed in a logic-based representation language, this issue is closely related to the consistency-based approach to diagnosis [Reiter 87, Hamscher *et al.* 92, Console *et al.* 94, Cordier 97]. However, both issues differ on several key points.

First, the main goal of the reasoning systems is to perform deductive reasoning with or without abnormal functioning conditions (assuming the latter situation by default): explaining faulty behaviors comes as a by-product, only. Second, because of the complexity and size of the targeted devices (e.g., a nuclear plant or a complex embarked system), we cannot assume that these ones can be represented in a complete way. Although we expect that as much knowledge as possible is encoded (in particular, knowledge from first principles), it is often impossible to envision all possible effects of all possible device failures. Indeed, when things go wrong in large-scale complex monitoring systems, the user is often confused and overwhelmed by various symptoms of bad behavior that seem unrelated, making it difficult for him (her) to understand the actual situation and problem, and to react in a fast and proper way (e.g., warning alarms went in series in most reported nuclear power plants incidents). Actually, many such symptoms are often indirect consequences of a single source problem and appear unrelated simply because of unforeseen interactions of components and processes, and thus of the incompleteness of the encoded knowledge. They often hide the real cause of the faulty behavior and/or the immediate action to undertake. In this respect, computing all diagnoses *directly* on the basis of an incomplete description of the device is not only computationally expensive but does neither necessarily clarify the situation nor provide the user with the actual explanation, since essential chains of failure can have been missed.

On the contrary, our reasoning system points out the main source of problems or the more serious one (hereafter, the *first failure*) to allow the user to assess the situation quickly and get a hint about the real problem that is occurring and of the action to undertake. It thus informs the user about the device failure that is most important according to a scale that is predefined with respect to the specific application; it can feature failures that should trigger immediate action or a preference for the source of a failure chain. To complete the picture, the reasoning system also gives the user *one* whole coherent interpretation of the situation, explaining also detected failures that are apparently not consequences of the first failure, at least on the basis of the available knowledge. If the user finds it necessary, he (she) can assess the situation again, using the same process in an iterative way after having fixed the first failure or simulated its repair (together with the other detected unrelated failures if he (she) decides it).

The proposed computational approach implements a form of nonmonotonic inference to allow reasoning by default when assuming the absence of device failures. The logic is carefully designed as to match the expressiveness required by this specific application. It exhibits both acceptable worst case and excellent empirical computational results even for very large scale applications. By itself, the computational approach is original from at least two points of view. First, it makes use of local search techniques while preserving deductive completeness. Second, it proves efficient for very large knowledge bases (*KB*s) thanks to new heuristics in the use of local search.

2 A Simple Useful Nonmonotonic Framework

Thanks to dramatic progress in solving extremely large propositional satisfiability problems, there is a growing interest in designing manageable propositional encodings of various forms of knowledge and problems [Selman *et al.* 97]. In this respect, the knowledge representation language proposed here is full clausal normal form (CNF) propositional logic. Thus, this does not mean that the actual knowledge must be in propositional form. For example, all the computational results in this chapter apply to instantiated counterparts of finite first-order *KB*s, where the size blow-up is empirically avoided thanks to a component-guided instantiation schema, as in various diagnostic systems [Hamscher *et al.* 92, Console *et al.* 94, Cordier 97].

The language is enriched with a finite set **AB** of McCarthy's prioritized Abnormality propositions [McCarthy 86], noted Ab_i, allowing rules of default reasoning to be expressed. Abnormality propositions are used to describe possible (but unexpected) faulty components. For instance, the rule asserting that, under normal circumstances, when the switch is on then the lights should be on is represented by the formula $switch\_on \wedge \neg Ab_1 \Rightarrow lights\_on$, and in clausal form by $\neg switch\_on \vee Ab_1 \vee lights\_on$. In this representation framework *à la* consistency-based diagnosis [Reiter 87], we also expect that a principle of parsimony does hold, i.e., only a limited number of Abnormality propositions are to be interpreted as *true* when the *KB* is queried (unless Murphy's law applies, of course). Ab_i propositions are expected to be *false* under normal operating circumstances of the device since they are intended to represent unexpected faulty conditions. This is thus a very specific use of Ab_i variables. We do not consider default rules where exceptions are expected to exist under normal circumstances, like « Typical humans are right-handed ».

This knowledge representation formalism is interpreted under a model-preference framework that matches our needs. First, let us recall that an *interpretation* assigns values from {*true*, *false*} to the propositional variables of the (enriched) language. A *model* of a formula *f* is an interpretation under which the formula *f* gets the truth value *true*. A model will be represented by the set of propositional variables that it satisfies, i.e., interprets to *true*. The Abnormality propositions Ab_i are prioritized according to a pre-order, translating the scale of importance of the various possible failures. The set **AB** is partitioned into *m* subsets $\mathbf{AB}_z$, called strata, s.t. $z \in \{1,...,m\}$. We say that $Ab_i \in \mathbf{AB}_r$ is lower than $Ab_j \in \mathbf{AB}_s$ (noted $Ab_i < Ab_j$) when $r < s$. We write $Ab_i = Ab_j$ when both propositions belong to a same stratum $\mathbf{AB}_s$. We also assume that the numbering of the Ab_i propositions follows the ordering of their stratum, i.e., when $Ab_i \in \mathbf{AB}_r, Ab_j \in \mathbf{AB}_s$ and $r > s$, then $Ab_i > Ab_j$ and $i > j$.

The knowledge engineer is thus expected to encode the default rules in such a way that more important device failures relate to lower Ab_i. Accordingly, models that are such that the lowest Ab_i they contain is as low as possible will be preferred. More formally, Let M_1 and M_2 be two models of *KB*.

Definition 2.1

M_1 is at least as preferable as M_2 (noted $M_1 \leq M_2$) iff

- *M_1 does not contain any Ab_i*
- *or both M_1 and M_2 contain at least one Abnormality proposition and $\forall Ab_i \in M_2$, $\exists Ab_j \in M_1$ such that $Ab_j \leq Ab_i$.*

Let us stress that $\leq$ allows us to compare any pair of models. Since $\leq$ is a pre-order, M_1 and M_2 are not necessarily identical when both $M_1 \leq M_2$ and $M_2 \leq M_1$.

Definition 2.2
M_1 is a preferred model of KB iff M_1 is minimal w.r.t. $\leq$.

Accordingly, nonmonotonic inference from *KB* is defined as follows. Let *f* be a literal, i.e., a propositional variable or its negation (the following is easily extended to clauses or formulas).

Definition 2.3
KB |~ f iff f is true in all preferred models of KB.

Such a preferential logic enjoys enough structural properties so that the following more convenient result can be established easily.

Theorem 2.1
Assume KB is a consistent set of propositional clauses and f is a literal. KB |~ f iff there exists a model M_1 of $KB \cup \{f\}$ and there does not exist a model M_2 of $KB \cup \{\neg f\}$ such that $M_2 \leq M_1$.

Let us stress that this logic differs from (semantical accounts of) prioritized circumscription [McCarthy 86] in the sense that the set of models is here partitioned into $m + 1$ equivalence classes, where m is the total number of strata in **AB** (roughly, only the lowest Ab_i proposition that is *true* in a model does influence the ordering relation; no refinement is made depending on the truth value of the higher Ab_j propositions, i.e., on less important failures).

This requirement is motivated by the nature of the targeted applications: most important failures should be exhibited in order to conduct a first assessment of the situation; other ones are often simply consequences or are considered secondary and treated afterwards (they should not hide the *real* problem). This also allows acceptable worst cases computational results to be obtained: this inference is polynomial under a number of calls to an NP-oracle that is logarithmic with respect to the number of strata in **AB**. Assume that *KB* is consistent and that *f* is a literal.

Theorem 2.2
Checking *KB |~ f* is in $P^{NP[O(\log m)]}$, where *m* is the total number of strata in **AB**.

The proof is given in Annex 2. When the Abnormality propositions are totally ordered, the worst case computational complexity is thus in $P^{NP[O(\log n)]}$, where *n* is the total number of Abnormality propositions in **AB**.

Interestingly enough, we propose a logically sound and complete decision procedure for *KB |~ f* (assuming that *KB* is consistent) that features new heuristics in the use of local search techniques. This will most often allow this decision problem to require a small constant number of calls to an empirically fast satisfiability checking procedure, only. Moreover, this number of calls (about 5, in case of device failures) is independent of the numbers of Abnormality propositions, clauses, propositional variables and strata in **AB**! But first let us recall how local search techniques work and how they can (sometimes) help us to prove inconsistency.

3 Local Search Techniques

Our proof-procedure will be based on powerful combinations of the logically complete so-called Davis and Putnam procedure [Davis *et al.* 62] (in short, DP) for propositional satisfiability checking and of local search techniques. Original heuristics about the use of local search techniques will allow us to detect the device failures and reduce the logarithmic number of calls to an NP-oracle to a small constant one (extremely often).

Most local search algorithms (mainly, GSAT and its variants [Selman *et al.* 92, 93, 97] [McAllester *et al.* 97]) perform a greedy local search for a satisfying assignment of a set of propositional clauses. The algorithms generally start with a randomly generated truth assignment of the propositional variables (i.e., assigns *true* or *false* to each propositional variable). Most of them change (« flip ») the assignment of the variable that leads to the largest increase in the total number of satisfied clauses. Such flips are repeated until either a model is found or a preset maximum number of flips is reached. This process is repeated as needed up to a preset maximum of times. To escape from local extrema, local search algorithms are provided with additional (e.g., random) moves. In the following, a variant procedure called TSAT is used [Mazure *et al.* 97a, 98b]. It makes use of a tabu list forbidding recurrent flips and appears competitive in most situations.

Although local search techniques prove efficient in showing the consistency of hard propositional clausal *KBs*, they are clearly logically incomplete in the sense that they do not cover the whole search space of interpretations and cannot thus directly prove that a formula is inconsistent.

4 Finding Contradictory Knowledge

Recently, it has been discovered that the trace of local search algorithms can allow one to determine the smallest inconsistent kernels very often when they fail to prove consistency within a preset amount of computing time [Mazure *et al.* 97c, 98a]. More precisely, for each clause, taking each flip as a step of time, the number of times during which this clause is falsified is updated. A similar trace is also recorded for each literal occurring in *KB*, counting the number of times it has appeared in the falsified clauses. When the local search technique fails to prove consistency, it appears extremely often that the most often falsified clauses form the smallest inconsistent kernels of *KB* if *KB* is actually inconsistent. Likewise, literals that exhibit the highest scores belong to these kernels. Independently, this ordering of literals has been proposed by [Crawford 93] as a branching heuristic in DP for solving hard random problems.

Let us stress that the validity of this experimental result depends on the size of the discovered probable kernels with respect to the size of *KB*. When the size of the set-theoretic union of smallest inconsistent kernels converges towards the size of *KB*, no significant result is obtained. Fortunately, in large *KBs* describing physical devices, inconsistency is often due to a limited number of contradictory rules and facts.

5 Inferring in an Efficient Way

Let us now come back to our initial problem. From now on, we assume that enough Abnormality propositions have been introduced into *KB* to keep it consistent under faulty conditions. How can the above techniques be used to compute $KB \mid\!\sim f$ in an efficient way and allow failures to be detected?

Inference will be computed according to Theorem 2.1: in order to infer f from *KB*, we need to find a model M_1 of $KB \cup \{f\}$ in such a way that there is no model M_2 of $KB \cup \{\neg f\}$ such that $M_2 \leq M_1$.

The complete proof-procedure is given in Annex 1. Let us simply describe its main features here.

The procedure contains two successive steps. In **Step 1**, we assume that the modeled device is working under normal conditions: $KB \mid\!\sim f$ is thus decided thanks to the complete satisfiability checks of $KB \cup \{f\}$ and $KB \cup \{\neg f\}$ with all Ab_i set to *false*. Any model that can be exhibited at this step is called a *normal-circumstances model* [Mazure *et al.* 97c].

Each satisfiability check is based on a combination of DP with a call to the local search procedure TSAT (but most other local search procedures could be used in place of TSAT). TSAT is run first, as it is expected to be efficient in case of consistency. Contrary to other applications of local search for solving SAT instances, it appears often efficient here to use a previously found model of *KB* as an initial interpretation for the first try. Also, all other factors being constant, we prefer a large number of tries over a large number of flips, in order to get a more significant trace. When TSAT delivers a model, the satisfiability check is over. When no model is found after a preset maximum computing time, a complete call to DP settles the (in)satisfiability issue. When the inconsistency is due to a limited number of clauses (which is often the case in this application domain, as explained above), DP (called *DP_focused_on_trace* in the proof-procedure) appears extremely efficient as we guide it by the ordering of literals obtained in the trace of the failed local search.

When no (normal-circumstances) models are extracted during the above **Step 1**, we enter **Step 2** in the proof-procedure, where faulty conditions are assumed.

According to our decision to focus on first failures, preferred models include the smallest possible Ab_i. Once again, we shall try to exhibit a model M_1 of $KB \cup \{f\}$ (resp. $KB \cup \{\neg f\}$) such that there is no model M_2 of $KB \cup \{\neg f\}$ (resp. $KB \cup \{f\}$) such that $M_2 \leq M_1$. Thus, we cannot have $Ab_j \leq Ab_i$ where Ab_i and Ab_j are the smallest Abnormality propositions of M_1 and M_2, respectively. The other Ab_k that are *true* in M_1 translate the other possible failures that a complete coherent interpretation of the situation indicates.

The powerful combination of DP and local search techniques will also be used here as a satisfiability checking procedure. However, the result of **Step 1** can be further exploited to reduce the worst case O(log *m*) calls to this checking procedure to a very small constant number of calls very often, which does neither depend on the size of the *KB*, nor on *m* and nor even on *n*.

Let us elaborate on this. **Step 2** is performed because no normal circumstances model does exist. The trace of the failed local searches (assuming all Ab_i being *false*) provides us with orderings on the clauses. Actually, if we focus on clauses with the highest scores that contain an Abnormality proposition, we are given the clauses of the smallest inconsistent kernels that can restore consistency under the assumption of faulty conditions (i.e., some concerned Ab_i in the clauses must be *true*). More formally,

Lemma 5.1
Let *C* be the set-theoretic union of all minimal inconsistent kernels of $KB \cup \{f\} \cup \bigcup(\forall Ab_i \in \mathbf{AB})\{\neg Ab_i\}$
(resp. $KB \cup \{\neg f\} \cup \bigcup(\forall Ab_i \in \mathbf{AB})\{\neg Ab_i\}$).
Any model of $KB \cup \{f\}$ (resp. $KB \cup \{\neg f\}$) contains at least one Abnormality proposition occurring in *C*.

In other words, when we were wrong in assuming the absence of faults, the heuristic about the trace of the local search gave us the smallest inconsistent kernels of *KB*, and thus the actual faults! This heuristic is most often accurate; it does not work mainly when a large part of the *KB* forms the minimal inconsistent kernels under the no-fault assumption. This worst case situation can only occur when Murphy's law applies really inordinately; i.e., most of all that can fail do actually fail.

Thus, we consider the list L of Ab_i propositions according to the orderings given by the trace of the failed local searches of **Step 1**. As an initial treatment, we sort the sub-list of Ab_i with the highest scores so that the Ab_i with the lowest *i* in this sub-list occurs first in the list L. Most probably, the first element of this list gives the Abnormality proposition that must be *true* in order to find a model of $KB \cup \{f\}$ or $KB \cup \{\neg f\}$. In a single fault circumstance, we are thus most probably given the explanation of the abnormality, i.e., the Ab_i that must be *true* in order to find a model of $KB \cup \{f\}$ or $KB \cup \{\neg f\}$.

We thus fix this first Ab_i proposition in the list L to *true* and check the existence of a model M_1 of $KB \cup \{f\}$.

1. If we find such a model M_1, we have to make sure that there is no model M_2 of $KB \cup \{\neg f\}$ s.t. $M_2 \leq M_1$. This last point can be decided thanks to a single satisfiability check of $KB \cup \{\neg f\} \cup \{Ab_1 \vee Ab_2 \vee \ldots \vee Ab_i \vee \ldots \vee Ab_r\}$,

where the last clause is formed with all Abnormality propositions Ab_r such that $Ab_r \leq Ab_i$. If we do not find such a model M_2, the procedure stops: $KB \mid\sim f$. If we find one, we have to iterate this process, in order to check the existence of a model M_3 of $KB \cup \{ f \}$ such that $M_3 \leq M_2$ (this existence exhibits a low probability, since this would imply that the heuristic at **Step 1** was not accurate). This process is finished only when one such model does not exist, or after the lowest stratum in **AB** has been considered.

2. If we do not find a model M_1, we check the existence of a model M_2 of $KB \cup \{ \neg f \} \cup \{Ab_1 \vee Ab_2 \vee \ldots \vee Ab_i \vee \ldots \vee Ab_r\}$, where the last clause is formed from all Abnormality propositions Ab_r such that $Ab_r \leq Ab_i$. If M_2 is found, we proceed as in case 1 (inverting the roles of f and $\neg f$, initially). If we do not find a model M_2, this implies that the heuristic result of **Step 1** was wrong. This occurs very rarely; in such a case, we restart **Step 2**, selecting the next Ab_i in the list as a candidate for the first failure (the procedure is logically complete and can end up considering the worst case where all Ab_i play this role, successively).

Most generally, selecting the lowest Ab_i in the list is often enough, as it represents the most probable actual first failure. Accordingly, in most cases of faulty conditions, the procedure will require a very small constant number of calls to a (in)satisfiability checking procedure: 4 or 5 (in case of failures) vs. 2 in the absence of device failures.

Theorem 5.1
The NonMonotonic_Inference procedure of Annex 1 is sound and complete with respect to the inference relationship of Theorem 2.1.

In the worst case, the above described procedure requires $O(n)$ calls to a complete satisfiability check, where n is the number of Abnormality propositions in **AB**. But it is a straightforward task to adapt the procedure in order to take the stratum concept into account and get $O(m)$ calls where m is the number of strata in **AB**. Moreover, we could include a dichotomic strategy in **Step 2** (see the proof of Theorem 2.2 in Annex 2), at least after the most probable failures have been considered. In this way, in the worst case, we could approach $O(\log m)$ calls as established by Theorem 2.2. However, since most of the times, the maximum first 5 calls will settle the $KB \mid\sim f$ issue, we have not included these features for clarity of presentation.

Let us also note that this procedure can be optimized in several other ways. For instance, successive calls to DP could avoid computing subtrees that have been considered in previous calls. More importantly, the connected component of the propositions occurring in $KB \cup \{ f \}$ could be delimited. Accordingly, only the corresponding subpart of KB would be considered in the proof-procedure.

Finally, let us stress that this way of computing inferences is not sensitive to the fact that it is defined for literals. It is a straightforward task to extend it to clauses.

6 Experimental Results

Let us just relate here a test (run on a 166 Mhz Pentium) that is a good representative of the size of the tested benchmarks and of the obtained computational results. For instance, we merged 5 large consistent *KBs* that are related to VLSI circuits [DIMACS 93] (namely, ssa-7522-{156, 157, 158, 159, 160}) with a small inconsistent Dubois' SAT instance [DIMACS 93]. The resulting inconsistent *KB* contains 15929 clauses that are not necessarily Horn, using 7041 variables and is intractable for standard improved DP techniques [Davis *et al.* 62]. We restored the consistency of the *KB* by introducing 1593 different Ab_i propositions in a random way inside the *KB* (however, making sure that *KB* becomes consistent and using at most one Abnormality proposition by clause). Each Ab_i was given a different stratum, making the problem computationally harder to solve (in the worst case). Moreover, we do not use previously found models of *KB* as initial interpretations of calls to TSAT. This *KB* remains intractable using standard DP procedures; using TSAT we show within 4.6 seconds CPU time that it is consistent.

We selected *f* in a random way, from the literals occurring in *KB* and used our algorithm in order to show whether $KB \mid\!\sim f$ or not.

The first step consists in checking whether there exists a normal circumstance model for *KB* that satisfies *f* or falsifies it. TSAT (with its weight option, a tabu length set to 10, 10 tries and 10000 flips) is run on the *KB*, with all Ab_i set to *false* and *f* set to *true*, during 1min01s30. It fails to deliver a model (indeed, no such model does exist). The trace obtained by this step is given in Figure 1. Running a DP optimized procedure on this trace allows us to prove that no such normal circumstances model does exist for *f*, using 0s94 CPU time. A similar work is performed with *f* set to *false*, spending 59s26 for TSAT and 0s93 on its trace by DP to show the absence of normal circumstances model for $\neg f$.

The trace of Figure 1 allows us to state that, *most probably*, Ab_4 is to be *true* in order for the *KB* to become consistent. Indeed, Ab_4 appears in the most often falsified clauses (less us stress that the fact that this is a very low Ab_i is purely accidental, and will not influence the efficiency results, most often). Within 39s35, TSAT allows us to find a model of $KB \wedge f \wedge Ab_4$. Then, we have to check that no model does exist for $KB \wedge \neg f \wedge (Ab_1 \vee Ab_2 \vee Ab_3 \vee Ab_4)$. This is done in one step, introducing the additional clause $Ab_1 \vee Ab_2 \vee Ab_3 \vee Ab_4$, using first TSAT and then DP to prove that no such model does exist. TSAT was run during 48s and the unique call to DP on the delivered trace was performed in less than 1.8 seconds.

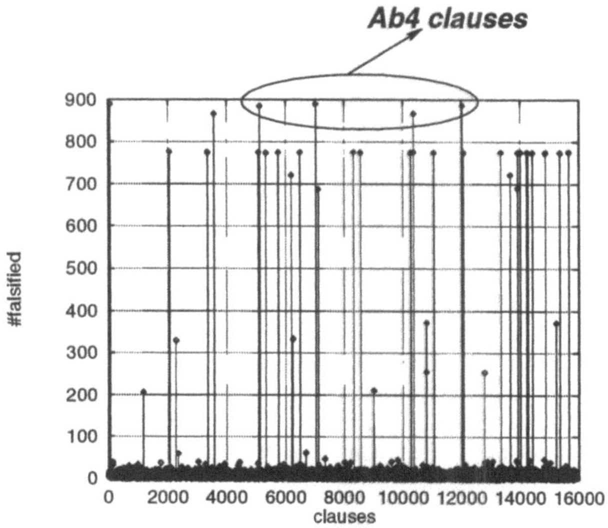

Figure 1. Experimental Illustration

7 Going Further

The goal of the proposed system was to reason notwithstanding possible device failures, obeying a preference for some explanations of these failures. In the introduction, we motivate when and why to focus on the first-failure and provide the user with *one* way to correct all lower level faults, at the risk of missing some alternative secondary faults with respect to the encoded knowledge. The reason were mainly the incompleteness of this knowledge that can anyway miss some causal links and an imperious goal of computational efficiency, due a need of quick answer about the situation of the modeled critical systems. In this respect, it should be noted once again that iterating this process after repairing (or simulating the reparation of) the detected faults does give a way to detect all actual device failures with respect to the encoded information. In this respect, we hope that the heuristics and approach proposed in this paper could offer a promising path for efficient alternatives to more standard computational techniques for consistency-based diagnosis, like the approaches based on ATMS [Hamscher *et al.* 92, Console *et al.* 94]. Accordingly, we agree with [Selman *et al.* 97] that an important challenge consists in finding adequate ways to encode first-order specialized knowledge into propositional form. Some promising but modest work in that respect is described in [Brisoux *et al.* 98].

The reader might have noticed that the use of a preorder between Abnormality propositions introduces additional considerations to be taken into account. Clearly, our system might allow one single Abnormality proposition to be featured out, while its stratum could contain several propositions explaining faults in different ways. However, exhibiting all Abnormal propositions inside this stratum that can explain the faulty situation can be achieved easily. Indeed, once we know one first failure, satisfiability checks can be run with all other elements of the stratum set to *true*, successively, while all elements of lower strata are set to *false*. The key point is that we already know the stratum of the candidates that can play the role of the first failures. When real-time deadlines are too strong to allow more than finding *one* first-failure, we are however able to provide the user with the list of all elements of the concerned stratum as candidates for alternative first failures. This information can be a really valuable one, especially when the strata are fine-grained or include various explanations of the same failure (which should however be avoided as much as possible at the time of knowledge encoding).

Acknowledgements

This work was supported in part by a "Contrat d'objectifs" and the Ganymède II project of the "Région Nord/Pas-de-Calais". It is a revised version of a preliminary work presented at the 3rd Int. FLINS Workshop on Fuzzy Logic and Intelligent Technologies for Nuclear Science and Industry, held in Antwerpen in 1998.

References

Brisoux, L., Saïs, L. and Grégoire, É. (1998), "Validation of knowledge-based systems by means of stochastic search," In: *Proc. DEXA workshop on verification, validation and integrity issues in expert and databases systems*, R.R. Wagner (ed.), Vienna, IEEE Computer Press, pp. 41-46.

Console, L., Friedrich, G. and Golumbic, M.C. (1984), Special Issue on Model-Based Diagnosis, *Annals of Mathematics and Artificial Intelligence*, **11**.

Cordier, M.-O. (ed.) (1997), *Proc. of the Eight Int. Worskhop on Principles of Diagnosis*, Mont-Saint-Michel, France.

Crawford, J.M. (1993), "Solving satisfiability problems using a combination of systematic and local search," In: Working Notes of the DIMACS Workshop on Maximum Clique, Graph Coloring and Satisfiability, 1993.

Davis, M., Logemann, G. and Loveland D. (1962), "A machine program for theorem proving," *CACM*, 7, pp.394-397.

DIMACS (1993), Second Challenge on Satisfiability Testing organized by the Center for Discrete Matt. and Computer Science of Rutgers University.

Grégoire, É. and Saïs, L. (1998), "Practical inconsistency management for critical-tasks decision support systems," in: D. Ruan *et al.* (eds.), *Fuzzy Logic and Intelligent Technologies for Nuclear Science and Industry*, World Scientific, Singapor, pp. 384-391,1998.

Grégoire, É., Mazure, B. and Saïs, L. (1998), "Logically-complete local search for propositional nonmonotonic knowledge bases," In: *Proc. 7th Int. Workshop on Nonmonotonic Reasoning*, Niemelä I. and Schaub T. (eds.), Trento, pp. 37-45.

Hamscher, W., Console, L. and De Kleer, J. (1992), *Readings in Model-Based Diagnosis*, Morgan Kaufmann.

Mazure, B., Saïs, L., Grégoire, É. (1997a), "Tabu search for SAT," In: *Proc. of the 14th Nat. Conf. on Artificial Intelligence (AAAI-97)*, Rhode Island, pp. 281-285.

Mazure, B., Saïs, L. and Grégoire, É. (1997b), "Checking several forms of consistency in non-monotonic knowledge-bases," In: *Proc. ECSQARU-FAPR97*, Lecture Notes in Computer Science 1244, Springer, pp. 122-130.

Mazure, B., Saïs, L. and Grégoire, É. (1997c), "An efficient technique to ensure the logical consistency of cooperative agents," *Int. Journ. of Cooperative Information Systems*, World Scientific, **6**(1), pp. 27-36.

Mazure, B., Saïs, L. and Grégoire, É. (1998a), "Boosting Complete Technique Thanks to Local Search Methods," *Ann. of Math. and Art. Int.*, **22**, pp. 319-322.

Mazure, B., Saïs, L. and Grégoire, É. (1998b), "System Description: CRIL Platform for SAT," In: *Proc. of the 15th Int. Conf. on Automated Deduction (CADE'15)*, Lecture Notes in Computer Science 1421, Kirchner C. and Kirchner H. (eds.), Springer, pp. 124-128.

McAllester, D., Selman, B. and Kautz, H. (1997), "Evidence for invariants in local search," In: *Proc. of the 14th Nat. Conf. on Artificial Intelligence (AAAI-97)*, pp. 321-326.

McCarthy, J. (1986), "Applications of circumscription to formalizing common-sense knowledge," *Artificial Intelligence*, **28**, pp. 89-116.

Reiter, R. (1987), "A theory of diagnosis from first principles," *Artificial Intelligence*, **32**, pp. 57-95.

Selman, B., Levesque, H. and Mitchell. D. (1992), "A new method for solving hard satisfiability problems," In: *Proc. of the 14th Nat. Conf. on Artificial Intelligence (AAAI-97)*, pp. 440-446.

Selman, B., Kautz, H. A. and Cohen, B. (1993), "Local search strategies for satisfiability testing," In: *Proc. 1993 DIMACS Workshop on Maximum Clique, Graph Coloring, and Satisfiability*.

Selman, B., Kautz, H. and McAllester, D. (1997), "Computational challenges in propositional reasoning and search," In: *Proc. of the 15th Int. Joint Conf. on Art. Intelligence (IJCAI-97)*, Nagoya, Japan, pp. 50-54.

Annex 1

The TSAT procedure is not described here (detailed descriptions can be found elsewhere [Mazure *et al.* 97a, 98b]). The following procedures are described using an imperative style and a Pascal-like pseudo-code. Obviously, the actual implementation avoids the many redundant computations that appear in this description.

```
Procedure Unit_Propagation ;      {used inside DP}
Input :  A KB in CNF
Output: A KB without unit clause or with an empty clause.
Begin
  while ((∃ unit clause ∈ KB) and ( not (∃ empty clause ∈ KB)))
  do
     l := a literal s.t. ( l ∈ c and c is a unit clause);
     KB := KB simplified by l ;
  enddo ;
  return KB ;
End ;
```

```
Procedure DP_focused_on_trace ;
Input : A KB in CNF, a list T of pairs <literal,score> built from the trace of TSAT
          computation ;
Output : true if KB is consistent,
         false otherwise ;
Begin
  KB := Unit_Propagation(KB) ;
  if (∃ empty clause ∈ KB) return false ;
  else if (KB = ∅) return true;
   else
    l := the literal with the greatest score in T and that belongs to KB ;
   return (DP_focused_on_trace(KB ∧ l, T) ∨ DP_focused_on_trace(KB ∧¬l, T));
   endif
End ; {Actually, this procedure is implemented in such a way that it considers first
the clauses that were most often falsified during TSAT search}

Procedure  Satisfiability_Test ;
Input :   A KB in CNF ;
Output : true if KB is consistent,  false otherwise ;
Begin
  if (TSAT(KB)) return true ;
  else
    trace := the trace delivered by TSAT ;
    return (DP_focused_on_the_trace(KB, trace)) ;
  endif ;
End ;

Procedure NonMonotonic_Inference ;
Input : A KB (assumed consistent) and f a literal ;
Output : true if KB |~ f,  false otherwise ;
Begin

{Step 1 : search for normal-circumstances models }
KB' := KB ∧ f ∧ { ∧¬Ab_i ( ∀ Ab_i ∈ AB) };
if (Satisfiability_Test(KB'))
  {A normal circumstances model exists, this model is necessary a preferred
   model}
  KB' := KB ∧ ¬f ∧ { ∧¬Ab_i ( ∀ Ab_i ∈ AB) };
  if (Satisfiability_Test(KB')) return false ;
  {Normal circumstances models exist for KB ∧ f and for KB ∧ ¬f,
   thus we cannot have that KB |~ f}
  else return true            {KB |~ f}
  endif
else
 KB' := KB ∧ ¬f ∧ { ∧¬Ab_i ( ∀ Ab_i ∈ AB) };
 if (Satisfiability_Test(KB')) return false ;
 {Normal circumstances models exist for KB ∧¬ f but not for KB ∧ f, thus we
  cannot have that KB |~ f}
```

```
else

{Step 2 : Search for preferred models that are not normal
          circumstances ones }

L := { Ab_s,...,Ab_l} ;
{L is the set of all Abnormality propositions that is sorted in decreasing order
 w.r.t. their scores in the trace of the various calls to TSAT during Step 1 }
 L := local-treatment(L);
 {the local treatment procedure sorts the first elements of L, i.e. the Ab_i with the
 highest scores, so that the Ab_i with the lowest i in this sub-list appears first in L}
proof := false ;
        {will be assigned true if its shown that KB |~ f}
while (not proof) and ( L ≠ ∅)
do
 lowest_Ab_j := the first element of L ;
 L := L \ {lowest_Ab_j } ;
 KB' := KB ∧ {∧¬ Ab_i ( ∀ Ab_i < lowest_Ab_j ) } ;
 KB' := KB' ∧ f ∧ lowest_Ab_j ;
 if (Satisfiability_Test(KB'))
    c := Ab_1 ∨Ab_2∨... ∨ lowest_Ab_j ;
    KB' := KB ∧ c ∧ ¬f ;
    if (Satisfiability_Test(KB'))
    then {There are preferred models for f and for ¬f, including lowest_Ab_j }
          Ab_r := the lowest Ab_i in the discovered model of ¬f ;
          L := { Ab_j s.t. Ab_j < Ab_r} ;
    else
        {there exists a model of KB f that has no preferable model that satisfies f,
         thus KB |~ f }
        proof := true ;
     endif
   endif
 enddo
return proof ;
endif
endif
End ;
```

Annex 2

Sketch of the proof of Theorem 2.2.

First, let us recall that **AB** = $\{Ab_1,\ldots,Ab_n\}$ is partitioned into m strata $\mathbf{AB}_1,\ldots,\mathbf{AB}_m$ such that when $Ab_i \in \mathbf{AB}_r$, $Ab_j \in \mathbf{AB}_s$ and $r < s$ then $i < j$.
Let **AB'** $\subset$ **AB**. The clause formed with all elements of **AB'** is noted $\vee$**AB'**. Let

KB be a consistent set of clauses and *f* be a literal. Clearly, checking whether *KB* $\cup$ {*f*} exhibits a model containing at least one member of **AB'** can be achieved thanks to the satisfiability check of *KB* $\cup$ {*f*} $\cup$ {$\vee$**AB'**}.

When *KB* $\cup$ {*f*} (or *KB* $\cup$ {$\neg f$}) contains a model without any Abnormality proposition, checking *KB* |~ *f* requires two consistency checks (where all occurrences of all Abnormality propositions have their value set to *false*). When no such model exists, we know that any model of *KB* $\cup$ {*f*} or *KB* $\cup$ {$\neg f$} contains at least one Abnormality proposition since *KB* is consistent and *f* is a literal. In this second case, we consider successive subintervals of $[\mathbf{AB}_1, \ldots, \mathbf{AB}_m]$ in a dichotomic way, modulo the notion of stratum. Let **AB'** = $[\mathbf{AB}_{inf} \ldots \mathbf{AB}_{sup}]$ (with *inf* = 1 and *sup* = *n*, initially). At each major dichotomic dividing step, **AB'** is split into two subintervals in the following way. Let *i* be the number of strata in **AB'**. We put the $\lfloor i/2 \rfloor$ first lower strata in $\mathbf{AB'}_{Lower}$ and the other remaining strata of **AB'** in $\mathbf{AB'}_{Upper}$ so that $\mathbf{AB'} = \mathbf{AB'}_{Lower} \cup \mathbf{AB'}_{Upper}$.
Let $\mathbf{AB'}_{Selected} = \mathbf{AB'}_{Lower}$.
Let A = *KB* $\cup$ {*f*} $\cup$ {$\vee\, \mathbf{AB'}_{Selected}$} and B = *KB* $\cup$ {$\neg f$} $\cup$ {$\vee\, \mathbf{AB'}_{Selected}$}.
The satisfiability of A and B are checked independently. Three cases can be distinguished.

1. When A (resp. B) is satisfiable while B (resp. A) is not satisfiable, then the procedure stops.
2. When both A and B are not satisfiable, the same consistency checks and analysis are undertaken with $\mathbf{AB'}_{Selected} = \mathbf{AB'}_{Upper}$.
3. Else, we proceed to the next major dichotomic dividing step, with $\mathbf{AB'} = \mathbf{AB'}_{Selected}$.

In the worst case, this procedure has to end when $\mathbf{AB'}_{Selected}$ contains the Abnormality propositions of one stratum; $\log_2 m$ major dividing steps are thus required, each of them making use of a constant (at worst 4) calls to a consistency check.

Artificial Intelligence Classifiers and Their Social Impact

J. Carlos Aguado Chao

Automatic Control Department (ESAII)
Polytechnical University of Catalonia (UPC)
C/ Pau Gargallo 5, 08028 Barcelona, Spain
jaguado@esaii.upc.es

This chapter is clearly divided into two parts. In the first one, the undeniable importance of classification in most of the goal-directed systems (we could spare the word intelligent only for human actions) will be briefly justified, and the main so-called Artificial Intelligence classification algorithms will be revisited. The possibilistic classifiers based on hybrid connectives will be more deeply presented, because of their novelty and power. After that, we will step out into unknown land and try to foresee by analogy some of the social changes that will be and are already being provoked by the new paradigms in information processing. Not surprisingly, there are no conclusions, because we are not concluding a process, but initiating it.

1 The Importance of Classification

The classification processes play a basic role in any intelligent behavior, and we can find main reasons for it. The first one is that before any decision can be made about any future action, it is crucial to determine what is the current situation. Only in a purely numerical system, which of course we humans are not, a decision can be taken straightforward, without any analogy to already known patterns. Thus in an artificial intelligence tool -a software program that tries to mimic human or natural goal oriented behavior- nearly always a classification precedes any practical action.

The second reason is related to our present social environment, and shows that long before any action is scheduled, classification can reveal itself crucial. The world all around us can be truly called an information society, and it makes plainly impossible to keep up with the new information continuously provided in any field of the human activity, specially within science and technology.

Centuries ago, it made sense to try to form an exhaustive compilation of books, like the great library of Alexandria. Those books were handwritten, and the procedure to collect them was as simple as taking possession of any of them noticed to be in the city, and give back a copy to the owner. The contents of these papers were mere transcriptions of spoken speeches, intended to be read aloud, and the society could be considered mainly oral. This feature implied that human interaction, and thus human reasoning was mostly directed to move and convince an audience. Ideas should be brightly and beautifully presented, and a dramatic lie was always preferred to a boring truth.

When books could finally be printed in unlimited number with a minimum effort, the social change was drastic. Printed words could get to more people, and they were no longer considered as mere substitutes of the oral speech. Any educated person could have its own copy of a book, read it not necessarily aloud and take his/her own conclusions. The sentences were now intended to be clear and convincing, not only by their beauty but by their logical argumentation. Even the everyday reasoning of educated persons, whose number could also increase faster, grew more ordered and logical.

Finally, nowadays books can be produced faster and not only printed ones. Electronically written words can reach the whole world as fast as they are created, and the economic cost is minimum, as long as the reader has access to a computer and a phone. But even paper printed information has overwhelmingly multiplied its volume. Curiously enough, in our current society, the superabundance of information has provoked a culture where image is more important than ever. So, in some sense, we are back in a world where the main goal in human communication is to impress and move with brightly presented sensations, just as it was it the earlier years of human history.

Anyway, images or written information, persons can no longer even read every word or look at every document ever comes their hands casually. We have to perform a first data selection into topics that we need or want to be informed about, and topics we have to ignore. This classification is not always guided by personal likes and dislikes, because any human organization can be considered as a member of an information ecosystem. Each of such entities has to recognize the useful information, its sources and which other entities it has to compete or collaborate with. That way, a second basic function of an Artificial Intelligence tool is to recognize by any given criteria the considered useful information. The next question is what type of algorithms can automatically learn to classify information.

2 Artificial Intelligence Classifiers

By classifying algorithm it will be understood, from now on, any technique is able to learn to separate observations in a number of classes. Obviously, this is a basic capability of intelligent behavior, and that is the reason why automatic classifiers are studied within the field of Machine Learning (ML), which is a branch of the Artificial Intelligence (AI) theory.

Now a brief exposition will be provided on the state of the art in Machine Learning, and then a new paradigm will be introduced, the logical connectives based classifying algorithms. A first partition in ML methods can separate them into symbolical and numerical ones. The former can be divide into classical inductive, classical deductive techniques and analogy based methods --case based reasoning--, and the latter into connectionist --artificial neural networks-- and evolutionary algorithms. We will look at their advantages and weak points and justify how our new type of algorithm can combine some of their best features.

2.1 Symbolical Classification Algorithms

This kind of techniques tries to model the process of human reasoning as it appears to ourselves, specially the construction of abstract concepts from concrete individuals and the later pattern recognition process, which applies these concepts to new individuals.

2.1.1 Classical Inductive Methods (Supervised and Unsupervised)

Inductive learning consists in the process of extracting knowledge from a set of observations. That process of going from the concrete to the abstract, from the particular to the general, presents a main disadvantage, which is the preservation of the falsehood but not the truth. In other words, if concrete observations do not fulfill a property, general classes will behave the same way, but properties satisfied by single individuals may not be fulfilled by generated classes. Inductive reasoning, nevertheless, lays on the very foundations of our culture and civilization, because it supports the Science: a scientific theory can perfectly reveal itself to be false, but we can never demonstrate its correctness, only admit and use it until a contra example is provided.

Classical inductive methods perform a search into a set of data by means of two operators, specialization and generalization. This learning process can be either supervised or not. In the former case, concepts are described by means of positive, and sometimes negative, examples of individuals which belong to them. This paradigm was implemented by Winston and a series of authors [Winston 70], [Hayes-Roth 77], [Vere 75], [Dietterich 81], [Mitchell 83]. The latter approach --clustering-- does not require examples, but single observations that are clustered according to certain similarities discovered in them; again Michalski [Michalski 80] introduced this point of view, and later [Fisher 87] presented COWEB, a hierarchical and incremental classification system able to perform bi-directional searches in the data. In order to see how this process can be carried out, let us take a look at ID3, the Inductive Dichotomizer Tree first sketched by Hunt and enhanced by Quinlan [Hunt 66], [Quinlan 86].

The basic idea in this classifier is the construction of a decision tree each of whose leaves represents a unique class meanwhile every inner node stands for a decision point which partitions the data according to the value they present for a given binary descriptor. This particular algorithm employs as a splitting criterion a measure from the Information Theory, the information content of a set of n objects C that belong either to the class P with a probability p_i, or to the class Q with a probability q_i:

$$I(C) = \sum_{i=1}^{n} (-p_i \log_2 p_i - q_i \log_2 q_i) \qquad (1)$$

This measure is calculated after and before partitioning the data on the values of a given descriptor, and at each step the algorithm chooses the descriptor producing a larger information gain, that is, the most relevant one. Main advantages of this method are the linear growth of its complexity with the number of descriptors, its ability to ignore irrelevant ones and the incremental nature of the computations that allows to leave out an add on later a number of descriptors.

So as to remark the common features to both supervised and unsupervised inductive classification techniques, let us recall that they all preserve falsehood but not truth, and many of them use criteria borrowed from Statistics, Information Theory or Mathematics (for instance, many of them employ some kind of distance from an individual to a given class). Most of them work with a hierarchical classification based on bivalued descriptors, and usually there are a number of parameters which need to be tuned out by the user -for instance, the threshold to be trespassed in order to consider an individual close enough to a given class in a clustering algorithm.

2.1.2 Classical Deductive Algorithms

Deductive techniques appeared in the eighties with the obvious goal to be truth preserving. Of course there is an implicit trap in this possibility, because the only case where a deduction is demonstrably correct is when it does not add any knowledge to our data base, but only expresses existing information in a different form. Is this real learning?

Anyway, this process can also be necessary in a given domain, and the so-called Explanation Based Learning [Mitchell 83] is a very active field in the Artificial Intelligence. These methods start typically from a domain theory, a base of rules and facts, and they enhance it by means of provided examples of a goal concept. STRIPS [Fikes 72] is an early example of these techniques, and Mitchell [Mitchell 86] introduced later a relatively domain-independent formalism.

All of these methods preserve truth, but they need a vast amount of domain knowledge, a complete and sound theory in order to start working.

2.1.3 Analogy Based Methods (Case Based Reasoning)

Analogic reasoning is based on the assumption that two objects that agree in a relevant aspect are likely to agree in others. It requires a store of analogous observations --sources-- in order to try to classify a new individual --target. Many times this inference is applied to problem solving, and this means that the proposed solution must be modified according to the new situation.

The goal of this case based reasoning can be to obtain a representation of the target including inferences carried out from different sources. This process can be organized [Kedar-Cabeli 88] in five stages:

- Access: finding one or more good sources close to the target, usually by means of similarity measures.
- Elaboration: available information from the source is modeled by means of derivation of attributes, relations or other chains of causes.
- Mapping: it is very unlikely that two individuals match exactly, so only selected inferred attributes are transferred from the source to the target.
- Evaluation: this is the stage where inferences are assessed, frequently resulting in repairing or refining the analogy.
- Learning: finally the acquired knowledge is stored in its final form.

2.2 Numerical Classification Algorithms

Up to this point, all the considered classifying techniques were based on symbolic reasoning, that is, the way human reasoning appears to us. The other great group of classifiers collects those which perform a numerical minimization, searching for the point where a mathematical function takes its minimum value. Nevertheless, the so far most popular member of this group of algorithms, artificial neural networks, was designed to model the deep behavior of the human brain.

One main difference worth noticing in this group of techniques is their capability of working together. For instance, a genetic algorithm can be employed to find an efficient architecture for a neural network in order to solve a determinate problem, or a neural network can be used, as we will see, so as to implement the parallel simulated annealing paradigm.

2.2.1 Connectionist Techniques (Artificial Neural Networks)

Our brain is a vast and very complex structure where multiply networked cells, called neurons, exchange electric signals. As early as the mid fifties, very simple models of human neurons were applied to perform some computations. But in the late sixties the book "Perceptrons" [Minski 69] by Minski and Papert showed that those structures were not enough even to learn the computation of the exclusive-or logical function. Only in the mid eighties neural networks regained popularity thanks to Hopfield, Rumelhart and McClelland [Hopfield 82], [Rumelhart 86], who proposed and compiled many improvements derived from the point of view of graph theory and distributed and parallel computation.

In a general form, a neuron is a simple processing element with many input connections and only one output connection. As soon as the input data is ready, which implies a synchronous operation, every term is multiplied by a corresponding weighting factor, a bias value is added and after a global summation a so-called activation function either linear --habitually-- or not is applied so as to yield the output.

The operation of a neural network, or forgetting its limited resemblance to our brain, the operation of a connectionist algorithm, presents two different stages: the training, when the weights are determined for each node --neuron--, and the pattern recognition, when a trained network is able to approximate any function. Again, back in the training stage, the learning can be either supervised or not, but a different network architecture needs to be employed in one or the other case.

Such algorithms have obtained and still are producing spectacular results, but also show important disadvantages: given a certain problem, there is no method to determine the architecture of the network that solves it, even though it has been proved that it has to exist; given a problem and a certain architecture, there is no theory to estimate the number of iterations needed to learn to approximate the goal function; given a problem and a network that has been successfully trained, there is no general way to explain *what* the networked has learnt, how it is able to approximate a function from the examples, or if its performance will also be satisfactory for new individuals.

2.2.2 Evolutionary Methods

Also in the line of numerical classifiers, we found a group of algorithms which are closely related to artificial neural networks but show more clearly their parallel and evolutionary nature. Of course, they were proposed to somehow overcome the unpredictability of neural networks, but so far they have not reached the popularity of the former.

Genetic Algorithms

Holland [Holland 92] proposed to apply evolution as it appears in natural populations to computation. His proposal was, in order to solve a given problem, to generate a population of possible solutions, the fitness of each one could be assessed; new individuals would be obtained by combination --crossover-- of the existing ones, and some of the worst ones --lower fitness-- would disappear; finally, from time to time, a mutation would assure the richness of the population in order to escape from local maximums of the fitness function and try to follow the global ones, balancing somehow the need to concentrate on best individuals and the need to explore new ways. The usual implementation of his ideas is by coding each solution as a fixed length binary chain.

Genetic computation is a very general technique and has been applied successfully to many optimization problems. Its main advantages are the parallel operation and the production of sub optimal solutions at any step of its operation, but only in very restricted conditions we can assure the ability to produce the optimal one. On the other hand, the necessity of a fitness function implies that this algorithm can only be employed for supervised learning.

Simulated Annealing

This final technique is very close to genetic algorithms, and even more predictable than them, but it has never been as well known. Obviously, this method is inspired by the annealing process which leads some physical materials to a new and purer crystallization. This is accomplished by modeling the effect of temperature and the whole process can be studied by the chains of Markov statistical theory, which can assure its convergence and, given some conditions, bind it within determined limits. Again the result is a method able to quickly produce approximate solutions to complex optimization problems, and again the main difficulty is to escape from local minima.

Every natural system tends to minimize its energy so as to reach a thermodynamic balance for every given temperature. It has been found out that the probability of every configuration --distribution of the elements in a system-- is:

$$P_T(c) = \frac{1}{z_T} e^{\frac{-E(c)}{kT}} \tag{2}$$

where c is the configuration, $E(c)$ its energy, T the temperature and the rest are constant terms. So, the compared probability for two given configurations is:

$$\frac{P_T(c_1)}{P_T(c_2)} = e^{\frac{-\Delta E}{kT}} \tag{3}$$

and that implies that when $\Delta E \geq 0$ then $P_t(c_2) \geq P_t(c_1)$.

On the other hand, traditionally it has been known a good procedure in order to avoid inner defects of a given material, for instance a metal, which consists in heating and then cooling it, but in a progressively slower way, so as to allow the system to reach its thermodynamic balance at each temperature. Obviously, avoiding inner defects means minimizing the energy of the system.

The computer implementation of this process presents two separate actions, a tempering process and a relaxation. The former is just a sequence of constant temperature cycles, with decrements progressively more separated in time. The relaxation algorithm, executed at each temperature, generates random elementary changes; the probability of accepting any of them is

$$p = \min(1, e^{\frac{-\Delta c}{T}}), \tag{4}$$

and we can observe that

$$T \to 0 \Rightarrow \begin{cases} p(\Delta c > 0) \to 0 \\ p(\Delta c < 0) \to 1 \end{cases}, \tag{5}$$

which implies that at very low temperatures only changes that decrement the system energy will be accepted, but at high temperatures changes which increment the energy can be permitted. That is the key to get out from local minima.

The theoretical foundation of this method relies on the statistical theory. It has been demonstrated that if 1) every state can be reached (ergodicity) and 2) the chains of Markov can be infinitely long, that is, we do not limit the number of state changes, then the algorithm is to converge towards the set of optimal states. Furthermore, other theoretical results help to choose the initial and final temperature, the decrement size and number of allowed transitions in order to get a desired sub optimal solution.

The main disadvantage of this kind of techniques is their slow functioning, but artificial neural networks can help to reduce this inconvenience. A Machine of Boltzmann is a particular Hopfield network whose neurons present binary outputs (0 or 1) and bi-directional, never reflexive, connections either positively or negatively weighted. Its configuration at any moment allows to compute the defined energy as

$$E = \sum_{I<J} w_{ij} S_i S_J , \tag{6}$$

which every neuron tends to minimize by computing in parallel

$$\Delta E_k = \sum_J w_{kj} S_J \tag{7}$$

and thus determining its output:

$$\begin{cases} \Delta E_k \geq 0 \Rightarrow S_k := 1 \\ \Delta E_k < 0 \Rightarrow S_k := 0 \end{cases}. \tag{8}$$

The ability of these networks to perform parallel computations allows us to carry out simulated annealing the way it happens in Nature, leading to solutions closer to the optimal ones in a reasonable time. At the same time, the annealing technique allows to escape from local minima. The concrete algorithm to solve an optimization problem would be: a) constructing a Machine of Boltzmann with as many binary output neurons as the necessary bits to encode every possible state; b) choosing initial weights and connections such that their cost function depends directly of the defined energy, and every of its minima is a possible state; and c) replace the standard Boltzmann's determination of the output of each neuron by:

$$p(S_k = 1) := \frac{1}{1 + e^{\frac{-\Delta E_k}{T}}} \tag{9}$$

2.3 Logical Connectives Based Algorithms

The last explored proposal to Machine Learning is a kind of inductive techniques that combines some of the best characteristics of both symbolic and numerical algorithms. In order to do so, they rely on the generalizing power of Fuzzy Logic [Zadeh 65], and the interpolation capability of logical hybrid connectives [Piera 87].

2.3.1 Hybrid Connectives

By linearly compensated hybrid connective we will mean the interpolation between a t-norm and a t-conorm by means of the β parameter such that $\beta = 0$ represents the intersection and $\beta = 1$ means the union. This parameter will -inversely- determine the exigency level of the classification, so it can be called *tolerance*.

2.3.2 An Implementation: *LAMDA*

The implementation of these possibilities assumed the form of *LAMDA* algorithm, which was developed by Josep Aguilar in collaboration with a series of authors (Ramon López de Mántaras, Núria Piera, Julio Waissman, ...) [Aguilar 81, 82, 96], [Desroches 87], [Piera 89], [Aguado 98 99] who from the eighties have been enhancing this original self learning classifying technique. The input data to this algorithm are in the form of a set of observations or individuals recorded as rows of a text file, each of them divided in different values, one for every descriptor or column. Every descriptor can be either quantitative or qualitative and, in this last case, each possible value is called attribute or modality of the descriptor.

The information treatment is very simple: in a first step every modality is acquired for the qualitative descriptors, and the range of values exhibited in the quantitative ones is determined for normalization purposes. A second stage, as soon as the universe of the data is known, consists in either loading the whole population to be treated or reading and treating the individuals one by one in an indefinite series.

From this moment on, a triple alternative is offered: we can classify the individuals according to a known and fixed set of classes, or we can start from a given set of classes which can be modified, or even we can start from zero and construct our own partition. Of course the last possibility is by far the most interesting one, as it comprehends the specific features of the other two and still adds up some other new characteristics.

Anyway, the basic assignment of an individual to a class follows the same process in spite of being or not the prelude to a possible modification or creation of a class. The schema is shown in Figure 1:

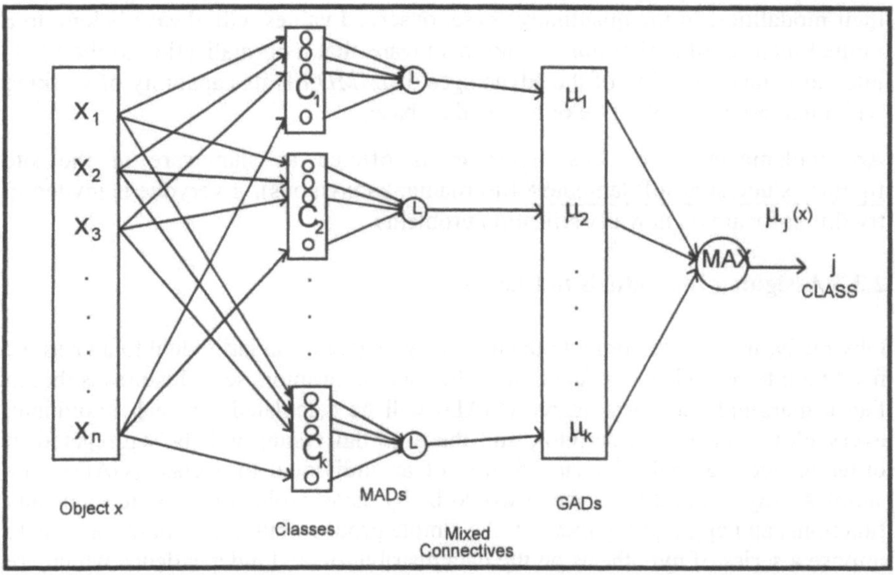

Fig. 1. Structure of classifiers based on hybrid connectives

where MAD and GAD stand for Marginal and Global Adequacy Degree, respectively, of an individual to a given class. It should be noticed that classifying structure remarkably resembles that of a single neuron in an artificial neural network, as can be seen in Section 2.3.6.

Not amazingly there are, as in the neural operation, a training stage when classes can be modified and created, and a pure pattern recognition step when we just want to assign individuals to fixed classes. Nevertheless these stages can be compatible and learning could go on forever.

So as to formalize the description of *LAMDA*, we can define an individual as a series of descriptors values $d_1, ..., d_n$ such that each d_j takes values from the either finite or infinite set D_j. We will call universe or context to the Cartesian product $U = D_1 \times D_2 ... \times D_j$. Thus, any object or individual is represented as a vector $x = (x_1, ..., x_n)$ from U, such that each component x_j expresses the value for the descriptor d_j in the object x. The subset of U gathering all these vectors will be called data base or population.

The type of the descriptors can be either qualitative or quantitative. In the former case we can still distinguish between ordered (i.e., D_j = {small, medium, large}) or non-ordered finite sets (like D_j = {Mathematics, History, Geography, Physics}), but, anyway, the possible values for these descriptors will be called their modalities. In the quantitative case, observed values will always belong to a limited interval of real numbers, and this means that a normalization to the [0, 1] interval is possible. One of the advantages of *LAMDA* is its capability of working with qualitative, quantitative or hybrid data bases.

An implementation of this algorithm is offered as shareware in the site ftp.upc.es/upc/arca/soft/lambda98 (username: anonymous). Everyone is invited to try this software on new classification problems.

2.3.3 Assigning Individuals to Classes

Obviously, in order to compute the adequacy degree of an individual to a class we need both to be coded as a function of the same descriptors set. This means that in fact a marginal adequacy degree (MAD) will be calculated for each individual, every class and each descriptor, and these partial results will be aggregated in order to get a global adequacy degree of an individual to a class (GAD). The simplest way to build this system would be by using probability distributions and functions, and aggregating them by the simple product, but that would force us to impose a series of hypothesis on the data distribution and independence which are too arbitrary.

As a consequence, possibility functions (we could see them as fuzzified probability functions) are applied, and we estimate on either a parametric or a non-parametric basis, that is, with or without imposing an *a priori* distribution, their parameters, in order to describe the classes. This means that *LAMDA* is a statistical criteria based inductive technique, but generalizes this kind of algorithms by using fuzzified distribution functions and combining them linearly by means of hybrid connectives.

In the standard implementation of *LAMDA*, our possibility functions are derived from distribution functions, mainly by leaving out the normalization constant. In the case of qualitative descriptors our estimation is exactly the maximum likelihood one for a multinomial distribution, and we need to store in a class just the observed frequencies of every modality. Nevertheless, quantitative descriptors are completely different, because any real number within an interval is a possible measure, instead of only one label from a limited set.

The main way for introducing fuzzy quantities in a system is to depart from a situation with binary variables and generalize them by admitting every intermediate value. In *LAMDA* the first fuzzified function was the binomial distribution, considering experiments where every in-between result between "success" (1) and "fail" (0) is possible. In order to estimate these distributions it was only necessary to store as the kernel of a class the observed average value of quantitative descriptors.

From this kernel we obtain the marginal adequacy degree for a given individual, class and descriptor; if the descriptor is qualitative, this degree is just the frequency of the observed modality in the class. On the other hand, for a quantitative descriptor, j, described in the class C_k by the parameter ρ_{kj}, when the observed value is x_j we calculate:

$$MAD(x_j, C_k) = \rho_{kj}^{\;x_j}(1-\rho_{kj})^{1-x_j} \tag{10}$$

where this function represents a non-parametric --that is, without imposing a determinate probabilistic distribution-- Bayesian estimation. This is the originary possibility function employed in *LAMDA*, but the algorithm is general enough to employ any other one.

A second function was proposed by Julio Waissman. Since our original function represents a probability distribution law only lacking a normalization constant, it can be seen as a fuzzy generalization of a probability law. Let us recall the simple case of tossing a coin, an experiment with two possible results, either equiprobable or not; arbitrarily we call one of these "success" and the other one "fail". After replicating the experiment n times, we know that the probability of obtaining k "successes", being p the probability of obtaining one in a single toss, is:

$$\binom{k}{n} p^k (1-p)^{n-k} \tag{11}$$

where the combinatorial number only expresses the different possible ordered sequences of "successes" and "fails". This was just the classical binomial probability law in the case of repeated experiments.

But the main difference between fuzzy and probabilistic measures is that the former are not applied to repeated experiments with unknown and categorical results, but to a single experiment which result can continuously range between 0 and 1. Thus, considering a measure that occurs only once, n = 1, we obtain a degree of success x with the probability

$$\rho^x (1-\rho)^{1-x} \tag{12}$$

This is the direct extension of a classical probability law, but it still retains an excessive tendency to a bipolarization "success-fail", identified to 1 and 0, which implies unstability and indefinition when our observations come close to 0.5. Of course this case is not habitual and *LAMDA* can work perfectly well relying on this function.

Julio Waissman's proposal was born by observing data around 0.5 and questioning why "success" always had the meaning of "observations around 1" --by symmetry, observations around 0 were also correctly considered. *LAMDA* builds classes from observed data, and that implies that a true "success" is observing a new individual close enough to the average of the class. Let us assume that we want to classify points in an n-dimensional space from knowing their coordinates; then each class can store the coordinates of the *center* of the class, which are the average of the assigned points. Thus the degree of success can be represented by the inverse of the distance of a new point to this center:

$$d_{kj} = \left| x_j - c_{kj} \right| \tag{13}$$

$$MAD(x_j, C_k) = \rho_{kj}^{\;1-d_{kj}} (1-\rho_{kj})^{d_{kj}} \tag{14}$$

We can consider this new function as a fuzzyfication of the law that already generalized a probabilistic distribution, because a new parameter can now range from 0 to 1. Within the developed software, the user has the possibility of working with the classical function of *LAMDA* or --default option-- this new one, or else, why not, an assumed normal --Gaussian-- distribution:

$$MAD(x_j, C_{kj}) = \frac{1}{\sigma_{kj}} e^{-\frac{1}{2}\frac{(x_j-\mu_{kj})^2}{\sigma_{kj}^{\;2}}} \tag{15}$$

Now the hybrid connectives appear, not restricted to the product of independent possibilities, but including it [Zadeh 65, 71], [Schweizer 83]. This means that there exist a wide variety of families of fuzzy connectives from where we can choose, and also we are able to select their tolerance and combine according to it a pair of a t-norm (intersection) and a t-conorm (union) [Sklar 77], [Piera 87], [Piera 88].

By proceeding this way, a global adequacy degree is calculated, for an individual to every class. Of course this GAD will rule the choice of the class for the individual to be assigned to. At this point of the classification process, most standard algorithms employ a minimum threshold which the membership needs to trespass in order to provoke an effective assignment; usually the user is expected to provide this threshold, but *LAMDA* is different, for it holds a Non-Informative Class, NIC, automatically constructed for a given universe, so as to model the maximum entropy, which means returning the same GAD for any individual. Whenever the object that is to be classified does not possess a membership for another class greater than its membership to the NIC it can not be assigned, otherwise the object will be assigned to the class which exhibits the maximum GAD, and the first found one if two or more of them produce the same result. The NIC always remains empty.

2.3.4 Modifying Classes

Obviously the representation of a class is expected to vary after a new individual has been assigned to it. The part of the kernel which describes the qualitative descriptors parameters will change this way: the frequency of the observed modality, maximum likelihood estimation, is incremented:

$$frec' = frec + \frac{1 - frec}{N+1} = \frac{N \cdot frec + 1}{N+1} \tag{16}$$

where N stands for the number of elements previously assigned to the class. On the other hand, the rest of frequencies are diminished according to:

$$frec' = frec + \frac{0 - frec}{N+1} = frec - \frac{frec}{N+1} = \frac{N \cdot frec}{N+1} \tag{17}$$

As far as quantitative descriptors as concerned, there are, in the current *LAMDA* implementation, three different sets of parameters that could be computed, depending on the assumed distribution to estimate. The original estimation function has but one single parameter for each descriptor and class, which after the assignment of a given observation, x_j, will be modified:

$$\rho'_{kj} = \rho_{kj} + \frac{x_j - \rho_{kj}}{N+1} \tag{18}$$

and this means that this description of the average individual will not vary when it is already equal to the new-come observation.

If we try the Waissman's proposed function, now we have to recursively estimate two parameters, the new center and its probability, by means of a new distance:

$$c'_{kj} = c_{kj} + \frac{x_j - c_{kj}}{N+1} \tag{19}$$

$$d'_{kj} = \left| x_j - c'_{kj} \right| \tag{20}$$

$$\rho'_{kj} = \rho_{kj} + \frac{1 - d'_{kj} - \rho_{kj}}{N+1} \tag{21}$$

And, finally, if we assume that our data follow the Gaussian distribution --supposition that results will support or refute-- we have to compute a new mean and variance:

$$\mu'_{kj} = \mu_{kj} + \frac{x_j - \mu_{kj}}{N+1} \tag{22}$$

$$\sigma^{2\prime}_{kj} = \sigma^2_{kj} + \frac{(x_j - \mu'_{kj})^2 - \sigma^2_{kj}}{N+1} \tag{23}$$

2.3.5 Creating Classes

The next question is considering which ones could be the more appropriate classes in order to perform a particular classification, and whether *LAMDA* is able to automatically construct them just the same as the NIC is constructed. The answer is yes, our algorithm can either learn its own classes or accept the ones provided by the user. In fact the Non-Informative Class plays a major role in the creation of any other class.

As it has been said before, an individual is assigned to the class for which its Global Adequacy Degree results greater. In the special case when this class turns out to be the NIC, this means that the individual is not close enough to any of the existent classes. Thus, two actions are possible, if we are carrying out a pure pattern recognizing, this individual is to remain unclassified, but, on the other hand, if we are on a learning stage, this situation implies that we need a new class which is suitable for the individual. The creation proceeds by temporally assigning the individual to the NIC and modifying it so as to form a brand new class; this process, however, in order to construct a class which is between the NIC and the new individual, treats the NIC as it held one individual instead of none. Otherwise, the newly created class would be so fixed on its only individual that *probably* will not accept any other, and that is bad politics for a learning stage. The current implementation of *LAMDA* keeps always this virtual element --taken from the NIC-- additional to every class, so as not to arbitrarily reduce only the weight of the first arriving element.

2.3.6 Parallelisms and Differences with Respect to Other Classifying Algorithms

Let us recall some of the characteristics of the classifying methods we have reviewed. Symbolic classifiers usually work with qualitative bivalued (true-false) descriptors, need some parameters arbitrarily supplied by the user, rely on some heuristic, need supervised learning and are generally slow and always falsehood preserving.

Numerical methods produce results difficult to understand and work only with quantitative descriptors. On the other hand, they perform either supervised --much more commonly-- or unsupervised learning for a given structure, but not both of them, and are iterative and slow methods where there is no clear mark of 'learning finished'. Moreover, artificial neural networks, the most popular among them, are highly unpredictable tools.

LAMDA's parallelisms with these techniques are quite obvious. *LAMDA is* an inductive method, with classification criteria generalized from Statistics, and shows a remarkably similar behavior to connectionist techniques. In fact, its classifying structure --remember Figure 1--remarkably resembles that of a single neuron in an artificial neural network, as can be seen in Figure 2.

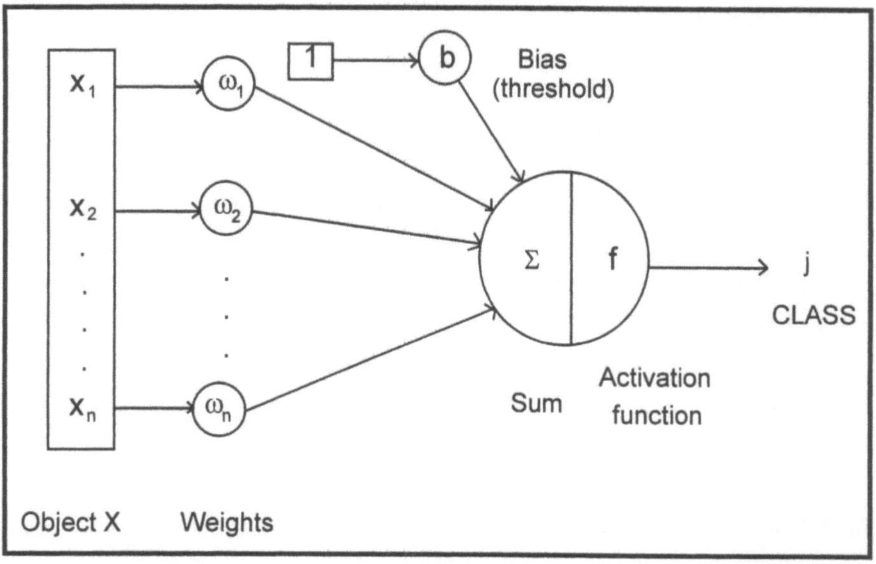

Fig. 2. Structure of a single perceptron

Here the class description (kernel) plays the role of the neuron weights and, consequently it is expected to vary in a learning stage. If we use neurons to classify individuals, we can be interested in finding the maximum output signal, that is computed as:

$$\max_{j}(f(b_j + \sum_{i} x_{ij} \cdot w_{ij})) \tag{24}$$

where b_j is the bias of every output neuron, f the activation function, x_{ij} the different inputs and w_{ij} the correspondent weights that are determined in the training stage. That can be perfectly equivalent to a particular case of *LAMDA*, which can assign an individual to the class that yields:

$$\max_{j}(\prod_{i} \rho_{ij}^{x_i}(1-\rho_{ij})^{(1-x_i)}) \equiv \max_{j}(\sum_{i}(1-x_i)\log(1-\rho_{ij}) + x_i \log \rho_{ij}) \tag{25}$$

where x_i are the different values for each descriptor and r_{ij} the parameters for every descriptor and class that are determined in the learning process.

Even more, if we try Radial Basis Function Neural Networks, the parallelism is even closer with *LAMDA* if the Gaussian distribution is employed, as shown in Figure 3.

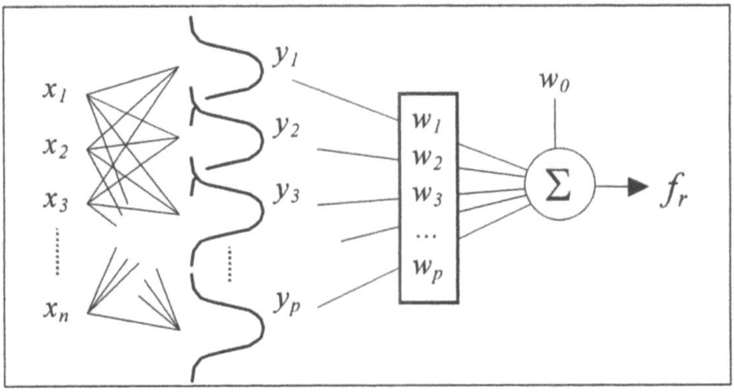

Fig. 3. RBF network architecture

Nevertheless, differences to these methods are usually advantages for *LAMDA*. The first one is generality. Standard classifiers working only on qualitative bivalued descriptors are generalized and overpowered by *LAMDA* and its capability to deal with both qualitative --bivalued or not-- and quantitative descriptors simultaneously.

On the other hand, our algorithm, not based on some heuristic but on generalized statistical functions, does not need a minimum user supplied minimum threshold, which is to be trespassed so as to assign an individual to class, as is the case in most of the statistical classifying algorithms. In *LAMDA* this minimum adequacy level is automatically determined as a function of the data base, and its value is that of the global adequacy for any individual to the maximum entropy class, the NIC.

As a third difference, the whole assignment process is not a classical statistical one, but fuzzy, thus more general. Additionally the overall exigency can be parametrized in order to produce every possible coherent classification for a given pair of logical connectives and a universe. Of course, when the user has enough information he/she can fix this exigency but *LAMDA* also can automatically explore the different values that yield every possible partition [Piera 91].

Comparing to neural networks, *LAMDA* does not search for minima in unknown functions, our learning process is not the result of "blind" weights combinations, but a coherent modification of the classes' description. This description is easily understandable because it is nothing more that the list of observed modalities and their appearance frequencies, for every qualitative descriptor, and the average value observed for every quantitative descriptor in the individuals belonging to this class.

Another major advantage of *LAMDA* with respect to usual classifying techniques is its sequential data treatment. We only need to store the present classes and deal with the individual that is to be classified, unlike the case based reasoning (CBR) algorithms which need to store a massive stock of information from the already processed individuals. This means that *LAMDA* is faster and demands but little memory; the counterpart is its dependence of the order in which the observations are provided. Nevertheless, we humans are not less order dependent in our learning processes.

And finally, the results. Among the qualitative advantages, these algorithms can generate a multiplicity of possible results, and include a post classification analysis of the obtained fuzzy partitions. Although the final result always is presented as a set of classical --crisp-- partitions *LAMDA* internally operates on fuzzy subjacent partitions that can provide some valuable information about the conceptual vicinity among either classes or individuals. [Aguilar 90]. From the quantitative point of view, results are similar to or slightly better than the ones provided by neural networks, as can be seen trying a public domain benchmark [Prechelt 94].

Table 1. RBF network results

Problem	Width	Centers	Training Data Mean	Training Data stddev	Validation Set mean	Validation Set stddev	Test Set mean	Test Set stddev	Test set Class. Error
Cancer1	5.126	40	0.024	0.081	0.017	0.063	0.026	0.070	1.72%
Diabetes1	0.597	29	0.129	0.175	0.133	0.186	0.173	0.226	25.00%
Heartc1	2.415	33	0.071	0.109	0.062	0.102	0.158	0.244	22.67%

Table 2. *LAMDA* results

Problem	Function	Connective	Tolerance	Validation Set Error	Min. Class. Error	Test set Class. Error
Cancer1	F_3	Frank	0	8.22%	0 %	0 %
	F_4	Probabilistic	0	8.22%	0.58%	0.58%
Diabetes1	F_2	Probabilistic	0.9434	23.00%	21.35%	25.00%
	F_4	Probabilistic	$1-\delta$	21.00%	21.35%	25.52%
	F_4	Frank	0.874	20.31%	21.35%	25.52%
Heartc1	F_4	Probabilistic	0	17.5%	13.33%	19.44%
	F_4	Frank	0.6898	17.5%	13.33%	19.44%

3 Social Impact of Information Technologies

The question is: where can these new information treatment tools lead us? When we look back in time it is relatively easy to perceive strong tendencies that provoked social changes. Nevertheless, if we try to look around us in our current environment, we can only appreciate a number of contender forces fighting each other to guide the social evolution. There are two main reasons for that contrast, the first one is the easiness to judge past actions: we know what happened first and we know what came later, so its is wise to assume that the latter was consequence of the former. The second reason is our tendency to simplify: if there were opposed forces and we know that one was the strongest, we can despise all the others.

Looking backwards, we can see that the real revolution leading us to the once mythic year 2000 has taken place within the information technologies. Nowadays, cars can not fly, the planets have not been colonized and most of our behavior is not very different (in the Western culture) from the way it was 25 years ago. Nevertheless, we can communicate "instantly" with someone in the other side of the world, and we can easily determine our physical location with only an error of few meters.

3.1 An Analogy, a Past Revolution

It is obvious that the information revolution is happening right now. But if we want to foresee its direction, one way could be the comparison with past technological, and consequently social, great changes. Let us go back to the first days of the Industrial Revolution, started by the use of the steam machine. If we admit the analogy between a motor and a computer, that revolution was caused by the flexibility in the generation of strength, meanwhile our revolution is being caused by the capability of managing information.

What was the evolution of the engines? Of course the early ones were big, inefficient and very expensive. Step by step they were progressively improved, and the combustion and the electric engines appeared. In the first industries, an enormous motor stood in the central point, and engineers tried to transmit its strength everywhere. When finally the engines were small and cheap enough, the distributed operation was born, and today we can find from 10 to 85 tiny electric engines in a single automobile.

From a social point of view, the first devices were constrained within the factory limits, and their impact on the people was merely indirect and economic. Of course, that does not imply a small effect, since an enormous exodus from the country to the cities took place in these years, and the human landscape underwent major changes.

Even so, the everyday life of the common people had its own revolution when motors entered their very homes. Combustion engines made automobiles possible, and electric appliances changed household forever. It is difficult to assess the derived social consequences, because of their magnitude. The middle-class was born, millions of people from poor families could live as only very rich people could have lived until then. They were able to be easily transported very far away, and they had their houses attended faster and with little physical effort. They had "mechanical servants", and the everything was transformed: the houses, the cities, the business, the social life, the family, the sexual relationships and roles ....

3.2 Possible Consequences

Trying to apply this analogy to our present situation, the first thing we have to remember is that information is a more abstract and general concept than strength. In fact is one of the most general and subtle concepts the human race has ever thought of. The parallelism with the Industrial Revolution can not be complete, and it is wise to expect the present change to be much deeper.

Anyway, the past stages of the computer's evolution could be assimilated to the ones of the engines. The early devices were enormous and expensive, and in order to employ them only in big industries a centralized model was imposed. When the machines finally were small and cheap the distributed systems could be born. Nevertheless the distribution of intelligence is, of course, more flexible than the distribution of the strength, and distributed computers can cooperate and use the addition of their capabilities or even more –emergent computation, systems where the union of simple elements brings upon a more powerful complex system.

If we try to extend further the analogy, we could wonder about the changes in the operation that represented the combustion and electrical motors. The best parallelism in the computation field could be the different trials to model the human brain. After all, we had domestic animals to inspire the engine constructors –as can be observed in the unit "steam horse" or "horse-power"–, but we knew the human brain as the only information processor. The original meaning of the word computer was "a person who makes calculations" but nowadays is always understood as "a device that processes information". Only recently some other natural behaviors have been considered intelligent or, at least, aimed at a purpose, and therefore, able to a find a solution to a problem. Instances of them could be collective intelligence of a colony of ants and the combined process of evolution and natural selection.

We could establish an arbitrary compared classification of intelligent techniques, its inspiration and engines, for instance:

1) First steam motors compared to conventional imperative programs, a model of the human behavior, simple orders to process data.
2) Combustion motors compared to the classical symbolical Artificial Intelligence, a model of "how the human reasoning looks", logical programming.

3) Electrical motors compared to artificial neural networks, modeling "how the human reasoning is", or at least the unconscious learning.
4) Hybrid combustion and electrical motors compared, for instance, to fuzzy logic and hybrid connectives based classifiers, mixing qualitative and quantitative data.
5) Other technologies, modeling some natural processes like genetic algorithms (natural selection in living populations) and simulated annealing (ordered crystal cooling processes).

We already have little computer processors, even based on fuzzy logic, in our homes, in our domestic appliances, but their still fail to model most of our reasoning flexibility. Many of the classical Artificial Intelligence programs are just in the situation of the centralized computers in the 1960's, they can work, but they are not practical. At that time, expensive computers could run a whole factory, but they were only efficient at small ones, and could not take over big industrial plants until the development of cheap distributed systems. Still nowadays we have bulky expert systems that can support, for instance, medical decisions, but they are very sparingly used. "Intelligent" techniques will not get into our everyday lives until they are distributed into light tiny systems applied everywhere.

3.3 Beyond any Analogy

Nevertheless, it was already said before that information is one of the most general systems humans can work with. And that leads us beyond any possible analogies. Probably even Humankind was born when the information concept was acquired. Some philosophers state that human intelligence appeared when for the first time we could look at the track of an animal and eliminate the immediate reaction. We became intelligent when we could get the information and "coldly" plan the best action to take.

Even more, the abstracter treatment of the information, the more effective behavior. Nowadays there are theories sustaining that the intelligence of modern humans was abstracter and more symbolic than the one of Neanderthal people. After all, their brain was at least as big as ours, we coexisted with them, we are still here and they are not.

Everything alive is information after all. We can look at the living creatures as a kind of genetic experiment, and that makes us nothing but the means to explore and perpetuate genetic codes.

But let us return to more practical issues. The capability to process more information implies a positive feedback which generates more data, and demands a greater processing capability. Even the "5th Framework Programme" of the European Commission issued an special program on the "Information Society Technologies" with a proactive initiative named "Universal Information

Ecosystems". This topic, considered as a future and emerging technology, intended to explore the possibility of modeling our society as an "ecosystem" of "knowledge entities". Every person or human organization is seen as an entity receiving information from many others, and deciding the more appropriate ones to communicate, collaborate or trade with. This interactive system is open, dynamic and global by nature, and artificial intelligence tools are necessary to guide these decisions and the implied automatic self-organization.

References

[Agell 94 a] "The Qualitative Perspective for Mobile Tracking", Agell, Aguado, Piera, proceedings of the SCS European Simulation Multiconference, pp. 529-534, June 1994.

[Agell 95 a] The chapter "Mathematical foundations of Qualitative Reasoning" within the book "Current Trends in Qualitative Reasoning and Applications", edited by N. Piera and the CIMNE, ISBN 84-87867-67-7 pp. 12-24, January 1995.

[Agell 95 b] "Fusing Qualitative Images (Ongoing Research)", Agell, Aguado, Piera, proceedings of the III IJCAI Workshop on Engineering Problems for Qualitative Reasoning, Montreal, Canada, August 1995.

[Agell 95 c] "A Qualitative Images Fusion", Agell, Aguado, Piera, proceedings of the IX International Workshop on Qualitative Reasoning, Amsterdam, Netherlands, May 1995.

[Aguado 98] *A Mixed Qualitative-Quantitative Self-Learning Classification Technique Applied to Situation Assessment in Industrial Process Control.* Ph. D. dissertation in the Polytechnical University of Catalonia. December 1998.

[Aguado 99] "Comparison of structure and capabilities between a non-standard classification technique and the Radial Basis Function Neural Networks", Aguado, Català, Parra, proceedings of the 13th European Simulation Multiconference (ICQFN 99),Vol II, pp 442-448, Warsaw, June 1999.

[Aguilar 81] *Estimation Recursive d'une Partition, Exemples d'Apprentissage et Auto-Apprentissage dans* R^N *et* I^N. J. Aguilar-Martín, M. Balssa and R. López de Mántaras. Qüestió. Vol. 5, nº 3, pp. 150-172. September 1981.

[Aguilar 82] *The process of classification and learning the meaning of linguistic descriptors of concepts.* J. Aguilar-Martín and R. López de Mántaras. Approximate Reasoning in Decision Analysis. pp. 165-175. North Holland, 1982.

[Aguilar 88] "Probabilistic and fuzzy relational semantics systems in propositional approximate reasoning", 18th Intern, Symp. on Multi-Valued Logic, IEE pp. 205-209, Palma de Mallorca (Spain), 1988.

[Aguilar 90] *Conceptual Connectivity Analysis by means of Fuzzy Partitions.* J. Aguilar-Martín, M. Martín and N. Piera. Information Processing and Management of Uncertainty in Knowledge Based Systems. Vol. 1, pp. 250-252 (extended abstract), 1990.

[Aguilar 96] *Data Based Inductive Knowledge Representation,* Aguilar, Aguado, Piera, pp. 1-6, proceedings of the International Panel Conference on Soft and Intelligent Computing, edited by the Technical University of Budapest, October 1996.

[Codina 94 a] "Neural Network Identification of Non Linear Systems Using State Space Techniques", Codina, Aguado, Fuertes, proceedings of the European Symposium on Artificial Neural Networks, pp. 13-18, Brussels, Belgium April 1994.

[Codina 94 b] "Capabilities of a Structured Neural Network Learning. Comparison with Classical Techniques", Codina, Aguado, Fuertes, proceedings of the IMACS International Symposium on Signal Processing, Robotics and Neural Networks, pp. 334-338, Lille, France, April 1994.

[Desroches 87] *Syclaire: Système de classification avec apprentissage et recoinassaince de formes. Manuel d'utilisation.* P. Desroches. "Research report" 87/9 del Centre d'estudis Avançats de Blanes. November 1987.

[Desroches 90] *Variation points in incremental conceptual clustering.* P. Desroches, N. Piera and J. Aguilar-Martín. "Technical report" of the LAAS n° 90254. July 1990.

[Dietterich 81] *Inductive learning of structural descriptions: evaluation criteria and comparative review of selected methods.* T. G. Dietterich and R. S. Michalski. *Artificial Intelligence*, 16, pp 257-294.

[Fikes 72] *Learning and executing generalized robot plans.* R. E. Fikes, P. E. Hart and N. J. Nilsson. *Artificial Intelligence*, 3, pp 251-258.

[Fisher 87] *Knowledge acquisition via incremental conceptual clustering.* D. H. Fisher. *Machine Learning*, 1, pp 267-283.

[Hayes 77] *Knowledge acquisition from structural descriptions.* F. Hayes-Roth and J. McDermott Proceedings of the IJCAI, Cambridge, Massachusetts, pp 356-362.

[Holland 92] *Adaptation in natural and artificial systems.* The MIT Press, Cambridge, Massachusetts.

[Hopfield 82] *Neural networks and physical systems with emergent collective computational abilities.* J. J. Hopfield. *Proceedings National Academy of Science*, USA, vol 79, pp 2554-2558.

[Hunt 66] *Experiments in induction.* E. B. Hunt. and P. J. Stone.New York: Academic Press.

[Kedar-Cabelli 88] *Analogical reasoning, chapter Analogy- from a unified perspective.* S. T. Kedar-Cabelli. Kluwer Academic Press.

[Michalsky 80] *Conceptual clustering: inventing goal-oriented classifications of structured objects.* R. S. Michalsky and R. Stepp III. *Machine Learning*, 1, pp 472-498.

[Minsky 69] *Perceptrons.* M. Minsky and S. Papert. MIT Press, Cambridge, Massachusetts.

[Mitchell 83] *Learning by experimentation. acquiring and refining problem solving heuristics.* T. M. Mitchell and P. Utgoff. *Machine learning: an artificial intelligence approach*, edited by Michalski, Carbonell and Mitchell, pp 163-190.

[Mitchell 86] *Explanation based learnng: a unifying view.* T. M. Mitchell, R. M. Keller and S. T. Kedar-Cabelli. *Machine Learning*, 1, pp 47-80.

[Piera 87] *Connectius de lògiques no estandard com a operadors d'agregació en classificació multivariable i reconeixement de formes.* N. Piera. Doctorate dissertation in the Universitat Politècnica de Catalunya. July 1987.

[Piera 88] *Mixed Connectives of Lineal Compensation.* N. Piera and J. Aguilar-Martín. "Technical report" of the LAAS nº 88301. October 1988.

[Piera 89] *LAMDA: An Incremental Conceptual Clustering Method.* N. Piera, P. Desroches and J. Aguilar-Martín. "Technical report" of the LAAS nº 89420. December 1989.

[Piera 90] *Variation points in pattern recognition.* N. Piera, P. Desroches and J. Aguilar-Martín. Pattern Recognition Letters. Vol. 11 pp. 519-524. August 1990.

[Piera 91] *Controlling Selectivity in Nonstandard Pattern Recognition Algorithms.* N. Piera and J. Aguilar-Martín. IEEE Transactions on S, M & C. Vol. 21, nº 1, pp. 71-82. January-February 1991.

[Prechelt 94] *PROBEN1 — Set of Neural Network Benchmark Problems and Benchmarking Rules.* L. Prechelt. Technical Report 21/94, Department of Computer Science, University of Karlsruhe, 1994.

[Quinlan 86] *Induction decision trees.* J. R. Quinlan. *Machine Learning,* 1, pp 81-106.

[Rumelhart 86] *Learning internal representations by error propagation.* D. E. Rumelhart, G. E. Hinton and R. J. Williams. *Parallel distributed processing: explorations in the microstructure of cognition. vol 1: Foundations.* Edited by Rumelhart and McCleland, the MIT Press, Cambridge, Massachusetts.

[Sklar 77] *Random variables, joint distributions and copulas.* A. Sklar. Kybernetica. Vol. 9 pp. 449-460, 1977.

[Schweizer 83] *Probabilistic metric spaces.* B. Schweizer i A. Sklar. Elsevier North-Holland, 1983.

[Vere 75] *Induction of concepts in predicate calculus.* S. A. Vere. Proceedings of the IJCAI, Cambridge, Massachusetts, pp 281-287.

[Winston 70] *Learning structural descriptions from from examples.* P. H. Winston. PhD thesis, MIT, Project MAC, Cambridge, Massachusetts.

[Zadeh 65] *Fuzzy sets.* Information and Control. Vol. 8, pp. 338-353, 1965.

[Zadeh 71] *Fuzzy sets as a base for a theory of possibility.* Fuzzy Sets and Systems. Vol. 1 (1), pp. 3-28, 1971.

Appendix: Quantitative versus Qualitative Descriptors

Our world, above the atomic level, appears quantitative but our representation of it always is qualitative, because of the limitation of our resources. A quantitative descriptor takes its values from the set of the real numbers; a qualitative one can only show a finite number of values. For instance, if we want to measure the length of a piece of metal, the result could theoretically be any positive real number, but our measuring and computing devices have a limited precision, finite in spite of how large it can be. Anyway, we always operate as if our tools were not limited, and we obtain results which better or worst approximate the reality.

Numerical learning techniques work on this hypothesis, and ignore qualitative descriptors, which differ from quantitative ones in the fact that the former cannot admit interpolation, but these methods always interpolate. Even though maybe under some restrictions numerical approximations could be enough in a practical sense, connectionist methods, the most commonly employed, present such an unpredictability that they can not guarantee either any useful results for a given problem or a reasonable time and structure to solve it. That is why some efforts have been made to try to find a more regular basis of functioning, like for instance the Fourier series [Codina 94a, 94b].

But even if predictable raw numerical process is not subtle enough to deal with most of the real life problems, where qualitative descriptors are introduced by human devices, due either to a limited, incomplete representation of data, or to a discrete operational state. Qualitative Reasoning is able to cope with incomplete data, to express categorical classifications and even speed up computations by adequating the quantity of information to the needed abstraction level in every situation. All this possibilities are reflected in [Agell 94, 95a, 95b, 95c].

But symbolic methods are not the solution either, since they lack the ability to quickly cope with apparently infinite-valued descriptors. We need to combine the best of both techniques, numerical self learning treatment and qualitative process, as *LAMDA* does [Aguilar 96]. We usually combine in our reasoning processes descriptors with different number of possible values; for instance, if we want a clock to indicate the time to have meal it can only need to have a resolution of about fifteen minutes, if it is used to announce the time of every class in a school it could need five minutes resolution, but if we want to control an athletic competition the resolution has to be greater, and if we want to predict physical phenomena we would ideally ask for an infinite resolution.

3.

RISK IN DECISION SUPPORT AND CONTROL SYSTEMS

Analysis Support System of Human Reliability with Linguistic Expressions

Takehisa Onisawa and Akio Ohmori

Institute of Engineering Mechanics and Systems, University of Tsukuba
1-1-1, Tennodai, Tsukuba, 305-8573, Japan
onisawa@esys.tsukuba.ac.jp

This chapter presents a human reliability analysis method with natural language based on the consideration that human reliability should be analyzed by linguistic terms rather than by numerical values and that analyst's subjectivity should be openly incorporated into the analysis. In this method, (1) linguistic expressions of information on human reliability are used, (2) the meaning of linguistic expressions is expressed by a fuzzy set defined on a unit interval [0, 1], (3) the parametrized operations of fuzzy sets are used in order to reflect analyst's subjectivity towards analyzed human tasks, (4) task dependence and task procedure are considered, which are peculiar to human reliability, and (5) analysis results are expressed linguistic expressions.

Many analysis results can be obtained according to analysts even if the analyzed task is the same since the results are dependent on analysts subjectivity. Therefore, it is necessary to discuss reliability in order to obtain a compromise between different analysis results. The chapter also presents the procedure for a compromise. The human reliability method including the procedure for a compromise is implemented on a personal computer as the analysis support system of human reliability. Finally, an analysis example is presented to explain the analysis support system.

1 Introduction

In the mathematical sense, reliability is generally characterized or measured by the probability that an entity can perform one or several required function(s) under given conditions for a given time interval [1]. According to the entities, the notion of time interval should be replaced by the notion of cycles, distance traveled, etc. There have been many applications of the probabilistic method to analyses of equipment reliability and system reliability [2,3]. It is said that by the probabilistic method it is possible to assess component reliability and system reliability rationally, i.e., objectively, since a failure probability or an accident probability is compared with probabilities of other rare events. The probabilistic method is applied to not only reliability analysis of hardware component, machinery and instruments but also that of a human operator. THERP [4] is famous as the probabilistic analysis model of human reliability and the probabilistic analysis of human reliability is applied to nuclear power plants [5]. The probabilistic approach is based on the assumption that a hardware failure, a system accident or a human error occurs at random. Furthermore, this approach assumes that a system state or a human performance is good or bad, and these states are mutually exclusive.

The assumptions, however, have many questions [6]. Concerning human reliability, the probabilistic approach has other problems. We usually have not enough amount of data on human errors to assess human reliability by probability. To solve the problem about the lack of data on human errors, various methods are used [7]. Nevertheless, expert judgements have been widely used and they will continue to be used to fill in the gaps left by data sources on human errors [7]. Human reliability is affected by many environmental factors, situations, and psychological stress of a human operator. These factors are called performance shaping factors [5]. Even if a human operator performs a task, he does not always perform it in the same way since a human being is distinct from machinery. Therefore, it is necessary to modify the basic error probability according to the performance shaping factors. The modification is usually performed by multiplying the basic error probability by numerical factors which are determined by expert judgements [5]. These numerical factors are not objective but subjective, i.e., fuzzy. Furthermore, even if human tasks are analyzed in detail, human reliability is not estimated precisely since human behavior is complex and unpredictable. In these situations, information on human reliability should be expressed with linguistic terms rather than with numerical values and human reliability should be analyzed roughly and fuzzily [8]. There have been many applications of fuzzy theory to the reliability engineering field including human reliability considering the above mentioned problems [9–18].

The chapter is based on the viewpoint that human reliability should be analyzed by linguistic terms rather than by numerical values and that analyst's subjectivity should be openly incorporated into human reliability analysis. The chapter also considers task dependence and task procedure for the following reasons. Human tasks are often complex and these tasks are usually dependent on each other. Therefore, human reliability analysis must imperatively take the dependence into account [7]. There are two kinds of dependences among consecutive similar tasks [5]. One is failure dependence and the other is success dependence. The latter is peculiar to human reliability. Furthermore, task procedure in human tasks is usually prescribed. In this situation the error in task procedure leads to the error of a given task even if a human operator performs the task itself. Therefore, the error in task procedure must be considered in the analysis. This is also peculiar to human reliability.

This chapter subtilizes the human reliability analysis approach by linguistic expressions. In this approach, (1) linguistic expressions of human reliability are used, (2) the meaning of the linguistic expression is expressed by a fuzzy set defined on a unit interval [0, 1], (3) the parametrized operations of the fuzzy sets are used in order to reflect analyst's subjectivity towards analyzed human tasks, (4) task dependence and task procedure are considered, which are peculiar to human reliability, and (5) analysis results are expressed by linguistic expressions. In this approach there is a possibility that many analysis results are obtained since analysis results are dependent on analyst's subjectivity. Therefore, it is necessary to discuss human reliability for obtaining a compromise between different analysis results. The chapter presents the procedures for obtaining a compromise between them. The present human reliability analysis method including the procedures for a compromise is implemented on a personal computer as the analysis support system of human reliability. The analysis support system is explained with the help of an example.

2 Human Reliability Method

2.1 Linguistic Expressions of Human Reliability

2.1.1 Subjective Measure of Unreliability

Numerical values such as human error probability are necessary to quantitative human reliability analysis. In practice, however, it is often difficult for even experts to estimate human reliability by numerical values since the amount of data on human error probability are limited as mentioned before. In these situations human reliability should be expressed by linguistic terms. For example, when experts have not enough amount of data about human reliability, its reliability estimate such as 10^{-4} [1/hr] is unreasonable and difficult. They should estimate it by linguistic expressions based on their experienced engineering judgements such as *human operator has rather high reliability* since in this situation fuzziness rather than randomness becomes important, and fuzziness in the estimation is expressed by linguistic terms.

This chapter expresses human reliability with linguistic terms in the form of reliability estimate and its fuzziness. Table 1 shows linguistic expressions of reliability estimate and Table 2 shows linguistic expressions of its fuzziness. "*Human reliability is standard*" means that its reliability is evaluated to be equal to expert's standard. "*Human reliability is high*" means that its reliability is evaluated to be higher than expert's standard.

The meaning of linguistic expression is represented by a fuzzy set better than by numerical values since the meaning is fuzzy by itself. This approach uses a fuzzy set to represent the meaning of linguistic expressions about human reliability, and the fuzzy set is defined on a unit interval [0, 1] which means subjective evaluation of unreliability. Therefore, a numerical value in the unit interval is not probability but subjective evaluation.

Table 1. Linguistic expressions of human reliability estimate and corresponding parameter values

Class	Expressions of Reliability Estimate	Parameter x_0 (Representative Value)
	(human operator has)	
1	no reliability	–
2	very low reliability	0.9 – 1.0 (0.95)
3	low reliability	0.7 – 0.9 (0.8)
4	rather low reliability	0.55 – 0.7 (0.625)
5	standard reliability	0.45 – 0.55 (0.5)
6	rather high reliability	0.3 – 0.45 (0.375)
7	high reliability	0.2 – 0.3 (0.25)
8	quite high reliability	0.1 – 0.2 (0.15)
9	extremely high reliability	0.05 – 0.1 (0.075)
	(human error is)	
10	next to impossible	0.0 – 0.05 (0.025)
11	Impossible	–

Table 2. Linguistic expressions of fuzziness of human reliability estimate and corresponding parameter values

Class	Expressions of Fuzziness	Parameter m
1	low fuzziness	2.0
2	medium fuzziness	2.5
3	rather high fuzziness	3.0
4	high fuzziness	3.5

The fuzzy set is called a *subjective measure of unreliability* in this chapter. The subjective measure of unreliability is a non-probabilistic reliability measure in the sense that it does not satisfy axiomatic laws of probability such as additivity. Eq. (1) is used as the membership function of subjective measure of unreliability [19].

$$F_S(x) = \frac{1}{1 + 20 \times |x - x_0|^m} \quad , \tag{1}$$

where x_0 and m are parameters, and $0 \leq x,\ x_0 \leq 1$. The parameter x_0 gives a maximal grade of $F_S(x)$ and the parameter m is related to fuzziness. The constant term 20 can be replaced by another one, e.g., 10, in the sense that almost the same analysis results are obtained in both cases [19]. In this chapter the constant term is assumed to be 20. Fig. 1 shows an example of a subjective measure of unreliability. It is interpreted that the smaller the numerical value of the abscissa, the higher the subjective evaluation of reliability. For example, the subjective evaluation of unreliability 1 means that a human operator is subjectively estimated

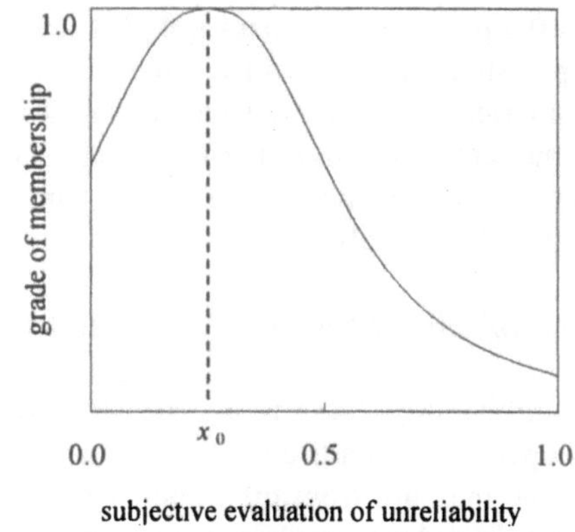

Fig. 1. Subjective measure of unreliability

to commit an error. On the other hand 0 means that he is subjectively estimated to perform a task without error. The middle point 0.5 means subjective standard evaluation of unreliability, that is, reliability is estimated to be equal to expert's standard. The ordinate means the grade of membership. Then the larger the numerical value of *m*, the fuzzier the subjective measure of unreliability.

The fuzzy set (1) has the following properties:

1. The fuzzy set is normal and convex [20]. This implies that it is easy to translate linguistic expressions as shown in Tables 1 and 2 to fuzzy sets.
2. $F_S(0) \neq 0$ and $F_S(1) \neq 0$. These imply that even if reliability is quite low, a human operator does not necessarily commit an error in a given task, and that even if reliability is quite high, a human operator does not always perform the task without error.

The parameter x_0 and m correspond to linguistic expressions of reliability estimate and its fuzziness, respectively, as shown in Tables 1 and 2. Especially, the subjective measure of unreliability in class 1 and that in class 11 are defined as follows:

$$\text{class 1} \ : \ F_S(x) = \begin{cases} 1 : x = 1 \\ 0 : x \neq 1 \end{cases}, \tag{2}$$

$$\text{class 11} : \ F_S(x) = \begin{cases} 1 : x = 0 \\ 0 : x \neq 0 \end{cases}. \tag{3}$$

Fuzziness is not taken into consideration for linguistic expressions in classes 1 and 11.

The correspondence of parameters x_0 and m to linguistic terms is defined subjectively. Correspondence relationships, however, are consistent in the following sense. The $x < 0.5$ part of the abscissa in Fig. 1 means higher reliability. Therefore, linguistic expressions such as *rather low reliability*, or *low reliability* do not correspond to numerical values of x_0 smaller than 0.5. And as mentioned above, the larger the numerical value of m, the fuzzier the subjective measure of unreliability. Once the correspondence relationships are defined, they are consistently employed through the analysis.

2.1.2 Opposite View of Subjective Measure of Unreliability

Subjective measure of unreliability $F_S(x)$ is the measure from the unreliability viewpoint that a human operator commits an error in a task. On the other hand, it is necessary to consider the measure from the reliability viewpoint that a human operator succeeds in a task in order to analyze success dependence as mentioned below. Success dependence implies that if a human operator succeeds in one task, he is likely to succeed in a consecutive similar task. The opposite view of subjective measure of unreliability should have the following features:

1. Estimation that a human operator does not make an error in a task corresponds to estimation that he performs the task without error.
2. Estimation that a human operator makes an error in a task corresponds to estimation that he does not perform the task successfully.
3. The relation between subjective measure of unreliability and its opposite view is monotonous.
4. Estimation of unreliability at "*standard*" is equal to estimation of reliability at "*standard*".

In this chapter, for simplicity, a function N: subjective evaluation of unreliability → subjective evaluation of reliability is defined as

$$N(x) = 1 - x \ . \tag{4}$$

When $F(x)$ is the subjective measure of unreliability, $N(F(x))$, which is the opposite view of the subjective measure of unreliability, is called the *subjective measure of reliability* and is defined by

$$N(F(x)) = 1 - F(x) \ . \tag{5}$$

The extension principle [20] is applied to Eq. (5). In this chapter the extension principle is applied to all operations of fuzzy sets.

2.2 Fault Tree Analysis by Subjective Measure of Unreliability

2.2.1 *AND* Operation and *OR* Operation of Subjective Measures of Unreliability

In this chapter human reliability is analyzed by the fault tree analysis. In the fault tree analysis basic operations are an *AND* operation and an *OR* operation.

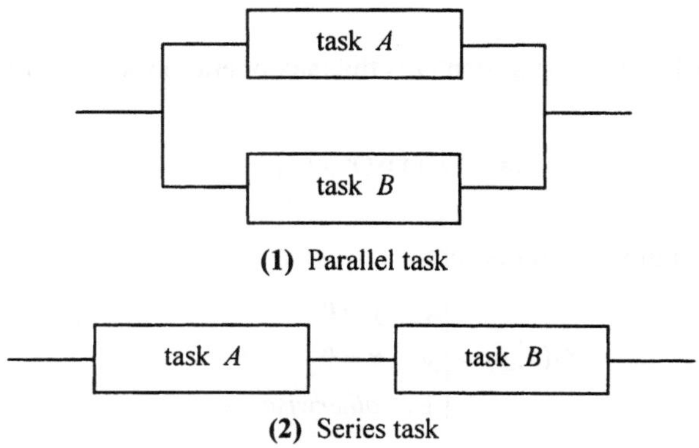

Fig. 2. Parallel task and series task

An analyzed whole task is divided into basic tasks. The process that errors in all tasks lead to the error of a whole task is expressed by an *AND*-gate in a fault tree. This whole task corresponds to a parallel task as shown in Fig. 2(1). On the other hand the process that an error in at least one of tasks leads to the error of a whole task is expressed by an *OR*-gate in a fault tree. This whole task corresponds to a series task as shown in Fig. 2(2). In the fault tree analysis by the subjective measure of unreliability the following parametrized t-norm and t-conorm are used as the *AND* and the *OR* operations of subjective measures of unreliability, respectively.

$$H(x,y)=\frac{1}{1+\left[\{(1-x)/x\}^{1/n_H}+\{(1-y)/y\}^{1/n_H}\right]^{n_H}}\quad , \tag{6}$$

where $0 < x, y \leq 1$, $H(0, y) = H(x, 0) = 0$, and nH is a non-negative parameter.

$$G(x,y)=\frac{\left[\{x/(1-x)\}^{n_G}+\{y/(1-y)\}^{n_G}\right]^{1/n_G}}{1+\left[\{x/(1-x)\}^{n_G}+\{y/(1-y)\}^{n_G}\right]^{1/n_G}}\quad , \tag{7}$$

where $0 \leq x, y < 1$, $G(1, y) = G(x, 1) = 1$, and nG is a non-negative parameter.

The operation H and the operation G are the Dombi t-norm and the Dombi t-conorm, respectively [21]. The operations H and G have the following properties [21]. The operation H becomes the drastic product when the parameter nH is infinite.

$$H(x,y)=\begin{cases} x, & y=1 \\ y, & x=1 \\ 0, & otherwise \end{cases}, \tag{8}$$

and the min operation when *nH* is zero.

$$H(x,y)=\min(x,y)\ . \tag{9}$$

On the other hand the operation *G* is the max operation when *nG* is infinite,

$$G(x,y)=\max(x,y)\ , \tag{10}$$

and the drastic sum when *nG* is zero

$$G(x,y)=\begin{cases} x, & y=0 \\ y, & x=0 \\ 1, & otherwise \end{cases}. \tag{11}$$

From the viewpoint of reliability analysis, the operations *H* and *G* can cover the range of reliability estimate from the most pessimistic estimate through the most optimistic estimate [19]. The parameters *nH* and *nG* can reflect analyst's subjectivity towards an analyzed task whether the task should be analyzed pessimistically or not. Parameter determination is mentioned below.

2.2.2 *AND* Operation and *OR* Operation of Subjective Measures of Reliability

The *AND* and the *OR* operations of subjective measures of reliability are obtained by Eq. (4), Eq. (6) and Eq. (7), where $x' = N(x)$ and $y' = N(x)$.

$$\begin{aligned} H'(x',y') &= N(G(x,y)) \\ &= \frac{1}{1+\left[\{(1-x')/x'\}^{n_G}+\{(1-y')/y'\}^{n_G}\right]^{1/n_G}}, \end{aligned} \tag{12}$$

where $0 < x', y' \le 1$, $H'(0, y') = H'(x', 0) = 0$, and *nG* is a non-negative parameter.

$$\begin{aligned} G'(x',y') &= N(H(x,y)) \\ &= \frac{\left[\{x'/(1-x')\}^{1/n_H}+\{y'/(1-y')\}^{1/n_H}\right]^{n_H}}{1+\left[\{x'/(1-x')\}^{1/n_H}+\{y'/(1-y')\}^{1/n_H}\right]^{n_H}}, \end{aligned} \tag{13}$$

where $0 \le x', y' < 1$, $G'(1, y') = G'(x', 1) = 1$, and *nH* is a non-negative parameter.

2.3 Task Procedure

Task procedure is usually prescribed in a task which consists of some basic tasks. In this case the error in task procedure must be considered in the analysis. For example, let us consider a task such as opening two valves A and B, and let us assume that according to a given procedure, a human operator must open the valve B after opening the valve A. If the operator opens the valve B at the beginning, the error in the procedure may lead to the error of the whole task. That is, even if the operator opens the valve A and the valve B successfully, if the operator makes an error in task procedure, the task is estimated to be error. This kind of error is called a procedure failure in this chapter. In the above example of opening valves, if the procedure failure is not considered, then only the error in opening valves leads to an error in the task. Although it is not necessary to deal with the procedure failure in the hardware reliability analysis, the procedure failure must be considered in the human reliability analysis since this is peculiar to human reliability.

We usually have not enough amount of data to estimate the probability of the procedure failure. Therefore, the procedure failure should be also estimated by linguistic terms rather than by numerical values. In this chapter linguistic terms about the procedure failure are assumed to be the same as the ones in Table 1 and Table 2. That is, the estimation of the procedure failure is expressed by reliability estimate about the procedure failure and its fuzziness. The meaning of linguistic terms of estimation of the procedure failure is expressed by a fuzzy set with the same membership function as that of the subjective measure of unreliability.

The subjective measure of unreliability considering the procedure failure is obtained by the operation G.

$$F_P = G(F, P) \quad , \tag{14}$$

where F is the subjective measure of unreliability of a given task without consideration of the procedure failure and P is a fuzzy set representing estimation of the procedure failure. This implies that the error in a given task occurs when the procedure failure happens or the error in a task itself happens.

2.4 Dependence Analysis

2.4.1 Dependence among Human Tasks

If consecutive tasks are similar, it is necessary to consider dependence among tasks in the analysis. Let us consider a parallel task in which errors of all basic tasks lead to the error of the whole task. When the failure dependence is estimated to exist among basic tasks, the error in one task has an influence on the error in another task. Therefore, the failure dependence must be dealt with since analysis results become optimistic without consideration of the failure dependence. In a series task, however, the failure dependence is meaningless in the analysis since the error of at least one of tasks leads to the error of the whole task. On the other hand, when the success dependence is estimated to exist among tasks in a series task, the success in one task has an influence on the success in another task. In this case, the success dependence must be dealt with since successes of all tasks lead to the success of a series task. In a parallel task, however, it is no use considering the success dependence since the success of at least one of tasks leads to the success of the whole task.

Knowledge on dependence situations is often obtained by experts who are familiar with the task based on their engineering experienced judgements. In this chapter dependence evaluation is expressed by linguistic terms in the form of dependence level estimate and its fuzziness. Table 3 shows linguistic terms about dependence level estimate referring to [5]. With regard to linguistic expressions of fuzziness of dependence level estimate the same terms as the ones in Table 2 are used.

For the sake of simplicity, the meaning of linguistic terms of dependence evaluation is expressed by a fuzzy set with the same membership function as the subjective measure of unreliability.

$$R(r) = \frac{1}{1 + 20 \times \left| r - r_{0i} \right|^m} , \tag{15}$$

where $0 \le r, r_{0i} \le 1$, and r_{0i} and m are parameters. With regard to numerical values of m, the same numerical values as shown in Table 2 are used. The parameter r_{0i} corresponds to level i (i = 1,2,3,4,5) in Table 3. Determination of the parameter r_{0i} is mentioned below. Especially, the fuzzy set representing dependence evaluation in level 1 and that in level 5 are defined as follows:

$$\text{level 1}: \quad R(r) = \begin{cases} 1 : r = 1 \\ 0 : r \neq 1 \end{cases} , \tag{16}$$

$$\text{level 5}: \quad R(r) = \begin{cases} 1 : r = 0 \\ 0 : r \neq 0 \end{cases} . \tag{17}$$

Fuzziness is not taken into consideration for linguistic expressions in levels 1 and 5.

Table 3. Linguistic terms about dependence level estimate

Level	Expressions of Dependence Level Estimate
1	complete dependence
2	high dependence
3	moderate dependence
4	low dependence
5	zero dependence

2.4.2 Dependence Considering Task Procedure

Let us consider tasks as shown in Fig. 3 and Fig. 4. Fig. 3 implies that the error of task $A(B)$ has an influence on that of task $B(A)$. On the other hand Fig. 4 means that the success of task $A(B)$ has an influence on that of task $B(A)$.

The failure dependence analysis and the success dependence analysis consider two situations [22]. One is the situation in which the procedure failure is considered as shown in Fig. 3(1) and Fig. 4(1). The other is the one in which the procedure failure is not considered as shown in Fig. 3(2) and Fig. 4(2). In both situations as far as the dependence level is not complete, the error or the success of task $A(B)$ does not always have an influence on the error or the success of task $B(A)$. The following two cases are considered: (*Case I*) Error(Success) of task $A(B)$ has an influence on error(success) of task $B(A)$, and (*Case II*) Error(Success) of task $A(B)$ has no influence on error(success) of task $B(A)$. Let F_A and F_B be subjective measures of unreliability of tasks A and B, respectively, and R and P be fuzzy sets representing estimation of the dependence level and that of the procedure failure, respectively.

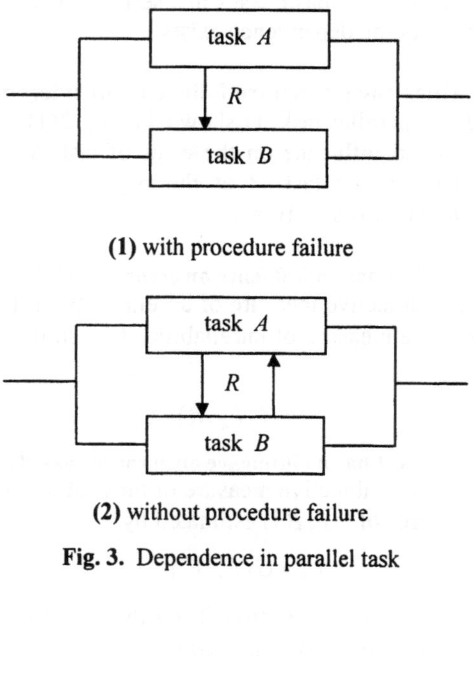

(1) with procedure failure

(2) without procedure failure

Fig. 3. Dependence in parallel task

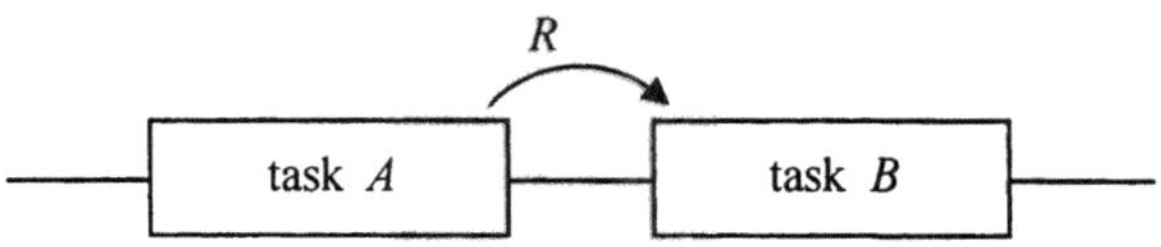

(1) with procedure failure

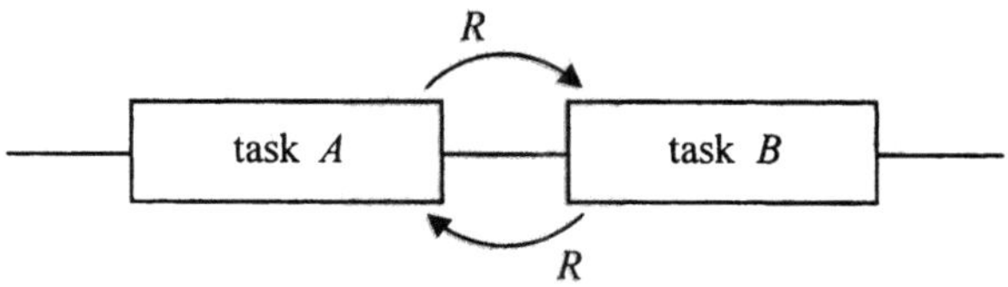

(2) without procedure failure

Fig. 4. Dependence in series task

(1) Failure Dependence

In the dependence analysis only a parallel task as shown in Fig. 3 is considered since the error of at least one of basic tasks in a series task leads to the error of the whole task whether or not the dependence exists.

(*a*) Situation in which the procedure failure is considered.

Let us consider a parallel task as shown in Fig. 3(1) and assume that the error of task A has an influence on the error of task B. Therefore, a human operator must perform the task A at the beginning. Human error of the procedure is applied to this situation.

***Case I*: Error of task *A* has an influence on error of task *B*.**

Let F'_1 be the subjective measure of unreliability of the whole task in this case. The subjective measure of unreliability F'_1 can be estimated by F_A and R, i.e.,

$$F_1' = H(F_A, R) \quad . \tag{18}$$

***Case II*: Error of task *A* has no influence on error of task *B*.**

The portion of the subjective measure of unreliability of task A, having no influence on the error of task B is estimated by

$$F_A = G(F_A', F_1') \quad , \tag{19}$$

where F'_A corresponds to the portion. The subjective measure of unreliability of the whole task in *Case II* is estimated by

$$F_1'' = H(F_A', F_B) \quad . \tag{20}$$

The subjective measure of unreliability of the whole task considering both *Case I* and *Case II* is estimated by

$$F_1 = G(F_1', F_1'') \cdot \quad (21)$$

Considering the procedure failure, the subjective measure of unreliability of the whole task in the situation (*a*) is estimated by

$$F_P = G(F_1, P) \ . \quad (22)$$

(*b*) Situation in which the procedure failure is not considered.

Let us consider a parallel task as shown in Fig. 3(2). In this situation it is not necessary to consider which of errors of task *A* or task *B* occurs first. Human errors in a parallel task without the procedure failure are applied to this situation.

Case I **: Error of task *A*(*B*) has an influence on error of the other task *B*(*A*).**

When the error of task *A* occurs first and the error of task *A* has an influence on the error of task *B*, the subjective measure of unreliability of the whole task is obtained by

$$F_{0A}' = H(F_A, R) \ . \quad (23)$$

On the other hand, when the error of task *B* occurs first and the error of task *B* has an influence on the error of task *A*, the subjective measure of unreliability of the whole task is obtained by

$$F_{0B}' = H(F_B, R) \ . \quad (24)$$

Considering Eq. (23) and Eq. (24) the subjective measure of unreliability of the whole task in *Case I* is estimated by

$$F_0' = G(F_{0A}', F_{0B}') \ . \quad (25)$$

Case II **: Error of task *A*(*B*) has no influence on error of the other task *B*(*A*).**

The portion of the subjective measure of unreliability of task *A*, having no influence on the error of task *B*, is estimated by

$$F_A = G(F_A', F_{0A}') \ , \quad (26)$$

where F'_A corresponds to the portion. And the portion of the subjective measure of unreliability of task *B* having, no influence on the error of task *A*, is estimated by

$$F_B = G(F_B', F_{0B}') \ , \quad (27)$$

where F'_B corresponds to the portion. The subjective measure of unreliability of the whole task in *Case II* is estimated by

$$F_0'' = H(F_A', F_B') \ . \quad (28)$$

The subjective measure of unreliability of the whole task considering both *Case I* and *Case II* in this situation is estimated by

$$F_0 = G(F_0', F_0'') \ . \quad (29)$$

Let us consider the case in which the dependence level is complete. In this case, $F'_{0A} = F_A$ and $F'_{0B} = F_B$ are obtained by Eq. (23) and Eq. (24), respectively. Then $F_0 = G(F_A, F_B)$ since F'_A and F'_B are equal to Eq. (3). It is found that in this case a parallel task is equal to a series task virtually.

(2) Success Dependence

In the success dependence analysis only a series task as shown in Fig. 4 is considered since the success of at least one of tasks in a parallel task leads to the success of the whole task whether or not the success dependence exists. In the success dependence analysis, opposite views of F_A and F_B, i.e., $N(F_A)$ and $N(F_B)$ obtained by Eq. (5), are used. Let $N(F_A)$ and $N(F_B)$ be S_A and S_B, respectively. The same procedure mentioned in (1) is performed in order to estimate the subjective measure of reliability $S(x)$ of a series task with success dependence where F_A, F_B, the operation H and the operation G are replaced with S_A, S_B, the operation H' and the operation G', respectively. Then $S(x)$, the subjective measure of reliability of a series task with success dependence, needs to be replaced with $F(x)$ using Eq. (30).

$$\begin{aligned} F(x) &= N^{-1}(S(x)) \\ &= 1 - S(x) \end{aligned} \tag{30}$$

$F(x)$ obtained by Eq. (30) is the subjective measure of unreliability of a series task with success dependence.

2.5 Linguistic Expressions of Analysis Results

Results of human reliability analysis are expressed by linguistic terms in the form of reliability estimate and its fuzziness. Let F be the subjective measure of unreliability obtained by the presented analysis method, and F_S be the subjective measure of unreliability with the membership function (1). Let α-cuts of F and F_S be $(F)_\alpha = (x_1(\alpha), x_2(\alpha))$ and $(F_S)_\alpha = (x_{S1}(\alpha), x_{S2}(\alpha))$, respectively. The distance between F and F_S is defined as

$$d(F_S, F) = \int_0^1 \left[\{x_1(\alpha) - x_{S1}(\alpha)\}^2 + \{x_2(\alpha) - x_{S2}(\alpha)\}^2 \right]^{1/2} d\alpha \,. \tag{31}$$

The parameter x_0 ($0 \le x_0 \le 1$) and m ($= 2.0, 2.5, 3.0, 3.5$) of F_S are selected so as to minimize the distance d. Table 1 and Table 2 are used in order to express results with linguistic terms according to the selected parameters x_0 and m.

2.6 Parameters

2.6.1 Parameters *nH* and *nG*

Let us consider a parallel task and a series task as shown in Fig. 2. And let reliability estimate of each basic task and its fuzziness be *standard* and *medium*, respectively, where basic tasks are assumed to be independent of each other. The parameters *nH* and *nG* are determined according to each analyst's reliability estimate of the parallel task and the series task, respectively. For example, if an analyst estimates reliability of a parallel task and that of a series task as *extremely high* ($x_0 = 0.075$) and *rather low* ($x_0 = 0.625$), these estimates lead to determination of parameters $nH = 3.51$ and $nG = 1.05$.

This estimation depends on analysts and analyzed tasks whether their human reliability should be evaluated optimistically or not. Therefore, the parameters *nH* and *nG* reflect analyst's subjectivity towards an analyzed task. These parameters play an important role in the analysis.

2.6.2 Parameter r_{0i}

(1) Parameter of Failure Dependence

Let us consider a parallel task shown in Fig. 3, where no procedure failure is considered. Let reliability estimates of basic tasks and their fuzziness be *standard* and *medium*, respectively. When the dependence level is estimated to be *complete*, in the situation of Fig. 3(1) reliability estimate of the whole task is *standard* and its fuzziness is *medium*. In the situation of Fig. 3(2), reliability estimate of the whole task is obtained using the operation G with the determined parameter *nG*. On the other hand, when the dependence level is estimated to be *zero*, reliability estimate is obtained using the operation H with the determined parameter *nH* in both situations of Fig. 3(1) and Fig. 3(2). Let x_{0i} be the parameter x_0 of the subjective measure of unreliability F_S of the whole task under the dependence level i, and let r_{0i} be the parameter r_0 of R representing the dependence level i, where $i = 1$ implies *complete dependence*, $i = 2$, *high dependence*, $i = 3$, *moderate dependence*, $i = 4$, *low dependence*, $i = 5$, *zero dependence*. The numerical value of the parameter x_{01} in the situation of Fig. 3(1) is equal to 0.5. And the numerical value of the parameter x_{01} in the situation of Fig. 3(2) is obtained by the operation G and Eq. (31). The numerical value of the parameter x_{05} is obtained by the operation H and Eq. (31). Numerical values of parameters x_{0i} ($i = 2, 3, 4$) in the situation of Fig. 3(1) are obtained by Eq. (21) and Eq. (31). In the situation of Fig. 3(2), parameter values x_{0i} ($i = 2, 3, 4$) are obtained by Eq. (29) and Eq. (31). The parameters r_{0i} ($i = 2, 3, 4$) are determined so as to satisfy the following.

$$(x_{01} - x_{0i}) : (x_{0i} - x_{05}) = \begin{cases} 1:3 & (i=2) \\ 1:1 & (i=3) \\ 3:1 & (i=4) \end{cases} . \tag{32}$$

That is, two kinds of parameters values (r_{0i} ; $i = 1, 2, 3, 4, 5$), which correspond to dependence levels in a parallel task shown in Fig. 3(1) and to that in Fig. 3(2), are obtained. When the parameter nH and nG are 2.88 and 0.41, respectively, the parameters r_{0i} ($i = 2, 3, 4$) are as follows:

$$\text{Fig. 3(1)} \quad \begin{cases} r_{02} = 0.67 & (x_{02} = 0.41) \\ r_{03} = 0.27 & (x_{03} = 0.33) \\ r_{04} = 0.08 & (x_{04} = 0.24) \end{cases}$$

$$\text{Fig. 3(2)} \quad \begin{cases} r_{02} = 0.73 & (x_{02} = 0.64) \\ r_{03} = 0.30 & (x_{03} = 0.48) \\ r_{04} = 0.10 & (x_{04} = 0.31) \end{cases}$$

The parameters r_{0i} ($i = 2, 3, 4$) are dependent on the parameters nH and nG, and the existence of the task procedure.

(2) Parameter of Success Dependence

In a series task shown in Fig. 4, the parameters of the dependence level are determined in a similar way, where no procedure failure is also considered. Let x_{0i} be the parameter x_0 of the subjective measure of unreliability F_S of the whole task under the dependence level i, and let r_{0i} be the parameter r_0 of R representing the dependence level i, where $i = 1$ implies *complete dependence*, $i = 2$, *high dependence*, $i = 3$, *moderate dependence*, $i = 4$, *low dependence*, $i = 5$, *zero dependence*. The success dependence parameters r_{0i} ($i = 2, 3, 4$) are determined so as to satisfy the following.

$$(x_{0i} - x_{01}) : (x_{05} - x_{0i}) = \begin{cases} 1:3 & (i = 2) \\ 1:1 & (i = 3) \\ 3:1 & (i = 4) \end{cases} \tag{33}$$

3 Procedure for a Compromise between Different Analysis Results

Many analysis results are obtained even if reliability of the same human task is analyzed since analysis results are dependent on each analyst's subjectivity. Some of analysis results can be classified into the same group in Table 1 since the present human reliability analysis approach expresses analysis results using linguistic terms. There still exists a problem, however, in which all analysts results are not classified into the same group. Therefore, analysts need to consider a compromise between different results. This section describes the procedure for reaching a compromise between different analysis results. The procedure is dependent on the difference degree between analysis results whether the degree is within one class in Table 1 or not.

3.1 Procedures in Small Difference Degree Case

If the difference degree between linguistic expressions of analysis results is within one class in Table 1, then the analysis results are defined to have a consensus. For example, there is a consensus between "*standard reliability*" and "*rather low reliability*" , where fuzziness of analysis results is not considered. In this case, an average of all analysis results is presented as the analysis result.

Let F_j (j = 1, 2, ..., n) be the subjective measure of unreliability obtained by jth analyst, where n is the number of analysts. The average of the results is defined by

$$F_{AV} = \frac{1}{n} \sum_{j=1}^{n} F_j \quad . \tag{34}$$

The F_{AV} is also expressed by linguistic terms in the form of reliability estimate and its fuzziness by the use of Eq. (31).

3.2 Procedures in Large Difference Degree Case

When the difference degree is more than two classes in Table 1, it is defined that there exists disagreement in analysis results. For example, there is not a consensus between "*standard reliability*" and "*low reliability*". In this case, the following procedures for obtaining a compromise are taken.

First of all, all analysis results are shown. Comparing other analysts results, analysts can recognize differences among them. It is necessary to select items which should be re-estimated in order to decrease the difference. The re-estimation of the item having large influence on analysis results is effective for decreasing the differences between the analysis results.

Let F and F_k be subjective measures of unreliability of a whole task and the k-th item, respectively. The influence degree is defined by

$$Id = \frac{d(\ F + \Delta F,\ F\ )}{d(\ F_k + \Delta F_k\, ,\ F_k)} \quad , \tag{35}$$

where ΔF is the change of reliability estimate of a whole task, i.e., analysis results, depending on the ΔF_k , i.e., the change of reliability estimate of the k-th item. When the item is reliability of a basic task or reliability of a task procedure, F_k is its subjective measure of unreliability. When the item is the dependence level, F_k is replaced by the subjective measure of unreliability of a parallel task or a series task considering the dependence.

The re-estimation order of the item is determined according to the influence degree Id. First of all, the item with the largest influence degree is selected. If analysts re-estimate the selected item, the re-estimated analysis results are shown. If there is still disagreement among the analysis results, the item with the second largest influence on the analysis result is selected. And after its re-estimation, the same procedure is repeated until a compromise is obtained or all estimate items are shown.

When a compromise is obtained, the procedures mentioned in 3.1 are taken.

When it is found that a compromise is not obtained even repeating the above mentioned procedures, the followings are suggested as a compromise:

- an average analysis result of all analysts results: F_{AV} defined by Eq. (34).
- an analysis result including all analysts results in the sense of their union: F_{CU} normalized by the convexity operation [8] on the union of F_J defined as follows:

$$F_U = \bigcup_J F_J \ . \tag{36}$$

  F_{CU} is interpreted as the largest part of the reliability evaluation that all analysts agree with, since F_{CU} includes all F_J's obtained by analysts [19].
- an analysis result including all analysts results in the sense of their intersection: F_{NI} normalized by the maximal grade of the intersection of F_J defined as follows:

$$F_I = \bigcap_J F_J \ . \tag{37}$$

  F_{NI} is interpreted as the common part of the reliability evaluation that all analysts agree with one another, since F_{NI} includes only the parts of F_J which all analysts agree with [19].

F_{AV} , F_{CU} and F_{NI} are expressed by linguistic terms in the form of reliability estimate and its fuzziness using Eq. (31).

4 Analysis Support System

The analysis support system is implemented on a personal computer based on the method described in Section 2 and Section 3. The system consists of two parts; a human reliability analysis part and a support part for obtaining a compromise.

4.1 Human Reliability Analysis Part

Figure 5 shows the user interface of the human reliability analysis part in the analysis support system.

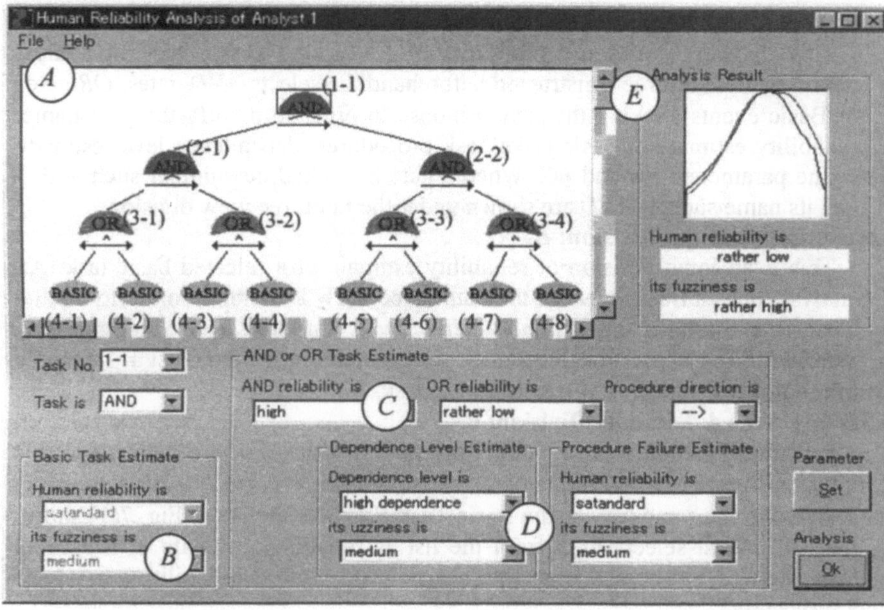

Fig. 5. User interface of human reliability analysis

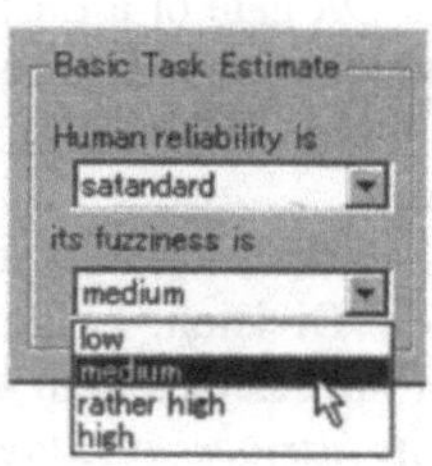

Fig. 6. Basic task estimate division

(1) reliability estimate of a parallel task **(2)** reliability estimate of a series task

Fig. 7. *AND* or *OR* task estimate division

Fault Tree View Division: ***A***

This division shows the fault tree of analyzed tasks. When an analyst modifies the fault tree constructed beforehand, he selects *AND* gates, *OR* gates or Basic events clicking them by a mouse in order to modify the gate name, reliability estimate of basic tasks, task procedures, dependence level estimate or the parameters *nH* and *nG*. When a gate is clicked, its number such as 1-1 and its name such as *AND* are shown under the fault tree view division.

Basic Task Estimate Division: ***B***

This is an input division of reliability estimate of a selected basic task. An analyst clicks a Basic event in the Fault Tree View Division in order to input a basic task estimate, e.g., *Human reliability is* ***standard*** *and its fuzziness is* ***medium***. Fig. 6 shows that an analyst selects ***medium*** from the list as fuzziness of reliability estimate of a selected basic task.

AND **or** ***OR*** **Task Estimate Division:** ***C***

Reliability estimates of a parallel task and a series task are inputted in this division in order to determine parameters values *nH* and *nG*. Each estimate is selected in the same way as the Basic Task Estimate Division. Fig. 7(1) shows that an analyst selects ***high*** from the list as reliability estimate of a parallel task. Fig. 7(2) shows that an analyst selects ***rather low*** as reliability estimate of a series task.

Dependence Level Estimate Division and Procedure Failure Estimate Division: ***D***

Dependence level estimate, task procedure failure estimate and their fuzziness estimate are inputted in this division. Each estimate is selected from the list after an *AND* gate, an *OR* gate or a Basic event is clicked by a mouse. The task procedure is inputted using the procedure direction input division above the Procedure Failure Estimate Division.

Analysis Result View Division: ***E***

This division shows the analysis result of a selected gate or a selected basic event in the Fault Tree View Division. Linguistic expressions of reliability estimate and its fuzziness of the selected item are also shown such as *Human reliability is* ***rather low*** *and its fuzziness is* ***rather high***.

Set **button:**

This button is pressed when an analyst modifies his estimation.

Ok **button:**

This button is pressed when an analyst finishes inputting each item estimate.

The analyzed fault tree is constructed from the top event through basic events by the following procedures using the user interface shown in Fig. 5.

The task number and its name such as *AND*, *OR* or Basic are inputted through the user interface. When the inputted name is *AND* or *OR*, parameters *nH* and *nG*, the dependence level estimate and the task procedure failure estimate are inputted. When the inputted name is Basic, its reliability estimate is inputted. When all inputs are finished, the analyzed fault tree is presented in the Fault Tree View Division.

4.2 Support Part for Compromise

Figure 8 shows the user interface of the support part for compromise.

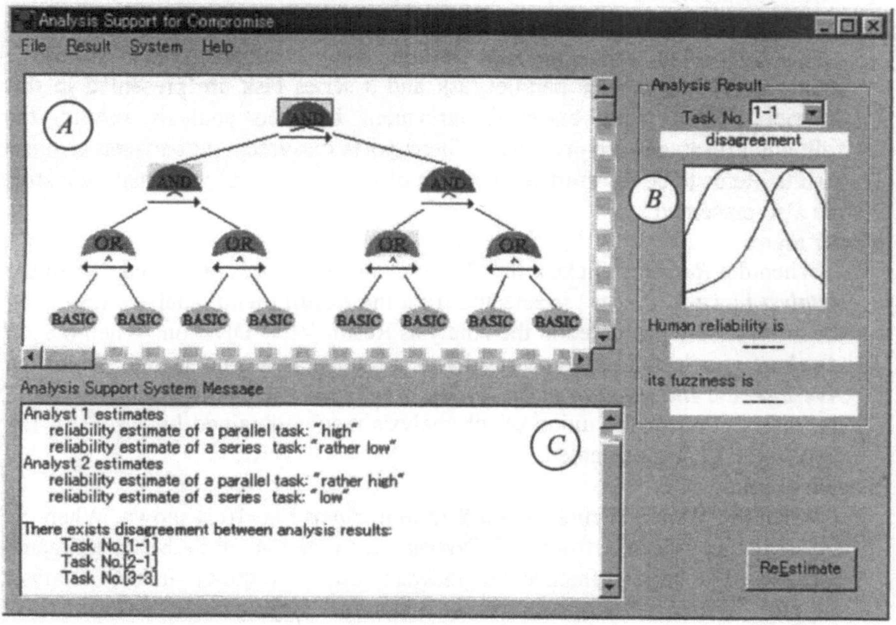

Fig. 8. User interface of analysis support for compromise

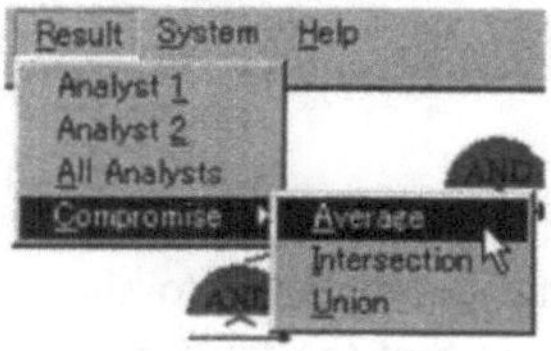

Fig. 9. Result menu

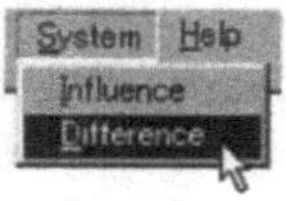

Fig. 10. System menu

Fault Tree View Division: ***A***

This division shows the fault tree of analyzed tasks.

Analysis Result View Division: ***B***

Analysis results according to the selected item in the ***Result*** menu mentioned below are presented.

Analysis Support System Message Division: ***C***

This division presents a message from the analysis support system. Analyst's reliability estimates of a parallel task and a series task are presented in this division. When there exists disagreement between analysis results, the following messages are presented: There exists disagreement between analysis results. Items to be re-estimated because of a large influence on analysis results are also presented.

Result **menu:**

When the ***Result*** is clicked, the ***Result*** menu as Fig. 9 is presented. When the ***Analyst j*** (j = 1, 2,.., n) is selected from the ***Result*** menu, analysis results by the analyst j are presented in the Analysis Result View Division. When the ***All Analysts*** is selected, analysis results of all analysts are presented. When the ***Average***, the ***Intersection*** or the ***Union*** is selected, the average analysis result, the intersection or the union of all analysis results obtained by Eq. (34), Eq. (36) or Eq. (37) is presented.

System **menu:**

When the ***System*** is clicked, the ***System*** menu as Fig. 10 is shown. When the ***Influence*** is selected from the ***System*** menu, the item to be re-estimated because of a large influence on analysis results is shown in the Analysis Support System Message Division. When the ***Difference*** is selected, task numbers which have disagreement between analysis results are shown in the Analysis Support System Message Division.

ReEstimate **button:**

When this button is pressed, an analyst re-estimates the item suggested by the analysis support system using the user interface shown in Fig. 5.

5 Analysis Example

This section presents an analysis example to show the feasibility of the present support system. Let us assume that the fault tree shown in Fig. 11 is constructed beforehand and that this tree is analyzed by two analysts. Tables 4 and 5 show their reliability estimates of each basic task, each procedure failure estimate and dependence level estimate. Let us assume that their fuzziness estimates of all items are ***medium***. Table 6 shows their reliability estimates of a parallel task and a series task. It is found that the analyst 1 is more optimistic than the analyst 2 about a parallel task. However, there is no telling which analyst is more optimistic about other items.

Table 4. Estimate of each item by Analyst 1

		Estimate Item of Analyst 1	
Task No.	Reliability Estimate	Dependence Level Estimate	Procedure Failure Estimate
task (1–1)	–	*low dependence*	*extremely high reliability*
task (2–1)	–	*high dependence*	*extremely high reliability*
task (2–2)	–	*high dependence*	–
task (3–1)	–	*zero dependence*	–
task (3–2)	–	*high dependence*	–
task (3–3)	*standard reliability*	–	–
task (3–4)	*standard reliability*	–	–
task (4–1)	*high reliability*	–	–
task (4–2)	*high reliability*	–	–
task (4–3)	*standard reliability*	–	–
task (4–4)	*standard reliability*	–	–

Table 5. Estimate of each item by Analyst 2

		Estimate Item of Analyst 2	
Task No.	Reliability Estimate	Dependence Level Estimate	Procedure Failure Estimate
task (1–1)	–	*moderate dependence*	*rather high reliability*
task (2–1)	–	*high dependence*	*rather high reliability*
task (2–2)	–	*low dependence*	–
task (3–1)	–	*low dependence*	–
task (3–2)	–	*low dependence*	–
task (3–3)	*high reliability*	–	–
task (3–4)	*high reliability*	–	–
task (4–1)	*high reliability*	–	–
task (4–2)	*high reliability*	–	–
task (4–3)	*high reliability*	–	–
task (4–4)	*high reliability*	–	–

Table 6. Reliability estimate of parallel task and series task

	Analyst 1	Analyst 2
Reliability Estimate of a Parallel Task	*quite high reliability*	*rather high reliability*
Reliability Estimate of a Series Task	*rather low reliability*	*rather low reliability*

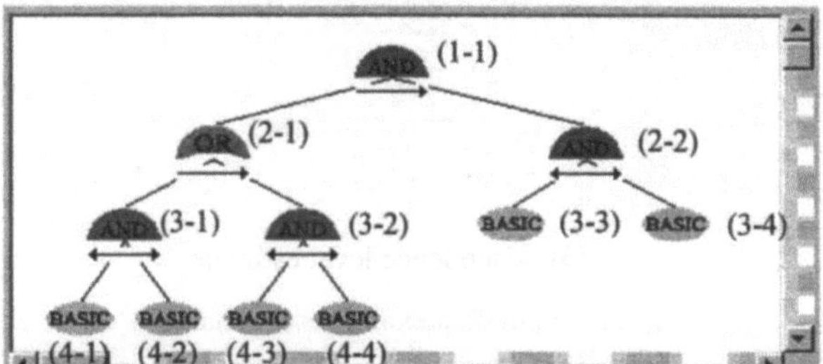

Fig. 11. Fault tree

Figure 12 shows examples of analyst 1's input data about reliability estimate of basic task (4–1), estimate of procedure failure (1–1), dependence level estimate (1–1), and reliability estimates of a parallel task and a series task. Fig. 13 shows two analysts results. It is found that the difference degree between the results is three classes in Table 1. It is necessary for two analysts to discuss human reliability of the analyzed task whether a compromise is obtained or not. Two cases are described in the following: One is the case that the analysts make a compromise with each other. The other is that they do not make a compromise with each other.

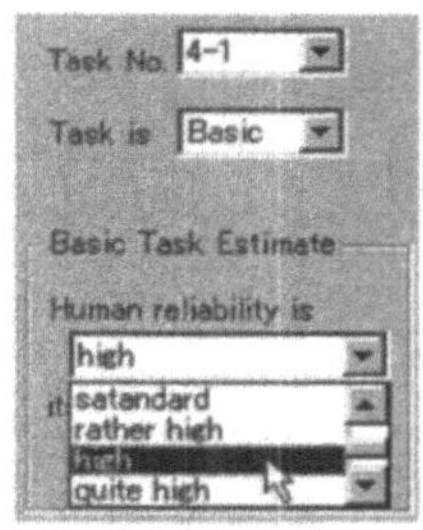

(1) reliability estimate of basic task

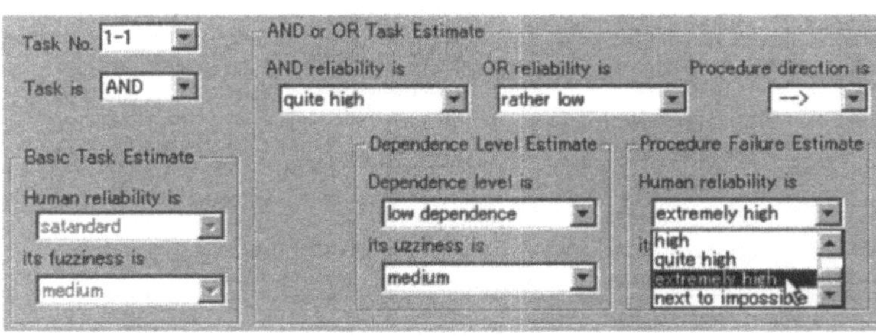

(2) procedure failure estimate

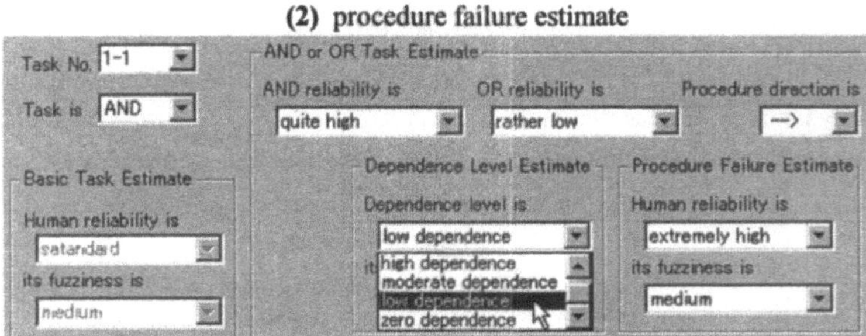

(3) dependence level estimate

Fig. 12. Input data examples of Analyst 1

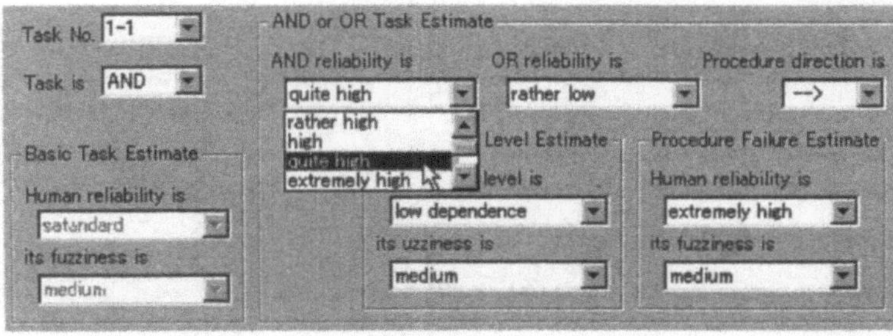

(4) reliability estimate of a parallel task

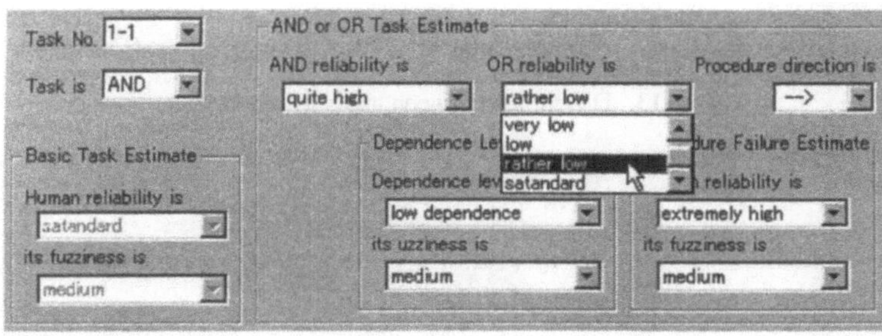

(5) reliability estimate of a series task

Fig. 12. Input data examples of Analyst 1 (continued)

(1) ***The analysts make a compromise with each other***

- **The 1st reestimation**

The analyst 1 selects the ***Difference*** menu. The system shows that there exists disagreement between the analysis results as shown in Fig. 14. Fig. 14 shows that the analyst 1 is in disagreement with the analyst 2 about estimates of task (1-1), (2-1), (2-2), (3-1), (3-2), (3-3), (3-4), (4-3) and (4-4) which are shown by rectangular marks. The analyst 1 selects the ***Influence*** menu. The system shows the item "task (1–1): procedure failure estimate", and suggests the analyst 1 to re-estimate it since it has the largest influence on the analysis result for the analyst 1. The analyst 1 re-estimates this item.

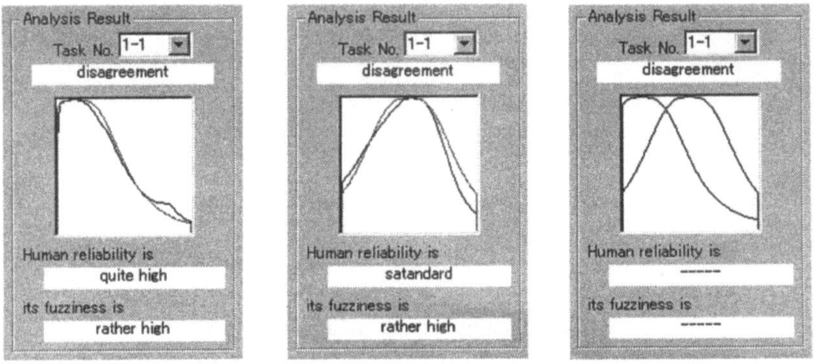

(1) analyst 1 **(2)** analyst 2 **(3)** all analysis results

Fig. 13. Difference between analysis results

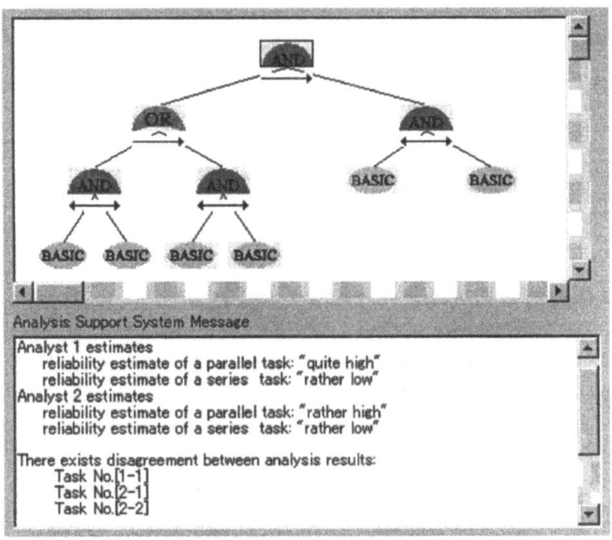

Fig. 14. Disagreement between analysis results

The system also shows the analyst 2 that there exists disagreement between the results and suggests him to re-estimate the item "task (1–1): procedure failure estimate" in the same way.

As the result of the re-estimation, two analysts modify the estimate of "task (1–1): procedure failure estimate" as shown in Table 7.

- **The 2nd re-estimation**

Figure 15 shows the difference between the analysis results after the first re-estimation. The system shows that there is still a disagreement between the analysis results. The system shows the item "task (1–1): dependence level estimate" and suggests to re-estimate this item since this has the second largest influence on the analysis results. And then two analysts re-estimate the suggested item. This process is repeated several times as shown in Table 7.

Table 7. Re-estimate of items

	Analyst 1	Analyst 2
The 1st Re-estimation	Task (1–1): procedure failure *extremely high → quite high*	Task (1–1): procedure failure *Rather high → high*
The 2nd Re-estimation	Task (1–1): dependence level no modification	Task (1–1): dependence level *Medium → low*
The 3rd Re-estimation	Task (2–1) : procedure failure *extremely high → quite high*	Task (2–1): dependence level No modification
The 4th Re-estimation	Task (2–1): dependence level no modification	Task (2–1) : procedure failure *Rather high → high*
The 5th Re-estimation	Task (3–1): dependence level *zero → medium*	Task (2–2): dependence level No modification

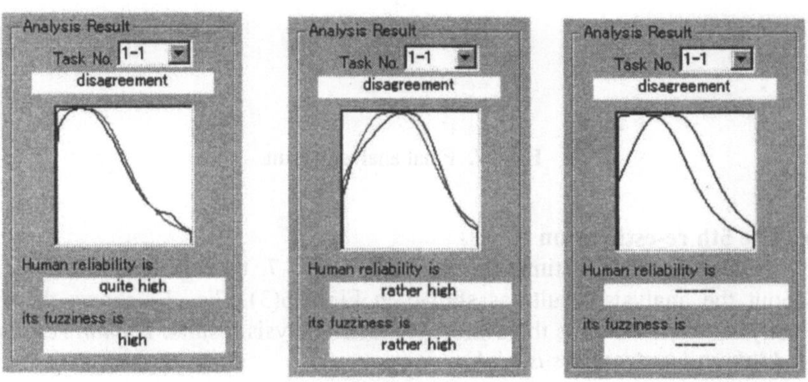

(1) analyst 1 **(2)** analyst 2 **(3)** all analysis results

Fig. 15. Difference between analysis results (after the 1st re-estimation)

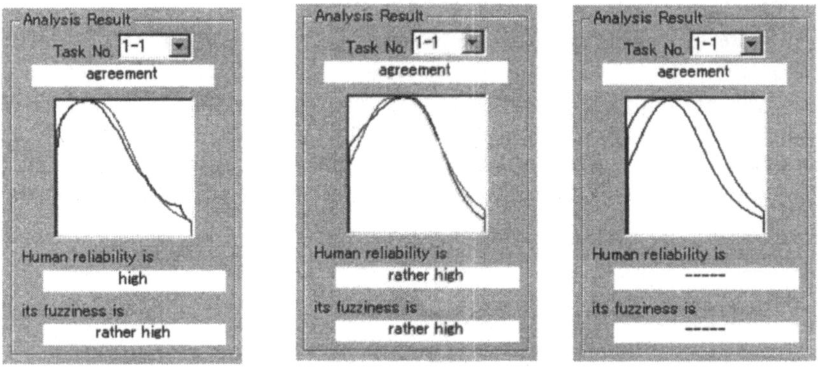

(1) analyst 1 **(2)** analyst 2 **(3)** all analysis results

Fig. 16. Difference between analysis results (after the 5th re-estimation)

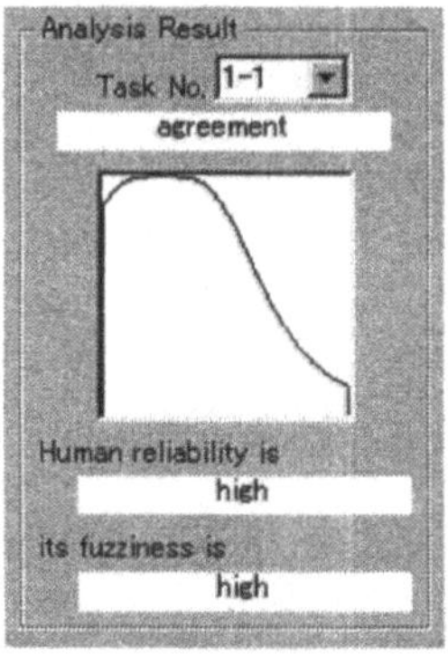

Fig. 17. Final analysis result

- **The 5th re-estimation**

After the 5th re-estimation shown in Table 7, the agreement is obtained about the analysis results as shown in Fig. 16(3). Fig. 17 shows the final analysis result which is the average of their analysis results: *Human reliability is **high** and its fuzziness is **high**.*

(2) *The analysts do not make a compromise with each other*

Let us assume that the analyst 1 modifies the estimation of suggested items as shown in Table 7 and that the analyst 2 does not modify them.

There is still a disagreement between analysis results as shown in Fig.18. The system presents that "The agreement of the analysis result is not obtained" and shows the following results as a compromise between two analysts results as depicted in Fig. 18 and Fig. 19.

(a) an analysis result by the analyst 1
"*Human reliability is* ***high*** *and its fuzziness is* ***rather high***."

(b) an analysis result by the analyst 2
"*Human reliability is* ***standard*** *and its fuzziness is* ***rather high***."

(c) an average analysis result of two analysts results
"*Human reliability is* ***rather high*** *and its fuzziness is* ***medium***."

(d) an intersection analysis result of two analysts results
"*Human reliability is* ***rather high*** *and its fuzziness is* ***medium***."

(e) a union analysis result of two analysts results
"*Human reliability is* ***rather high*** *and its fuzziness is* ***high***."

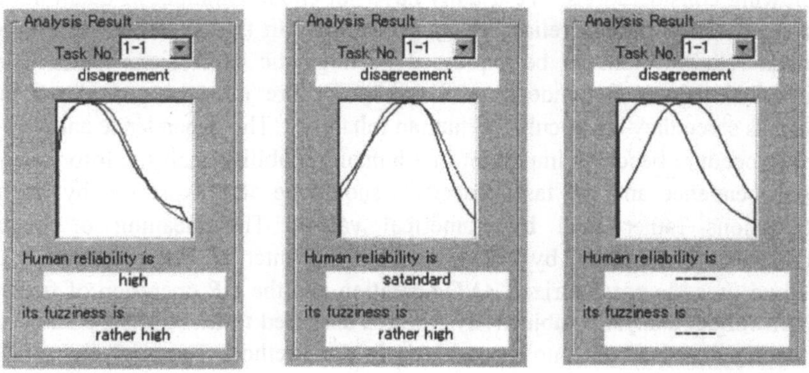

(1) analyst 1 **(2)** analyst 2 **(3)** all analysis results

Fig. 18. Difference between analysis results

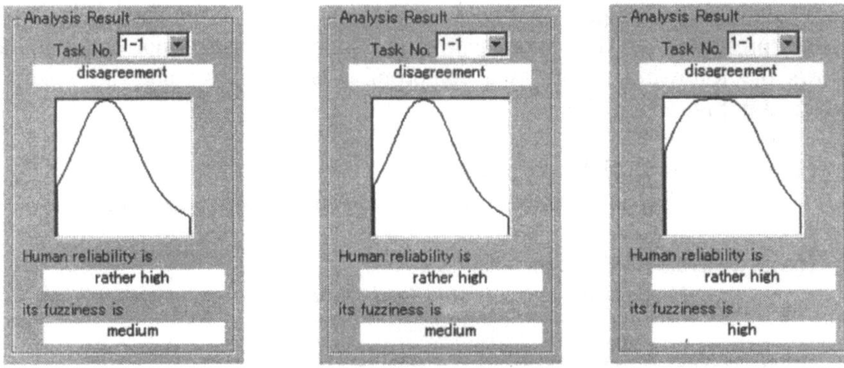

(1) an average analysis result **(2)** an intersection analysis result **(3)** a union analysis result

Fig. 19. Compromise between two analysis results

6 Conclusions

This chapter presents the analysis method of human reliability using linguistic expressions and the analysis support system. The presented method is based on the following considerations. We usually have not enough amount of data on human errors to assess human reliability by probability. In this situation, estimation of human reliability should be expressed by linguistic expressions rather than by numerical values. Dependence and task procedure must be considered in the analysis since they are peculiar to human reliability. The dependence analysis with task procedure becomes important in a human reliability analysis. Information on the dependence and the task procedure should be also expressed by linguistic expressions rather than by numerical values. The meaning of linguistic expressions is expressed by a fuzzy set on a unit interval. The presented analysis method uses the parametrized *AND* operation and the *OR* operation of fuzzy sets which reflect analyst's subjectivity towards analyzed tasks. Analyst's subjectivity is openly incorporated into the analysis in this method. The analysis results are expressed by linguistic expressions.

Many analysis results are obtained even if the same tasks are analyzed since analysis results are dependent on each analyst's subjectivity towards analyzed tasks. Therefore, it is necessary to consider a compromise between analysis results. The procedures for obtaining a compromise are also presented. The analysis support system is implemented on a personal computer based on the present analysis method and the procedures for a compromise.

The analysis example shows that analyst's subjectivity towards analyzed tasks can be reflected by the method and that analysis results are comprehensible for discussion of human reliability since linguistic expressions are employed in human reliability analysis. The analysis support system takes the procedures for a mutual compromise and gives suggestions.

The present analysis support system is useful to the human reliability analysis with linguistic expressions.

References

1. A. Villemeur, Reliability, Availability, Maintainability and Safety Assessment, Volume 1, Methods and Techniques, John Wiley & Sons, Chichester (1992).
2. K.B. Misra, Reliability Analysis and Prediction, A Methodology Oriented Treatment, Elsevier, Amsterdam (1992).
3. P. Kafka and J. Wolf eds., Safety and Reliability Assessment – an Integral Approach –, Elsevier, Amsterdam (1993).
4. A.D. Swain, A Method for Performing a Human Factors Reliability Analysis, Report SCR-685, Sandia Corporation (1965).
5. A.D. Swain and H.E. Guttmann, Handbook of Human Reliability Analysis with Emphasis on Nuclear Power Plant Applications, NUREG/CR-1278 (1983).
6. T. Onisawa, System Reliability from the Viewpoint of Evaluation and Fuzzy Sets Theory Approach, in: T. Onisawa and J. Kacprzyk, Eds., Reliability and Safety Analyses under Fuzziness, pp. 43-60, Physica-Verlag, Heidelberg (1995).
7. A. Villemeur, Reliability, Availability, Maintainability and Safety Assessment, Volume 2, Assessment, Hardware, Software and Human Factors, John Wiley & Sons, Chichester (1992).
8. K.J. Schmucker, Fuzzy Sets, Natural Language Computations, and Risk Analysis, Computer Press, Rockville, MD, (1984).
9. K.S. Park, Fuzzy Apportionment of System Reliability, IEEE Transactions on Reliability, R-36, pp. 129-132 (1987).
10. T. Onisawa and Y. Nishiwaki, Fuzzy Human Reliability Analysis on the Chernobyl Accident, Fuzzy Sets and Systems, Vol. 28, pp.115-127 (1988).
11. D. Singer, A Fuzzy Set Approach to Fault and Reliability Analysis, Fuzzy Sets and Systems, Vol. 34, pp. 145-155 (1990).
12. R. Kenarargui, Even-tree Analysis by Fuzzy Probability, IEEE Transactions on Reliability, R-40, pp. 120-124 (1991).
13. K.Y. Cai, C.Y. Wen and M.L. Zhang, Posbist Reliability Behavior of Typical Systems with Two Types of Failure, Fuzzy Sets and Systems, Vol. 43, pp. 17-32 (1991).
14. T. Washio, H. Takahashi and M. Kitamura, A Method for Supporting Decision Making on Plant Operation Based on Human Reliability Analysis with Fuzzy Integral, Proc. Of 2nd International Conference on Fuzzy Logic and Neutral Networks, Vol. 2, pp. 841-845 (1992).

15. T. Onisawa and K.B. Misra, Use of Fuzzy Sets Theory: Applications, in: K.B. Misra, Ed., New Trends in System Reliability Evaluation, pp. 551-586, Elsevier, Amsterdam (1993).
16. T. Onisawa and J. Kacprzyk, Eds., Reliability and Safety Analyses under Fuzziness, Physica-Verlag, Heidelberg (1995).
17. Special Issue on Fuzzy Methodology in System Failure Engineering, Fuzzy Sets and Systems, Vol. 83, No. 2 (1996).
18. E. Kerre, T. Onisawa, B. Cappelle and I. Gazdik, Chapter 12 Reliability, in: Vol. 4, Fuzzy Sets in Decision Analysis, Operations Research and Statistics, ed. R. Slowinski, in: The Handbooks of Fuzzy Sets Series, eds., D. Dubois and H. Prade, Kluwer Academic Publishers, Massachusetts (1998).
19. T. Onisawa, Subjective Analysis of System Reliability and its Analyzer, Fuzzy Sets and Systems, Vol. 83, pp. 249-269 (1996).
20. D. Dubois and H. Prade, Fuzzy Sets and Systems: Theory and Applications, Academic Press, New York (1980).
21. M. Mizumoto, Pictorial Representations of Fuzzy Connectives, Part I: Cases of t-norms, t-conorms, and Averaging Operators, Fuzzy Sets and Systems, Vol. 31, pp. 217-242 (1989).
22. A. Ohmori and T. Onisawa, Human Reliability Analysis by Linguistic Expressions – With Consideration of Dependency among Tasks and Task Procedure –, Journal of Japan Society for Fuzzy Theory and Systems, Vol. 10, No. 5, pp. 947-956 (1998) (in Japanese).

A Cognitive and Formal Terminology for Descriptive Parameters in Concurrent Real-Time Distributed Software Systems

Atoosa Jalashgar

Department of Automation, Building 326, Room 024
Technical University of Denmark
DK-2800 Lyngby, Denmark
aj@iau.dtu.dk

This chapter is about the variety, difficulty and importance of characteristics of Concurrent Real-Time Distributed (CRTD) software systems, with special focus on failure, reliability and robustness analyses. Concurrent real-time related factors common to distributed as well as centralised systems are described in detail. The focus here is merely on task scheduling and inter-task communication. In that regard, problems encountered as a result of improper choice of design parameters are covered. Next, the characteristics of distributed systems are addressed. Limitations of software analysis methods in accounting for parameters involved in the design and analysis of CRTD systems are discussed. An approach to be used as an alternative framework is suggested.

1 Introduction

The policies chosen in accounting for factors related to requirement and design specifications of Concurrent Real-Time Distributed (henceforth CRTD) software systems play a crucial role in the availability of the constructed system. Knowledge about the consequences of these policies plays an equally crucial role in the failure, reliability and robustness analyses of the system.

In traditional software reliability analysis considerable emphasis is put on developing test methods. When it comes to CRTD systems, any test-based knowledge about the reliability of the software is insufficient knowledge. The main reasons are the combinatorial explosion of the test variants and the difficulties in simulating particular system states, or conditions. Furthermore, the main focus in traditional reliability methods is on failure effects, the number of which is already limited (as being revealed by tests) and the importance of which is often assumed to be mutually equal.

Consequently, in order to be able to identify possible deficiency patterns and thus to assess the true reliability and robustness of CRTD systems, two complementary tasks need to be accomplished. The first is to develop more realistic and adequate system models as input to the reliability estimators. The second is to develop estimators that allow such inputs (Fig. 1). The focus in this chapter is on the first task.

Being able to develop more suitable models of the characteristics of CRTD systems call for descriptive methodologies that need to be not only detailed but also systematic in their approach.
Knowledge about the functional, behavioural and structural complexity of software systems is usually expressed in terms of requirement and design specifications. System designers often tend to forget about the fact that the requirement and design specification models need also to be understandable to other individuals, e.g., the system analysts. Therefore, there is always a high possibility that this knowledge has not been systematically expressed. Conclusively, the analysts will need the descriptive methodologies to develop an instruction, a structured frame that allows detailed requirement and design specifications to be organised and described in an unambiguous and systematic manner. The desirable goal by applying the frame is then to form a new model of the system, ready to be used in the associated failure, reliability and robustness analyses.

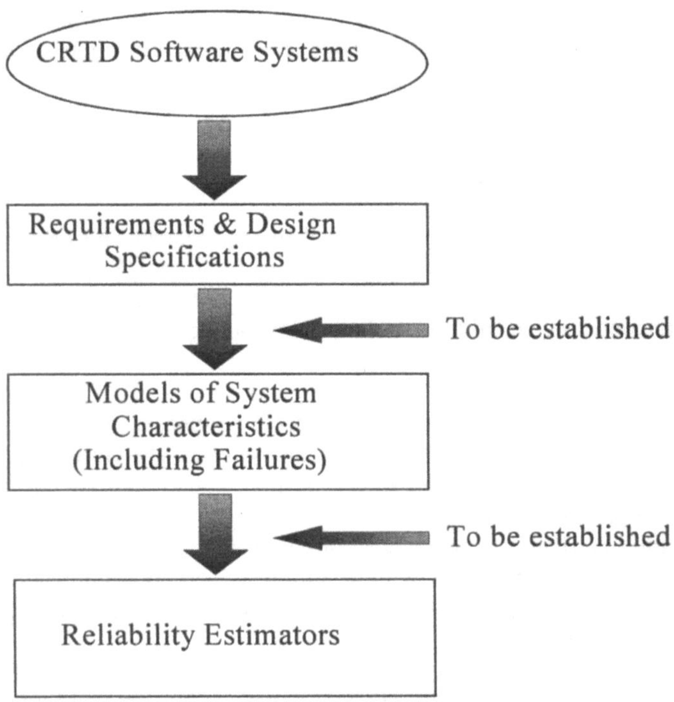

Figure 1. The complementary tasks to maintain the system availability of Concurrent Real-Time Distributed software systems

Developing such a complex but structured frame for CRTD systems is not an easy task, as it requires comprehensive amounts of knowledge about various kinds of architectures, each built upon distinct kinds of requirement and design decisions. The fact that it also requires help from and combination of various kinds of descriptive methodologies makes the job even more difficult.

The chapter is divided into four sections. The first is about concurrent real-time systems. The second copes with the subjects related to distributed software systems. These two sections can be read independently of each other. The last two sections deal with CRTD systems as an integrated software environment. The third section discusses failure and reliability analysis in more detail. The last section furnishes some first steps towards the formation of an analysis frame, based on a combination of cognitive modelling, object-oriented techniques and industrial formal methods.

2 The Characteristics of Concurrent Real-Time Systems

This section is about the descriptive parameters of concurrent real-time systems. To begin with, two major topics involved during the system construction, namely task scheduling and inter-task communication are elaborated. In that respect, some design strategies are explained. Problems related to the choice among these strategies are equally addressed.

2.1 Task Scheduling

In those software environments where most tasks are periodic and independent of each other, and the execution times have well-defined upper bounds, it is possible to schedule the tasks *statically* by assigning each task an explicit execution time. This approach is called *synchronous* execution.[1,2] Based on the number of tasks and their processing time an interval to be triggered by the timer interrupt is defined. The interval contains a number of "slots" each assigned to a task. If all tasks require the same "start-to-run" frequency (compared to each other), each slot is identical to its task's execution time. If different frequencies are required the tasks with lower frequencies will not receive CPU time in all intervals, and the tasks with too large execution time will be split up into subtasks and distributed over several intervals. However, whether the start-to-run frequencies are the same or different, each task is executed at a constant frequency (Fig. 2). Synchronous execution is used in systems with no concurrency.

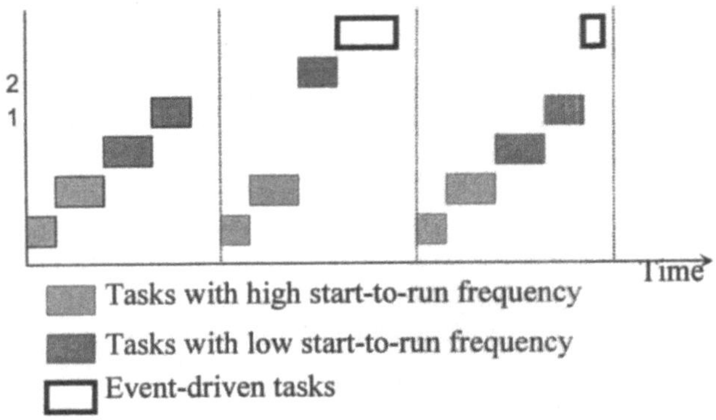

Figure 2. Synchronous task execution

When a task execution depends on several conditions, each combination must be explicitly expressed in the task's code. Such a condition could be the availability of a certain resource or a certain action (performed by another task) the task needs in order to proceed to its next state. In applications with plenty of complicated patterns of event and time dependencies between the tasks, and races for common resources, task scheduling must be carried out dynamically and asynchronously. This is done in terms of parallel and simultaneous task execution, where logically runnable tasks are allowed to be temporarily suspended. The strategy is also called *preemptive scheduling* as opposed to *run-to-completion* methods used in sequential structures. Task execution is based on preemptive scheduling in all concurrent, also known as *multitasking*, software systems.

Several algorithms based on preemptive scheduling have been developed.[1,2,3,4] Because today's concurrent systems usually use two or more of these algorithms, two of them are described in the following. This will at the same time give the reader a more clear picture of the level of complexity and hence the number of failure possibilities in connection with task scheduling.

2.1.1 Round Robin Scheduling

This method makes the assumption that all tasks are equally important. Each task is assigned a time slot in which it is allowed to run. If the task is still running at the end of the slot, the CPU time is given to another task. The CPU will also switch if the task has blocked (due to unavailability of a specific resource, for example) or finished before the slot has elapsed. The implementation of Round Robin is simple and performed by maintaining a list of runnable tasks, starting from the head of the list and putting the current task on the tail of the list, when its time slot runs out (see Fig. 3).

A : Previous task
B : Current task
C : Next task

Figure 3. Round Robin scheduling with the list of runnable tasks

As indicated, the length of the time slot is constant for all tasks. The definition of this length is crucial in Round Robin, as each task switch requires a certain amount of time. A too short slot causes many task switches as well as a decline in CPU efficiency. A too long slot may cause poor response to short CPU time requests.

2.1.2 Priority Scheduling

In true real-time applications there is always one or several tasks with very strict time constraints, due to their close interaction with the external environment. One example is real-time process control systems. Here the efficiency of multitasking must be sacrificed in order to meet deadlines for real-time tasks. Round Robin scheduling is therefore not a very good choice in such systems. The respect for external factors is the main reason for using Priority scheduling, where each task is assigned a priority and the runnable task with the highest priority is granted the CPU time. To prevent high-priority tasks from running too long, the scheduler may decrease the priority of the currently running task at each timer interrupt.

In practical real-time applications, a combination of Round Robin and Priority scheduling is often used.[5] Here, the tasks are grouped into priority classes, so that Priority scheduling is used among classes and Round Robin scheduling is used among tasks in each class (Fig. 4). Furthermore, the priorities are usually adjusted among lower priority classes, either statically or dynamically.

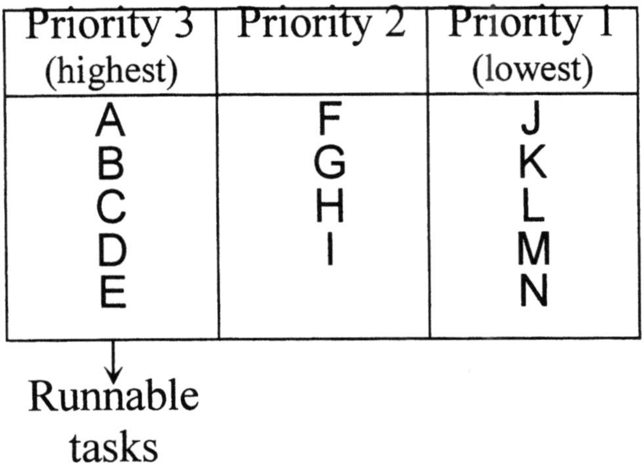

Figure 4. A scheduling algorithm with three priority classes

2.2 Inter-Task Communication

Decisions related to task scheduling view the tasks mostly as independent entities manifesting independent sets of behaviours or actions. Nevertheless, all tasks in multitasking real-time applications depend on each other's behaviours to be able to perform their own. This inter-dependency leads to *communication* among the tasks.

In many multitasking systems, apparently-independent tasks often share resources such as data storage (main memory or a shared file) and I/O devices. It could happen, that two or more try to access a resource at the same time. In such a case, the final state of the resource will depend on precisely when which task runs. This could suddenly become a very serious problem, if a third task with a time constraint is waiting for a certain state of the resource supposed to be established by the output of one of the two resource-competing tasks! Here, we have an example of a true inter-dependency, where parallel progress of tasks *depends on* each other's behaviour, and they therefore need to be mutually synchronised.

Situations where two or more tasks compete accessing a common resource are called *race conditions*.[1,2,4] Detecting race conditions in dynamically configured systems that at the same time are real-time, is itself a very challenging effort. Soon we will see that solutions to race conditions indeed give rise to some other serious problems!

A task in a concurrent environment is mainly busy taking care of internal computations and other matters that do not lead to race conditions. As indicated above, however, the task sometimes needs to access a common resource. The part of a task's program where accessing a shared resource is expressed is often called the *critical section.* If no two tasks are ever in their critical sections at the same time, then race conditions are avoided. The idea is called *mutual exclusion* and has been realised in many ways. The simplest implementation is, of course, that a task disables all interrupts at the start of its critical section and enables them at the end.[4] This approach is an unwise solution, since it gives the control of interrupts, belonging to the system executive, to application tasks. Three feasible methods are discussed in the following.

2.2.1 Check & Set

In this method the memory bus is locked in order to set a flag, after making sure that it was down just before the alteration. The memory bus is unlocked after setting the flag or detecting that some other task has set it. In the latter case, the Check & Set procedure starts all over again. Usually, it is implemented as a single hardware instruction by a task that wants to enter its critical section (Fig. 5).

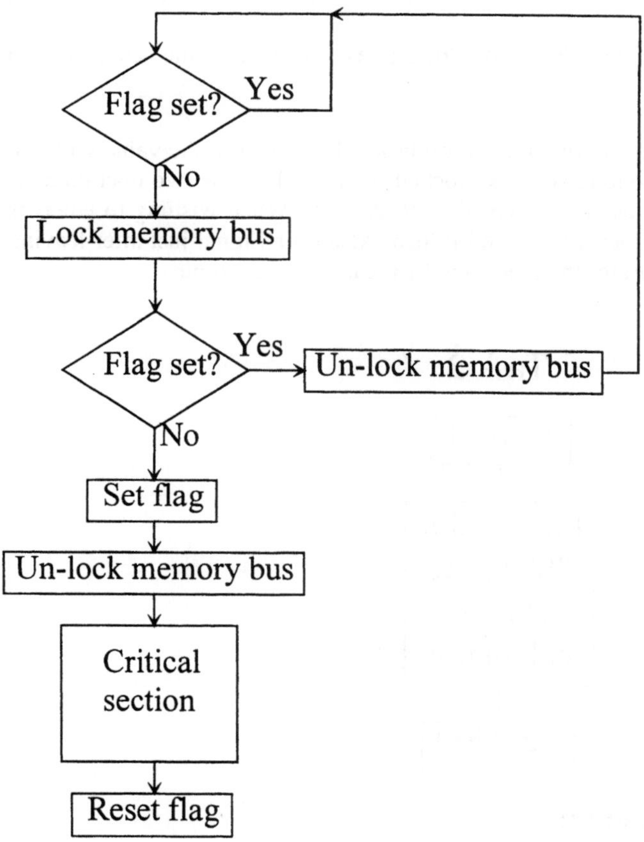

Figure 5. The Check & Set algorithm

2.2.2 Semaphores

The Check & Set approach is simple and effectively realises mutual exclusion, but there is one major problem attached to it: Valuable processor time is wasted due to the busy-waiting nature of the approach. With semaphores, the task blocks instead of busy-waiting.

The idea of using semaphores as a means for mutual exclusion was originally introduced by E. W. Dijkstra in 1965.[6,7,8] Since then, many other algorithms relying on the basic idea have been developed.

A semaphore itself is just an integer variable. What makes it special is the modification and interpretation of its value. Usually, two operations are associated with a semaphore:

- *Wait* checks the value, decrements it if greater than 0 or forces the executing task to block if not.
- *Signal* increments the value, and allows a task blocked on the semaphore to finish its Wait.

After a Signal on a semaphore with several tasks blocked on it, the value will still be 0, but there will be one fewer task blocked on it. Each of the two operations is normally implemented as an "indivisible" system call. When wishing to enter its critical section, a task does a Wait, and it then executes a Signal just after leaving the section. Fig. 6 illustrates the preceding by means of an example.

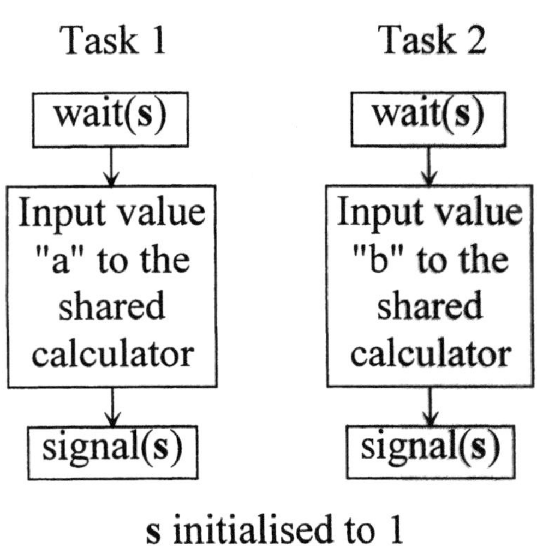

Figure 6. Mutual exclusion by using semaphores

2.2.3 Message Passing

Semaphores provide an efficient way to prevent race conditions among concurrent tasks sharing a common resource. Semaphores can also be used to synchronise tasks that are dependent on the result of each other's behaviours. Of course, semaphores themselves need to be protected from mutual exclusion somehow! This is not a problem in centralised systems with a common memory, since they can protect semaphores by putting them in this memory and protecting them by the Check & Set algorithm. In distributed multitasking systems, however, there is no common memory present, which makes our solution to race conditions inapplicable. Additionally, semaphores cannot be used for exchange of information among the tasks.

The kind of inter-task communication, where tasks can get protected access to common resources, be mutually synchronised, and exchange information can be established based on the concept of *message passing*.[1,2,4] Here, two operations Send and Receive are used as the communication means, usually with three parameters: the identity of the sender or receiver, a message and a semaphore. As far as the mutual exclusion and synchronisation are concerned, Send and Receive work exactly as Signal and Wait, respectively.

The most basic discrepancy between message passing methods is whether the communication is synchronous or asynchronous. In the former, the pair of operations is executed at the same time instant. This means that both tasks must be ready to communicate for the operations to complete, otherwise the "faster" task must block until its partner is ready to receive or send the message. In the latter, the sender does not wait for the receiver and just delivers the message to a buffer and continues. The receiver, however, must wait for the arrival of the message.

For each of the two major groups of message passing many possible implementations exist, the further exploration of which is, however, prevented by the scope and space limitations of this chapter.

2.3 Problems to Cope with

The area of concurrent real-time software systems is very rich with regard to the number of methods and techniques available. The methods described in this chapter are basic and simple. Over the years, a great number of improved versions, combinations and new approaches have been invented. Many have attracted supporters who claim their favourite approach is the best. Due to that and due to the needs and demands allied with various kinds of systems, people use different kinds of design algorithms, some of which may not seem to be the best choices. As a system analyst, however, one must be aware of these possibilities in order to obtain a good understanding of system behaviours and their immediate or future impacts. The fact that many details about the behaviours of a system will never be found, explicitly or implicitly, in the corresponding requirement and design specification reports makes a solid systems understanding even more crucial, when the system, designed and implemented already, becomes subject to failure and reliability analyses.

As stated previously, several algorithms based on preemptive scheduling have been developed, and the scheduler of most multitasking systems uses different algorithms for different groups of tasks. How a group of tasks is scheduled influences itself the choice of a procedure to avoid race conditions. For example, choosing Check & Set method may be a simple and sufficient approach in many systems but has a serious drawback for priority-scheduled tasks in others. Here, a high priority task 1 can risk looping in its Check & Set routine forever, if it is a lower priority task 2 that is in its critical section.

In this case, using semaphore operations may seem to be a much better choice, as task 1 simply blocks instead of busy-waiting, so that task 2 can get some CPU time to reset the flag and come out of its critical section. Unfortunately, however, it is not so difficult to make tiny mistakes with serious impacts while using semaphores. An example is shown in Fig. 7 in pseudo C.

Put is the function of a crane at the head of a long conveyer belt, and Remove is the function of another crane at the end of the belt. Turn is used for mutual exclusion in the case where the belt is full and Put is executed. Not_Full and Not_Empty are used to indicate when the belt is completely full (Put blocks) and when it is completely empty (Remove blocks), respectively. Now, consider what happens if the first two Wait calls in Put are reversed in order, and the belt eventually becomes full: Put blocks after resetting Turn. Next time Remove is executed, it finds Turn reset already, so it blocks too. Both functions are now blocked, forever!

The bug addressed above leads us to the problem of *deadlock*, still subject to research and one of the main topics in numerous literatures.[9] What makes deadlock interesting is specially, that it occurs as a consequence of granting exclusive access to shared resources, to avoid race conditions! In fact, such a discriminating nature of mutual exclusion approaches gives rise to another serious problem, namely *starvation*. This is the situation where tasks continue to run indefinitely but without any progress. Detecting deadlocks and starvations in complex concurrent systems is not easy at all, and neither is finding remedies for them. Nevertheless, both have a great influence on the reliability estimations of the systems.

```
semaphore Turn = 1;
semaphore  Not_Full = N;
semaphore Not_Empty = 0;

Put ()
{ int item;
  while (TRUE)
   { Wait (Not_Full);
     Wait (Turn);
      Put_Item (item);
      Signal (Turn);
      Signal (Not_Empty);
   }
}


Remove ()
{ int item;
  while (TRUE)
   { Wait (Not_Empty);
     Wait (Turn);
      Remave_Item(item);
      Signal (Turn);
      Signal (Not_Full);
   }
}
```

Figure 7. A bug in mutual exclusion using semaphore operations

3 The Characteristics of Distributed Systems

Design of software in distributed systems involves accounting for many significant factors such as functional and operational flexibility, redundancy and transparency in software and hardware, and system configuration and inter-task communication. Alternative solutions for each of these factors are, among other things, influenced by the increasing degree of system optimisation and automation, and often result in considerable amounts of system variety and complexity.

A universal (and correct) definition of a distributed system is several, relatively autonomous units collaborating and interacting with one another through a network in order to attain an overall objective defined by the designer or user of the system.

The characteristics of distributed systems are multiple remote units, interconnections among the units and the overall shared state they together are supposed to establish and maintain.[12] Within the automation industry, distributed systems allow the use of small and cost-effective computers, close connections to the plant and therefore a more efficient automation job. They allow different kinds of components, software tools and versions, and management policies to be adopted. They can grow over a large range of sizes. Last but not least, their units usually do not all crash at once! Nevertheless, the characteristics of such systems also necessitate addressing issues like possible dependencies between failures occurring in each unit, unreliable communication between the units due to concurrency and the existence of a network, and increased system administration.

Compared to centralised systems, all features of distributed systems introduce, in one way or another, new sources of deficiencies and failure patterns. Failure analysis of these systems is therefore much more complicated than that of centralised systems.

In the following, typical requirement specification parameters associated with distributed software systems are addressed. It is then explained how these can influence the design parameters. The main stress in this respect is put on the central executive of the system, the *manager*.

3.1 Issues Involving Requirements

Associated with the construction of distributed systems are requirements involving major issues such as flexibility, system configuration, inter-task communication, naming and transparency.[13]

Basically, three kinds of flexibility can be addressed:

- **Functional** flexibility, meaning that it is possible to assign new roles or functions to the system components without changing their operation. An example could be a new and different response to the interrupt from an alarm device.
- **Operational** flexibility, meaning that it is possible to define new ways of operation for the system components without modifying their functions. An example here could be to replace the low-pass filter in an electronic establishment with a more advanced one.
- **Topological** flexibility, meaning that it is possible to change the physical as well as the logical structure of the system. This could be to add new

computers or to change the communication patterns between the software units.

Functional, operational and topological flexibility can be time-independent and dynamic, if it can be implemented at any time and without system shut down.

Generally, distributed automation systems operate in real-time to guarantee the attainment of time-critical functions. This means that minimising response time is much more important than maximising throughput. Guarantee criteria with respect to time constraints have a great influence on making decisions about task scheduling mechanisms that are followed during system configuration.

Distributed systems usually perform many tasks the progress of which can be either concurrent or inter-dependent. Creating the proper environment for smooth progress of parallel as well as inter-dependent tasks calls for an evaluation of various resource management and synchronisation strategies, in order to account for race conditions, deadlock and starvation. The need for exchange of data in addition to synchronisation signals, however, advocates only for message passing as the communication mechanism. Although this along with the distributed resources of distributed systems excludes considering methods such as semaphore operations and monitors, it creates some other issues to consider. Associated with communication by message passing are the choices between synchronous and asynchronous communication, between various data formats, between channel connection patterns (one-to-one, one-to-many, etc.), and finally between addressing forms (direct, mailbox, etc.). Nevertheless, even when all choices are carefully made and implemented, there will still be behaviours that are impossible to anticipate, due to the existence of an unreliable network, for example.

Another important topic is naming. As stated previously, most distributed systems have complex physical as well as logical structure. Furthermore, their structure changes due to a flexible functional, operational and topological architecture, and due to the use of redundant logical and physical components. These factors make the choice of a proper, flexible and robust naming strategy essential and difficult.

When all decisions about the system flexibility, system configuration, inter-task communication and naming are made, the designer(s) then must figure out what kinds of information should be visible and in which extent, and how they should be exchanged internally between the system components and externally between the system and its users. This leads to the issue of transparency, which opens for many interesting possibilities, among them incipient sources of failures! Fig. 8 sums up the above-mentioned issue.

3.2 Issues Involving Design

Once the requirement specifications with respect to the main issues addressed above are formulated, it is time to design the software system accordingly. During this process, it is important to bear in mind not to end up designing the wrong system! In other words, all the definitions, choices, specifications and resulting models should be *valid* with regard to the requirements.

The first step is to identify the units and then to define their respective purpose and functions. One good way to start this is to group the software structure in three main categories: Input/Output units, Application units and Executive units (Fig. 9).

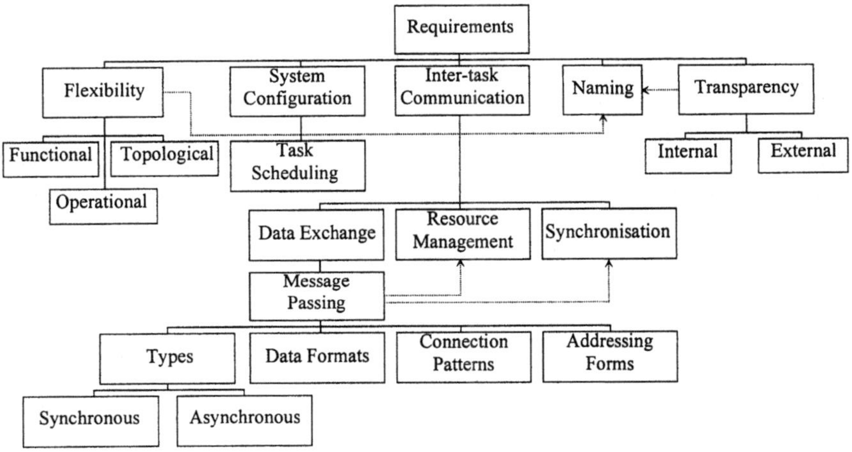

Figure 8. Issues involved in the requirement specifications of distributed systems

I/O units are assigned plant-dedicated I/O functions such as collecting sensor-data and actuating. All other plant-dedicated functions, like assessment and manipulation of plant-data, are performed by the Application units. Functions the realisation of which is necessary for the proper behaviour and communication of the I/O and Application units are implemented by the Executive units. These functions are plant/application independent. The manager of a distributed system consists only of Executive units. The I/O and Application units communicate with each other through the manager.

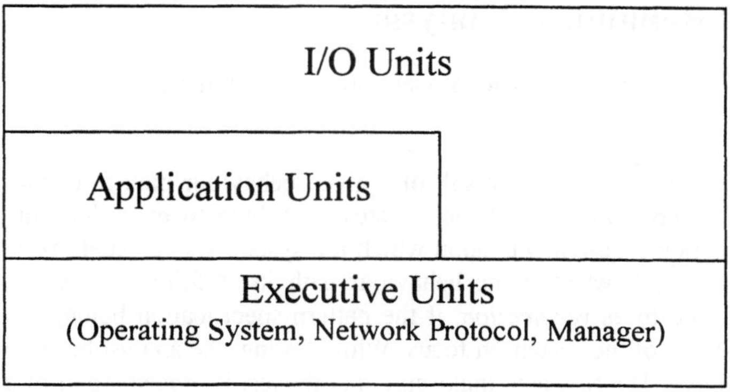

Figure 9. Input/Output, Application and Executive units in the software structure of distributed systems

While the static part of I/O and Application units (i.e., their *programme*) depends mainly on what purpose, functions and capabilities (not necessarily valid!) they are assigned to, their dynamics depends almost entirely on the decisions made by the manager. Once the distributed system is activated, the control of the internal behaviours and communications, i.e., making programmes into *tasks* (or *processes*), is handed over to the manager, making every statement about the global states of the system enormously difficult.[14]

The core job of the manager is to establish the proper means for the I/O and Application units, so that their parallel or inter-dependent progress altogether complies with the design specifications and the progress results contributes to the attainment of the requirement specifications. The manager establishes such means in terms of services it offers to the units. Consider the operational flexibility being one of the requirements in the distributed automation system of a nuclear power plant. The manager should comply with its actual design specification by, for example, supporting a replacement of a PID controller with an adaptive controller without affecting the function of the controller. Another requirement, e.g., involving transparency, could result in services giving an Application unit the impression of being a part of a centralised system.

In practical implementations of true distributed systems, each machine is provided a manager. The only difference between these managers lies in the interface facilities they use to establish a contact with one another. Such a facility could be a list over all already-introduced managers, a list over all local I/O and Application units, etc.

4 Failure and Reliability Analysis

As indicated in Introduction, traditional software analysis techniques rely mostly on test methods.

Analysts execute tests to compare the result of a given behaviour path to certain end states. This is called *validation*, if these states contribute to or include the overall goal and function requirements upon which the system was created. Tests are also executed to check whether the behaviour path itself follows a certain pattern. This is referred to as *verification*, if the pattern specification belongs to the design specifications of the system in focus. While testing, the analyst has two major problems to solve. The first is to make sure that the test is indeed a part of a validation or verification task. In other words, the analyst must gain deep knowledge about the system requirement and design specifications. The next is to make sure that the test is feasible to carry out, meaning that ample knowledge is obtained about the actual selected states (in terms of data values) that define or influence the system behaviour. This knowledge is indeed vital to a successful validation and verification of the behaviour.

In CRTD systems with many concurrent tasks, each state of the system is a result of not only the independent behaviour of a single task, but also of inter-task dependencies that are reflected in the communication among the tasks. Hence, each state of the system is practically defined by all states of all tasks that constitute the system itself. Having the concurrent and often real-time nature of these tasks in mind, the simulation of selected system states during the test will be extremely difficult, if not impossible.

In addition to the profound reliance on test methods, another limitation of traditional software reliability techniques is their implicit tendency of only considering failure *effects* and not failure *sources* in the corresponding reliability estimators. One evidence is that reliability and availability measures such as mean-time-between-failures and probability of failure over a given interval, are deduced from estimates of elapsed time between occurrences of faulty behaviour and not of failures that caused those behaviours. The failure effects, already a limited group as being identified by tests, are in most cases also considered to have identical weights.[10]

Such characterisations of software systems may be sufficient when dealing with systems of relatively simple architecture. CRTD systems inherit the features of two complex software structures. Revealing a group of their behaviours and assigning them the same importance is not adequate for claiming to take the requirement and design parameters in CRTD systems into account.

In fact, one main reason for choosing a distributed solution over a centralised one is to increase fault tolerance of the system-plant combination by means of proper job assignments and redundancy. However, this will be impossible to obtain without an effective and systematic anticipation of possible failure modes, particularly those implanted as a consequence of using a distributed software architecture! In fact, the ability of pinpointing units, operations and functions to be replicated or placed in certain locations will be feasible only through an analysis of failure modes and patterns.

To highlight the role of original sources of failure, consider a message loss as a faulty behaviour in a network-based environment. This fault can be attributed to the sender, receiver or communication channel, depending on from where the actual failure causing that behaviour originates. If the loss is due to a noise-corrupted signal, the channel is to be inspected. If the loss is due to buffer overflow, the receiver or the sender must be checked. Nevertheless, the origin of the loss has an impact on the reliability estimate of the system, because mean-time-between-failures is different for the communication channel, receiver and sender.

The ability of actually distinguishing between various kinds of failures is also of great pertinence to discuss. For instance, consider two processors in a CRTD system where one has crashed and the other is performing slowly. Other processors in the system can safely perform actions on behalf of the crashed processor, because it cannot perform at all. However, the others cannot take over the jobs of the slow processor, because its actions might not be consistent with those performed on its behalf by others. Nevertheless, unless the communication form is synchronous or the crash is of a detectable character, revealing the difference between the two processors is almost impossible.

5 An Analysis Frame: Some First Steps

In order to maintain the functional and operational availability of CRTD systems in accordance with the requirement and design specifications, and in order to develop more detailed and systematic models of failures in such environments, the approach suggested here is to form an analysis frame by developing and using *cognitive, object-oriented* and *formal* models of requirement and design specifications as the first step.

The importance of a valid design model with respect to the requirements is needless to stress. Nevertheless, the statement seems to sound obvious because we are dealing with homogeneous establishments (hence the term *system*), with strong interconnections and interdependencies. Some of these *bounds* explain the relationship between the objectives and functions of the system units. Others explain the relationship between a particular function and the operation modes of the unit. There are also patterns describing the relationship between these modes and the communication with other units inside the system. By cognitive models we simply mean models the use of which force us to take all these bounds between different characteristics of the system into consideration.

Regardless of the specification phase and its level of detail, the resulting models will not be good enough, however, if they just display knowledge about system properties and their mutual interactions. Especially when it comes to complex structures, it is equally crucial how the acquired knowledge is displayed.

The object-oriented school of thought is built upon three concepts of information organisation: 1) objects and attributes, 2) wholes and parts, and 3) classes and members. The first deals with the object and its properties. The second views the object as a homogeneous compound of objects. The third describes different classes of objects, and the membership relation between an object and a certain class.

Clearly, cognitive and object-oriented reasoning are closely related to one another. In fact, it can be said that the former is stronger in describing the relationships between the attributes of an object, whereas the latter is more powerful in describing the relationships between the objects (in the same or different systems, and in the same or different classes). Furthermore, due to their focus on *organisation*, object-oriented methods represent knowledge in a way that is more understandable and therefore more useful, not only for a computerised but also for a human user.

The above suggests that a model created by exploiting the features of cognitive and object-oriented methodologies is a good model. However, it is still not good enough! To be able to carry out the validation and verification tasks, the associated models need also to represent knowledge in an *unambiguous* way.

Dealing with the complexity issue in an unambiguous manner, especially with regard to the dynamic characteristics of systems, is the prime idea behind formal methods. In recent years, considerable attention within telecommunications and embedded systems has been paid to certain formal specification, validation and verification techniques with a high practical application. Some of these techniques have, more or less, adopted object-oriented principles in their resulting models.[11] They are believed to have a complementary role in the trio of cognitive, object-oriented and formal modelling of requirement and design specifications of CRTD systems. Referring to former statements in this chapter, the trio is therefore also believed to be good enough as an *input* to software reliability estimators.

6 Conclusion

This chapter has addressed important issues involved in coping with complex software structures in general, and with the software of Concurrent Real-Time

Distributed (CRTD) systems in particular. Additionally, the focus has been put on failure and reliability analysis. In that regard, the properties of traditional software analysis techniques have been discussed. It is argued that in order to maintain the availability of CRTD structures in accordance with the requirement and design specifications, two complementary tasks are needed: to develop more realistic and adequate system models as input to the reliability estimators, and to develop estimators that allow such inputs. The pertinence of the first task, being the topic of this chapter, is demonstrated through an elaboration of essential issues involved during requirement and design specifications of CRTD systems. The relevance of the next task is implied by addressing important limitations of traditional software analysis.

As far as the first task is concerned, the chapter has also featured some suggestions about where to start, by recommending to take advantage of cognitive modelling, object-oriented principles and industrial formal techniques.

The approach suggested is an attempt to represent the true properties of CRTD systems in a more explicit and structural way, so that they can be both valid and verifiable.

Acknowledgment

The author wishes to acknowledge Senior Software Designer Harald Thunem, Nokia Telecommunications A/S, Denmark, for useful comments on this chapter.

References

1. Peterson J., Silberschatz and A. and Galvin P., *Operating System Concepts, 4th Edition*, Addison-Wesley, 1993.
2. Deitel H., *Operating Systems, 2nd Edition*, Addison-Wesley, 1990.
3. Bach M. J., *The Design of the Unix Operating System*, Prentice Hall, 1987.
4. Tanenbaum A. S., *Operating Systems: Design and implementation*, Prentice Hall, 1987.
5. Dayan P. S., *The OS-9 Guru, 1st Edition*, Galactic Industrial Limited, 1992.
6. Ben-Ari M., *Principles of Concurrent Programs*, Prentice Hall, 1982.
7. Hansen B., *The Architecture of Concurrent Programs*, Prentice Hall, 1977.
8. Hansen B., *Operating System Principles*, Prentice Hall, 1973.
9. Martin J. M. R., *The Design and Construction of Deadlock-Free Concurrent Systems*, PhD Thesis, University of Buckingham, UK, 1996.
10. Jalashgar A. & Stølen K., "Failure Analysis of Real-Time Systems", *Safety and Reliability*, Volume 1, Lydersen S., Hansen G. K., Sandtorv H. A., 601-608, A.A.Balkema/Rotterdam/ Brookfield, 1998.
11. Bræk R. & Haugen Ø., *Engineering Real-Time Systems: An Object-Oriented Methodology Using SDL*, Prentice Hall, 1993.

12. Sloman, M. & Kramer, J. 1987. *Distributed Systems and Computer Networks*. Prentice Hall. ISBN 0-13-215849-3.
13. Jalashgar, A. *et al.* 1991. *Design af manager til distribuerede proceskontrolsystemer*. Final report, Institute of Electronic Systems, Aalborg University.
14. Mullender, S. 1993. *Distributed Systems*. Second Edition. Addison-Wesley. ISBN 0-201-62427-3.

Safety Licensable and High Speed Programmable Digital Controllers Providing any Required Control Behaviour

Peter Vogrin and Wolfgang A. Halang

Faculty of Electrical Engineering, FernUniversität
D-58084 Hagen, Germany
E-mail: wolfgang.halang@fernuni-hagen.de

Abstract. The design principle of control algorithms working with set-point pre-processors, called SPP controllers, is described. Set-point pre-processors calculate internal set-point graphs of controlled variables in such a way that very high controller gains are attainable and, thus, stability is increased. The behaviour of SPP controllers is normally much closer to the "best physically possible" controller performance, and much more predictable than the one of conventional control structures. Furthermore, otherwise conflicting design objectives can nearly all be achieved, e.g., stability, safety, high speed, small energy consumption, or steady and harmonic temporal controller output values. Since SPP control algorithms allow for extremely fast reactions, their implementation in form of sequential software could be counter-productive by ruining the speed advantages. This problem was solved by approximating the algorithms by rule base tables allowing to implement them on safety licensable inference controllers, which are characterised by fuzzification with analogue-to-digital converters and inference by look-up in tables containing fuzzy rule sets. An inference controller consists of a few elements, only. Thus, it is reliable, safe, verifiable, cheap and small. Owing to the simplicity of both its hardware and software, safety licensing of the controller is facilitated. With regard to software, this can easily be carried out by inspecting the table content. The controller is very fast, with its speed mainly determined by the EPROM access time, and works almost jitter-free. Operating in a strictly cyclic fashion, the controller exhibits fully predictable real time behaviour. Its hardware operation is supervised by a fail-safe logic immediately initiating an emergency shut-down in case of a malfunction.

1 Introduction

With regard to the following two main reasons, the performance of state-of-the-art controllers employed in safety critical systems is often very unsatisfying:

1. Owing to the (limited) possibilities for safety licensing and the guidelines of the licensing authorities, the design principles of control equipment have to be quite elementary. In particular, according to the "*List of Type Approved Programmable Electronic Systems*" [1] published by TÜV Rheinland, *it is prohibited to use PID or other more complex control algorithms in safety related applications.*

2. Another detriment to present controllers is their inherent lack of two aspects of speed:
   - Owing to their mathematical models, control systems react slowly after modifications of their set-points, and after unforeseen events, such as disturbances, defects and alterations of parameters of the technical processes being controlled.
   - Loop execution time.

These restrictions to the design of safety related controllers often lead to great disadvantages: As supervisors of safety critical functions have to be designed without using complex control algorithms, they may be constructed out of low complexity and well proven industrial standard equipment, only. For instance, dead-zone controllers and self-acting controllers working with mechanical or electromechanical components such as springs or cable pulls with bobs making use of the bobs' gravitational attraction. In these cases, plant safety is normally ensured by some laws of nature. The use of such simple equipment is to enhance the availability of safety related systems. On the other hand, particularly mechanical components are subject to aging processes. Due to this property, the service intervals of the safety related equipment must be fairly short. Furthermore, state-of-the-art safety systems usually perform very simple actions, such as emergency shut-down, in response to controller failures or malfunctions. Moreover, the speed of technical processes is slowed down by the inherent slowness of controllers. These detriments may lead to reductions of the inherent safety of controlled systems. A considerable improvement of controller performance is only attainable if at least one of the following properties of controller design is enhanced:

1. The major need for the design of enhanced controllers for safety critical systems is the use of complex and dynamic algorithms, which is not approved by the licensing authorities, yet. A safety licensable controller needs to be developed being able to execute complex control algorithms. Such a controller must have the following two ostensibly contradictory properties: on one hand, it must be able to perform complex operations, and on the other one, its software and hardware must be so simple that safety licensing becomes possible.

2. The loop execution times have to be kept short.

3. Safety and economic reasons often make it desirable to design mathematical control models which allow for more determinable and faster reactions.

The organisation of this chapter is as follows. First the basic structure of a control system is described. In order to improve the three controller properties mentioned above, we then introduce the design principle of controllers with set-point pre-processors which are advanced mathematical control algorithms and, furthermore, we shall develop the design principle of an inference controller. The advantages of these two controller types as compared to conventional controllers are shown by a case study and finally some conclusions are drawn.

2 Basic Structure of a Control System

Fuzzy controllers and most conventional controllers - for instance, PID or multiloop control systems - have the same fundamental structure as displayed in Figure 1. In the general case, a multiloop control system has k controlled variables and β actuations. A special case of this is a single loop feedback system. Apart from the fact that the sensors measure the states of the technical process instead of controlled variables, state control systems have the same structure. Moreover, it is possible to design complex automation systems with several such structures.

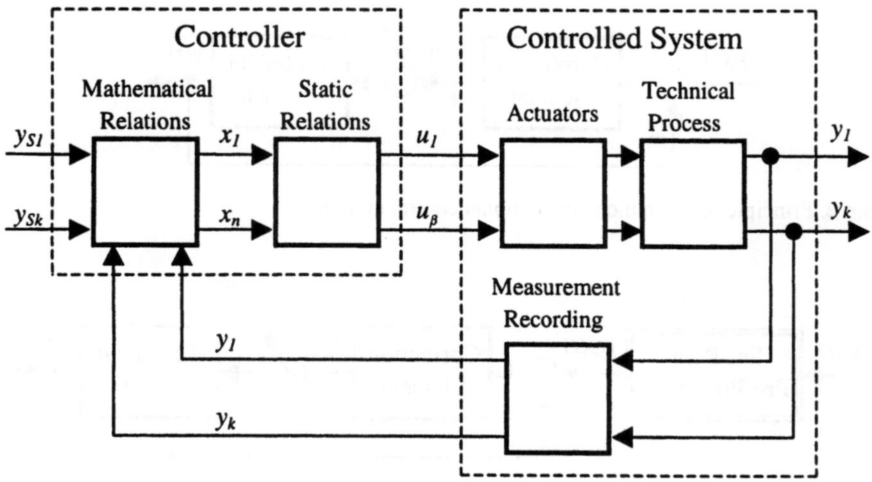

Fig. 1. Basic structure of a typical control system

The mathematical relations comprise at least one comparator. The comparator of a conventional controller usually calculates the control errors $e_j(t) = y_{S\,j}(t) - y_j(t)$ as deviations. The comparator of a controller working with a set-point pre-processor normally calculates the differences between the internal set-point graphs and the controlled values. In this case, the internal set-point graphs are calculated by a set-point pre-processor. Furthermore, dynamic controllers require differentiator and integrator elements.

3 Controllers with Set-Point Pre-Processors

The design principle of controllers working with set-point pre-processors called SPP controllers are described in this chapter. Its set-point pre-processor calculates the internal set-point graph in such a way that the maximum difference between the set-point graph and the real position is much smaller than the maximum control error of a conventional controller. Furthermore, the temporal graph of this difference is steady and harmonic. Thus, higher controller gain is possible and, therefore, feedback control systems become more stable. Another great advantage of SPP controllers is the continuous temporal behaviour of their actuations. Moreover, compared with conventional controllers such as PD or PID, the influence of disturbances on controller performance is very small.

The only differences between the design principle of a conventional controller of a single loop feedback system as shown in Figure 2 and the one of a SPP controller as depicted in Figure 3 are that a SPP controller contains a set-point pre-processor and that its correction element gains are higher.

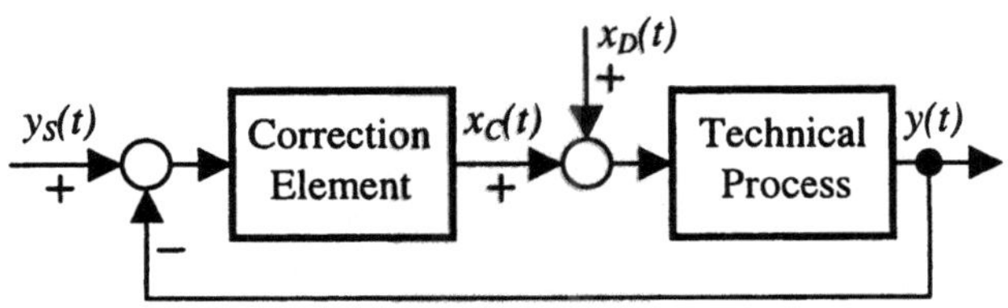

Fig. 2. Principle of a common conventional control system

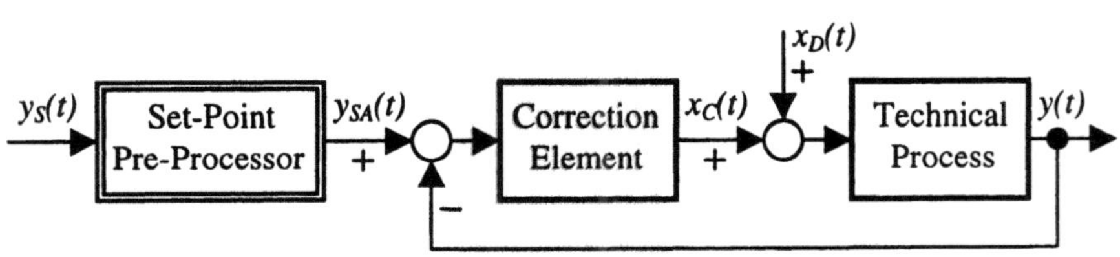

Fig. 3. Principle of a controller working with a set-point pre-processor (SPP controller)

The set-point pre-processor input $y_S(t)$ is the desired temporal graph of the controlled value. Its output is the internal set-point graph $y_{SA}(t)$. The actuations of a real controller are always limited. The set-point pre-processor calculates $y_{SA}(t)$ assuming that the necessary actuation controlling the technical process $x_C(t)$ is within the range of possible actuations of the control system. Thus, the controlled value $y(t)$ is, in principle, able to follow the internal set-point graph $y_{SA}(t)$, and the difference $y_{SA}(t) - y(t)$ is very small all the time if the controller's amplifications are high enough. On the other hand, the controller may be unstable, or at least, there may be a performance degradation if the amplifications of the controller are too high. An efficient way to find the most convenient parameters is to compare the performance of some controllers with different amplifications.

Generally, larger amplifications in the controller lead to smaller differences between the controlled values and the internal set-point graphs if the amplifications are selected appropriately. Or in other words, the higher the gains of the controller are the more predictable and perhaps the better is its performance. On the other hand, higher gains lead to greater requirements for the hardware and software of the controlled system. Higher gains in the controller leads to the following necessary changes of properties of the controlled system: The time difference between the alteration of the controller's input values and the subsequent modification of its output values must be smaller. Thus, the loop execution time and the measuring times of the sensors have to be shorter. Furthermore, the actuator's reactions have to be faster after actuation modifications, and the sensors have to be more accurate.

In practice, it is not possible to design a controller with the theoretical "best physically possible" performance. To this end, amplifications would have to be infinitely high, and the controller's loop execution time, the sensor's measuring time, and the reaction time of the actuators had to be zero. Moreover, the mathematical model considered by the design of the set-point pre-processor and the sensor output values must be fully accurate. All these characteristics are not achievable. However, the only practicable way to improve the controller behaviour is the modification of its mathematical model, if the controller's loop execution time is short enough.

As will be shown in the following example, there is often a big gap between the real behaviour of a conventional controller and the "best physically possible" controller performance. Using the difference $y_{SA}(t) - y(t)$ instead of the control error $y_S(t) - y(t)$ as static controller relation inputs leads to the great advantage that the behaviour of the controlled system is much more predictable as the one of a conventional controller. Moreover, the real performance of a properly designed SPP controller is usually much closer to the "best physically possible" controller behaviour. Thus, the priorities by the controller design are almost freely selectable, for instance:

- Stability and Safety: The control system must be stable even if the technical process is highly disturbed, e.g., by a force, by a torque or by a temperature input.
- In order to reduce production costs, the speed of the technical process should be as fast as possible.
- For reason of reduction of expenses and ecological damages, the energy consumption of the technical process has to be small.
- The temporal graph of the actuations should be steady and very continuous.

3.1 Example: Control of a Sleigh on a Horizontal Line

As displayed in Figure 4, position control of a sleigh on a horizontal line is assumed in this very elementary example. We shall only show the basic design principle of controllers working with set-point pre-processors. The design and the behaviour of a conventional PD controller (PD controller 1) is compared with the ones of two PD controller working with a set-point pre-processor (SPP controllers 2 and 3).

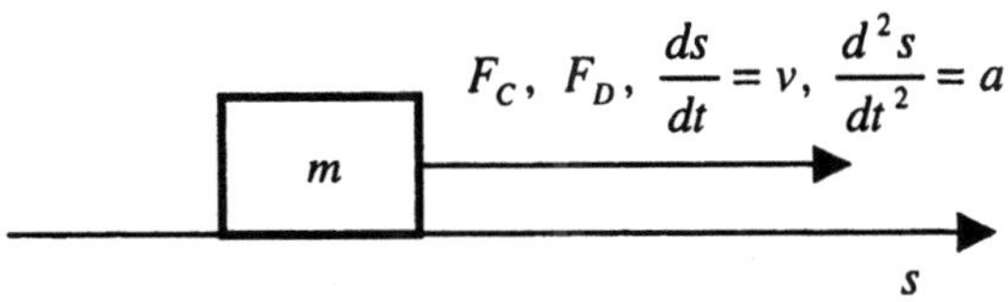

Fig. 4. Sleigh on a horizontal line

The mathematical model is characterised by the following differential equation:

$$F_C(t) + F_D(t) = m \bullet \frac{d^2 s(t)}{dt^2} = m \bullet a(t)$$

The total force F_C + F_D and the speed of the sleigh v have to be zero in the stationary state. Discrete degrees of freedom of robots often have similar properties. The control problem is to control the position of the sleigh s by adjusting the force F_C. The controlled system has an integrative behaviour. Thus, the performance of the best PD controller is better than the performance of the best PID controller, if the disturbing force F_D is zero.

3.1.1 Design of the Conventional PD Controller 1

In this example, the force controlling the sleigh $F_C(t)$ is limited between −25 N and 25 N. It was tried to design controllers with the best performance assuming that the maximum amount of the set-point jump is 100 m. Moreover, the control errors due to disturbing forces $F_D(t)$ should be as small as possible. The control deviation $s_S(t)$ − $s(t)$ of the conventional PD controller 1 may be very large. It ranges between −100 m and 100 m. $F_C(t)$ is about ($s_S(t)$ − $s(t)$) multiplied by the proportional amplification P. Thus, P has to be rather small, otherwise the controlled system will be unstable. The chosen mathematical algorithm of the correction element is described by the following equation:

$$F_D(t) = P \bullet (s_S(t) - s(t)) + PD \bullet v(t) = 3.5\ \text{Nm}^{-1} \bullet (s_S(t) - s(t)) - 3\ \text{Nsm}^{-1} \bullet v(t)$$

These assumptions lead to some great disadvantages, viz., inherent tardiness after a set-point jump and after a sudden change of $F_D(t)$. Another detriment is the large permanent control error if a permanent force disturbs the sleigh.

3.1.2 Construction of the SPP Controllers

As described above, controller design objectives are nearly freely selectable. In this example, we consider the desired temporal graph of the force controlling the sleigh $F_A(t)$. The task of the set-point pre-processor is, therefore, the computation of the internal set-point graph $s_{SA}(t)$ by assuming the desired temporal graphs $s_S(t)$ and $F_A(t)$, whereas s_S is the set-point of the controlled value.

There is a static relationship between the position of the sleigh $s(t)$ and the force accelerating it. So, lag or dead time elements are not contained in the mathematical model of the controlled system. We do not consider occurring disturbing forces (F_D = 0). The sleigh mass is m =1 kg. These considerations lead to a simple equation of the set-point pre-processor's calculator:

$$s_{SA}(t_1) = s_{SA}(t_0) + \frac{1}{m} \bullet \int_{t0}^{t1} \left\{ \int_{t0}^{t1} F_A(t) \bullet dt \right\} \bullet dt$$

Set-Point Pre-Processor of the SPP Controller 2

The set-point pre-processor calculates the fastest internal set-point graph physically possible by considering that the controlling force is within the controller's actuation range ($-25 \text{ N} \leq F_C(t) \leq 25 \text{ N}$). Therefore, $F_A(t)$ is either the maximum or the minimum of the controlling force $F_C(t)$ or zero. This property has the great disadvantage that the controller has small force reserves if the real technical process is different from the one considered by the design of the controller, e.g., due to disturbances or aging processes. The internal set-point graph $s_{SA}(t)$ is shown in Figure 5.

Set-Point Pre-Processor of the SPP Controller 3

The design principles of the SPP controllers 2 and 3 resemble each other. The only difference is that $F_A(t)$ is limited between −6.25 N and 6.25 N. Thus, the theoretical reserve force of the SPP controller 3 is 25 N − 6.25 N = 18.75 N. On the other hand, as displayed in Figure 5, the time after which the internal set-point graph $s_{SA}(t)$ reaches the set-point $s_S(t)$ are twice as long.

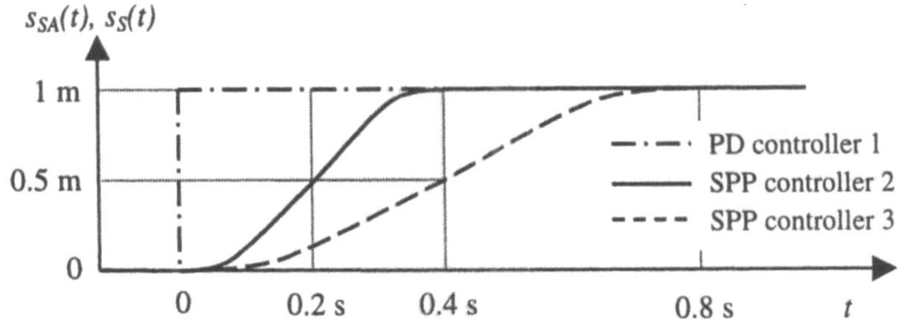

Fig. 5. $s_{SA}(t)$ or $s_S(t)$, respectively, leading to a set-point jump from $s(t) = 0$ to 1 m

Correction Element of the SPP Controllers

Generally, the design principle of controllers working with set-point pre-processors is to distinguish between the desired temporal graph of the controlled value (in this example, $s_S(t)$) and its internal set-point graph ($s_{SA}(t)$). The mathematical algorithm of the set-point pre-processor is designed in such a way that the difference between the internal set-point graph $s_{SA}(t)$ and the actual controlled value $s(t)$ is rather small. Due to this property, it is both possible and essential that the amplifications of the controller are high. Owing to the small difference $s_{SA}(t) - s(t)$, it is possible to design controllers with high amplification, even though their actuations are limited. Additionally, also in order to decrease this difference ($s_{SA}(t) - s(t)$) the controller's amplification has to be large. The temporal graph $s_{SA}(t) - s(t)$ and, thus, the performance of the controller primarily depends on the controller's proportional amplification P_P. Therefore, SPP controllers should be designed according the following procedure:

1. Design of the set-point pre-processor. The controlled values should be able to follow the internal set-point graphs even if the controlled technical process is disturbed by any anticipated fault.
2. Coarse determination of the controller's proportional amplification P_P. Afterwards, its differential and integral amplifications P_D and P_I are set. Meaningful values of P_P, P_D and P_I mutually affect each other.

These considerations lead to following actuation function of the SPP controllers:

$$F_D(t) = P_P \bullet (s_{SA}(t) - s(t)) + P_D \bullet v(t) = 250\ \text{Nm}^{-1} \bullet (s_S(t) - s(t)) - 26\ \text{Nsm}^{-1} \bullet v(t)$$

Their proportional amplification P_P is 250 Nm^{-1} instead of $P = 3.5\ \text{Nm}^{-1}$ by the conventional PD controller 1. Or in other words, in this example, the proportional amplification of the best controller working with a set-point pre-processor is 71.4 times larger than the proportional amplification of the best conventional PD controller.

3.1.3 Performance of the Controllers

As shown in Figures 5 and 6, at the moment t_l (in this example, $t_l = 0$) a set-point jump immediately leads to a sudden change of the conventional controller's control error. At the time t_l the control deviation jump $s_S(t_l) - s(t_l)$ corresponds to the amount of the set-point jump (in this example, 1 m). Moreover, also the actuation $F_C(t_l)$ jumps at this time as displayed in Figure 7. With regard to the stability of the controlled system, the amplification of the controller has to be small, particularly if the maximum amount of the set-point jump is high. Therefore, conventional controllers are usually slow and not very stable. The aperiodic border case is assumed in the design of the conventional PD controller 1. Thus, the sleigh reaches its desired values $s_S(t)$ without overshoots if it is not greatly disturbed. However, only small enlargements of the controller's amplifications result in big overshoots, and still, the time after which the set-point is reached the first time is much longer than the one for an SPP controller which exhibits a minute overshoot, at most.

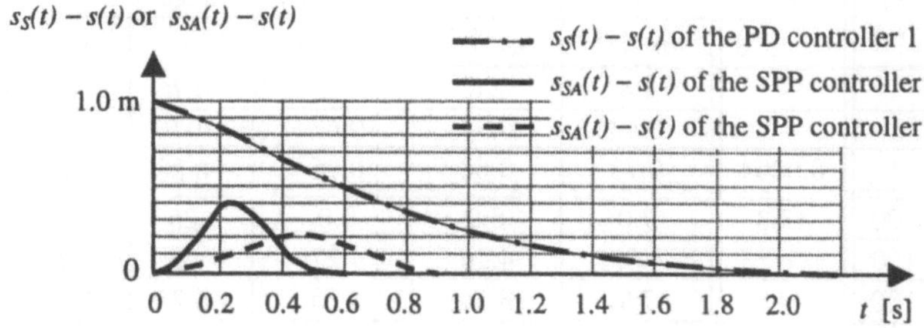

Fig. 6. Control errors after a set-point jump from zero to $s = 1$ m

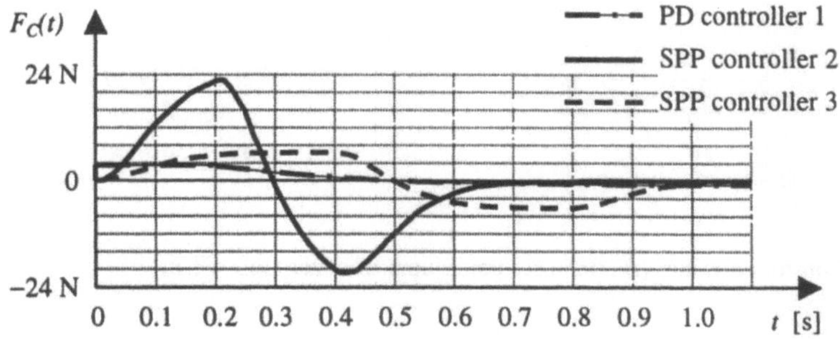

Fig. 7. Controlling force $F_C(t)$ after a set-point jump from zero to $s = 1$ m

In contrast to the control deviation of a conventional controller, the SPP controller's temporal graphs of its difference $s_{SA}(t) - s(t)$ are continuous. The sleigh position is not able to exactly follow the internal set-point graph due to the controller's finite proportional amplification ($P_P = 250$). Therefore, the actuations $F_C(t)$ of a SPP controller are very continuous. In this example, their actuation graphs $F_C(t)$ have roughly the shape of one sine wave. The positive force halfwave speeds up the sleigh towards the set-point. Afterwards, the negative force halfwave decreases the sleigh's speed and guides it to the desired position. The temporal graph of the controlling force $F_C(t)$ and, therefore, its maximum and minimum amounts can be well determined. Due to these properties, the maximum amount of $F_C(t)$ and the reaction speed of SPP controllers can be much larger than the ones of conventional controllers as depicted in Figure 8 and Table 1. Nevertheless, the sleigh reaches its desired position without overshoots.

Table 1. Comparison of the performance of the PD controller 1 **(1)**, the behaviour of the SPP controller 2 **(2)**, and the performance of the SPP controller 3 **(3)** after a set-point jump from zero to $s_S = 1$ m or to $s_S = 100$ m, respectively. s_B is the amount of the first overshoot. t_D is the time after which the output error is smaller than 0.1 m (t_E : $|\Delta s| < 0.01$ m; t_F: $|\Delta s| < 0.001$ m)

type	s_S [m]	overshoot s_B [m]	$\|\Delta s\| < 0.1$ t_D [s]	$\|\Delta s\| < 0.01$ t_E [s]	$\|\Delta s\| < 0.001$ t_F [s]	range of F [N] F_{C-}	F_{C+}
(1)	1	-	1.4	2.6	3.6	–0.76	3.5
(2)	1	-	0.43	0.56	0.67	–20.92	22.92
(3)	1	0.0004	0.73	0.88	0.96	–6.28	6.27
(1)	100	13.3	7.0	8.0	9.0	–25	25
(2)	100	-	4.03	4.16	4.27	–25	25
(3)	100	-	7.9	8.14	8.35	–6.25	6.25

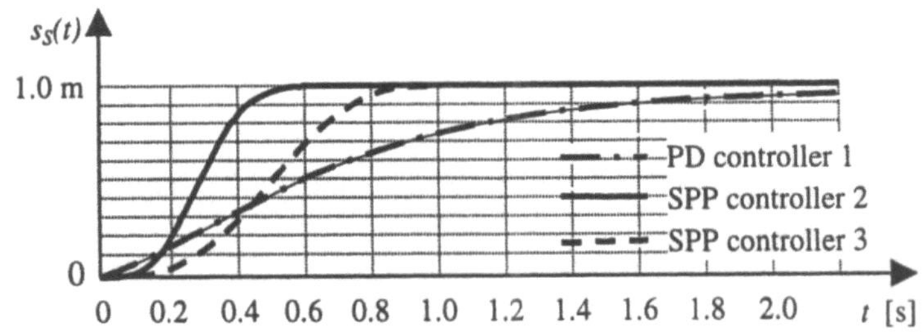

Fig. 8. Controlled value $s(t)$ after a set-point jump from zero to $s = 1$ m

The actuation domains are dissimilar after a set-point jump from zero to $s = 1$ m:

PD controller 1: $-0.76 \text{ N} \le F_C(t) \le 3.50 \text{ N}$

The maximum value of the set-point jump must be considered in the design of the controller (in this example 100 m). Thus, the amplification of a conventional controller has to be small due to the stability of the controlled system.

SPP controller 2: $-20.92 \text{ N} \leq F_c(t) \leq 22.92 \text{ N}$
It was tried to design a controller which is as fast as possible. The SPP controller 2 is about 5 times faster than the conventional PD controller 1. On the other hand, it has only a small reserve to prevent disturbances, e.g., due to permanent disturbing forces or large fluctuation of the technical process.

SPP controller 3: $-6.28 \text{ N} \leq F_c(t) \leq 6.27 \text{ N}$
In this example, the set-point pre-processor is designed in such a way that the controller has a large force reserve. Thus, this controller behaves much better than the SPP controller 2 if the controlled system is essentially disturbed. Even so, the SPP controller 3 is approximately 3 times faster than the PD controller 1. This controller design may be a good compromise between safety and speed.

4 Safety Licensable Inference Controller

Generally, it is possible to subdivide a single controller into the two elements mathematical relations and static relations as shown in Figure 1. This property often leads to essential simplification of controller design. One main difference between the individual controller types is their element static relations. For instance, the static relation of a conventional controller is a calculator, the one of a fuzzy controller is a fuzzy algorithm. Common fuzzy controllers and also most conventional controllers normally consist of a very complex hardware and, particularly, software. For instance, a fuzzy algorithms is the software of a computer calculating the controller output values (actuations). Therefore, it was nearly impossible to construct advanced error-free controllers or error-free safety critical systems. Another disadvantage of these controllers is their inherent lack of speed.

In order to avoid the detriments described above, the architecture of a programmable and safety licensable inference controller called "IF controller" is developed. The static relations of an IF controller are easily implemented in rule base tables. IF controllers can be described by many fuzzy algorithms with singletons as output membership functions and, in addition, with rectangular input membership functions, which do not overlap each other.

As shown in Figure 9, we design a fuzzy controller with a *condition interface* producing *fuzzy equivalents* of several input variables. These are then subjected to an inference engine co-operating with a rule base. The outputs from the *inference engine* are *fuzzy results* provided to an *action interface*, which finally performs process actuation and defuzzification.

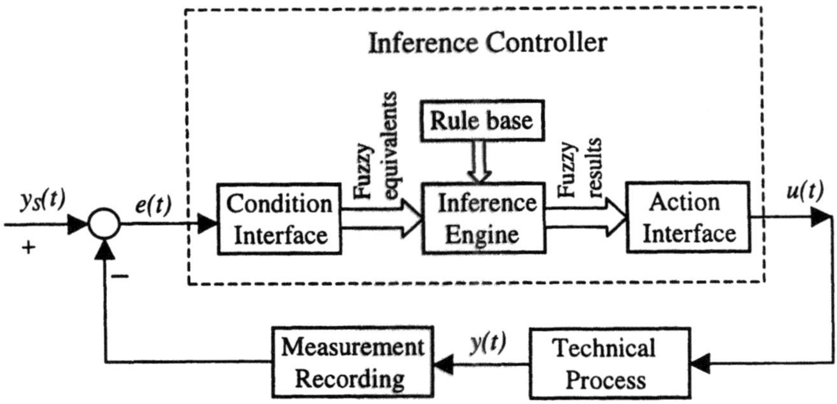

Fig. 9. Block diagram of a programmable inference controller (IF controller)

As inputs several analogue signals, such as angle of rotation and speed, are provided to the controller in form of control errors, i.e., differences between actual measured values and desired values. For reasons of algorithmic simplification in the controller, and to use proven hardware as widely as possible, these differences are determined in analogue form with operational amplifiers.

The main component of the controller is the inference engine. It operates under a strictly cyclic regime, such as a programmable logic controller. In contrast to the latter, however, each loop execution takes exactly the same amount of time, because the same operations are carried out in every iteration. Thus, the controller's real-time behaviour is fully deterministic and easily predictable.

Every loop execution comprises three steps: (1) input data generation in the condition interface (A/D converter), (2) inference by determining appropriate control rules, and (3) control actuation in the action interface (D/A converter). These steps as well as the overall operation cycle are strictly synchronised with a system clock.

Condition Interface

The control errors are fed into the condition interface, whose function is to fuzzify the input values. As depicted in Fig. 10, we subdivide the domains of the input variables in intervals. As the input values are given in analogue form, the most simple and straightforward way to directly obtain the corresponding fuzzy values in appropriate digital coding is to employ analogue-to-digital converters. Thus, the condition interface reduces to a set of linear or non-linear A/D converters, whose number equals the number of input variables. The fuzzy variables with rectangular membership functions are used as classifications of the input data. In this example, the temperature deviation is subdivided in the values "much too cold", "too cold", and so on. Such classifications directly correspond to the human way of thinking and reasoning about control and are, therefore, very realistic [2].

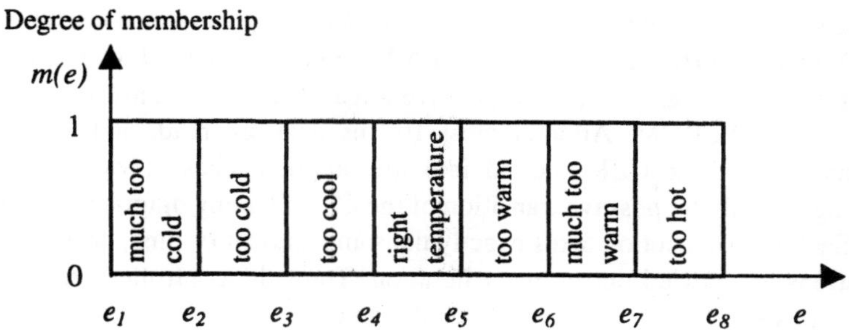

Fig. 10. Typical input membership functions

Inference Engine

The inference engine's rule base consists of one table for each output variable. These tables consist of if–then rules. Such tables are most easily implemented as random access memories which, for safety reasons, should be readable only, i.e., as ROMs, PROMs, or EPROMs [3]. As displayed in Fig. 10, the degree of membership of each classification is either 0 or 1. Moreover, strict one classification per condition interface input fires at the same time. Therefore, the output signal of an A/D converter signifies the current classification of the input value. These binary codes are concatenated to form addresses of table entries, i.e., memory locations, containing digital equivalents of the desired actuations. This hardware measure implements the steps: aggregation, inference and accumulation.

If the controller inputs (sensor outputs) are given in appropriate digital values, A/D converters are not necessary (Fig. 11). D/A converters are not needed if the actuators have digital control inputs. It is possible to design IF controllers that consist of inference engines only. An input latch needs to prevent jitter on address lines. The state of inputs is sampled and latched to provide the 16 address bits of a 16 bits wide 64 Kwords EPROM. The read out data represent the output value associated with the given inputs. Latches hold the output values until new ones are provided.

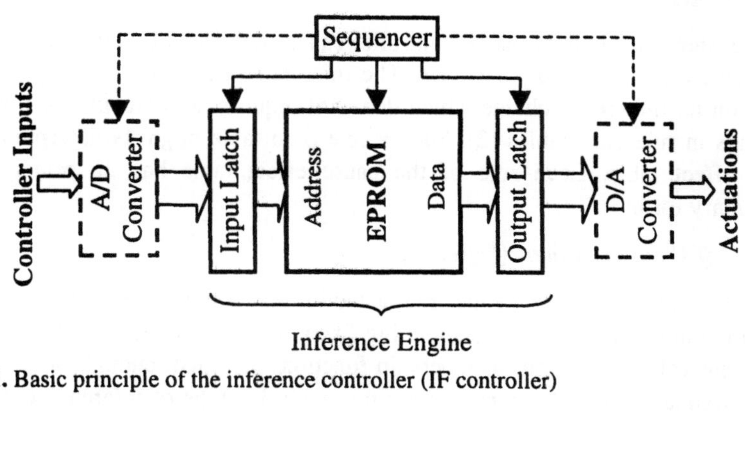

Fig. 11. Basic principle of the inference controller (IF controller)

The sequencer is implemented in a simple form of a clock generator consisting of an IC (7400) and a 5MHz quartz producing a 5 MHz clock signal CLK. The input latches latch the input signals with the positive edge of the CLK, thus generating an address for the EPROM. After at most 100 ns data are read, and with the negative transition of the CLK entered into the output latches. They latch the output data again with the positive transition of the CLK. Timing diagram is shown in Fig. 12. Each loop execution takes exactly the same amount of time, because the same operations are carried out in every iteration. Thus, the controller's real-time behaviour is fully deterministic and easily predictable.

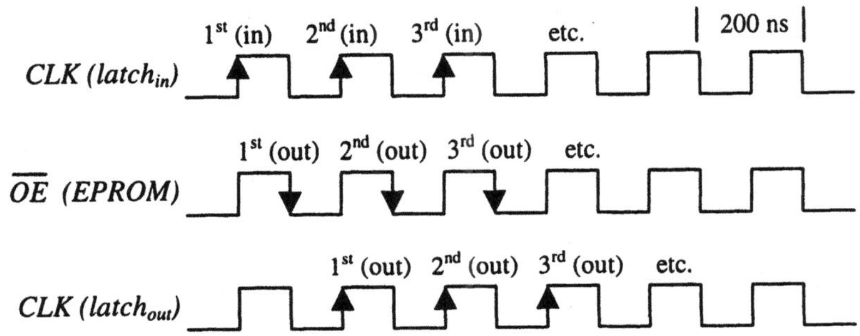

Fig. 12. Time diagram of the controller prototype's sequencer

Action Interface

The EPROM read-outs are finally provided to the action interface. In the design presented here an explicit defuzzification step is avoided by storing digital equivalents of the desired actuations in the rule base tables (EPROM). Hence, the action interface reduces to a standard digital-to-analogue converter for each output signal to be generated by the controller.

Basic Principle

The transfer behaviour of the IF controller can be described by the static relations between its input and output values. The inference operation of IF controllers is based on an inference scheme which does not require any computations, but just look-ups in rule base tables [2]. Such a rule base table might be interpreted as a cause effect table. Each rule of the cause effect table has, in principle, the elementary form:

if {"*cause* $_i$"} ***then*** {"*effect* $_j$"}

A digital equivalent of the desired actuation is assigned to any conjunctive combination of the values corresponding fuzzy input variables can assume. Thus, if the intervals of the input membership functions are small enough, it is possible to approximate any static control algorithm with this type of inference controller.

Moreover, dynamic controllers can be composed of IF controllers, integrators and differentiators. The number of words required in the EPROM corresponds to the product of the numbers of input membership functions, which may be large. However, there are possibilities to reduce the EPROM storage space required, e.g., cascading of IF controllers, IF controller with unequally spaced input intervals, and analogue inputs with adjustable center points and bandwidths. In practice, it turns out that the EPROM storage space required is often very small. Therefore, the IF controller presented here allows to solve difficult control tasks very easily and transparently.

Safety Aspects

The design of an IF controller containing a few well-proven industrial-standard components is very elementary and straightforward and, thus, reliable. Nevertheless, a breakdown of this hardware modules is not impossible, although not very likely. Fail-safe supervision is a possibility to improve the safety of the IF controller hardware. In case of a controller hardware malfunction, this supervisor generates a signal which can be used to perform an emergency shut-down of both the controller and the controlled technical process. As a consequence of this needs, the supervisor must be implemented in a fail-safe logic.

Owing to the inherent safety of the controller's hardware, the main task of a controller's safety proof is the verification of the correct implementation of the rule base tables. The contents of cause effect tables is the only application "software" implemented in IF controllers. Since cause effect tables must be kept in read only memories due to safety requirements, generally this kind of software takes on the form of firmware, for which complete test and diverse back translation are two rigorous software licensing methods.

Diverse back translation is essentially informal, easily conceivable, and immediately applicable without any training [4]. Thus, it is extremely well-suited to be used by people with the most heterogeneous educational backgrounds. The ease of understanding and use inherently fosters error-free application of the method. Hence, it is easy to validate specifications by both social consensus and complete test. Specified operations can be directly interpreted and executed by a machine without requiring complicated transformations and, therefore, verification of correct implementation. The only semantic gap is the coding between the levels relating to humans and to the machine. Each desired actuation (effect) corresponds to its digital equivalent stored in a certain memory location of the rule base table. Therefore, diverse back translation leads back in one easy step from machine code to problem specifications. For this architecture, the effort required to utilise diverse back translation for the safety licensing of control software is by several orders of magnitudes smaller than for verifying programs running on von Neumann computers. Thus, the need for generally difficult software verification simply disappears.

It easily possible to generate one or two dimensional graphs of the static relations of an IF controller with the readouts of its random access memory (EPROM). The use of such graphs may simplify the inspection of the rule base tables, especially if the number of rules of the controller's cause effect table is large.

5 Example

We consider **position control of a robot with two degrees of freedom**. The robot arm rotates on a horizontal plane around the pivot with the angular velocity ω. The distance r between the burden and the pivot of the robot is changeable with the velocity v. Thereby, the axis of the translation movement runs through the pivot. The control problem is to control the position of the burden with the mass m_L, which is located at the end of the robot arm, by adjusting the force F_C and the torque M_C.

5.1 Design of a Conventional Robot Controller 1

The principle of non-linear decoupling of systems as presented in [5] is used as design principle for this controller. Its controlling force $F_C(t)$ and its controlling torque $M_C(t)$ are not limited. It was tried to find the parameters of the fastest controller that leads to the assumption of the desired values without considerable overshoots by considering that $m_L = 20$ kg. Thus, the conventional controller 1 is described by the following equations for the distance r and the angle of rotation φ as functions of time:

$$F_C(t) = 70\,\text{Nm}^{-1} \bullet (r_S(t) - r(t)) - 119\,\text{Nsm}^{-1} \bullet v(t) - \{70\,\text{Nm}^{-1}\text{s}^2 \bullet r(t) - 25\,\text{Ns}^2\} \bullet \omega^2(t)$$

$$M_C(t) = \{17.67\,\text{kgm}^2 - 50\text{kgm} \bullet r(t) + 70\text{kg} \bullet r^2(t)\} \bullet [1\,\text{s}^{-2} \bullet (\varphi_S(t) - \varphi(t)) - 1.75\,\text{s}^{-1} \bullet \omega(t)] +$$
$$+ \{140\text{kg} \bullet r(t) - 50\text{kgm}\} \bullet v(t) \bullet \omega(t)$$

5.2 Design of an SPP Robot Controller 2

The conventional robot controller 1 and the SPP robot controller 2 compensate for the centrifugal force and the Coriolis torque of the robot. Thus, the translation dynamic controller only affects the dynamics of the robot arm translation, and the dynamics of the robot rotation is only influenced by the rotation dynamic controller. As shown in Figure 13, apart from the facts that the dynamic controllers working with a set-point pre-processor, and that the control algorithm is approximated by an IF controller, the SPP robot controller 2 has the same design as the robot controller 1. Its set-point pre-processor calculates the internal set-point graphs of the controlled values $r_{SA}(t)$ and $\varphi_{SA}(t)$ by assuming that the necessary actuations controlling the robot are within the ranges of possible actuations of the control system. Thus, the position of the burden $r(t)$ and $\varphi(t)$ is, in principle, able to follow the internal set-point graphs, and the differences $r_{SA}(t) - r(t)$ and $\varphi_{SA}(t) - \varphi(t)$ are very small at all times. Generally, if the mathematical model of the process to be controlled is assumed to be accurate in the design of the set-point pre-processor, design objectives, such as speed, stability, small energy consumption, can all be achieved close to the physical limits.

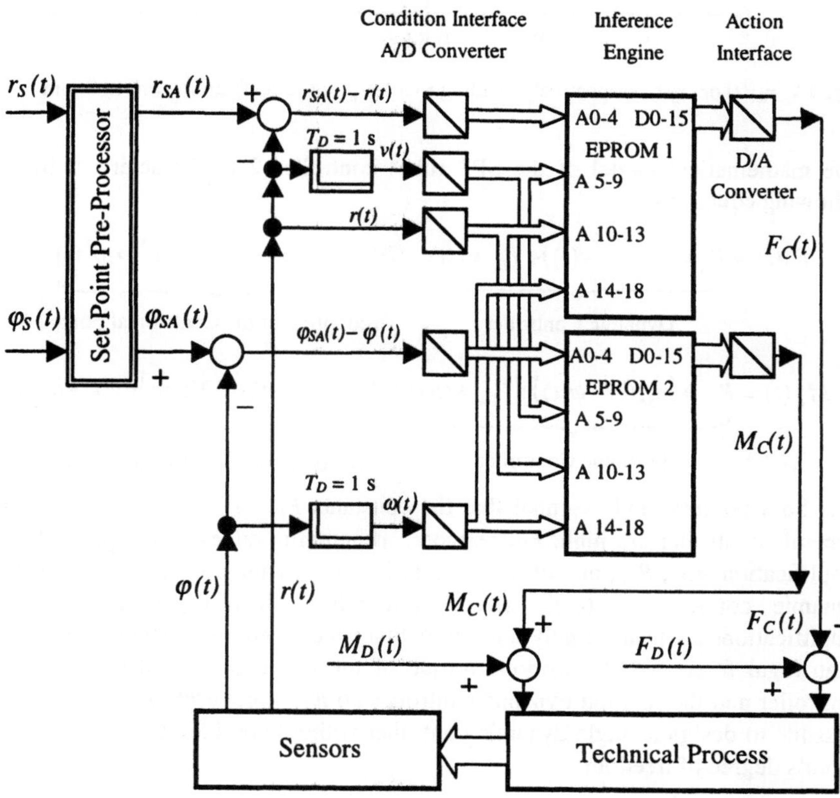

Fig. 13. Principle of the SPP robot controller 2

With this example, we shall show that SPP controllers exhibit excellent performance under any type of stress, even if the technical process to be controlled is insufficiently understood when constructing the set-point pre-processor. Thereby, the following simplifications are made: All non-linearities of the mathematical process model are not considered. Furthermore, the disturbing force $F_D(t)$ and the disturbing torque $M_D(t)$ are not considered, and the influences of the controllers' differential parts on the actuations are not considered. Thus, $r_{SA}(t)$ and $\varphi_{SA}(t)$ have similar shapes. An example of $r_{SA}(t)$ is shown in Figure 14. The internal set-point graph $r_{SA}(t)$ of the set-point pre-processor is described by the following equation for the desired actuation ($-100\ \mathrm{N} \le F_{Cd}(t) \le 100\ \mathrm{N}$) as function of time:

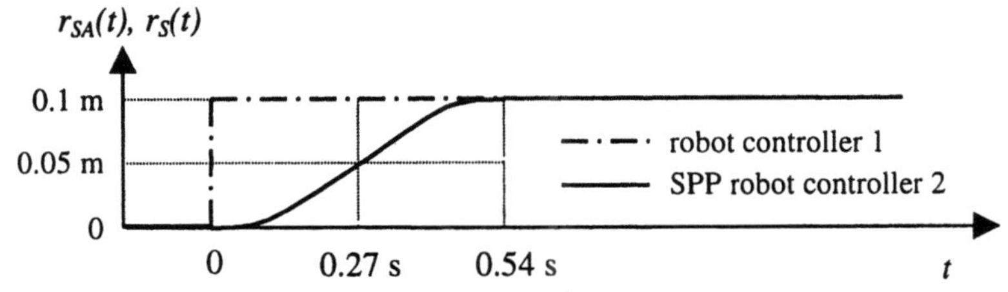

Fig. 14. $r_{SA}(t)$ or $r_S(t)$, respectively, leading to a set-point jump from $r(t) = 0$ to 0.1 m

The mathematical model of the SPP robot controller 2 is characterised by the following equations:

$$F_C(t) = \underbrace{P_{Pr} \bullet \left(r_{SA}(t) - r(t)\right) + P_{Dr} \bullet v(t)}_{\text{Dynamic Controller}} \underbrace{- \left\{70\ \mathrm{Nm^{-1}s^2} \bullet r(t) - 25\ \mathrm{Ns^2}\right\} \bullet \omega^2(t)}_{\text{Decoupling of the Centrifugal Force}}$$

$$M_C(t) = \underbrace{P_{P\varphi} \bullet \left(\varphi_{SA}(t) - \varphi(t)\right) + P_{D\varphi} \bullet \omega(t)}_{\text{Dynamic Controller}} + \underbrace{\left\{140\ \mathrm{kg} \bullet r(t) - 50\ \mathrm{kgm}\right\} \bullet v(t) \bullet \omega(t)}_{\text{Decoupling of the Coriolis Torque}}$$

It is both possible and essential that the constants P_{Pr}, P_{P_φ}, P_{Dr} and P_{D_φ} of the dynamic controller are high. Furthermore, meaningful values of the proportional amplification, e.g., P_{Pr}, and of the differential amplification, e.g., P_{Dr}, of a single dynamic controller mutually affect each other. However, otherwise these amplifications are nearly freely selectable. Since the controller compensates for the centrifugal force and the Coriolis torque of the robot, the translation dynamic controller and the rotation dynamic controller do not affect each other. Thus, it is possible to design a single dynamic controller without consideration for the other robot's degree of freedom.

The compared controllers are optimised for a simultaneous set-point jumps from zero to $r = 0.1$ m and $\varphi = 1$ rad. The rotation and the translation set-point jumps need about the same build-up periods and start at the same time to provide for the most difficult case to control. It was tried to find the parameters of the best controller with the properties described above. They are:

$P_{Pr} = 5800\ \mathrm{Nm^{-1}}$, $P_{P_\varphi} = 4000\ \mathrm{Nm}$, $P_{Dr} = -1260\ \mathrm{Nsm^{-1}}$ and $P_{D_\varphi} = -450\ \mathrm{Nms}$

In this example, the proportional amplifications of the SPP controller 2 are more than 80 times larger than the proportional amplifications of the conventional controller 1.

As described above, the real position of the burden is very close to $r_{SA}(t)$ and $\varphi_{SA}(t)$. Thus, the burden motion is of utmost continuity. It ranges very steadily within the first and last 10 % of its route after a set-point jump. Its acceleration is zero at the start instant of the movement. After the starting period, the robot ranges with high velocity in the centrepiece of its motion. Finally, it moves steadily and without overshoots to the desired position during the last actuation period. The temporal graph of robot movements should have exactly this shape. Such internal set-point graphs have, in addition, the benefit that the controllers' actuations are also very continuous..

As displayed in Figure 13, the dynamic controllers and the decoupling of the controller (centrifugal force and Coriolis torque) are designed as safety licensable IF controllers. The rule base table controlling the translation is stored on EPROM 1, and the rule base table controlling the rotation is stored on EPROM 2. A section of the rule base table of EPROM 1 is shown in Table 2.

Table 2. Part of the rule base table (cause effect table) stored on EPROM 1

Cause				*Effect*
if	***and***	***and***	***and***	***then***
interval of the input ω	interval of the input r	interval of the input v	interval of the input $r_{SA} - r$	actuation F_C (N)
30	11	5	28	122.69
30	11	5	29	234.05
30	11	5	30	591.33
30	11	6	0	–570.18
30	11	6	1	–212.90
30	11	6	2	–101.54
30	11	6	3	–45.86

For the signals $r_{SA}(t) - r(t)$, $v(t)$, $\varphi_{SA}(t) - \varphi(t)$ and $\omega(t)$ A/D converters with unequally spaced input intervals featuring higher precision around zero as shown in Figure 15 are selected. Due to the unequally spaced input intervals of the analogue-to-digital converters, under the given conditions the EPROM storage space required is by orders of magnitude less than the storage space required for a controller with equally spaced input intervals. The SPP robot controller 2 contains two 16 bits wide 0.5 Mwords EPROMs. An additional effective way to reduce the storage space required is cascading of IF controllers (rule base tables). For instance, it is also possible to design the SPP robot controller 2 with one 16 Kwords EPROM, two 1 Kwords EPROMs and one 0.5 Kwords EPROM, only.

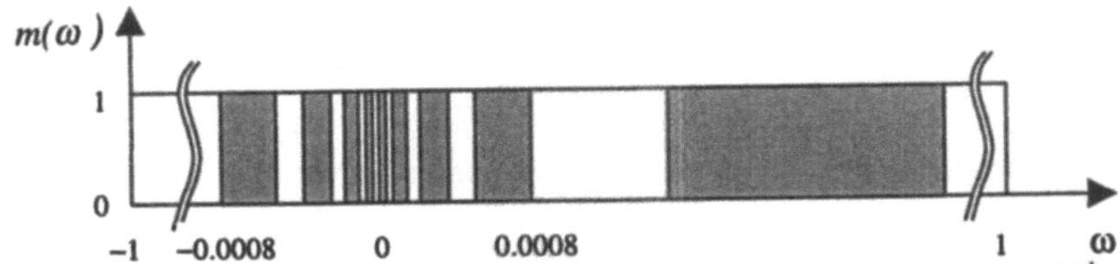

Fig. 15. Typical unequally spaced input intervals (e.g., angular velocity ω in $rads^{-1}$)

It was tried to design the best controllers by considering that $m_L = 20$ kg. In Table 1, we compare the performance of the conventional robot controller 1 **(1)** with the performance of the SPP robot controller 2 **(2)** under a number of circumstances: mass of the burden $m_L = 20$ kg **(a)**; $m_L = 0$ kg **(b)**; $m_L = 50$ kg **(c)**; $m_L = 20$ kg by considering damping due to springs **(d)**; $m_L = 20$ kg by assuming coincident disturbing impulses $F_D(t)$ and $M_D(t)$ **(e)**.

Table 3. Behaviour after simultaneous set-point jumps from zero to $r_S = 0.1$ m and $\varphi_S = 1$ rad **(a-d)**, and after disturbing impulses (100 N and 100 Nm) with the impulse width 1 s **(e)**

simulation	translation movement						rotation movement					
	r_O (10^{-3} m)		t_r (s)		$\Delta r_{(t=\infty)}$ (10^{-3} m)		φ_O (10^{-3} rad)		t_φ (s)		$\Delta\varphi_{(t=\infty)}$ (10^{-3} rad)	
	(1)	(2)	(1)	(2)	(1)	(2)	(1)	(2)	(1)	(2)	(1)	(2)
(a)	0.09	0	5.8	1.21	0	0	0.16	0	9.7	1.21	0	0
(b)	0	0	9.3	1.49	0	0	0.08	0	7.8	1.27	0	0
(c)	3.7	3.8	11.6	1.40	0	0	0.32	0	10.7	1.34	0	0
(d)	0	0	never	never	59	1.7	0	0	never	never	113	0.4
(e)	570	13.2	10.48	0.96	0	0	3340	26.3	12.08	0.41	0	0

Performance is expressed in terms of the amount of the first overshoot r_O (φ_O), the time t_r (t_φ) after which the output error is smaller than 0.0001 m (rad), and the permanent control error $\Delta r_{(t=\infty)}$ ($\Delta\varphi_{(t=\infty)}$).

The desired actuation functions of the SPP robot controller 2 are approximated by a safety licensable IF controller. It works very well and shows an excellent performance. This example reveals that the SPP robot controller 2 has, in addition to the ones mentioned in the Summary and the Introduction, many advantages as compared to the conventional robot controller 1:

- The time after which the control error is smaller than 0.0001 m after a translation set-point jump is between 4 and 8 times shorter, and the time after which the control error is smaller than 0.0001 rad after a rotation set-point jump is between 6 and 8 times shorter.

- A permanent force disturbing the robot arm leads to an about 80 times smaller control error, and a permanent disturbing torque leads to a 200 times smaller control error.
- The permanent control error due to damping because of springs $\Delta r_{(t=\infty)}$ is about 35 times smaller, and $\Delta\varphi_{(t=\infty)}$ is more than 200 times smaller.
- The overshoot after a force impulse disturbing the robot arm is 43 times smaller, and a disturbing torque impulse leads to an about 120 times smaller overshoot. Furthermore, the length of the period of time between the time after which the disturbing force F_D (torque M_D) vanishes after an F_D (M_D) impulse and the time after which the control error is smaller than 0.0001 m (rad) is about 10 times (30 times) shorter.

6 Conclusions

SPP controllers make use of an advanced mathematical control algorithm. As described above and in Table 4, the use of SPP controllers normally leads to essential performance improvements, even if the mathematical model of a technical process is very coarse and inaccurate assumed by the design of the controller. In contrast to the latter IF controllers are hardware structures. The expenditure to design IF controllers is reasonably small. Nevertheless, IF controllers approximate control algorithms with sufficient precision. Especially if the structure of an IF controller is customised to the mathematical model of a given controller, then the memory requirements for the corresponding rule base table often become surprisingly small. The IF controller consists of a few relatively simple, industry standard hardware modules. Hence, safety licensing of the hardware can follow well-understood and long-established procedures. The main task of a safety proof is to verify a rule set's correct implementation. For this, inspection or diverse back translation is used. The method is essentially informal and immediately applicable without any training. So, people with heterogeneous educational backgrounds are able to use diverse back translation very effectively due to its ease of understanding. A prototype of the IF controller with a 64 Kwords EPROM has already been build. Table 5 summarises the benefits achieved with IF controllers. The SPP robot controller 2 described in this paper has the advantages of both, of SPP controllers and of IF controllers.

Table 4. Comparison of the properties of SPP controllers and the ones of present controllers

Properties of **SPP controllers**	Properties of **present controllers**
The control behaviour can be determined almost freely within the physical limits.	The control behaviour can be determined only coarsely within small ranges.
The objectives of controller design can	Often the main design problem is the

be selected almost freely, e.g., stability, speed, reduction of expenses and ecological damages due to a technical process.	stability of control systems, and it is not possible to meet other design objectives as well.
The controller gain is very high. Thus, the stability of control systems against disturbances and changes of technical processes are extremely high. Furthermore, the process speed attainable is high.	The controller gain is small. Control system are, therefore, very sensitive to disturbances and parameter changes of the processes to be controlled, and control systems react slowly.
The temporal graphs of actuations are steady and continuous.	The actuation shapes are unsteady. This leads to clear performance degradation.

Table 5. Comparison of the properties of IF controllers and the ones of present controllers

Properties of **IF controllers**	Properties of **present controllers**
They are licensable for the highest safety requirements, viz., Safety Integrity Level 4 of IEC 61508, although they can approximate any arbitrarily complex control algorithm.	According to the authorities, it is not admitted to use PID or other complex control algorithms in safety related applications.
The use of appropriate, complex control algorithms enhances the inherent safety of technical processes to be controlled.	Very simple safety techniques lead to reductions of the inherent safety of controlled systems.
Operating in a strictly cyclic fashion, IF controllers exhibit fully predictable real-time behaviour.	The time needed for operation is usually not predictable. Thus, proper real-time behaviour cannot be guaranteed for these controllers.
Their loop execution times are extremely short, e.g., the one of the prototype is 0.2 microseconds.	The loop execution times normally exceed 1000 microseconds.

The use of IF controllers for safety related functions eliminates the disadvantages due to the restrictions of the licensing authorities, e.g., TÜV [1]. Furthermore, the use of SPP controllers promises a clear improvement of controller performance in many application cases.

References

1 Melchers, *List of Type Approved Programmable Electronic Systems (PES)*, Version 3.2, TÜV Rheinland Sicherheit und Umweltschutz GmbH, Institute for Software, Electronics, Railroad Technology (ISEB); Germany, 1998.

2 W.A. Halang, G.-H. Schildt, M. Colnaric, *A Fuzzy-Logic Based Programmable Electronic System for the Control of Safety-Critical Processes*, FernUniversität in Hagen, Faculty of Electrical Engineering, Germany; Vienna University of Technology, Institute for Automation, Austria; University of Maribor, Faculty of Technical Science, Slovenia, 1997

3 W.A. Halang, M. Colnaric, P. Vogrin, *Safety Licensable Inference Controller*, Proceedings 6th Zittau Fuzzy Colloquium, Hochschule für Technik, Wirtschaft und Sozialwesen Zittau/Görlitz, Germany, 1998.

4 H. Krebs, U. Haspel, *Ein Verfahren zur Software-Verifikation*, Regelungstechnische Praxis rtp, 26, 73-78, Germany, 1984

5 H. Hoyer, E. Freund, *The Principle of Non-linear Decoupling of Systems with Application to Industrial Robots*, rtp, 3, pp 80-87, and 4, pp 116-126, Germany, 1980.

4.

RISK IN ENERGY AND ENVIRONMENTAL SYSTEMS

Fuzzy Price Forecasting and Energy Contracting in Competitive Electricity Markets

Władysław Mielczarski[*,] and Grażyna Michalik-Mielczarska[**]**

[*] Systems Research Institute, Polish Academy of Sciences
Ul. Newelska 6, 01-447 Warsaw, Poland
Email: `Wladyslaw.Mielczarski@pse.pl`

[**]Energy Market – Consulting
Ul. Jaszuńskiego 10, 90-231 Łódź

1 Introduction

Of many applications, the fuzzy sets approach in cases involving human decisions seems to be the most promising. Electricity markets have brought new challenges to researches as varying prices and uncertain volume of energy production result in risk to be taken when contracting energy. The implementation of fuzzy approach allows the price pattern determination and the evaluation of the risk taken for various market attitudes [1]. The data used in research comes from the Australian Electricity Market, which seems to the most advanced electricity market with 5-minute dispatch prices.

2 Analysis of Spot Prices

Energy that has not been contracted in vesting or bilateral hedge contracts is traded on the spot market. The spot market determines half-hourly market clearing prices based on supply offers submitted by participating generators and aggregate electricity demand for the trading period. Participants provide their bids in 10 price bands. The spot prices, generator dispatch and electricity demand are determined through a central dispatch process. Subject to system and other operating constraints, generators are dispatched in merit order, that is with the lowest priced generator dispatched first – to balance supply and demand. The marginal price band dispatched in a given trading interval sets the market-clearing price for that interval.

Two periods have been choosen to present the analysis of spot prices and energy demand:

1. from July 1994 until May 1997 – Victorian Electricity Market operation (VicPool)
2. from May 1997 until January 1998 – National Electricity Market operation (NEM)

The analysis of spot prices over the 3 years of the Victorian Market (VicPool) operation shows that the average spot prices decreased significantly, despite some price oscillations and a slight increase in Pool demand (about 3%), as seen in Fig. 1.

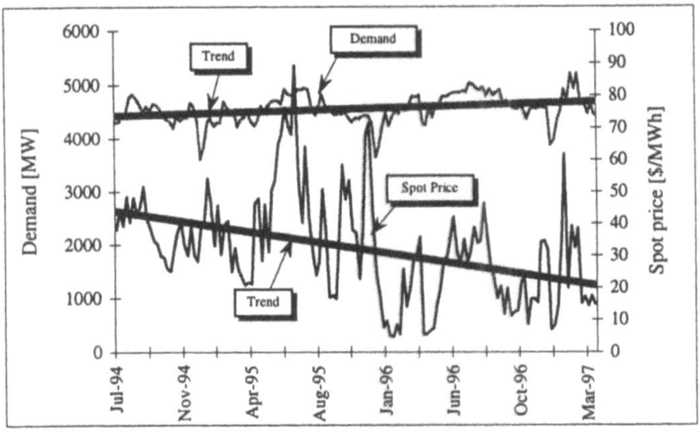

Fig. 1 Average weekly demand and spot prices along with their trends in VicPool (1994 ÷ 1997)

After the introduction of the National Electricity Market (NEM) in May 1997, spot price oscillations were significantly reduced, preserving the falling tendency of spot prices, as shown in Fig. 2.

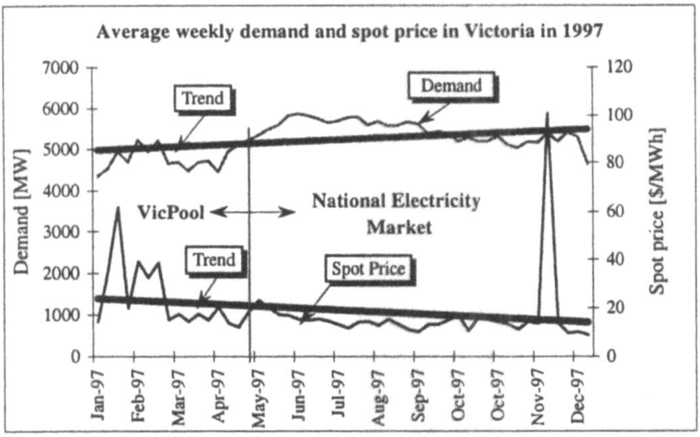

Fig. 2 Average weekly demand and spot prices in VicPool and NEM in 1997

2.1 Correlating Spot Prices

When carrying out the analysis of spot prices, it is important to find the correlation between the energy prices and other factors that can have a possible impact on their level. Spot prices can be influenced by a range of factors, including electricity demand, available generation capacity, energy transfer, and excess of generation. Out of these factors, the electricity demand has the most significant influence on spot prices; therefore, the analysis carried out for the VicPool and the National

Electricity Market (NEM) focuses on the investigation of the functional relations between these two variables.

Average spot prices and demand as functions of time

Correlation of average spot prices and electricity demand in 48 half-hourly intervals calculated for Mondays and Tuesdays in May 1997 is shown in Fig. 3. It can be observed that the spot price patterns reflect the patterns of energy demand quite accurately. As can be expected, the highest prices occur during evening peak hours, that is between 4:00pm and 8:00pm (36 ÷ 40 intervals), while the lowest prices between 3:00am and 5:00am (6 ÷ 10 intervals).

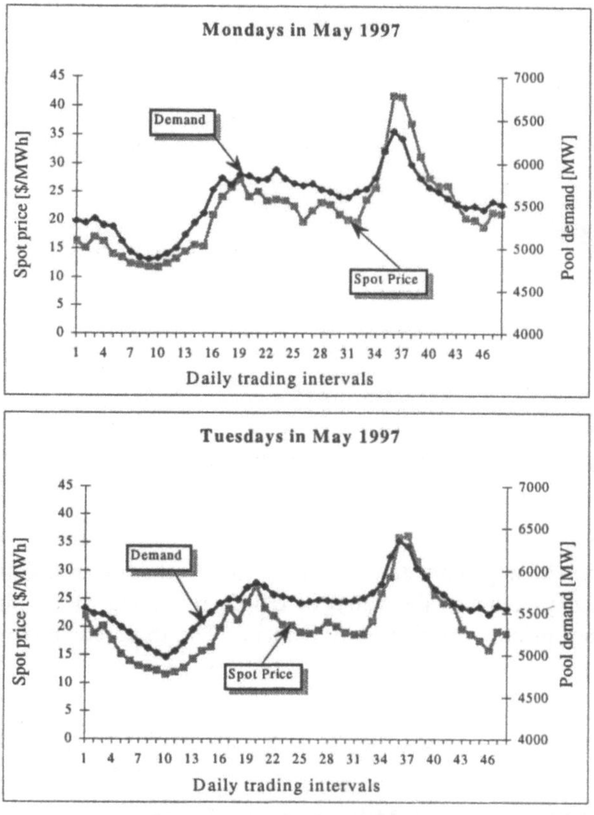

Fig. 3 Average spot price and electricity demand for Mondays and Tuesdays in May 1997

Average spot prices as a function of demand

Spot prices can be directly expressed as a function of pool demand. This method of spot prices analysis allows drawing the trend lines and investigating the accuracy of trend approximations. The trend relations between spot prices and pool demand for Mondays and Tuesdays in May 1997 are shown in Fig. 4. The data used to obtain these relationships are the same as demonstrated in Fig. 3 for the time dependent functions. It can be noticed that the linear trends can approximate quite accurately a functional relationship between spot prices and pool demand.

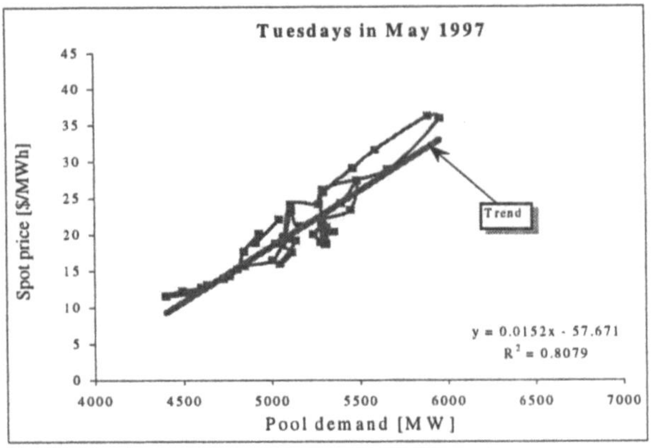

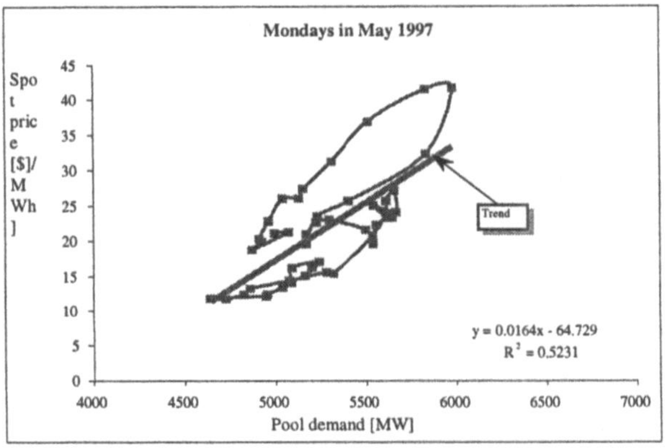

Fig. 4 Average spot prices as a function of pool demand for Mondays and Tuesdays in May 1997

Coefficient R^2, the measure of approximation accuracy, for Mondays is slightly above 0.5, while for Tuesdays reaches the value of 0.842, reflecting the strong correlation between the spot prices and electricity demand. The trend equation is expressed as:

$$y = mx + b$$

where y – spot price [$/MWh], x – demand [MW], m, b – coefficients.

Average Spot Prices as a Function of Demand Forecast

To assist market participants with their bidding strategies and ensure the adequate security of the power system, comprehensive information is provided both to a Market Operator and market participants, including a demand forecast, network constraints, expected aggregate plant capacity, forecast excess generation conditions, and projected system reserve conditions. This scheme is called PASA that stands for the Projected Assessment of System Adequacy. The Market Operator provides market participants with two types of forecasts:

1 Short-term PASA, which is produced each day and covers 2 ÷7 days ahead
2 Medium-term PASA which is produced weekly and covers two years period

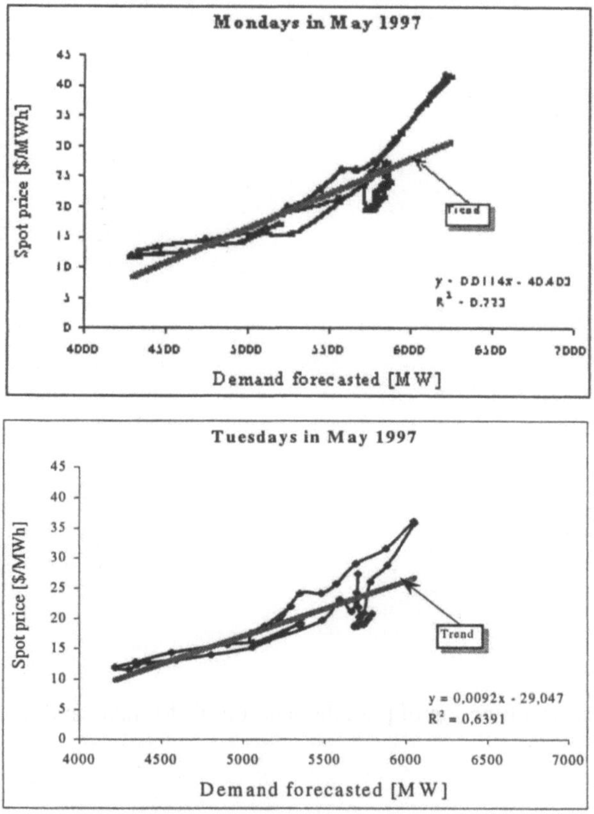

Fig. 5 Relations between average values of spot prices and the demand forecast for Mondays and Tuesdays in May 1997

A load forecast is prepared, taking into account historical load data, days of the week, public holidays, seasons (summer, winter), expected system conditions, and current weather forecast data. Although the actual demand differs from the prediction (mainly due to changing weather conditions), the forecasted demand provided in the short-term PASA assists substantially market participants in their bidding strategies. Therefore, it is essential to determine the relationships between spot prices and the demand forecast provided by the short-term PASA. Fig. 5 shows the correlation between average values of spot prices and a demand forecast for Mondays and Tuesdays in May 1997.

As can noticed, there is a strong correlation between demand forecast and spot prices - the coefficient R^2 is larger than 0.63. The good correlation between the forecasted and actual demand justifies the application of a demand forecast for predicting spot prices.

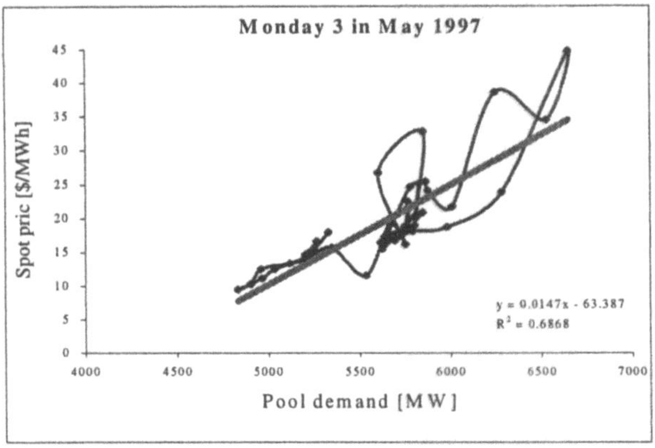

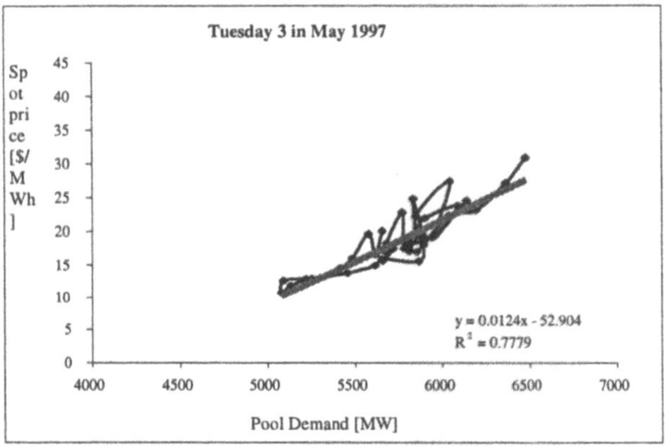

Fig. 6 Temporary spot prices as a function of pool demand for 3rd Monday and 3rd Tuesday in May 1997

Temporary spot prices as a function of demand

The spot price trends presented in Fig. 4 and 5 have been derived for average values of spot prices and average values of pool demand. However, spot price forecast should be carried out for particular days, so the relationships between average values of spot prices and pool demand can be only employed as auxiliary parameters. Therefore, it is essential to determine the trends for particular days. The analysis carried out both for the NEM shows the good correlation between the temporary spot prices and pool demand. Fig. 6 illustrates these relationships for 3rd Monday and 3rd Tuesday in May 1997.

2.2 Selection of the Best Trend Lines

A linear trend line is one of the most frequently used for statistical analysis, since it usually combines simplicity with adequate accuracy. However, other trend functions may also fit accurately, thus to compare different types of approximations four commonly used trend functions have been chosen:

1. Linear trend line

$$y = ax + b \tag{1}$$

2. Exponential trend function

$$y = a\exp(bx) \tag{2}$$

3. Power trend function

$$y = ax^b \tag{3}$$

4. Logarithmic trend function

$$y = a\ln(x) - b \tag{4}$$

The analysis of trend accuracy has been carried out for three months: May, June and July 1997. The illustration of this analysis is provided for two selected trends:

- 'Wide' – represents large deviations from the trend line (an example given for Monday 3, May 1997)
- 'Narrow' – represents small deviations from the trend line (for Tuesday 1, May 1997).

The Wide and Narrow patterns of spot price distribution as a function of pool demand are shown in Fig. 7 and 8.

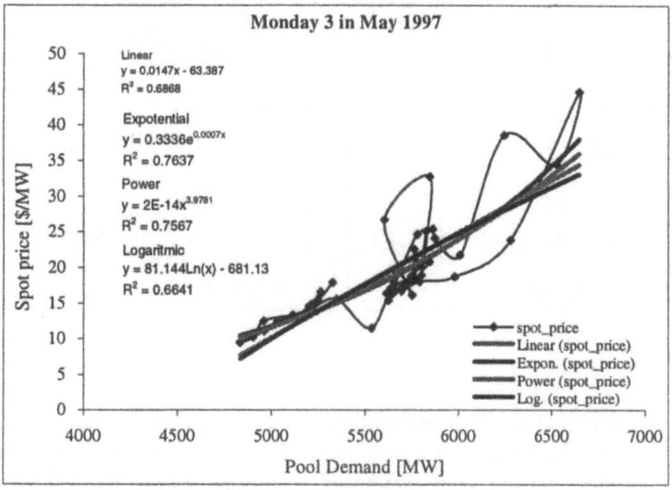

Fig. 7 Trends of spot prices as a function of demand on Monday 3 in May 1997 – Wide pattern

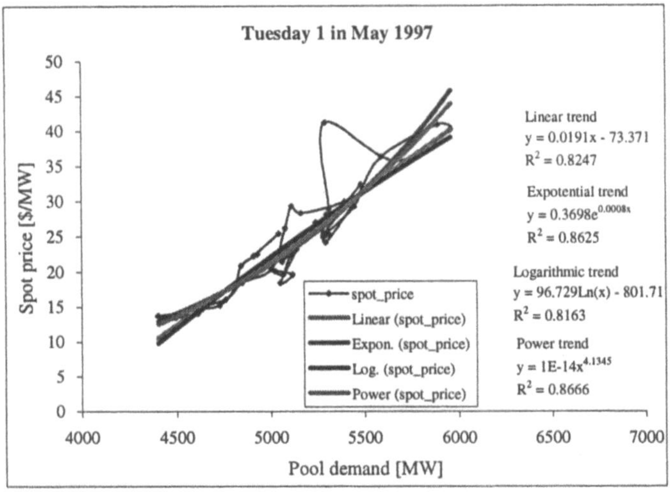

Fig. 8 Trends of spot prices as a function of demand on Tuesday 1 in May 1997 – Narrow pattern

Approximations by various trend functions are presented in Table 1.

Table 1 Approximations by various trend functions

Trends	Wide	Narrow
Linear	$y = 0.0147x - 63.387$ $R^2 = 0.6868$	$y = 0.0191x - 73.371$ $R^2 = 0.8247$
Exponential	$y = 0.3336e^{0.0007x}$ $R^2 = 0.7637$	$y = 0.3698e^{0.0008x}$ $R^2 = 0.8635$
Power	$y = 2E - 14x^{3.9781}$ $R^2 = 0.7567$	$y = 1E - 14x^{4.1345}$ $R^2 = 0.8666$
Logarithm	$y = 81.144\ln(x) - 681.13$ $R^2 = 0.6641$	$y = 96.722\ln(x) - 801.71$ $R^2 = 0.8163$

When trend equations are used for spot price approximation, the spot price forecast for a given day can be calculated. An example of forecasts for Monday 1 and Tuesday 1 in May 1997, resulting from the trend equations given in Table 1, is shown in Fig. 9.

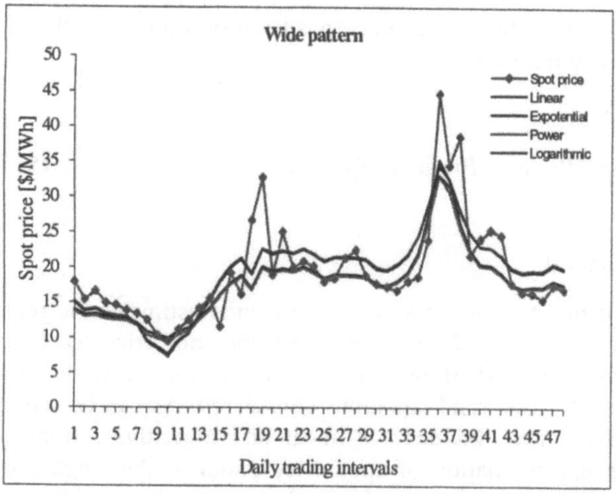

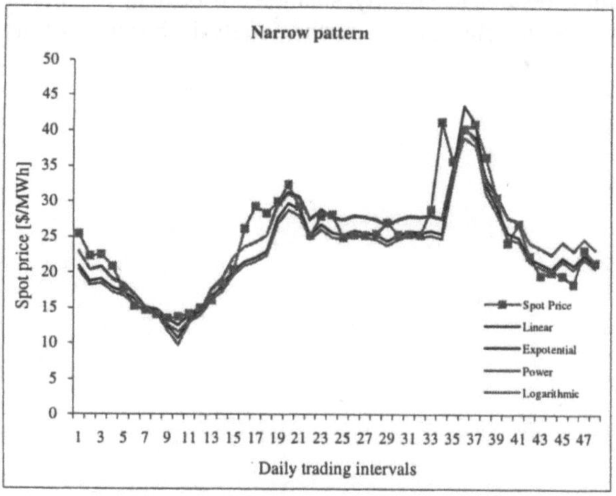

Fig. 9 Spot prices and price forecasts based on trend line approximations – Wide and Narrow patterns

The analysis of spot price forecasts based on different trend line approximations does not provide the apparent answer what type of approximation is the most suitable for the Wide, Narrow, or both patterns. The values of parameters R^2 for the four trends considered for both the Wide and Narrow patterns are quite similar, and therefore should not be used as the only criteria for the selection of the best trend. In addition, the parameter R^2, expressed as the squared differences between a trend line and a spot price pattern, seems not to be the most suitable measure for market participants to optimize their bidding policy. From a market participants' point of view, it is not essential how much the predicted spot price varies from the real price. The more important problem is to minimize the cumulative differences between the actual spot prices and the prices predicted, taking into account their positive and negative values. Obviously, the smaller the values of a cumulative difference, the better spot price prediction, and thus the smaller the risk resulting from uncertainty of spot price forecasts.

Therefore, a new measure of the prediction accuracy has been introduced, called a Daily Prediction Index (DPrI), and defined as:

$$DPrI = \sum_{i=1}^{48} (ActualSpotPrice_i - PredictedSpotPrice_i) \quad (5)$$

where 'i' denotes a trading interval.

When the *DPrI* is positive, it means that the spot forecast underestimates the real spot price. On the other hand, when the *DPrI* is negative, the spot price forecast overestimates the real spot price. Fig. 10 shows the *DPrI* values for the Wide and Narrow patterns of spot prices for the trend equations considered. As can be seen, the smallest values of the *DPrI* are obtained for linear and logarithmic trends, providing the most accurate approximation of spot price patterns. Although the logarithmic trend approximation results in slightly smaller errors than the linear trend, the linear equation has been selected for further analysis because of its simplicity.

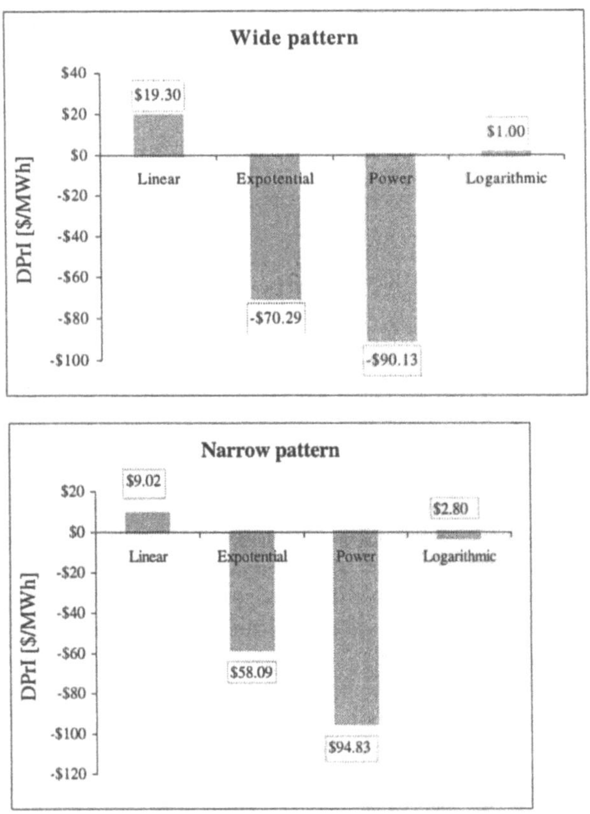

Fig. 10 Daily Prediction Index (*DprI*) values for the Wide (Monday 3, May 1997), and Narrow (Tuesday 3, May 1997) patterns

3 Fuzzy Function of Spot Price Patterns

Spot prices are influenced by a number of parameters, including the day of a week, a season, as well as the actual and forecasted demand. The analysis of spot prices focuses on the development of characteristic parameters required for examination and forecast of spot price variations. The data used for analysis comes from the National Electricity Market (NEM) in Australia and covers the period starting from the beginning of May 1997 until the end of December 1997.

3.1 Trend Lines

A linear trend line can be used for characterization of a functional relation between a spot price and its parameters, including energy demand. The vital advantage of linear trend lines is its simplicity. However, the linear trend line approximation does not reflect price deviations. Graphically, these deviations are represented by the points located above and below the trend line. In order to include the spot price deviations into the analysis, two trend borderlines, called the upper- and lower-trend lines or the upper- and lower- margins, have been proposed. Now, the functional relation between a spot price and its parameters can be expressed as:

$$\tilde{y}_{upper}(i) = mx(i) + b_{upper}(x) \tag{6}$$

$$\tilde{y}_{trend}(i) = mx(i) + b_{trend}(x) \tag{7}$$

$$\tilde{y}_{lower}(i) = mx(i) + b_{lower}(x) \tag{8}$$

where:

$\tilde{y}_{upper}(i)$, $\tilde{y}_{trend}(i)$, and $\tilde{y}_{lower}(i)$ – upper, main, and lower trend lines of spot price approximations

$x(i)$ – characteristic parameters, for example energy demand

b_{upper}, b_{trend} and b_{lower} – offsets of the upper, main, and lower trend lines

m – slope of the upper, main, and lower trend lines

i – trading interval

In relations (6), (7) and (8) it is assumed that the slope of trend lines is constant, while offsets vary. This assumption leads to three parallel trend lines, as shown in Fig. 11:

- Main trend line that reflects the average value of spot prices (in the middle)
- Two parallel trend lines on the both sides of the main trend line that define the upper and lower boundaries (margins)

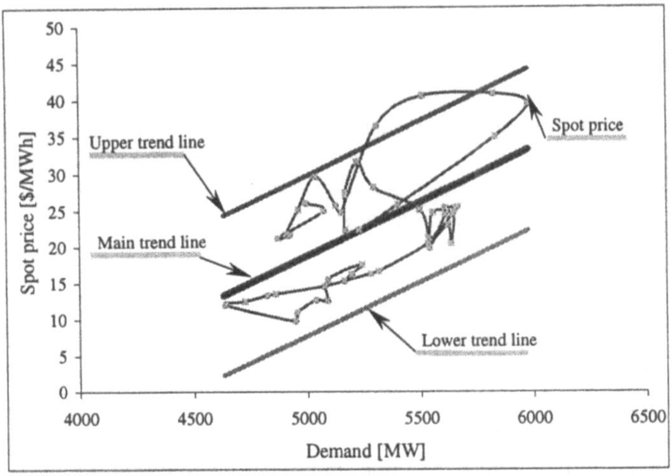

Fig. 11 Linear trend lines (upper, lower and main) of the relation between spot prices and electricity demand

In an example shown in Fig. 11, the upper and lower boundaries (trend lines) are shifted up and down by 20%, respectively. This shift is obtained by setting the following offsets:

$$b_{upper} = 1.2\, b_{trend} \tag{9}$$

$$b_{lower} = 0.8\, b_{trend} \tag{10}$$

As can be seen from Fig. 11, in the case presented all spot prices located below the main trend line are situated above the lower trend line. On the other hand, the upper trend line does not cover all spot prices located above the main trend line. The upper and lower offsets, determined by Eq. (9) and (10), can be considered as the levels of spot price deviations. General relationship for the spot price approximation can be expressed by:

$$\tilde{y}_k(i) = f\left(m, b_{trend}, b_{upper}, b_{lower}, x(i)\right) \tag{11}$$

for k = 1, 2 and 3, defining the main, upper, and lower trends, respectively; m, b_{trend}, b_{upper}, b_{lower}, i, and $x(i)$ – as defined for Eq. (6) ÷ (8).

3.2 Definition of Market Participant Attitudes

One of the main issues of spot price approximation is the proper selection of upper and lower boundaries. Because the slope of the three trend lines is the same, it is sufficient to determine only two offsets: b_{upper} and b_{lower}. Ideally, b_{upper} and b_{lower} should be determined in such a way that the space between them contains all spot price data. However, this approach may lead to very conservative evaluation of spot price deviations. In practice, a few spot price deviations outside the boundary lines may not affect significantly the trading balance when the trend assessment is used for the decision-making processes. The determination of the acceptable trend boundaries depends, in a great degree, on the attitude of decision-makers to risk-taking. As can be expected, more conservative market players tend to include more spot price data between the boundary lines. It has been assumed that all decision-markers are classified into four categories, according to their attitudes to risk-taking. Table 2 presents these categories and the corresponding levels of trend line boundaries.

Table 2 Categories of market players with different attitudes to risk-taking

Attitude of decision-makers to risk-taking	Amount of data contained between lower and upper boundaries
Very Conservative	95%
Conservative	90%
Neutral	80%
Risk-seeking	70%

When the proposed four approximation levels are applied to the upper and lower trend lines (margins) of spot price data, different values for b_{lower} and b_{upper} are obtained. The graphical representation of trend boundaries is shown in Fig. 12.

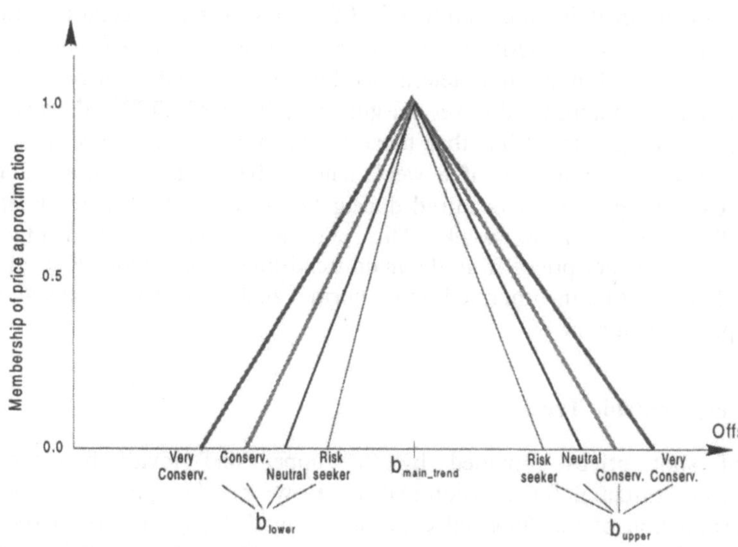

Fig. 12 Fuzzy filter applied to determine the boundaries for data deviations

In Fig. 12, the delta-shaped function, called a ' fuzzy filter', expresses the preciseness of the trend line approximation. Parameters b_{lower} and b_{upper} are determined in such a way that the area between boundaries contains the prescribed number of spot price data, according to the levels given in Table 2. This creates a set of data with members of 95%, 90%, 80%, or 70% family. When a data point lies exactly on the main trend line, the membership function of this point is equal to one, and the trend line provides the precise approximation. When a spot price is located either above or below the main trend line, its membership function decreases according to the distance from the main trend line. When a spot price is located below the lower or above the upper boundaries, its membership function is equal to zero.

Generally, the values of parameters b_{lower} and b_{upper}, expressed in \$/MWh, represent the boundary offsets on the spot-price axis. Although these parameters are important for decision-makers, several other parameters are also essential, including:

- The maximum and minimum prices for a given span of filter
- The number of trading intervals and the values of spot prices when spot price data is within the approximation margins
- The ratio of energy cost, sold or purchased, given by filter approximation to the total daily cost of trading
- The risk involved when decisions related to energy contracting are based on the filter approximation

Examples of filter approximations for a span of 80%, reflecting a Neutral attitude of market players to risk taking, for Mondays and Tuesdays are shown in Fig. 13.

As can be seen in Fig. 13 (a), for Mondays the distances between the upper and lower margins, determined by the family of 80%, show some fluctuations in different months (around \$15 ÷ \$20). When a filter with a span of 80% is applied, 80% of spot prices are within the limit determined by the upper and lover borders. Fig. 13 (b) shows that the upper and lower margins of spot prices for Tuesdays are generally closer to the main trend line than those for Mondays, because spot price oscillations for days in the middle of a week tend to be smaller. However, a significant increase in spot prices occurred during two days in the last week of November 1997, called a 'black week'. This unusual increase resulted in a substantial departure of spot prices from the main trend line. It is obvious that this unusual event should not be incorporated into a normal analysis of spot prices and requires the separate treatment.

3.3 Cost of Energy outside Trends

The limits of spot prices provided by the upper and lower boundary approximations are essential for decision-makers. However, these limits do not allow the determination of the financial exposure of market participants to spot price violations. Therefore, to enable the assessment of this exposure, it is also necessary to consider the cost of electrical energy when spot prices are outside the boundaries determined by the upper and lower trend lines.

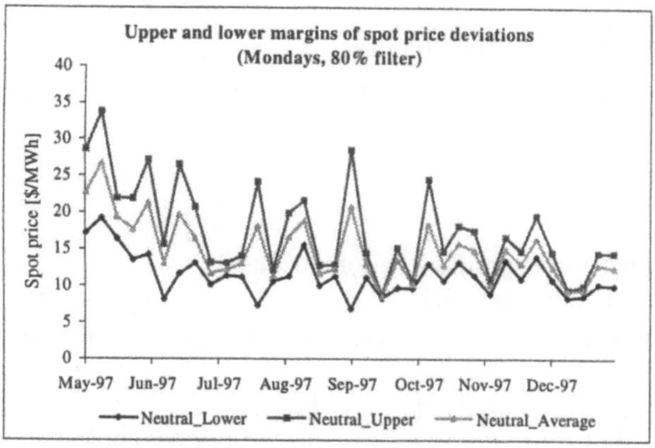

(a)

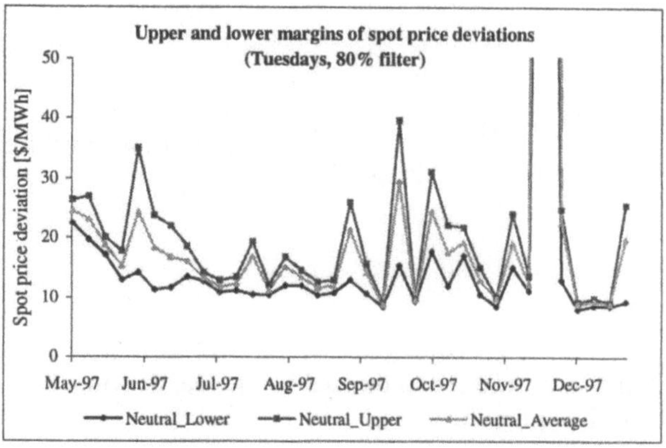

(b)

Fig. 13 Upper and Lower margins (80%) of spot price deviations along with the main trend of spot prices for a manager with a Neutral attitude to risk taking on Mondays (a) and Tuesdays (b)

Two different approaches have been applied in order to determine the additional costs that may be incurred when spot prices are above or below the upper or lower boundaries, respectively:

- *Uniform demand* – in this approach, it is assumed that the energy demand pattern remains uniform during the whole day
- *Real demand* – in this approach, the cost of energy is modified by the shape of real demand, which is low during the nights, high during business hours, and reaches the maximum during the evening

Uniform demand

For the uniform demand, the daily cost of the additional cost exposure when spot prices are above or below the upper or lower estimates is expressed as follows:

$$DAE = \frac{\sum_{i=1}^{48} [y(i) - \tilde{y}(i)] \times Sign[y(i) - \tilde{y}(i)]}{\sum_{i=1}^{48} y(i)} 100\% \qquad (12)$$

where:

DAE – daily additional cost exposure – uniform demand

$y(i)$ – spot price in the trading interval 'i'

$\tilde{y}(i)$ – upper or lower approximation of spot prices in the trading interval 'i'

$Sign[y(i) - \tilde{y}(i)_{upper}] = \begin{cases} 1 \text{ when } [y(i) - \tilde{y}_{upper}(i)] \geq 0 \\ 0 \text{ when } [y(i) - \tilde{y}_{upper}(i)] < 0 \end{cases}$ for prices above the main trend line

or

$Sign[y(i) - \tilde{y}(i)_{lower}] = \begin{cases} 1 \text{ when } [y(i) - \tilde{y}_{lower}(i)] \leq 0 \\ 0 \text{ when } [y(i) - \tilde{y}_{lower}(i)] > 0 \end{cases}$ for prices below the main trend line

Real demand

When the daily cost of additional exposure is coordinated by the real demand pattern, the following relationship applies:

$$DAE^{coord} = \frac{\sum_{i=1}^{48} [(y(i) - \tilde{y}(i))D(i)] \times Sign[y(i) - \tilde{y}\ (i)]}{\sum_{i=1}^{48} y(i) \times D(i)} 100\% \qquad (13)$$

where

DAE^{coord} - daily additional market participants' exposure - coordinated demand

$D(i)$ – Buyer's demand in a trading interval 'i'

$y(i)$, $\tilde{y}(i)$, and $Sign[y(i)$ - $\tilde{y}(i)$] - as for Eq. (12)

Generally, electrical energy market participants can be classified into two categories: Buyers and Sellers. For both of them it is essential to consider the cost of electrical energy when spot prices are outside the limits determined by the boundary trend lines; however, they focus on the spot market prices from the different perspective.

Buyers

Energy Buyers include Distribution Businesses, Retailers, and Customers that are allowed to purchase electrical energy in a wholesale market. These market participants are mostly interested in the price deviations above the upper trend lines. Significant price deviations can increase a market participants' exposure to the price volatility and consequently increase the risk involved in a market game. Uncertainty of spot price deviations may lead to a conservative behavior of decision-makers, resulting in hedge contracts with strike prices, as a rule, higher than spot prices. It should be noted that when the price of energy bought through various types of contracts is higher than the spot price, in practice it means the losses in revenue for Buyers.

Sellers

As energy Sellers are considered Generators producing electrical energy for the pool. They mainly focus on obtaining high revenue, which is a direct function of spot prices. These market participants are primarily concerned with the energy prices being below the lower trend lines. Obviously, the lower the spot prices, the smaller revenue the Sellers can obtain from energy generation. Electrical energy producers can protect themselves from the market exposure to low prices by entering the different types of hedge contracts with energy Buyers. It is essential for Generators to know the predicted spot prices as precisely as possible to enable them an adequate determination of contract strike prices in every trading interval. Generally, the lower the price expected, the lower the strike price in the contracts acceptable by energy producers. Although entering contracts may lead to smaller revenues when spot prices are above the strike prices agreed, the contracts protect energy producers against the elimination from the market game.

3.3.1 Spot Prices above the upper Trend Margin

Daily costs of the additional market exposure when spot prices are above the upper estimates are mainly of interests of energy Buyers. According to (12), for the uniform demand these costs can be calculated as:

$$DAE_{upper} = \frac{\sum_{i=1}^{48} \left[y(i) - \tilde{y}_{upper}(i)\right] \times Sign\left[y(i) - \tilde{y}_{upper}(i)\right]}{\sum_{i=1}^{48} y(i)} 100\% \qquad (14)$$

where:

DAE_{upper} – daily cost of an additional Buyer exposure when spot prices are above the upper estimates

$y(i)$ – spot price in the trading interval 'i'

$\tilde{y}_{upper}(i)$ – upper-trend approximation of a spot price in the trading interval 'i'

$$\mathrm{Sign}\left[y(i) - \tilde{y}(i)_{upper}\right] = \begin{cases} 1 \text{ when } \left[y(i) - \tilde{y}_{upper}(i)\right] \geq 0 \\ 0 \text{ when } \left[y(i) - \tilde{y}_{upper}(i)\right] < 0 \end{cases}$$

When applying the real demand approach, the daily cost of an additional market exposure can be calculated from the following relationship:

$$DAE_{upper}^{coord} = \frac{\sum_{i=1}^{48}\left[\left(y(i) - \tilde{y}_{upper}(i)\right)D(i)\right] \times Sign\left[y(i) - \tilde{y}_{upper}(i)\right]}{\sum_{i=1}^{48} y(i) \times D(i)} 100\% \qquad (15)$$

where:

DAE_{upper}^{coord} – daily additional Buyers' exposure – coordinated demand

D(i) – Buyer demand in the trading interval 'i'

$\tilde{y}_{upper}(i)$, y(i), and $Sign\lfloor y(i) - \tilde{y}(i)_{upper} \rfloor$ – as in Eq. 14

Fig. 14 shows the differences in a market exposure between the uniform and real-shaped demand approaches for the four categories of market participants. As can be seen, these differences are relatively small, with the largest value of 2.3% for the Buyer with the Conservative attitude to risk taking. Therefore, the further analysis of a market exposure will be carried out only for a real-shaped demand.

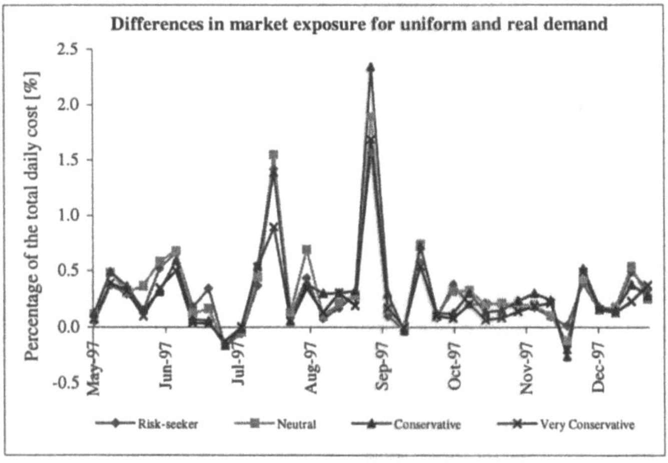

Fig. 14 Differences in market exposure between the uniform and real demand when spot prices above the upper margins (Mondays)

Fig. 15 shows the percentage of the daily energy cost of an additional market exposure for electricity Buyers with the Risk-seeking and Neutral attitudes on Mondays. Although an additional market exposure for Buyers with the Risk-seeking attitude can reach significant values for some periods (as for example 34% in September 1997), the average value calculated over 8 months (from the beginning of May 1997 until the end of December 1997) accounts for 9.6%. The maximum exposure of a Buyer with the Neutral attitude accounts for 31.5% with an average value equal to 8%.

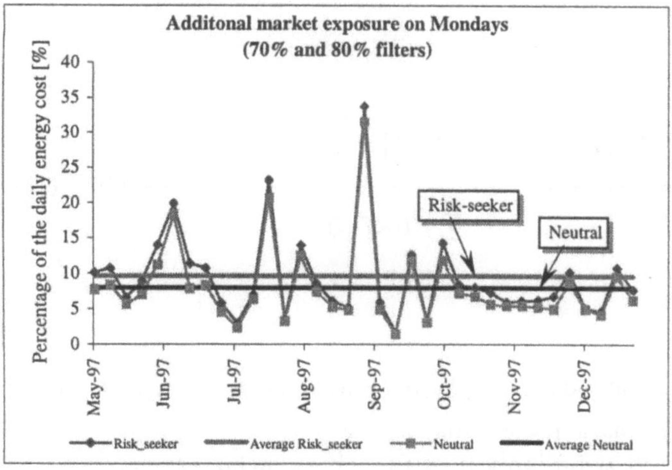

Fig. 15 Additional market exposure for Buyers with Risk-seeking and Neutral attitudes when spot prices above the upper margins (Mondays)

3.3.2 Spot Price below the Lower Estimates

The daily cost of additional market participants' exposure to prices below the lower estimates can be calculated according to Eq. (12) for uniform demand and (13) for the real-shaped demand as follows:

Uniform demand

$$DAE_{Lower} = \frac{\sum_{i=1}^{48} \left[y(i) - \tilde{y}_{lower}(i)\right] \times Sign\left[y(i) - \tilde{y}_{lower}(i)\right]}{\sum_{i=1}^{48} y(i)} 100\% \qquad (16)$$

where:

DAE_{lower} – daily cost of additional market exposure when prices below the lower estimates

$y(i)$ – spot price in the trading interval 'i'

$\tilde{y}_{lower}(i)$ – lower-trend approximation of spot price in the trading interval 'i'

$$Sign\left[y(i) - \tilde{y}(i)_{lower}\right] = \begin{cases} 1 \text{ when } \left[y(i) - \tilde{y}_{lower}(i)\right] \leq 0 \\ 0 \text{ when } \left[y(i) - \tilde{y}_{lower}(i)\right] > 0 \end{cases}$$

Real demand

$$DAE_{lower}^{coord} = \frac{\sum_{i=1}^{48} \left[(y(i) - \tilde{y}_{lower}(i)) D(i) \right] \times Sign\left[y(i) - \tilde{y}_{lower}(i) \right]}{\sum_{i=1}^{48} y(i) \times D(i)} 100\% \quad (17)$$

where:

DAE_{lower}^{coord} – daily additional Buyers' exposure - coordinated demand
$D(i)$ – Buyer demand in the trading interval 'i'
$\tilde{y}_{lowerr}(i)$, $y(i)$, and $\mathrm{Sign}\left[y(i) - \tilde{y}(i)_{lower}\right]$ – as in Eq. 12

Spot prices below the lower estimates affect both the electricity Buyers and Sellers. In this case, Buyers can purchase cheap electrical energy on the spot market. However, when the estimates have been applied to evaluate strike prices in hedge contracts, and the strike prices agreed is above the actual spot prices, the electricity Buyer has to compensate Generators for the difference. These differences can represent the loss of possible revenues for electricity Buyers in relation to the revenues that would have been obtained, if they had contracted less energy in hedge contracts.

Spot prices below the estimates do not affect directly electricity producers when they have contracted the bulk of generation in hedge contracts. However, low prices may result in non-scheduling of units when bidding prices are too high. Generally, producers with one-way hedge contracts should not generate energy in trading intervals with low prices.

3.3.3 Distribution of Additional Costs

In the decision-making processes, it is important to have information not only on the average value of the additional exposure, but also on the distribution of additional costs. Despite various patterns of the cost distribution for days of a week, several common elements can be distinguished. First, a Very Conservative market participant may expect that 80 ÷ 90% of spot price deviations above the upper limits are within 0 ÷ 5% of the daily total cost. On the other hand, market participants with the Risk-seeking attitude may expect 40 ÷ 50% of price deviations within the range of 5 ÷ 10%. In addition, they should expect that only 10% of spot price deviations are within 0 ÷ 5% of the total daily cost.

An example of the distribution of additional costs for market participants with different attitudes to risk taking when spot prices are above the upper estimates is shown in Fig. 16 (a), and when spot prices are below the estimates in Fig. 16 (b).

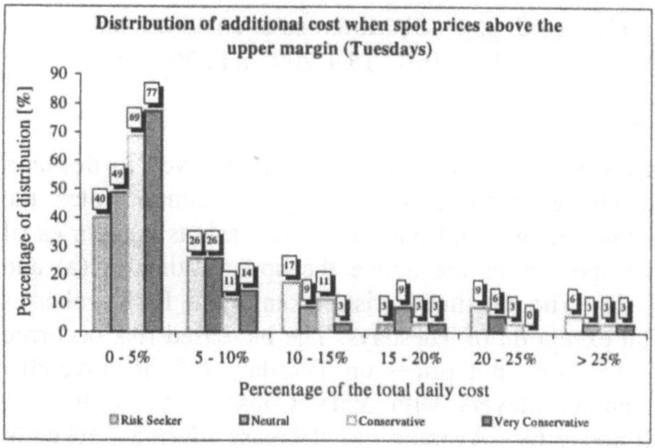

(a)

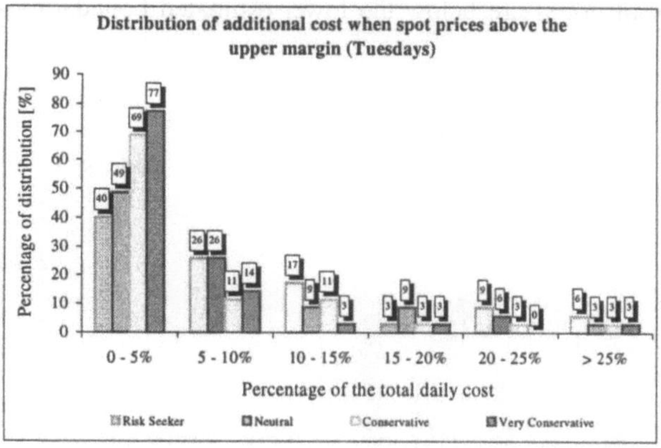

(b)

Fig. 16 Distribution of additional costs when spot prices above the upper estimates (a) and below the lower estimates (b) - Tuesdays

3.4 Risk Taken by Market Participants

The information on additional costs and their distribution is essential for decision-making processes. However, the most important parameter is the risk related to the different attitudes of market participants to a market game. In general, the risk is calculated as a product of the cost incurred and the probability of an event. In the spot market analysis, the risk can be calculated as follows:

$$\begin{aligned} &Risk = \\ &= C_{d:0-5} w_5 + C_{d:5-10} w_{10} + C_{d:10-15} w_{15} + \\ &+ C_{d:15-20} w_{20} + C_{d:20-25} w_{25} + Max_{\cos t} P_{d>25} \end{aligned} \tag{18}$$

where:

w_5, w_{10}, w_{20} and w_{25} – weighting coefficients equal to 0.05, 0.10, 0.15, 0.20 and 0.25, respectively

$C_{d:0\text{-}5}$, $C_{d:5\text{-}10}$, $C_{d:10\text{-}15}$, $C_{d:15\text{-}20}$, $C_{d:20\text{-}25}$ – distributions of an additional cost in bands $0 \div 5\%$, $5 \div 10\%$, $10 \div 15\%$, $15 \div 20\%$, and $20 \div 25\%$

The relation (18) calculates the maximum risk that may be involved in decision making based on the four characteristic attitudes. Fig. 17 demonstrates the maximum risk as the percentage of the total daily cost for various categories of market participants when the spot prices are above the upper estimates (a) and below the lower estimates (b). The maximum risk taken by a Risk-seeker is generally below 12%, with an exception of Tuesdays. The increased risk occurred on Tuesdays is a result of very high spot prices on Tuesday, 25th of November 1997. The risk taken by market players with Very Conservative attitude to electricity contracting is still above 6%. Average risk decreases during weekends, when spot prices are small – see Fig. 17 (a). Similar patterns of an average risk have been obtained when spot prices are below the lower estimated boundaries - see Fig. 17 (b).

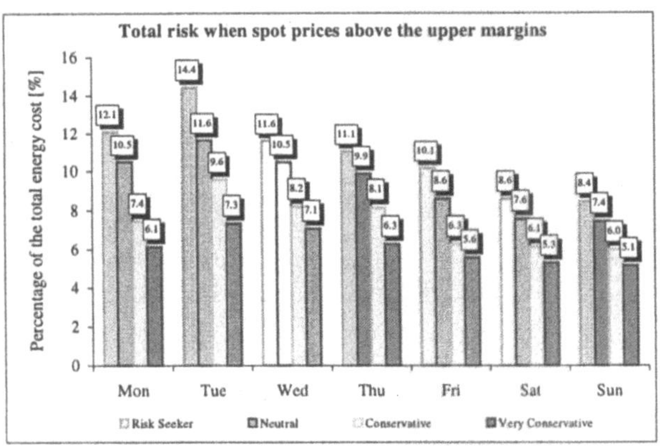

(a)

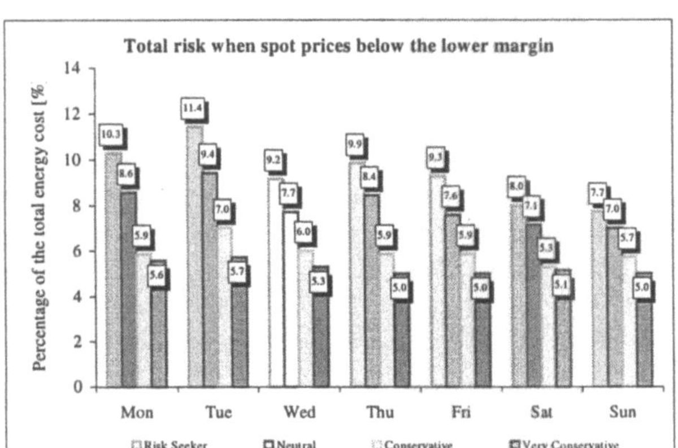

(b)

Fig. 17 Total risk when the spot prices above the upper (a) and estimate below the lower (b) estimates

4 Contracting Electrical Energy Using Fuzzy Set Approach

The analysis discussed above involved direct relationships between daily demand and daily spot prices for the eight months in 1997 of the National Electricity Market operation. Fuzzy patterns, additional costs, and the risk involved when using fuzzy patterns with various attitudes of decision-makers to the risk taking were calculated as an ex-post analysis. Such an analysis is essential for spot price pattern identification. However, the fuzzy patterns developed for particular days of a week can be applied into a prediction of spot prices in the coming days.

4.1 Estimate Categories

Several categories of estimates can be involved in a decision-making process. Three relationships have been selected to estimate patterns of spot prices for days of a week:

Weighted-sum-type estimate

$$\mu_i(x) = \alpha_{i-1}\mu_{i-1}(x) + \alpha_{i-2}\mu_{i-2}(x) + \alpha_{i-3}\mu_{i-3}(x) + \ldots + \alpha_l\mu_l(x) \tag{19}$$

where:

$\mu_i(x)$ – fuzzy estimate of spot prices for a given attitude of market players

$\mu_{i-1}(x), \mu_{i-2}(x), \mu_{i-3}(x), \ldots, \mu_1(x)$ – fuzzy patterns of spot prices in previous days

$\alpha_{i-1}(x), \alpha_{i-2}(x), \alpha_{i-3}(x), \ldots, \alpha_1(x)$ – coefficients of importance

- *Max-type of the estimate*

$$\mu_i(x) = \max\{\alpha_{i-1}\mu_{i-1}(x) + \alpha_{i-2}\mu_{i-2}(x) + \alpha_{i-3}\mu_{i-3}(x) + \ldots + \alpha_l\mu_l(x)\} \tag{20}$$

where

$\mu_i(x)$, $\mu_{i-1}(x), \mu_{i-2}(x), \mu_{i-3}(x), \ldots, \mu_1(x)$, $\alpha_{i-1}(x), \alpha_{i-2}(x), \alpha_{i-3}(x), \ldots, \alpha_1(x)$ as defined for Eq. (19).

- *Min-type of the estimate*

$$\mu_i(x) = \min\{\alpha_{i-1}\mu_{i-1}(x) + \alpha_{i-2}\mu_{i-2}(x) + \alpha_{i-3}\mu_{i-3}(x) + \ldots + \alpha_l\mu_l(x)\} \tag{21}$$

where

$\mu_i(x)$, $\mu_{i-1}(x), \mu_{i-2}(x), \mu_{i-3}(x), \ldots, \mu_1(x)$, $\alpha_{i-1}(x), \alpha_{i-2}(x), \alpha_{i-3}(x), \ldots, \alpha_1(x)$ as defined for Eq. (19).

The number of patterns used to estimate the future behavior of spot prices depends on a priori assumption supported by a market participant's experience and an initial spot price analysis. For this analysis, 'i' equals to 5 has been selected. That means that the data from four weeks have been applied to estimate the spot prices in the fifth week and this method is called a four-step filtering procedure.

Weighted-sum Type Estimate

Fuzzy filters used for representation of spot price deviations have three characteristic parameters: an average value, upper limit, and lower limit. A four-step filtering procedure based on the weighted-sum type estimate is defined as below:

$$\tilde{y}_l^k(i) = f\left(\hat{m}^k, \hat{b}_{trend}^k, \hat{b}_{upper}^k, \hat{b}_{lower}^k, x^k(i)\right) \qquad (22)$$

where:

$\tilde{y}_l^k(i)$ – estimated spot price for 'i' trading interval in 'k' trading days for the main trend, and upper and lower margins.

l – main, upper or lower estimates

This provides the main trend estimates and two (upper and lower) margins as follows:

$$\tilde{y}_{main}^k(i) = \hat{m}^k x^k(i) + \hat{b}_{trend}^k$$

$$\tilde{y}_{upper}^k(i) = \hat{m}^k x^k(i) + \hat{b}_{upper}^k$$

$$\tilde{y}_{lower}^k(i) = \hat{m}^k x^k(i) + \hat{b}_{lower}^k$$

where:

k – day of estimates, starting from the first week of June 1997, i – trading interval

$$\hat{m}^k = \alpha_1^{k-1} m^{k-1} + \alpha_1^{k-2} m^{k-2} + \alpha_1^{k-3} m^{k-3} + \alpha_1^{k-4} m^{k-4}$$

$$\hat{b}_{trend}^k = \alpha_2^{k-1} b_{trend}^{k-1} + \alpha_2^{k-2} b_{trend}^{k-2} + \alpha_2^{k-3} b_{trend}^{k-3} + \alpha_2^{k-4} b_{trend}^{k-4}$$

$$\hat{b}_{upper}^k = \alpha_3^{k-1} b_{upper}^{k-1} + \alpha_3^{k-2} b_{upper}^{k-2} + \alpha_3^{k-3} b_{upper}^{k-3} + \alpha_3^{k-4} b_{upper}^{k-4}$$

$$\hat{b}_{lower}^k = \alpha_4^{k-1} b_{lower}^{k-1} + \alpha_4^{k-2} b_{lower}^{k-2} + \alpha_4^{k-3} b_{lower}^{k-3} + \alpha_4^{k-4} b_{lower}^{k-4}$$

Since there are no preferences for a specific day to determine values of coefficients of importance, they have been assumed equal to 0.25 for the four-step estimating procedure. This procedure is similar to a four-step averaging procedure. Because the National Electricity Market started at the beginning of May 1997, the first estimate was applied at the beginning of June 1997.

Using the relations (14) ÷ (17), three trends for each day of a week have been constructed for a period from June to December 1997. The analysis of additional costs of the market exposure based on a four-step filtering procedure for Mondays is shown in Fig. 18. Despite some deviations in an additional cost, the average values of an additional exposure is about 9% for the Risk-seeker, while the average additional exposure for Very Conservative attitude is below 5% - Fig. 18 (a). A similar increase is observed for spot prices below the estimate, see Fig. 18 (b).

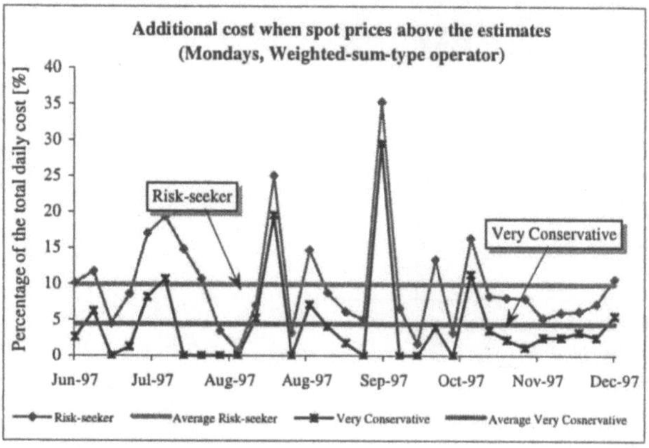

(a)

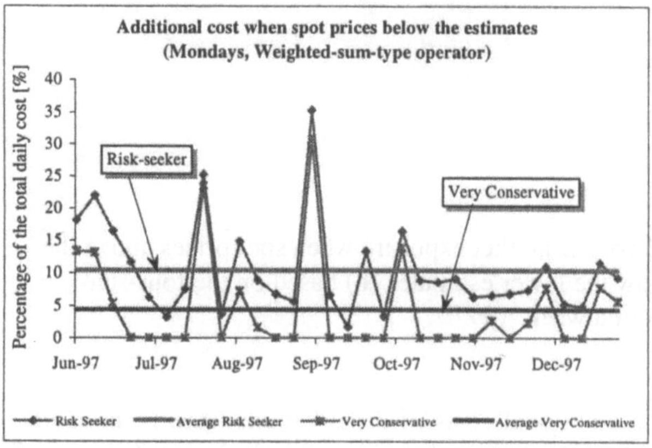

(b)

Fig. 18 Additional cost of market exposure for managers with very conservative and risk-seeking attitudes when spot prices above the upper (a) and below the lower estimates – Mondays, 1997

Average additional costs of a market exposure for days of a week when prices are above and below the estimates are shown in Fig. 19. The average additional cost for market participants with Risk-seeking attitude is below 12%, while a Very Conservative participant should expect an additional cost below 6%. In general, the additional cost expected is smaller when prices are below estimates, which results from smaller price variations.

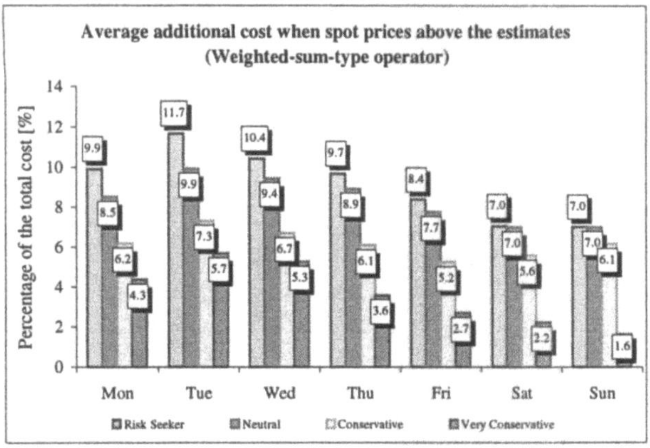

(a)

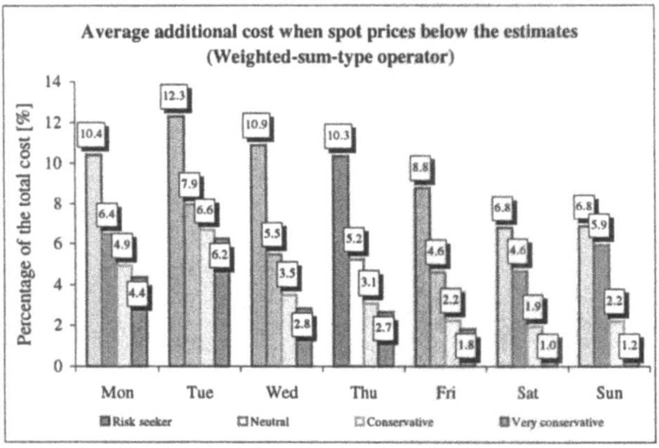

(b)

Fig. 19 Average additional cost of market exposure when spot prices above the upper estimates (a) and below the lower estimates (b) based on the four-step filtering procedure for days of a week, 1997

The distribution of possible additional costs of a market exposure is an essential parameter in decision-making processes. An example of this distribution for Mondays is shown in Fig. 20. Additional costs are distributed exponentially with some exception for market participants with Risk-seeking and Neutral attitudes.

Average total risk when spot prices above the upper estimates based on the four-step filtering for days of a week is shown in Fig. 21. It can be observed that the total risk decreases during non-working days due to low electricity prices.

Max-type Operator

When the max-type fuzzy operator is employed to determine patterns of spot prices, the trends of the expected spot prices are derived using general relationship (22) in which:

$\tilde{y}_l^k(i)$ – estimated price for 'i' trading interval in 'k' trading day for the main trend and two margins (upper and lower). The parameters are computed as follows:

$$\hat{m}^k = \max\left\{\alpha_1^{k-1} m^{k-1} + \alpha_1^{k-2} m^{k-2} + \alpha_1^{k-3} m^{k-3} + \alpha_1^{k-4} m^{k-4}\right\}$$
$$\hat{b}_{trend}^k = \max\left\{\alpha_2^{k-1} b_{trend}^{k-1} + \alpha_2^{k-2} b_{trend}^{k-2} + \alpha_2^{k-3} b_{trend}^{k-3} + \alpha_2^{k-4} b_{trend}^{k-4}\right\}$$
$$\hat{b}_{upper}^k = \max\left\{\alpha_3^{k-1} b_{upper}^{k-1} + \alpha_3^{k-2} b_{upper}^{k-2} + \alpha_3^{k-3} b_{upper}^{k-3} + \alpha_3^{k-4} b_{upper}^{k-4}\right\}$$
$$\hat{b}_{lower}^k = \max\left\{\alpha_4^{k-1} b_{lower}^{k-1} + \alpha_4^{k-2} b_{lower}^{k-2} + \alpha_4^{k-3} b_{lower}^{k-3} + \alpha_4^{k-4} b_{lower}^{k-4}\right\}$$

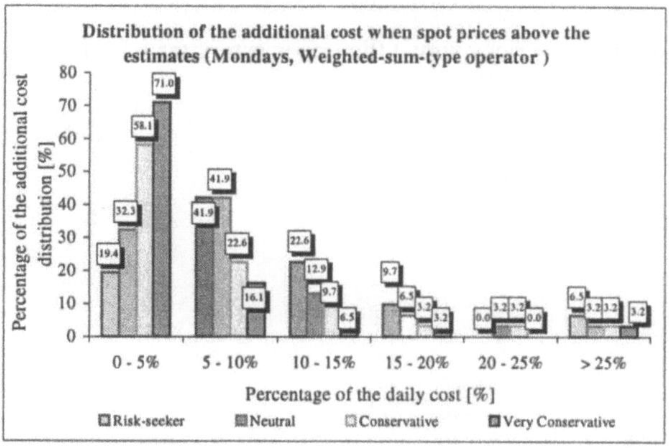

(a)

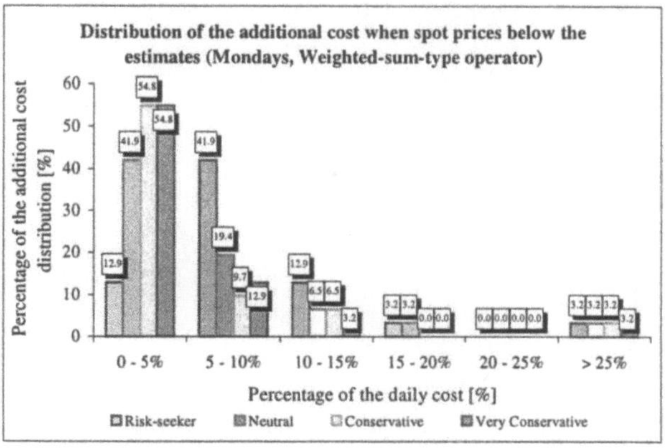

(b)

Fig. 20 Distribution of additional costs when spot prices above the upper estimates (a) and below the lower estimates (b) based on the four-step filtering procedure for Mondays in 1997

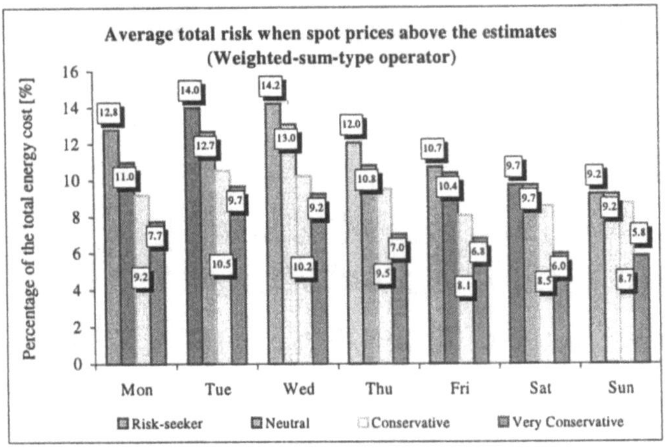

Fig. 21 Average total risk when spot prices above the upper estimates based on the four-step filtering for days of a week, 1997

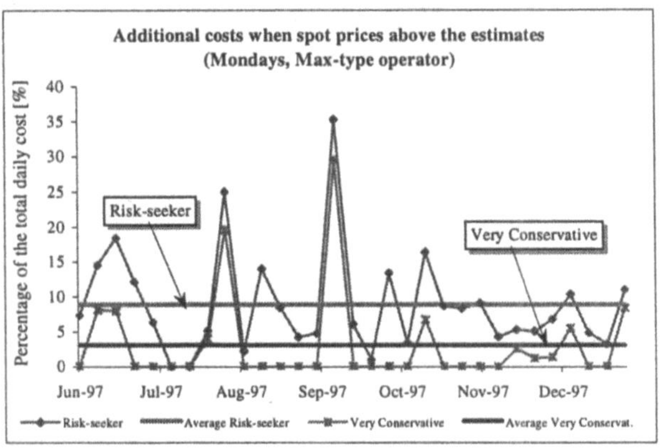

(a)

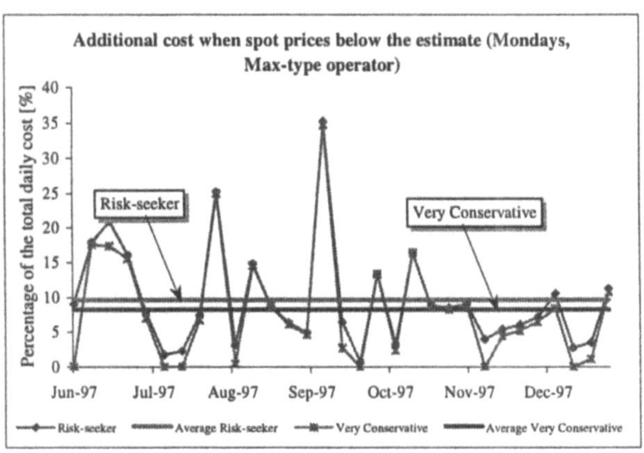

(b)

Fig. 22 Additional costs of the market exposure for the max-type operators, when spot prices above the estimate (a) and below the estimate based on the max-type operators on Mondays, 1997

An application of max-type operators to estimate patterns of spot prices results in the additional costs of market exposures, as shown in Fig. 22. When a max-type operator is applied with a combination of a Very Conservative attitude of the market participant, an additional cost is often equal to zero. This is an effect of wide margins of a spot price trend. It can be concluded that a low level of an additional cost can be obtained by widening trend boundaries, i.e. an introduction of more uncertainty.

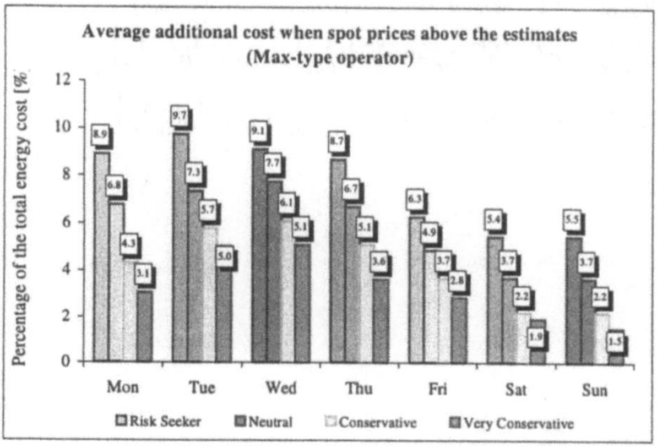

(a)

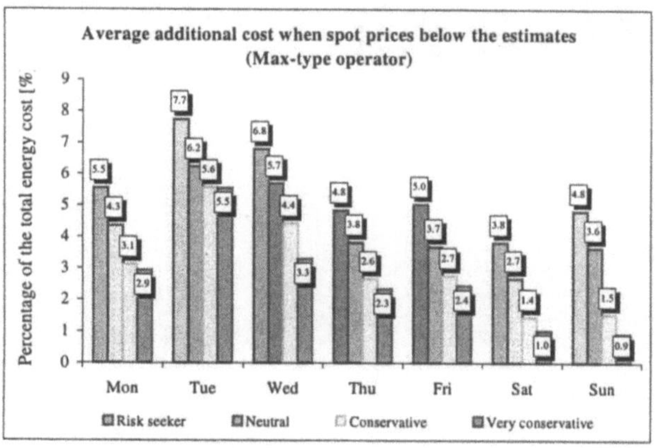

(b)

Fig. 23 Additional costs for the market exposure when spot prices above the upper estimate (a) and below the lower estimate (b) based on the max-type operator for days of a week in 1997

Average additional costs of market exposure for days of are week are shown in Fig. 23. The highest additional cost is observed on Tuesdays - the influence of the 25th November 1997 high prices. This cost decreases gradually towards the end of a week. The total risk of market exposure is shown in Fig. 24 with the lowest risk observed on Saturdays and Sundays.

Min-type Operator

When a min-type operator is used to determine patterns of spot prices, the main trend and two boundaries are calculated using the general relationship (22), which parameters are defined as follows:

$$\hat{m}^k = \min\left\{\alpha_1^{k-1} m^{k-1} + \alpha_1^{k-2} m^{k-2} + \alpha_1^{k-3} m^{k-3} + \alpha_1^{k-4} m^{k-4}\right\}$$

$$\hat{b}_{trend}^k = \min\left\{\alpha_2^{k-1} b_{trend}^{k-1} + \alpha_2^{k-2} b_{trend}^{k-2} + \alpha_2^{k-3} b_{trend}^{k-3} + \alpha_2^{k-4} b_{trend}^{k-4}\right\}$$

$$\hat{b}_{upper}^k = \min\left\{\alpha_3^{k-1} b_{upper}^{k-1} + \alpha_3^{k-2} b_{upper}^{k-2} + \alpha_3^{k-3} b_{upper}^{k-3} + \alpha_3^{k-4} b_{upper}^{k-4}\right\}$$

$$\hat{b}_{lower}^k = \min\left\{\alpha_4^{k-1} b_{lower}^{k-1} + \alpha_4^{k-2} b_{lower}^{k-2} + \alpha_4^{k-3} b_{lower}^{k-3} + \alpha_4^{k-4} b_{lower}^{k-4}\right\}$$

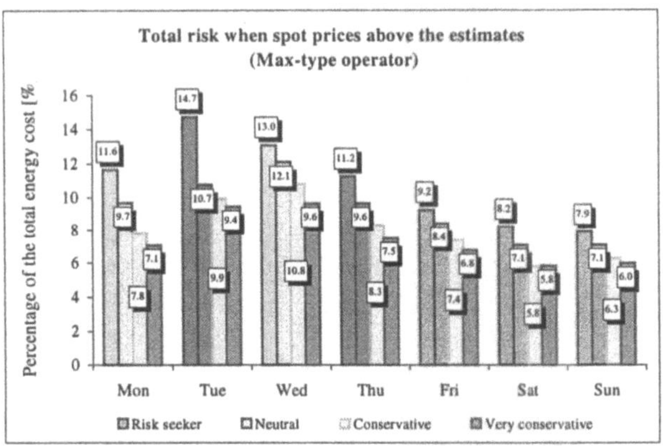

(a)

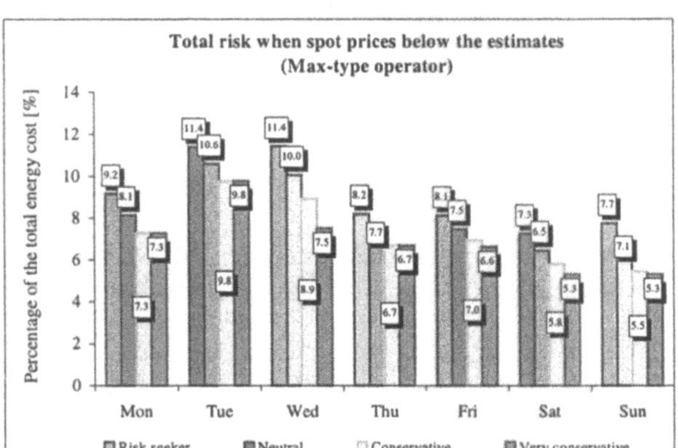

(b)

Fig. 24 Total risk of market exposure when spot prices above the upper estimate (a) and below the lower estimate based on a max-type operator for days of a week in 1997

Additional costs of the market exposure for Risk-seeking market participants that result from the differences between the upper and lower boundaries of trend lines and actual spot prices are shown in Fig. 25. It can be observed that, although an additional daily cost of the market exposure may reach the value above 30%, the average cost calculated for seven months of the National Market operation is below 9% for market participants with Risk-seeking attitudes.

Average additional costs of a market exposure for the min-type operator calculated for days of a week between June and December 1997 are shown in Fig. 26. Fig. 27 shows the total average risk for days of a week for four attitudes of market participants to risk taking.

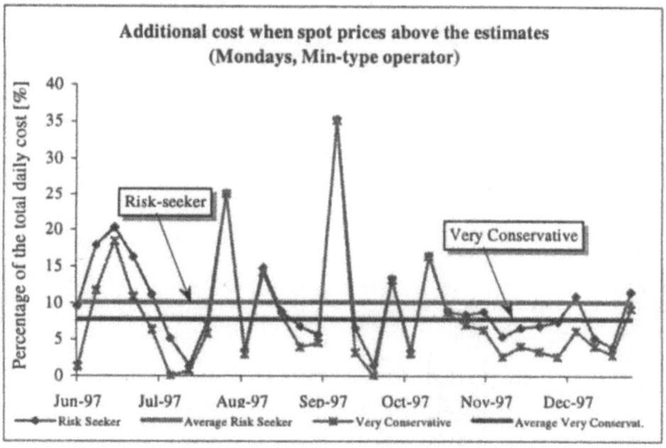

(a)

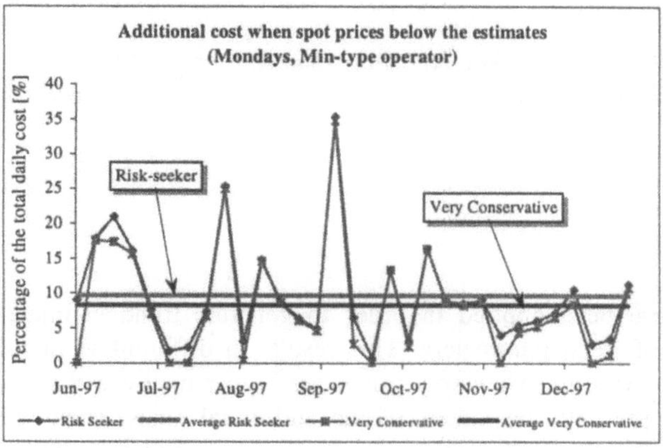

(b)

Fig. 25 Additional costs of market exposure when spot prices above the upper estimate (a) and below the lower estimate based on the min-type operator on Mondays in 1997

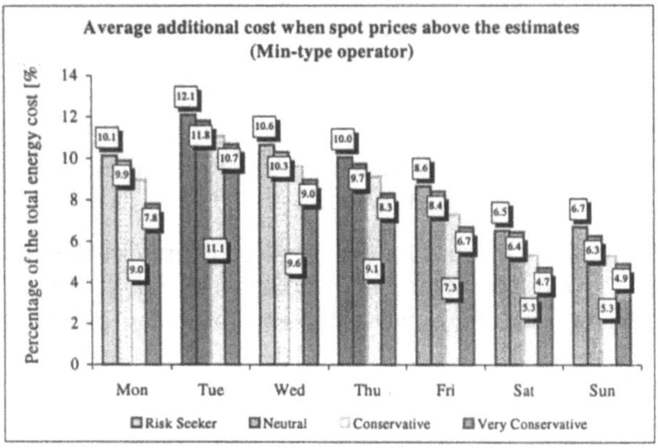

(a)

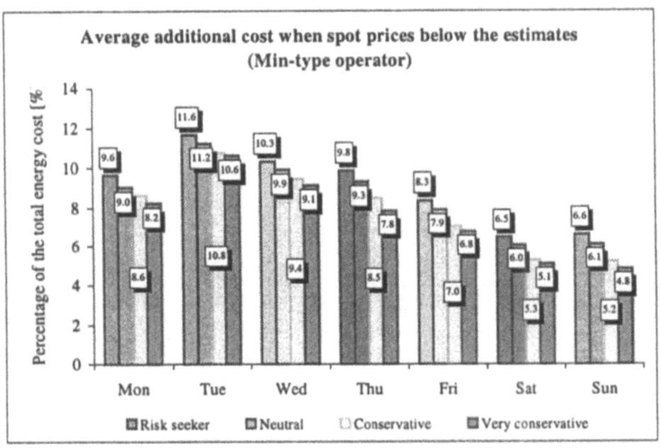

(b)

Fig. 26 Average additional costs of market exposure when spot prices above the upper estimate (a) and below the lower estimate (b) based on the min-type operator for days of a week in 1997

4.2 Comparison of Operators

Three fuzzy operators have been applied in order to generate trend estimates, providing various values of trend parameters. This results in different values of additional costs of market exposure when spot prices are higher or lower than the margins defined. Fig. 28 presents the values of the additional costs when spot prices are above the margin estimated for three fuzzy operators and four trading attitudes of market participants to risk taking.

When a Risk-seeking attitude is applied, the differences between three operators are insignificant. A Risk-seeking market participant accepts spot price boundaries when 70% of spot prices are within these boundaries and the main trend line. When Conservative and Very Conservative attitudes to spot trading are implemented, significant differences occur between the costs' estimated using various operators. Weighted-sum-type and maximum-type operators show lower market exposure comparing to a minimum-type operator. When market participants employ a min-type approach to a spot price estimation, their exposure to additional costs is the largest. It can be noticed that a min-type operator is insensitive to market attitudes, providing similar values of a market exposure independently on market participants' attitudes.

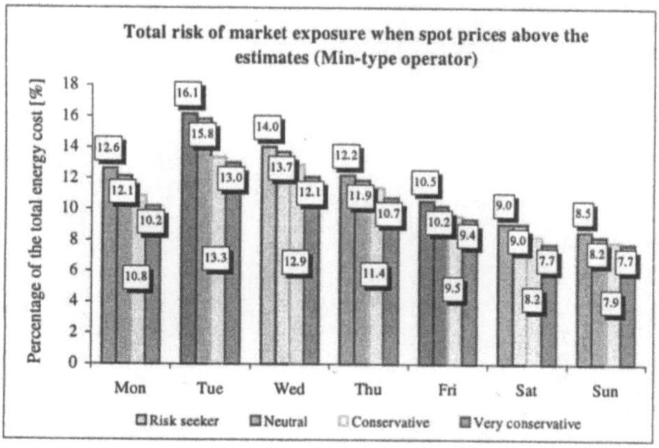

(a)

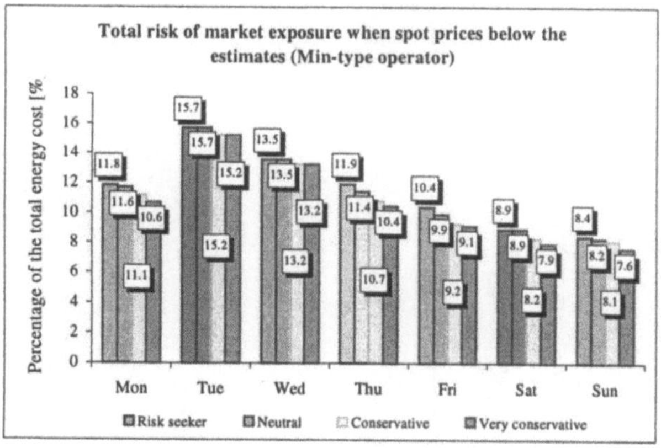

(b)

Fig. 27 Total risk of market exposure when spot prices above the upper estimate (a) and below the lower estimate (b) based on the min-type operator for days of a week in 1997

Fig. 29 demonstrates the additional costs when spot prices are below the lower estimate. These costs can be treated as the overpaid for electrical energy when a strike price in a hedge contract is based on trend estimates. Then, electricity Buyers have to compensate electricity Sellers the difference between the actual spot price and the strike price agreed. A min-type operator is practically insensitive to managers' attitudes. This operator also provides the highest value of additional costs. However, the values of additional costs when spot prices are below the estimates are usually smaller than in a case when spot prices are above the estimates. This can be explained by smaller price variations within a low-price range. Given that an average spot price is around $15 ÷ $20 and the lowest possible

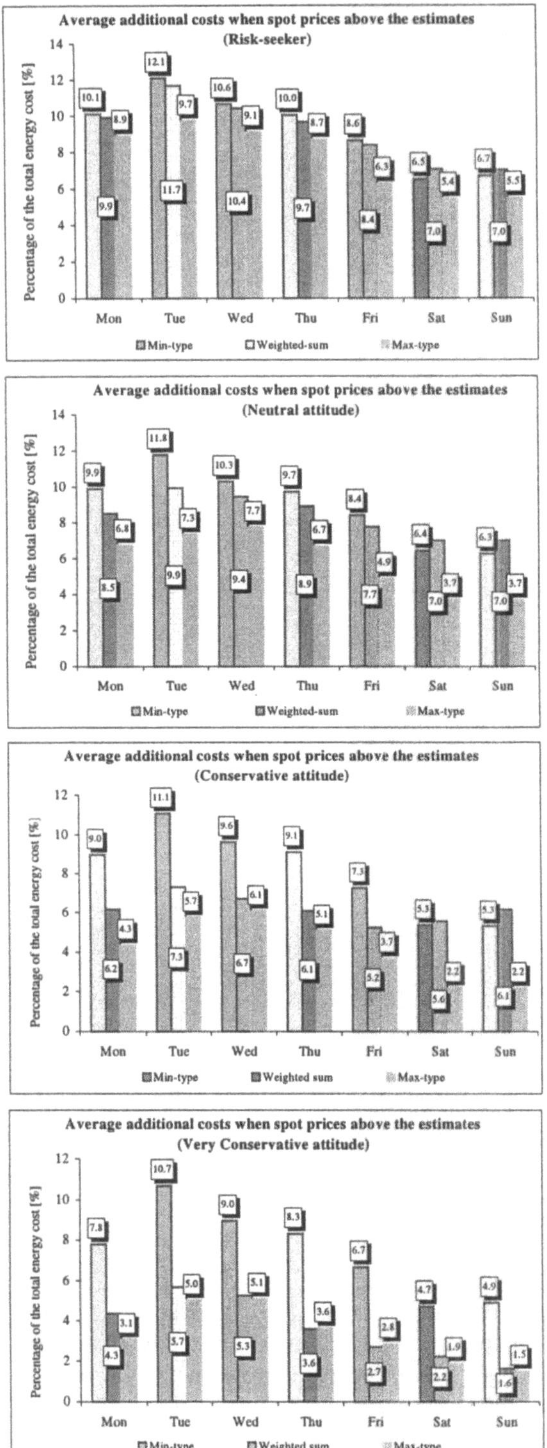

Fig. 28 Average additional costs of market exposure when spot prices are above the upper margins for market managers with different attitudes to risk taking

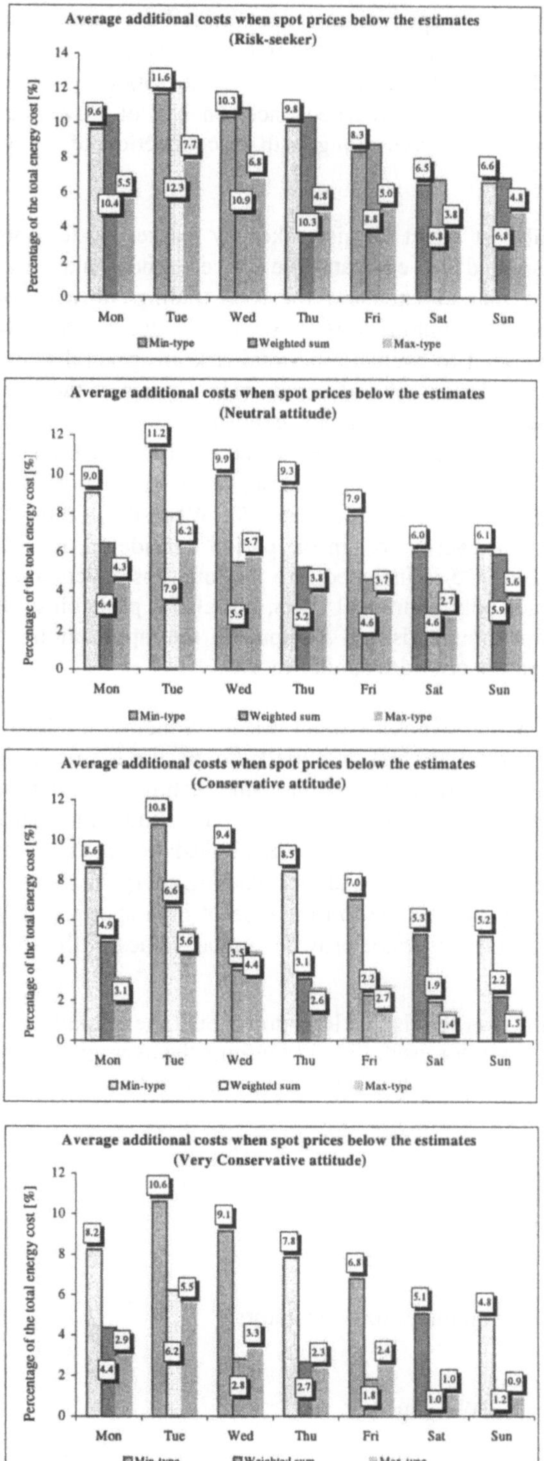

Fig. 29 Average additional costs of market exposure when spot prices are below the lower margins for market managers with different attitudes to risk taking

spot price is equal to zero, a decrease in spot prices cannot exceed the value of an average spot price. On the other hand, in a high price region, spot prices may reach up to VoLL (the Value of Loss Load), which is currently equal to $5000 per MW. Up to date, this highest spot price has been reached once, on 25th of November 1997, when abnormal high demand occurred along with disconnection of a few transmission lines.

Average values of the additional cost affect the risk taken by market participants when various operators are applied to estimate the price trends and their boundaries. The total risk of a market exposure for four trading attitudes of decision-makers, calculated as the maximum risk, is shown in Fig. 30. It should be noted that the fuzzy operators applied to estimate the total risk are generally less sensitive than those applied in calculations of additional costs. Values of the total risk are usually higher than values of an additional cost exposure. This is caused by two main factors. First, a total risk has been calculated as the maximum risk taking into account the highest limit of probabilities. The additional costs are classified in five bands, starting from 0 to 25% with the step equal to 5%. When the additional cost is higher than 25% of the total energy cost in the period considered, its real value with the probability equal to 0.25 is included into the total cost calculation. Second, the total risk includes not only additional costs, but also a probability of their appearance. When the maximum risk is computed, the application of weighted-sum and max-type operators provides similar results.

The total risk of an additional market exposure when prices are below the lower margins is shown in Fig. 31. Generally, the risk of a market exposure is smaller in this case due to smaller spot price deviations below the values estimated. It can be observed that a min-type operator implies the largest risk. The assumption that a trend boundary is a minimum from the previous four weeks leads to small values of differences between the main trend and its boundaries, leaving a large number of spot price data outside the defined regions. Like in the case of high spot prices, a weighted-sum operator and a max-type operator provide similar values of the total risk.

It is also necessary to compare the average risk taking into account the market managers' attitude and the fuzzy operator used. Fig. 32 presents the average values of the total, maximum risk. The notation used in these figures is as follows:

- R_S – Risk-seeking attitude
- N – Neutral attitude
- C – Conservative attitude
- V_C – Very Conservative attitude

These four attitudes are combined with three fuzzy operators:

- Minimum-type operator (Min)
- Weighted-sum-type operator (Weighted)
- Maximum-type operator (Max)

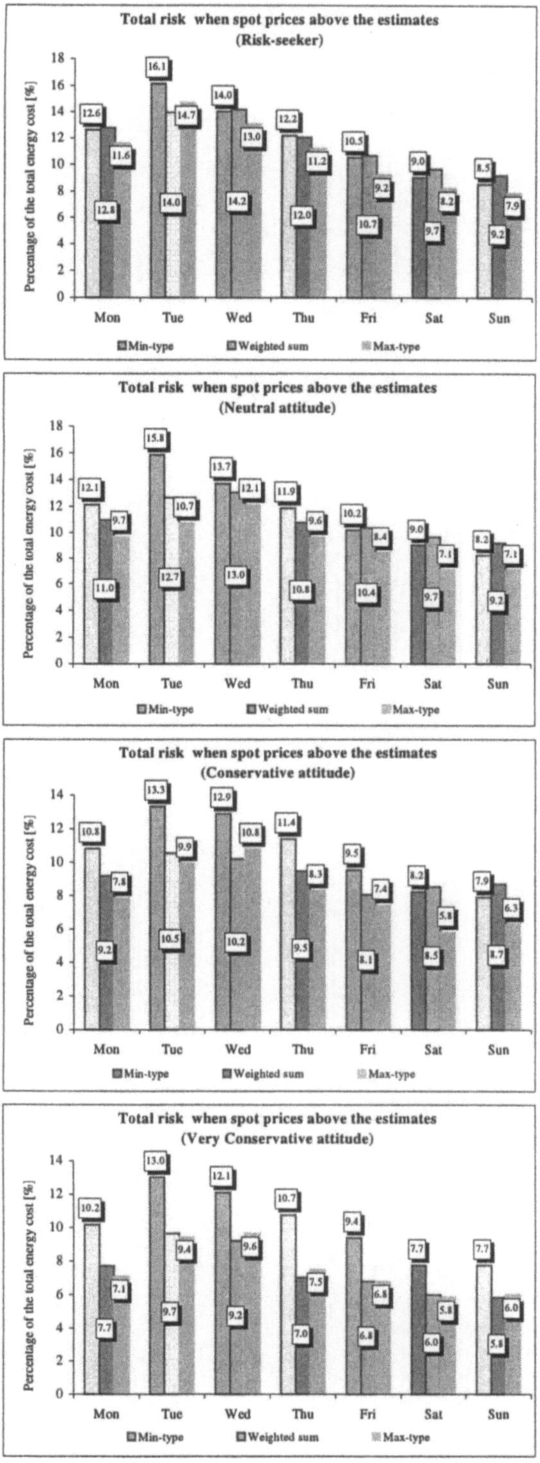

Fig. 30 Total risk of market exposure when spot prices above the upper margins for managers with different attitudes to risk taking, days of a week, 1997

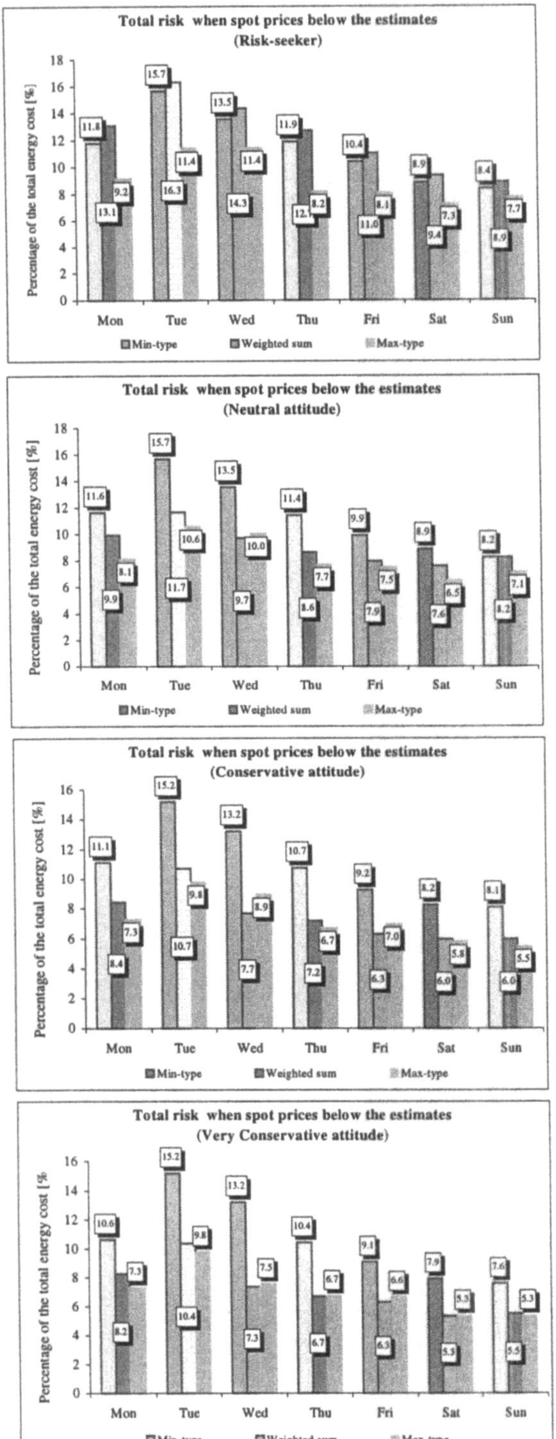

Fig. 31 Total risk of market exposure for managers with different attitudes to risk taking when spot prices below the lower estimates, days of a week, 1997

If the spot price estimates were applied from June to December 1997, market participants would be subject to the total risk of an additional market exposure about 10%. There are not significant differences resulting from various fuzzy operators applied to compute the trend estimates. The most significant difference caused by various market attitudes is approximately 4%. Such values for the maximum market exposure seems to be acceptable for a majority of market participants. These encouraging figures can justify the further studies with other parameters including an energy transfer level in order to reduce the total risk involved with the uncertainty of the price estimation.

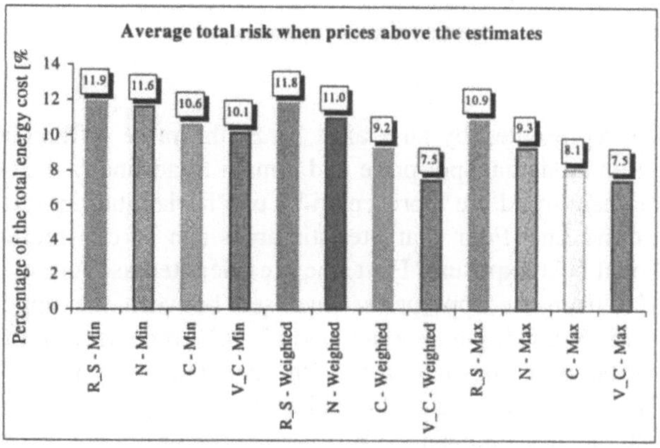

(a)

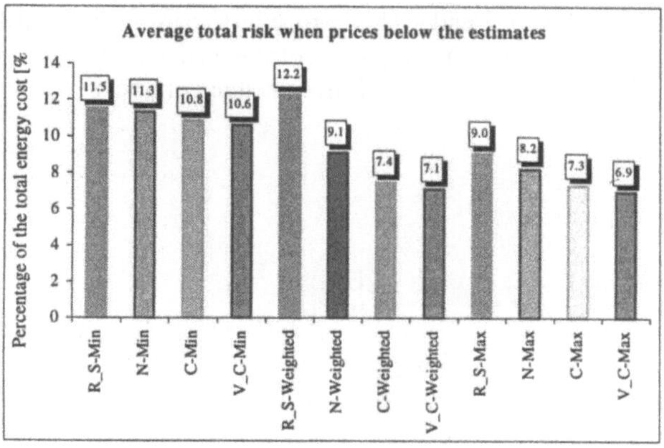

(b)

Fig. 32 Average total risk exposure when spot prices are above the estimates (a) and below the estimates (b) for different market participants' attitudes and different fuzzy filters applied

4.3 Costs between upper and lower Limits and outside the Margins

The additional costs of a market exposure have been calculated as the total cost of energy above the estimates for trading periods when spot prices are outside the upper margins, as well as for trading periods when spot prices are below the lower boundaries. This method of calculation allows the direct comparison of both the different attitudes of market participants to risk taking and the different fuzzy operators applied. However, this approach does not provide the detailed information on the portion of the costs incurred within boundaries i.e., the difference between the boundaries and the estimates, and outside of the upper and lower margins defined. The more detailed approach requires the breaking of the additional costs into two main groups:

- electricity costs when spot prices are between the main trend line and the boundaries
- electricity costs when spot prices are outside the boundaries

In Fig. 33, the main trend is represented by an x-axis. Thus, the price difference represents the distance between an actual spot price and a main trend line (x-axis). Therefore, boundaries of the main trend are represented by two horizontal lines set above and below the main trend line. Four characteristic areas can be considered when determining an additional cost exposure. First, the area denoted as 'A+' that represents the costs incurred from the spot prices enclosed between the upper margin and the main trend line. Second, the area denoted as 'A–', representing the costs when spot prices vary between the main trend line and the lower margin. Next, the area denoted as 'B+' that represents additional costs when spot prices are above the upper margin. Last, the area denoted as 'B–' where spot prices are below the lower margin. The above method of calculation is especially useful when a market participant is considering a hedge contract with a strike price equal to the upper or lower margin of the main trend. This procedure can provide the information on the costs incurred from shifting of a strike price to the trend margins and can determine additional costs that result from hedge contracts.

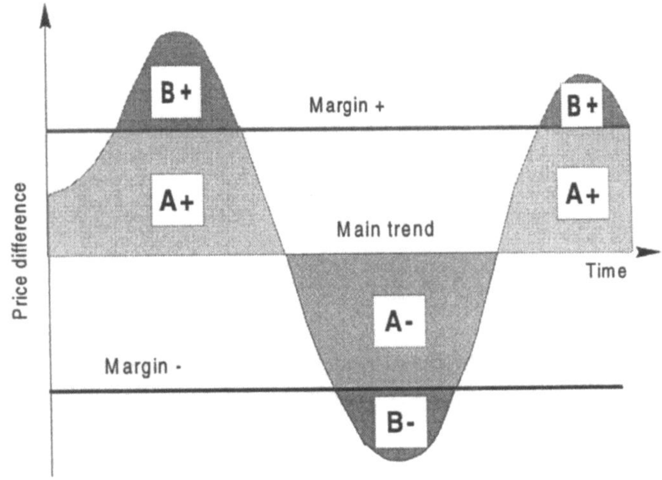

Fig. 33 Portion of the costs incurred within the boundaries and outside of the upper and lower margins

Average values of additional costs for four market participants' attitudes are shown in Fig. 34. The notation used is as follows:

RS_BM – costs when spot prices are Between the Margins for a Risk-seeking attitude
RS_OM – costs when spot prices are Outside the Margins for a Risk-seeking attitude
N_BM – costs when spot prices are Between the Margins for a Neutral attitude
N_OM – costs when spot prices are Outside the Margins for a Neutral attitude
C_BM – costs when spot prices are Between the Margins for a Conservative attitude
C_OM – costs when spot prices are Outside the Margins for a Conservative attitude
VC_BM – costs when spot prices are Between the Margins for a Very Conservative attitude
VC_OM – costs when spot prices are Outside the Margins for a Very Conservative attitude

When spot prices are high, Risk-seekers may expect an additional cost that is composed of two parts:

- Cost of energy purchased when spot prices are between the price estimated and the upper margin
- Cost of energy purchased when spot prices are above the upper margin (outside the margin)

The values presented in Fig. 34 (a) mean that a Risk-Seeking energy seller, who accepts the 70% upper margin and is ready to enter a hedge contract with an upper strike price determined by an upper estimate, should expect about 7% of the total revenue to be paid back to an energy Buyer. On the other hand, a Very Conservative energy producer, who is willing to enter the hedge contract when the upper estimate margin covers 95% of spot price deviations above the main estimate, should expect to pay back less than 2% of his revenue to an energy Buyer. However, a very high upper strike price may not be accepted by an energy Buyer who may prefer to enter a hedge contract with a Risk-seeking energy producer who is ready to offer a lower strike price, based on the 70% upper margin. Fig. 34 (a) can also provide essential information for energy producers who negotiate One-way Hedge Contracts. When determining a fee paid by a Buyer, a Seller should take into account all the additional costs expected. A Seller has to decide how much of additional costs should be covered by this fee.

Fig. 34 (b) presents the values of additional costs calculated as a loss of possible revenues when spot prices are below the level estimated. If a Risk-seeking energy Buyer accepts 70% lower margin, he may expect to pay back to a Generator about 3.5% of the total energy cost. On the other hand, a Very Conservative energy Buyer may try to enter a hedge contract assuming the lowest margin of price estimate. In this case, he will return to an energy Seller less than 0.5% of the total cost. However, it is possible that a Seller would not accept the hedge contract on higher strike price than resulting from his attitude, Fig. 34 (b) can provide information on the costs protected by hedge contracts and the loss of possible revenue from low spot prices.

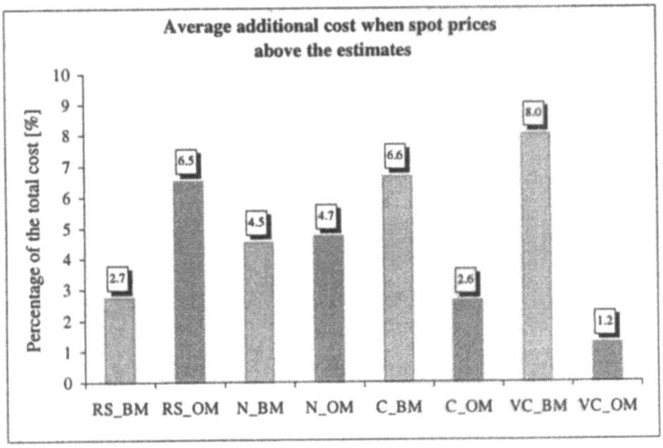

(a)

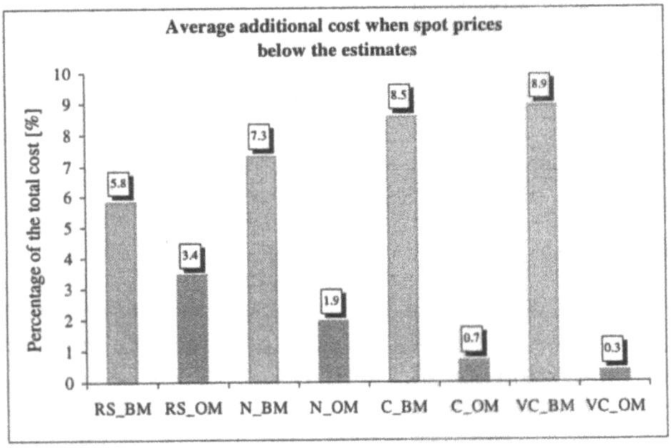

(b)

Fig. 34 Average values of additional costs when spot prices above the estimates (a) and below the estimates for different market participants' attitudes

Fig. 34 can be useful in negotiation of Collar/Floor-Ceiling hedge contracts, providing information on energy costs within and outside the Collar. This figure presents the post ante analysis of spot prices from May 1997 to December 1997 and additional costs calculated when the spot price estimate is based on energy demand. However, for market participants, it is more important to know a market exposure and the risk when spot prices are forecasted. Therefore, three fuzzy operators have been employed to compute the prediction of spot price trends. The results of an additional cost analysis are shown in Fig. 35 when spot prices are above their estimates. In general, the expected additional cost that is calculated on the trend estimate preserves a similar pattern and proportion to the post-ante analysis. This means small values between the margins and high values outside the margins for managers with Risk-seeking attitude, and high values between the margins and low values outside the margins for Very Conservative market participants.

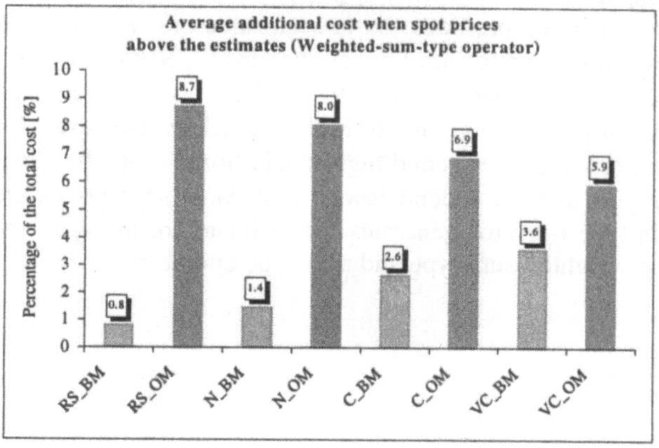

(a)

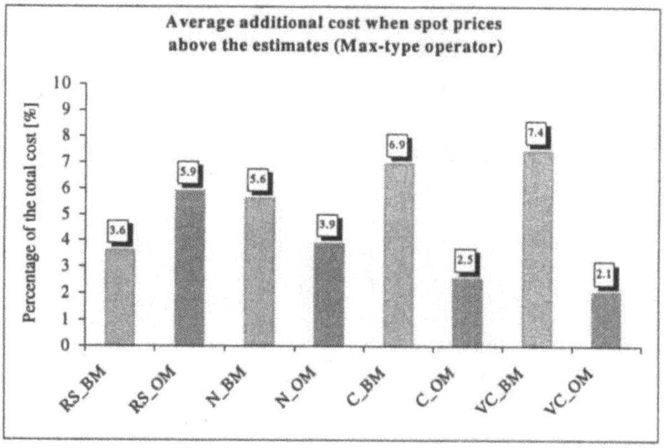

(b)

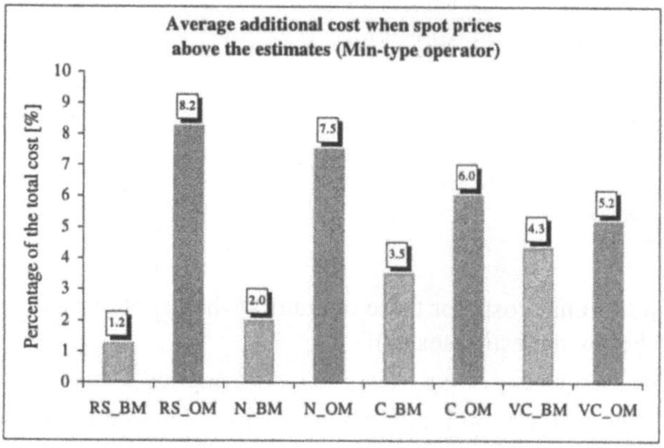

(c)

Fig. 35 Average values of additional costs when spot prices above the estimate by a weighted-sum-type operator (a), max-type operator (b) and a min-type operator (c)

Essential information can be obtained by a comparison of the three fuzzy operators presented in Fig. 36. First, a weighted-sum-type operator provides the estimate margins with the highest expected costs when spot prices are above the estimate (a). On the other hand, this operator provides the lowest values of the expected costs outside the margins when spot prices are low, as seen in (b). Second, a max-type operator provides the estimate margins with the highest cost between the margins and the lowest cost outside the margins for high spot prices. However, for low spot prices this operator provides the second highest additional cost when spot prices are between the margins and the second lowest cost when spot prices are outside the margins. A min-type operator generates the estimates resulting in the additional costs between the weighted-sum-type and max-type operators.

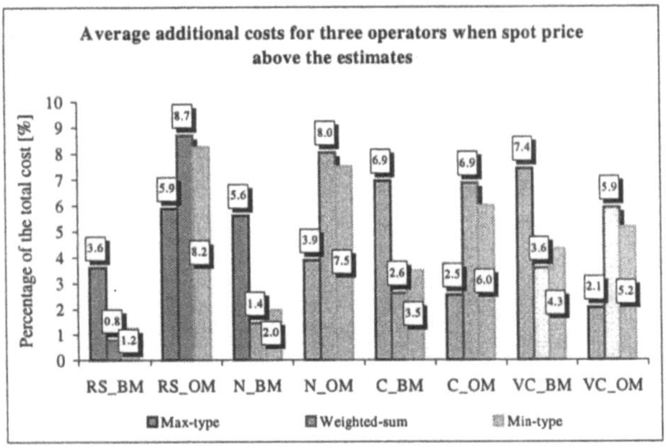

(a)

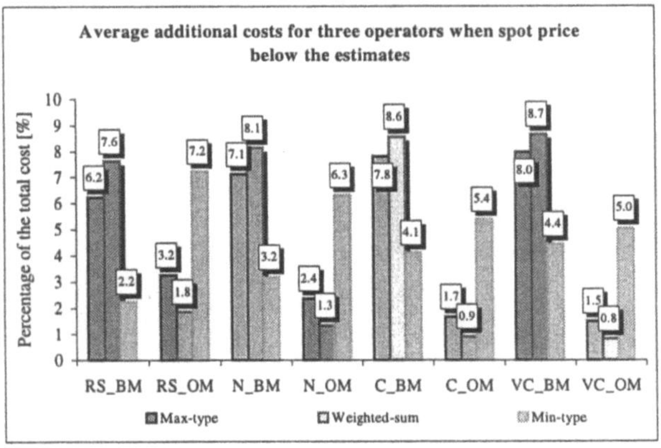

(b)

Fig. 36 Average values of additional costs for three operators when spot prices above the estimates (a) and below the estimates (b)

The analysis of average values of the total risk is shown in Fig. 37. The total risk when the estimates are based on three operators has a similar pattern to the additional costs expected when spot price below and above the margin estimated. However, the differences in the total risk are less significant when various market attitudes are applied to energy contracting.

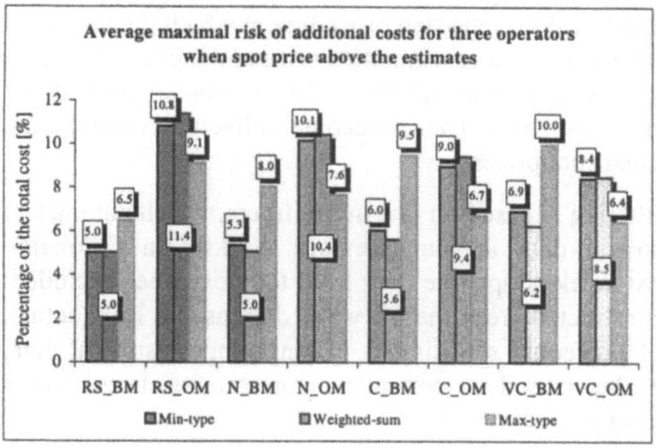

(a)

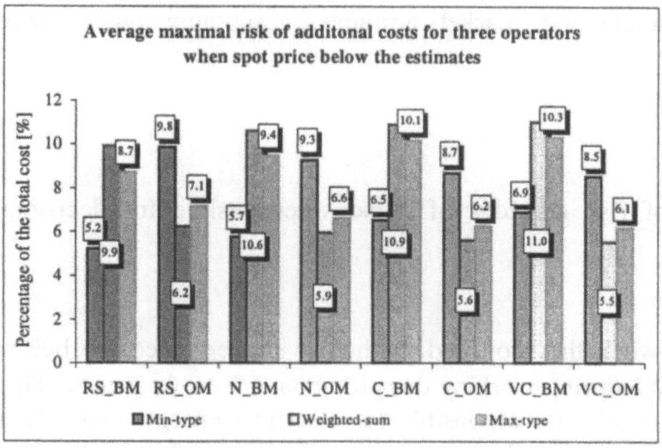

(b)

Fig. 37 Average maximum risk of additional costs for three operators when spot prices above the estimates (a) and below the estimates (b)

5 Conclusions

The fuzzy approach to spot price estimation has been implemented in an evaluation of electrical energy contracts. Such contracts are basic tools in reducing the risk of electrical energy trading. The better methodology applied to contract risk evaluation, the more stable behavior of market participants, and thus the more stable electricity markets.

The application of fuzzy logic techniques to electricity contract evaluations requires the use of fuzzy operators. Of many existing fuzzy operators, three operators have been chosen, including:

- Weighted-sum-type estimate
- Max-type of the estimate
- Min-type of the estimate

The estimate of contract risk is based on four-step procedure in which the historical records from the past four weeks are used to estimate a price and a contract risk in the 5th week. The four-step procedure was selected after a detailed analysis of energy demand and spot price patterns. This procedure reflects accurately the seasonal changes in demand and spot prices.

An idea of fuzzy four-step filtering is based on four fuzzy filters, which parameters are derived from the historical data in four previous weeks. An estimation procedure moving to the next week adopts the data from the next week included, while removing the filter constructed from the 5th week data as the least actual. This allows the reduction of procedure sensitivities to long range historical data. Three fuzzy operators have been used to compute parameters of fuzzy filters applied in the four-step filtering procedure.

The proposed procedure is simple an effective. It allows the precise estimation of risk and cost involved in contracting electrical energy. The computations were carried out for four characteristic market attitudes, combining three fuzzy techniques:

- Fuzzy filtering
- Fuzzy operators
- Linguistic definitions of risk attitudes of persons responsible for decision-making

Additional costs expected when the proposed method is implemented are below 10%. They vary from 8 ÷ 9% during working days to about 5% in weekends. This means that a decision-maker who is responsible for electrical energy contracting can have the confidence when using the proposed methodology that spot prices deviations outside the estimated ranges do not entail more than 10% of the total daily costs.

The comparison of fuzzy operators for various market attitudes shows that differences in results for risk seeking decision-makers are insignificant. The additional cost are the lowers for Max-type fuzzy operators for Conservative and Very Conservative attitudes. This means that there is no universal operator providing the best results in every case. The operators have to be adjusted to the market attitude of a decision-maker.

Reference

Michalik G., „*Fuzzy Logic Approach to Contracting Electrical Energy in Competitive Electricity Markets*", PhD thesis, Monash University, Melbourne, Australia, 1999.

System Modelling in Monitoring and Supervising System of Distributed Energy and Heat Supply Complex

Jiri Bila and Tomas Brandejsky

Czech Technical University in Prague,
Zikova 4, 166 07 Prague 6 - Czech Republic
Bila@fsid.cvut.cz

Comparing works which simulate buildings and their energy sources, this chapter turns attention to the monitoring of energy and heat supply system. The energy and heat supply complex (EHSC) consists of hydraulic turbine, electric boilers, heat pumps, solar collectors and many other needed facilities (expanders, compressors, armatures). As a prototype of such complex has been considered the energy and heat supply system of functional and building area in Herbertov - South Bohemia - the first ecological building-site in Czech Republic.

Though the main objective of the development of the Monitoring and Supervising System (MSS) has been motivated pragmatically (optimisation of electrical energy consumption), the analysis and design "time-space" was used also for application of non traditional approaches, methods and technologies (e.g,. - Object Modelling Technique (OMT) for analysis of the interaction between EHSC and MSS, Fuzzy Qualitative Modelling for human operator "What/IF" analysis and LONWORKS HW and SW environment for realisation of the MSS.

In the chapter only the problems and results related to Fuzzy Qualitative Modelling are emphasised.

1. Fuzzy Qualitative Modelling of Processes

In recent years there have emerged many inspiring ideas in discussions about means and objectives of Qualitative Modelling. Not reminding great persons from this field (there are really many and their names and results are collected, e.g., in book [Piera Carreté and Singh]), we try to explain how methods of qualitative modelling influenced our approach to modelling and simulation and how these methods will be used in the following sections.

Attractive metaphor which could illustrate one of points of view in qualitative modelling memorandum (programme) is known from processes of human capability evaluation (e.g., in testing of persons which aspire for a job or for a special function): in receiving tests is usually not required solutions of problems (which form a substance of function in a considered job) but solutions of some artificial problems (e.g., from IQ tests) and of problems which verify professional qualification of an applicant. Extrapolating this metaphor into the field of engineering artefacts, e.g. in the field of computation programmes, we consider the task of the type "How to verify correct function of a programme without computations of experimental examples?" Or in the field of decision problems – "How to find correct decision without detailed solution of the problem?" The last case could evoke memories of magic thinking. But from the practice of human operators which governs large-scale systems we know that such decisions are applied.

There is no necessary to speak about application of pre-formed actions which had been prepared in some "time off" and the whole decision problem was concentrated on their application in the right moment. This is no way of Qualitative Modelling. Much more better is to speak about pre-formed pre-models [Sugeno and Yasukawa, Bila].

In Qualitative Modelling we try to use special macro-variables (sometimes very hardly measurable) and to apply special methods for their processing. In many quantitative approaches there are obtained some numerical results (sometimes after difficult and time consuming computation phase) and they are interpreted as a qualitative results (sometimes by very simple but correct ways). One of objectives of qualitative modelling is to avoid such computational processes and to achieve the final result by its own proper method. Qualitative Models (and pre-models mentioned above) are often understood as not completely defined models. Their completeness is acquired during their application.

Among such meanings moves the concept of Qualitative Modelling (Qualitative Models) and we agree that it represents rather set of concepts than a one individual concept. From this point of view the distance between qualitative concepts, fuzzy concepts and fuzzy qualitative concepts is not large.

The application of Fuzzy Qualitative Modelling Methods introduced in following sections serves for purposes of a monitoring and supervising system. Qualitative Model is used for modelling of situations "What/IF" and in conjunction with a special expert system works for decreasing of electrical energy consumption in distributed energy and heat supply complex.

2. The System of Algebraic Operations with Fuzzy Linguistic Variables versus Rule-Based Approach

There will be emphasised the advantage of our approach for conditions, when some physical laws of the modelled system are known and the rule-based approach offers great amount of rules.

It is possible to write a classical solution method (Defuzzification-Algebraic operation-Fuzzification) for addition operation as

$$C = F(D(A) + D(B)), \tag{1}$$

Example 1: Considering observations of a system is fuzzy variable A described as

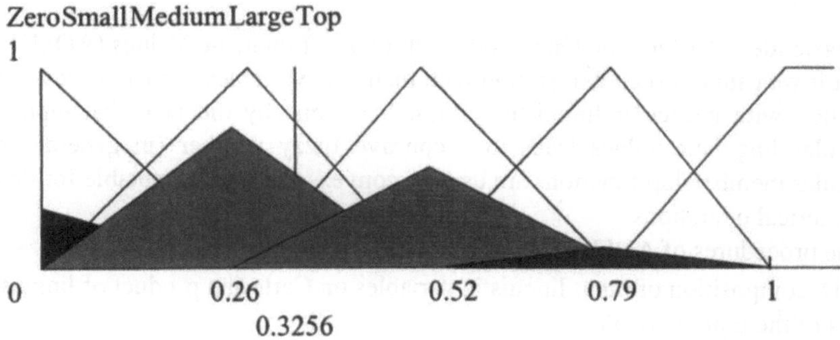

Figure 1: Variables A and B.

$$A = B = \left\{ \frac{0.3}{zero}, \frac{0.7}{small}, \frac{0.5}{medium}, \frac{0.1}{l\arg e}, \frac{0}{top} \right\}, \tag{2}$$

C variable has the identical structure and membership functions.
The result of formula (1) is

$$C = \left\{ \frac{0}{zero}, \frac{0}{small}, \frac{0.514}{medium}, \frac{0.485}{l\arg e}, \frac{0}{top} \right\}, \tag{3}$$

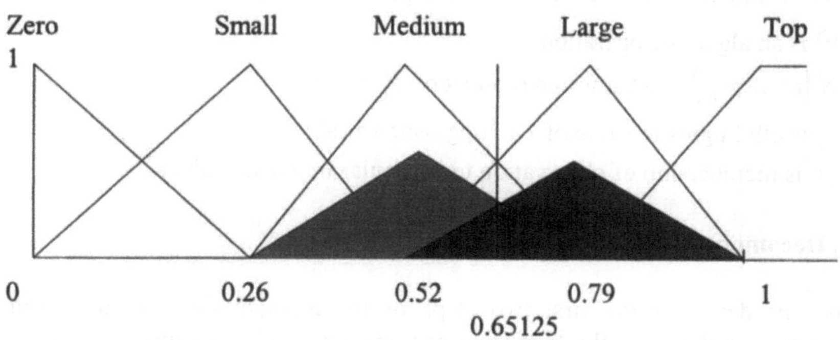

Figure 2: Variable C

3. Principles of Formation of Algebraic Operations with Fuzzy Linguistic Variables

The basic idea of Algebraic Operations with FUzzy Linguistic Values (AOFULV) is that if we cannot execute operation with linguistic variables, we can execute this operation with particular linguistic values. It is done by the fact, that union of particular linguistic values leads to a concave fuzzy number (in general), but particular membership functions are usually convex, and thus acceptable for doing of numerical operations.

The procedures of AOFULV consists of the following steps:

- Decomposition of input linguistic variables on Cartesian product of linguistic values of the input variables.
- Transformation of particular linguistic values into fuzzy numbers.
- Computation of algebraic operations with fuzzy numbers.
- Aggregation of final fuzzy numbers (results of the algebraic operations).
- Backward-transformation into linguistic variables.

The equation (4) summarises this steps

$$\mu_k^R = \max\left\{f_k^R \cap \left[\Theta\left(f\left(K\left(e_1,\ldots,e_m\right)\right)\right)\right]\right\} * \min\left\{\mu\left[K\left(e_1,\ldots,e_m\right)\right]\right\}, \tag{4}$$

where

μ_k^R is a membership value of the result to *k*th linguistic value of resulting variable R

f_k^R is a membership function of *k*th linguistic value of the variable R

Θ is an algebraic operation

$K\left(e_1,\ldots,e_n\right)$ is a Cartesian product of $e_1 \times \ldots \times e_n$

e_k is *e*th linguistic value of k-th linguistic variable

μ is membership of observation to particular linguistic values

3.1. Decomposition of Input Linguistic Variables

This part describes the first two steps of the method: the decomposition of linguistic variables and the formation of Cartesian product of linguistic values of the input variables.

Def.1: Let us have a function q with an arguments $q\left(a_1,\ldots,q_n\right)$. The arguments a_i are linguistic variables with particular linguistic values a_i^j. Thus

Cartesian product C is n-dimensional and we can write $C = a_1 \times \ldots \times a_n$ and element of C is $\left\{a_1^{j_1}, \ldots, a_n^{j_n}\right\}$.

Corollary 1: Cartesian product C contains $j_1 * \ldots * j_n$ elements different to each other. Linguistic values of particular linguistic variables give $j_1 * \ldots * j_n$ combinations, if each combination contains only one element from each variable. Thus Cartesian product C contains all combinations of linguistic values of arguments (linguistic variables).

3.2. Transformation of Particular Linguistic Values into Fuzzy Numbers

There are many methods for calculation with linguistic variables. The basic one is based on defuzzification, making an algebraic operation with real numbers and backward fuzzification. The main disadvantage of this method is a complete loss of information about distribution of uncertainty. Thus this method is good (and simple) in situations, where linguistic variables are projection of crisp data.

In situations, where is needed to achieve information about distribution of uncertainty exist three basic ways. Especially from control applications it is well known method based on rewriting of algebraic relations into fuzzy rules. The main disadvantage of the method is a sensitivity of rule system to definition of linguistic values (arguments) and the necessity to design the structure of resulting linguistic variable on the base of fuzzy rule structure, e.g., if you change the operation, you must change the structure of the resulting variable.

The first method of computing with linguistic variables developed by authors, see [Bila and Brandejsky], is based on the transformation of pairs of linguistic values into singletons, the execution algebraic operations with singletons and on backward transformations into resulting linguistic variable.

The presented method uses transformation of membership function and particular membership value of the linguistic value into fuzzy number. Fuzzy numbers are the best way of computing of algebraic operations from the point of achieving of information about distribution of uncertainty. The aggregation of information in the membership function and membership value is based on application of T-norm. In examples of the method is used T-norm in the form of multiplication.

$$\forall x \subset U : \mu_n(x) = \mu_f(x) * \mu , \tag{5}$$

where

μ_n is a membership function of created fuzzy number

μ_f is a membership function of transformed linguistic value

μ is a membership of an observed situation to transformed linguistic value

3.3. Computation of Algebraic Operations with Fuzzy Numbers

The method based on application of fuzzy numbers has some limits. The membership functions formed in the previous step must be convex.

Theorem 1: Let be given a membership function of linguistic value μ_f, membership of an observed situation to this linguistic value μ and T-norm T. Then membership function μ_n calculated by (5) is convex, if the membership function μ_f is convex.

Proof: Let t be T-norm and convex fuzzy membership function A is described as:

$$\forall \lambda \in \langle 0,1 \rangle : A(\lambda x + (1-\lambda) y) \geq A(x) \wedge A(y), \tag{6}$$

Then if T transforms convex membership function μ_f into non-convex membership function μ_n then T-norm T must satisfy

$$\begin{aligned} &\exists k, l, m, x : k < l < m, \\ &T(k, \mu_f(x)) > T(l, \mu_f(x)), T(l, \mu_f(x)) < T(m, \mu_f(x)) \end{aligned} \tag{7}$$

This presumption is in contrast with monotony condition of T-norms.

The use of fuzzy numbers brings the some limitations of acceptable operations, as classical calculations with fuzzy numbers.

Theorem 2: Let symbol Θ denote in the AOFULV method multiplication, then exists *neutral element* in the form of linguistic variable with the only linguistic value with nonzero membership, which is defined as a singleton in the point 1 with membership 1.

Theorem 3: Let symbol Θ denote in the AOFULV method addition, then exists *neutral element* in the form of linguistic variable with the only linguistic value with nonzero membership, which is defined as a singleton in the point 0 with membership 1.

3.4. Aggregation Operation

The next step of AOFULV method is the aggregation of particular results. The method described in [Brandejsky and Bila] or [Brandejsky *et al.*] uses aggregation before projection to result variable. It is different to our previous method [Bila and Brandejsky] and methods applied in rule-based approach, which uses an opposite order. The use of aggregation of particular results before the projection is better, because some particular result can support each other and thus decreases information loss given by final backward transforming (by granulation).

3.5. Backward Transformation into Linguistic Variables

The backward transformation works with a measure of linguistic similarity. It is linguistic similarity between membership function of linguistic value which membership is calculated and the intersection of this function with aggregated membership function (calculated in the previous step). There are many methods for calculation of linguistic similarity. The suggested method uses relation

$$\mu : \max(\mu_u \cap \mu_r), \tag{8}$$

Where μ represents resulting membership of particular linguistic value; μ_n means membership function of the linguistic value; μ_r is the union of partial results – fuzzy numbers.

Application of the defined AOFULV method improves computational efficiency in cases of problems that would have to be described by many rules. A procedural definition of non-trivial algebraic operations will be illustrated.

Simple examples of AOFULV design and application will be introduced and compared with rule-based results and even with non-fuzzy computation results.

3.6. Examples of AOFULV Capabilities

Example 2: This example contains the same condition as in Example 1, but the result is calculated by AOFULV method.

$$C = \left\{ \frac{0.376}{zero}, \frac{0.519}{small}, \frac{0.7}{medium}, \frac{0.518}{l\arg e}, \frac{0.5}{top} \right\}, \tag{9}$$

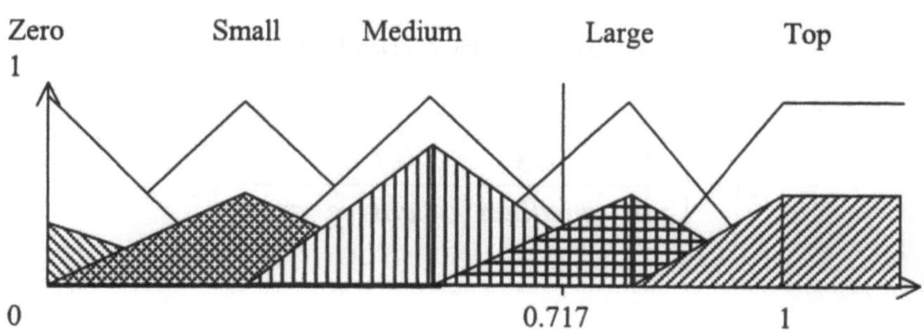

Figure 3: Variable C

Presented result is more acceptable than result achieved in Example 1, because it processed the information about distribution of uncertainty in a better way.

Example 3: TGV intensively brakes in order to stop. The degree of membership of linguistic values are for deceleration (medium = 0.3, intensive = 0.7) and for velocity (medium = 0.7, high = 0.5). Membership functions of fuzzy linguistic variables V, A, S are defined as:

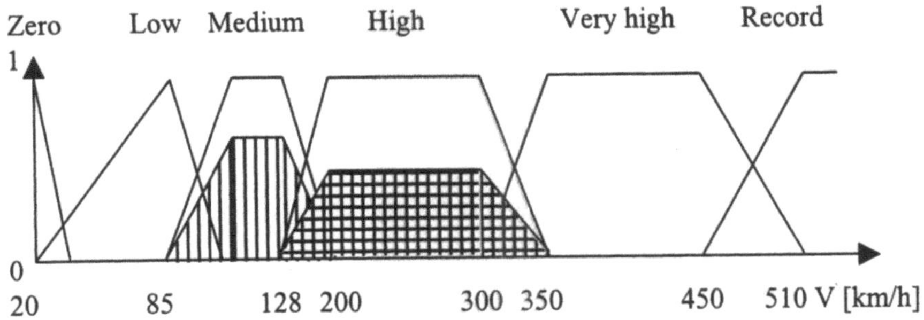

Figure 4: Variable V - Velocity

$$V = \left\{ \frac{0}{zero}, \frac{0}{low}, \frac{0.7}{medium}, \frac{0{,}5}{high}, \frac{0}{very\ high}, \frac{0}{record} \right\}, \tag{10}$$

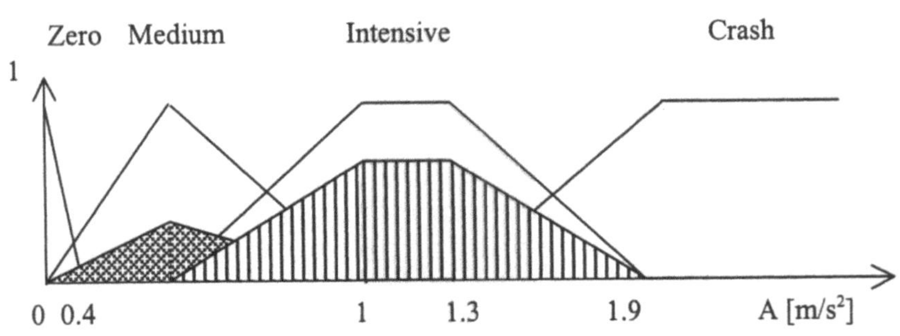

Figure 5: Variable A - Deceleration

$$A = \left\{ \frac{0}{zero}, \frac{0.3}{medium}, \frac{0.7}{intensive}, \frac{0}{crash} \right\}, \tag{11}$$

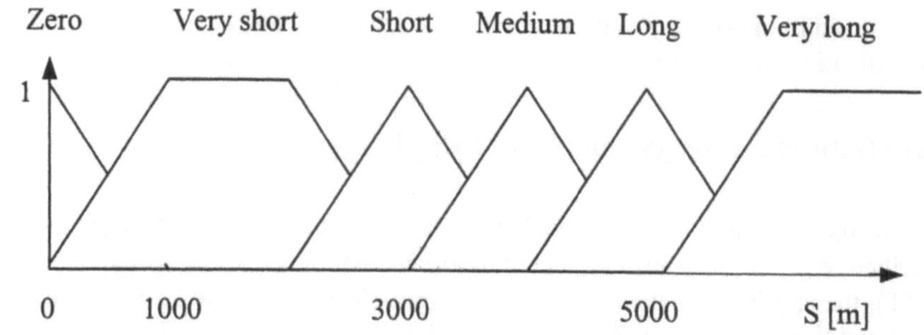

Figure 6: Variable S - Stopping distance

Classical formula which contains solving procedure is

$$S = \frac{v^2}{2a}, \tag{12}$$

We use this procedure in the place of algebraic operator Θ (form (4)). AOFULV method than produces result

$$S = \left\{ \frac{0.5746}{zero}, \frac{0.7}{very\ short}, \frac{0.5}{short}, \frac{0.4984}{medium}, \frac{0.4419}{long}, \frac{0.3853}{very\ long} \right\}, \tag{13}$$

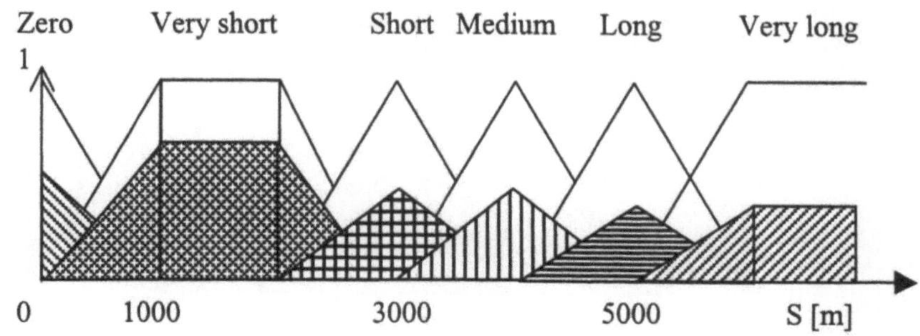

Figure 7: Variable S

Note: The calculation introduced above represents a situation when human observer estimated velocity and deceleration without measurement and he computed by AOFULV method stopping distance. And as it can be seen he achieved no bad result:

Stopping distance for TGV which goes by 150 km/hour and breaks by deceleration 0.7 m/s^2 is 721 m. Stopping distance of TGV which goes by 250 km/hour and breaks by deceleration 0.4 m/s^2 is 6024 m. Using COA defuzzification for fuzzy set S we obtain distance SD = 2942 m.

4. Distributed Energy and Heat Supply Complex

Monitoring system for Distributed Energy and Heat Supply Complex was described, e.g., in [Bila et al.], including the methodology of the system analysis (OMT), the applied HW methodology (Lonworks technology and Echelon Neuron chips) and the role of Fuzzy Qualitative Model within this system. The considered energy and heat supply complex consisted of hydraulic turbine, electric boilers, heat pumps, solar collectors and of many other needed facilities (expanders, compressors, armatures). As a prototype of such a system has been chosen the energy and heat supply system of one of the ***first ecological building areas*** in Czech Republic in Herbertov - South Bohemia. (The functional structure of energy and heat supply complex in Herbertov is illustrated in Figure 8.)

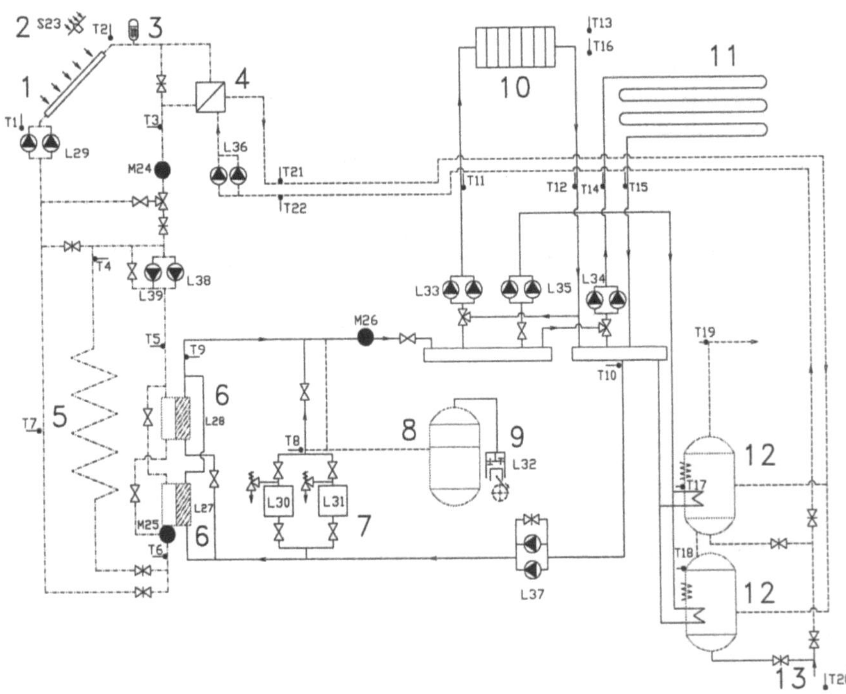

Figure 8: Herbertov's Energy and Heat Complex

Herbertov's Power Source System [Broz] consists of two parts. One part is employed as a system power source, and the other is utilised in consumption of energy produced by the system's power source. The main source of the system's power is provided by a 120 kW hydraulic turbine. This turbine utilises out the water energy from the nearby river, and in that way provides necessary electricity. As a form of prevention two electric boilers (7 in Fig.8) with a four cubic meter expander (8 in Fig. 8) and a compressor (9 in Fig. 8) are implemented to the system. The boilers act as alternative sources of energy.

There are certain constrains that should be considered. For example outside temperature, which depends on the season, could cause the water source, in our case nearby river, to freeze, and therefore constrains the pumping ability. To transcend this problem there is a glycol circuit (5 in Fig. 8) in the river and two heat pumps (6 in Fig. 8) installed into the system. There are two ways how to operate these pumps. We could do this manually, simply by an operator, or by differential switching, which is done automatically. During the summer season, but also in winter, we can introduce solar collector (1 in Fig. 8) and use it for more efficient heating of utility water. These solar collectors have surface area of one hundred square meters, while the absorption surface conveys fifty square meters. At the winter cycle, part of the energy produced by the power source, described above, is consumed for heating of the main building (10, 11 in Fig. 8). In case of non-sufficient temperature in building (T13, T14) and both the heat pumps are running /6/, the additional electric boilers (L30, L31) are switched on. In a case that the temperature of warm service water (T18) is not adequate the regulator that compares temperature differences (T2 and T18) activates two heaters (12) connected to the water piping system (13). These two heaters have a power of 20 kW. The state of the complete system is described with thirty-nine variables which are connected with basic components. Namely temperatures, twenty-two of them (T01 - T22), one solar IR radiance (S23), three flows (M24 - M25), and thirteen logical variables (L27 - L39).

4.1. Application of Novel Operations in Fuzzy Qualitative Modelling

Let us consider the computation of power

$$Q_c = \Delta \upsilon S k, \tag{14}$$

of heat pumps in case that the exact values of temperature difference $\Delta\upsilon$, the surface of expander **S** and quotient of heat transfer **k** are not known. The qualitative computation there is interpreted as a multiplication of three fuzzy variables and the result of this multiplication is a new fuzzy set which is simply interpretable.

The granulations of linguistic variables $\Delta\upsilon$, S, k and Q_C are illustrated in Figures 9, 10, 11 and 12.

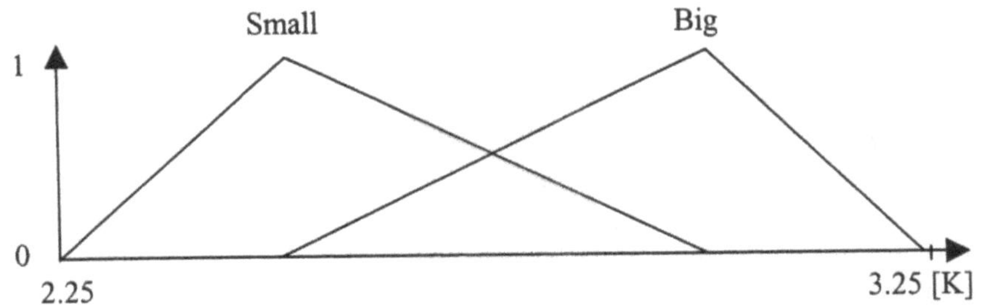

Figure 9: Variable $\Delta\upsilon$

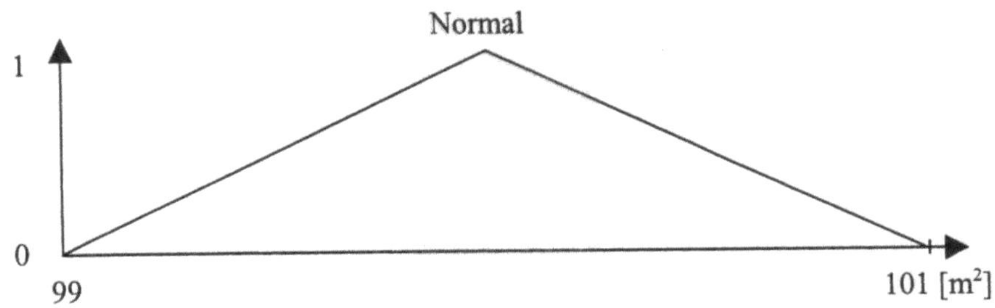

Figure 10: Variable S

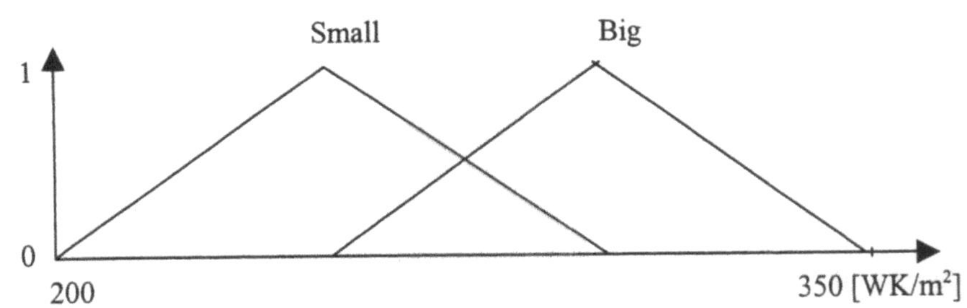

Figure 11: Variable k

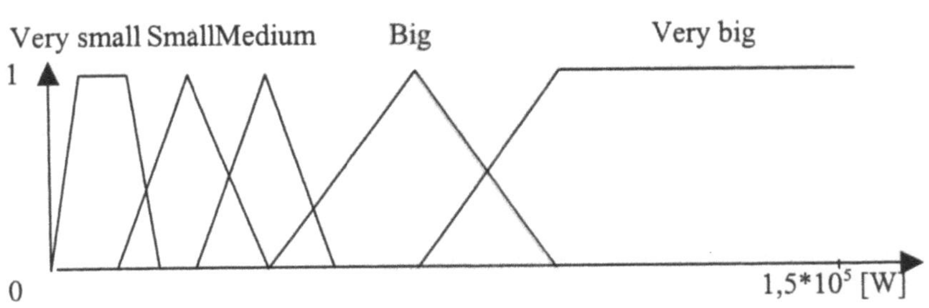

Figure 12: Variable Q_C

The computation of power of heat pump by AOFULV multiplication there is performed for conditions:

$$\Delta\upsilon = \left\{\frac{0.3}{small}, \frac{0.7}{big}\right\},$$
$$S = \left\{\frac{1}{normal}\right\}, \qquad (15)$$
$$k = \left\{\frac{0.8}{small}, \frac{0.2}{big}\right\}$$

and the result of operation (4) is:

$$Q_c = \left\{\frac{0.161}{very\ small}, \frac{0.442}{small}, \frac{0.657}{medium}, \frac{0.532}{big}, \frac{0.197}{very\ big}\right\}, \qquad (16)$$

This "translates" a human operator as "Power of the heat pump is medium. May be - big."

Comparing this result with the result achieved by multiplication of fuzzy numbers (of normal formal type [a, b,α,β])

$$Q_C = \Delta\upsilon Sk = "big * (normal * small)", \qquad (17)$$

there is obtained fuzzy number $\mathbf{Q_c}$

$$\mathbf{Q_c} = [(7{,}53.10^4), (7{,}53.10^4), (3{,}112.10^4), (2{,}375.10^4)]. \qquad (18)$$

Respecting fuzzy numbers **big**($\mathbf{Q_c}$) = [(9,5.10^4), (9,5.10^4), (2,1.10^4), (2,08.10^4)] and **medium**($\mathbf{Q_c}$) = [(7,0.10^4), (7,0.10^4), (1,0.10^4), (1,0.10^4)] (from Fig. 12.) and computing distances d($\mathbf{Q_c}$, **big**($\mathbf{Q_c}$)) and d($\mathbf{Q_c}$, **medium**($\mathbf{Q_c}$)) according to form [Leitch and Shen].

$$d(A,B) = [\ [1/2[2(a-b-c+d) - \alpha - \beta + \gamma + \delta]\ ]^2 + [1/2[c + d - a - b]\ ]^2\ ]^{1/2}, \qquad (19)$$

where A, B are fuzzy numbers A = [a, b, α, β] and B = [c, d, χ, δ] there are obtained values.

$$d(\mathbf{Q_c}, \mathbf{big(Q_c)}) = 2{,}1.10^4 \text{ [W]}, d(\mathbf{Q_c}, \mathbf{medium(Q_c)}) = 1{,}87.10^4 \text{ [W]}. \quad (20)$$

Fuzzy number $\mathbf{Q_c}$ from (18) is approximated by linguistic value **medium($\mathbf{Q_c}$)**, rather than by linguistic value **big($\mathbf{Q_c}$)**.

4.2. Fuzzy Qualitative Model in MSS

The developed Fuzzy Qualitative Model (FQM) is applied in Monitoring and Supervising System (MSS) especially for the support of control actions of human operator in situations with uncertainty. In this context FQM works in the mode "What / IF " and allows to operator to map important variants of control action consequences.

Though the role of human operator in MSS of Herbertov complex is reduced to a few control actions (nothing to say about alarm actions), the results of these control action activities are in some cases surprisingly useful.

The example which was solved in previous section illustrates one of situations which is impossible to identify exactly. The only one "measurable" variable (in relation (14) is the temperature difference $\Delta \upsilon$ but its quantity is relatively small. Both the other variables in (14) - the quantity of actual active surface **S** of heat pump and the quantity of actual heat transfer k - are uncertain and changes according to situation in river (where the heat pump is situated). However - for human operator is not difficult to evaluate the actual quantity of these variables in linguistic terms "normal, small, big."

Let us suppose that operator has for its control actions a small list of control rules and one of them is of the type:

IF ⟨((The *power* of heat pumps is **medium** or **big**) AND(The *Load* of the complex is **Standard$_1$**)AND(*Sunshine* is **Average**))⟩ ⇒ ⟨(THEN (*Heaters* T12 are allowed to be switched **Off**)⟩. (21)

Conditions which concerns of the *Load* and the *Sunshine* are easy to evaluate.

For evaluation of the first condition which concerns of the heat pump *power* there is necessary to do a small observations. If the operator finds observations as they have been given in our example and FQM computes linguistic value of power - the operator switches **off** the heaters T12.

Respecting the dynamics of the complex - this described control action is executed (assuming standard mental disposition of human operator) for ***half of an hour earlier*** than it could be done by automatic control system (as a reaction to increase of radiator temperature t_{11}).

Accepting proportions in situation which has been described above there is possible to estimate (roughly) the saving of electric energy as a consequence of the described human operator control action.

Direct consequence of the operator control action is saving of 40 kW hours. Let us suppose that such a situation repeats (on an average in year) five times a week. It means - 260 times per year. It represents saving of 10 400 kW hours per year. Though the influence of this saving on financial budgets for complexes of Herbertov type depends on national currencies, the ecological merit of this saving there is not necessary to emphasise.

It was only a small demonstration of application of FQM in MSS in case of static (not dynamic) situation. The complex works in most of its operation time in steady state mode and therefore it was possible to reduce fuzzy modelling of dynamic processes to minimum. There are for about thirty similar situations in which FQM helps to human operator.

5. Conclusions

In this chapter we introduced Fuzzy Qualitative Model with novel definition of algebraic operations. Merits of operations developed within the frameworks of this definition will be appreciated especially in modelling of special class of ill-defined systems. These systems are governed by laws which are known (in general) from some special science disciplines (e.g., Physics) but in application are mixed with strong uncertainties. Approaching such systems from the side of pure rule-based description there is necessary to construct lot of rules (sometimes) and to rewrite into them the knowledge structuralised before in known laws.

This is very inefficient. Our proposed approach opens a compromise way of fuzzy modelling which utilises information contained in structure of classical law description.

Besides a theoretical contribution to fuzzy modelling there has been in this chapter demonstrated procedural characteristic of a few operations constructed according to introduced novel procedural definition.

Application of Fuzzy Qualitative Model has been demonstrated within environment of Monitoring and Supervising System of one Heat and Energy Supply Complex.

Acknowledgements

This research has been conducted at the Department of Instrumentation and Control Technology, Faculty of Mechanical Engineering and at the Department of Automation in Transportation and Telecommunications, Faculty of Transportation Sciences (both CTU Prague, Czech Republic) and has been supported partially by Research Project of MŠMT ČR No.: 504/99:212 2 000 9 and partially by Grant Agency of Czech Republic within the framework of Grant Project 101/98/0336.

Bibliography

[Bila] Bila, J.: The Development of Qualitative Models for Knowledge Based Process Control Support Systems. In: Proc. of Int. Conference on SYSTEMS, Analysis, Control and Design, Lyon, France, July 4-6, 1994. Vol. 2, pp. 265-276.

[Bila and Brandejsky] Bila, J. and Brandejský, T.: An Alternative Approach to Theory of Fuzzy Numbers. In: Proc. of The Third European Congress on Intelligent Techniques and Soft Computing, *EUFIT 95*, Aachen, Germany, August 28-31, 1995. Vol.1, pp. 245-249.

[Bila et al] Bila, J., Broz, K., Jirovsky, V. and Rodic, H.: Monitoring System for Distributed Energy and Heat Supply Complex. In: Proc. of the Fifth IBPSA Conference on Building Simulation '97, Prague, CR, September 8-10, 1997. Vol. II, pp. 237 - 245.

[Brandejsky and Bila 98] Brandejsky, T. and Bila J.: Novel Algebraic Operations with Fuzzy Linguistic Variables based on the Theory of Fuzzy Numbers. In: Proc. of the Third Scientific - Technical Conference with International Participation - Process Control '98, Kouty nad Desnou, CR, June 7 - 10, 1998. Vol. 1, pp. 56 - 59.

[Brandejsky et al.] Brandejsky, T., Bila, J., Broz K.: Fuzzy Qualitative Modelling of Distributed Energy and Heat Supply Copmplex. In: Helena Szczerbinska (ed): Modelling and Simulation: A Tool for the Next Millennium. SCS, Warsaw, Poland. Pp. 391-394.

[Broz] Broz, K.: A Multiple Energy Source for Heating. In: Proc. of Regional Consultations for „RIO + 5" World Congress on Sustainable Development. Tallin, Estonia, January 25-27, 1997. 9 ps.

[Leitch and Shen] Leitch, R. and Shen, Q.: Fuzzy Qualitative Simulation. IEEE Trans. on Syst., Man and Cybernetics, No.4., 1993. pp. 1038 - 1061.

[Piera Carreté and Singh] Piera Carreté N., Singh M.G. eds: Qualitative Reasoning and Decision Technologies. In Procs. Of the IMACS Int. Workshop, CIMNE, Barcelona, 16-18 June, 1993.

[Sugeno and Yasukawa] Sugeno, M. and Yasukawa, T.: A Fuzzy-Logic-Based Approach to Qualitative Modelling. IEEE Trans. on Fuzzy Systems, Vol. 1, No. 1., 1993. pp. 7 - 31.

Evidential Reasoning in Nuclear Waste Management

Athena Tocatlidou

Department of Information and Communications Systems
University of the Aegean
83200 Karlovassi, Samos, Greece
atocat@aegean.gr

The problem of the management and disposal of nuclear waste is a complex decision making procedure dealing with different types of uncertainty, and relies to a considerable extend to domain experts judgment for the formation of rules and the relevant knowledge. The present work proposes a system that provides a generic and flexible way to model and support that decision. Evidence theory and fuzzy sets offer convenient formalisms to handle the uncertain information, and the evidential support logic mechanism is used to capture expert reasoning. For the problem of updating the existing knowledge is described an algorithm that extends fuzzy information in the presence of new evidence. An index related to the conceptual distance of the two fuzzy entities can be used as a parameter in the algorithm.

1 Outline of the Nuclear Waste Management Problem

Nuclear power is a secure source of energy, with an enormous potential as a substitute for conventional forms of power, it is inexhaustible, and with the exception of major accidents, clean for the environment and less harmful to the public health, producing a considerably small amount of waste. The disadvantages associated with this form of energy, include the risks associated with nuclear proliferation, the maintenance of the different types of nuclear reactors, the range of problems related to radiation contamination in the case of serious accidents, and the problem of disposal of the produced wastes, making nuclear power at present less attractive than conventional alternatives.

This last problem of radioactive waste disposal represents one of the major concerns for all countries with a nuclear program. Most of the existing high level waste is stored temporarily, a situation that needs a long term solution either because most of the facilities are reaching their limits, or because there exist serious concerns about public safety.

The design and operation of a radioactive waste management system is a very complex problem [Weinberg 1972, Thorne 1993]. The appropriate repository must ensure that wastes are contained in a way that warrants against entry into the biosphere for very long time periods, that radiation can be contained, and that the heat emission can be adequately handled. This complexity is reflected at the absence of final regulatory criteria for the issue in many countries.

There have been different proposals for the management of the waste, including some exotic ones, such as the placement of the dangerous material in the space or under the Antarctic ice sheet, but the focus of all organizations with a waste management problem has become some form of geological disposal. Such a scheme must make sure that, until the radioactivity decreases to very low levels, and this period can be decades, centuries or millennia, the material does not enter the biosphere. The most important route for this is groundwater, and its paths that can be altered by geological faulting, seismic activity, or event future human interventions [Roxburgh 1987, Walker 1983]. To minimize the possibilities of any risk the repository must be situated in geological formations that will have

- a minimal exposure to groundwater.
- geochemical and mineralogical properties that cause retardation of the radionuclides entry into the water.
- the ability to cope with the heat without destroying the surrounding geological structure.
- strength and stability to ensure safe containment of the waste.

Additionally the repository system must be

- deep enough to be inaccessible to animals and plants.
- located in areas with low probability of seismic activities.
- reasonably stable under glaciation conditions.

A typical approach for a geological disposal, is to lock the waste inside metal rods, sealed within multiple layers of thick wallet steel and other materials. The whole structure is then buried and monitored for the necessary amount of time [Gallegos and Bonano 1993].

A widely used method to assess the suitability of a certain site to serve as a waste repository is the prediction of the future state of the system given the occurrence of future scenarios, and a consequence analysis [Voegele, Younker and Alexander 1989]. This involves the investigation of the impact that different changes and events will have on the safety and integrity of the repository system. The scenarios are generic, i.e., without reference to any particular site characteristics, and consist of a list of the phenomena that could cause a repository failure, and as such they set the criteria against which to assess potential repository sites [Bonano, Cranwell and Davis 1989, Gallegos and Bonano 1993].

For the selection of an appropriate site, the approaches followed, despite differences due to regulations, geographical characteristics, the host medium etc., mainly consist of:

(i) Procedures for predicting the future state of the disposal system.

(ii) Models and computer systems to simulate the way the repository the waste and its environment will evolve.

(iii) Uncertainty and sensitivity analyses for the system [Helton 1993].

An important characteristic of a decision support system aiming to provide assistance for the above activities, is its ability to provide a high degree of flexibility regarding its modeling mechanisms, enabling that way the decision maker to formulate his/her description of the phenomena, as well as the rules/models defining a specific problem or domain.

One fundamental aspect of this flexibility is the provision for uncertainty handling. Although there is no general agreement on a classification of the different uncertainty types, there is a widely accepted distinction between stochastic uncertainty and vagueness, the first type involving information about known events with uncertain occurrence, while the second refers to entities resisting exact classification.

In the present paper is described a decision support system dealing with a representative of a class of problems, namely the management of information in the presence of risk. Following a brief presentation of the theoretical background and the modeling mechanism, is discussed the problem of extending the existing knowledge base in the light of new evidence. For this is proposed an algorithm based on simple set operations, and therefore applicable to both numeric and non numeric domains.

2 Theoretical Background and Design of the System

The standard mathematical theory of probability is based on the following axiomatic foundations:

Given a space X, and if by 2^X is denoted the set of all subsets of X, for each event E in 2^X a probability function is defined, associating with it a real number $p(E)$ satisfying the following principles:

(P1) $p(E) \geq 0$

(P2) $p(X) = 1$

(P3) For a countable set $\{E_1, E_2, \ldots, E_n\}$ of mutually exclusive events, the joint probability is given as

$$p(E_1 \cup E_2 \cup \ldots \cup E_n) = p(E_1) + p(E_2) + \ldots + p(E_n)$$

The adoption of this formalism for the treatment of uncertainty implicitly accepts the following two assumptions:

(a) The events described are clearly defined, and only their occurrence is uncertain.

(b) If the occurrence of an event is assigned a probability p, then the probability that the event will not occur must be equal to $(1-p)$.

In many cases experts are not comfortable with this artificially imposed precision.

The mathematical theory of evidence [Shafer 1976, Wu, Apostolakis and Okrent 1990] proposes an alternative approach introducing belief functions, and relaxing the third axiom of probability theory. According to this theory, given a space X and its power set 2^X a belief function

$$Bel : 2^X \rightarrow [0,1]$$

is defined such that

(E1) $Bel(\varnothing) = 0$

(E2) $Bel(X) = 1$

(E3) For every collection $\{E_{1,}, E_2, \ldots, E_n\}$ of events in X the joint belief is given by

$$Bel(E_1 \cup E_2 \cup \ldots \cup E_n) \geq \sum_i Bel(E_i) - \sum_{i<j} Bel(E_i \cap E_j) +$$
$$+ (-1)^{n+1} Bel(E_1 \cap E_2 \cap \ldots \cap E_n)$$

The belief $Bel(E)$ assigned to the event E represents all the evidence in support of that event, while the evidence that does not support its complement $\overline{E}$, given by $1 - Bel(\overline{E})$ is the value of the dual plausibility function $Pl(E)$ for the event E.

In the case of an event E and its complement it follows from axiom (E3) that

$$Bel(E \cup \overline{E}) = 1 \geq Bel(E) + Bel(\overline{E})$$

and therefore

$$Bel(E) \leq 1 - Bel(\overline{E})$$

The consequence of this relaxation of the additivity axiom (P3) makes necessary that for each event E two pieces of knowledge are needed: The belief to the truth of the event and the belief to the truth of its complement.

In this work is proposed a knowledge-based approach to the decision making problem of assessing the performance of a site to serve as a long term disposal repository for nuclear waste. The decision is based on a consequence analysis of various post-closure scenarios concerning climatic and geological phenomena occurring during the containment period specified by the different countries regulations.

The effects of these phenomena over a site are described using a hierarchical structure, which links each possible scenario with the phenomena describing it, and the later with the affected site parameters (Figure 1). A prototype system based on the above assumptions, and using an illustrative set of scenarios has been developed [Baldwin, Martin and Tocatlidou 1995].

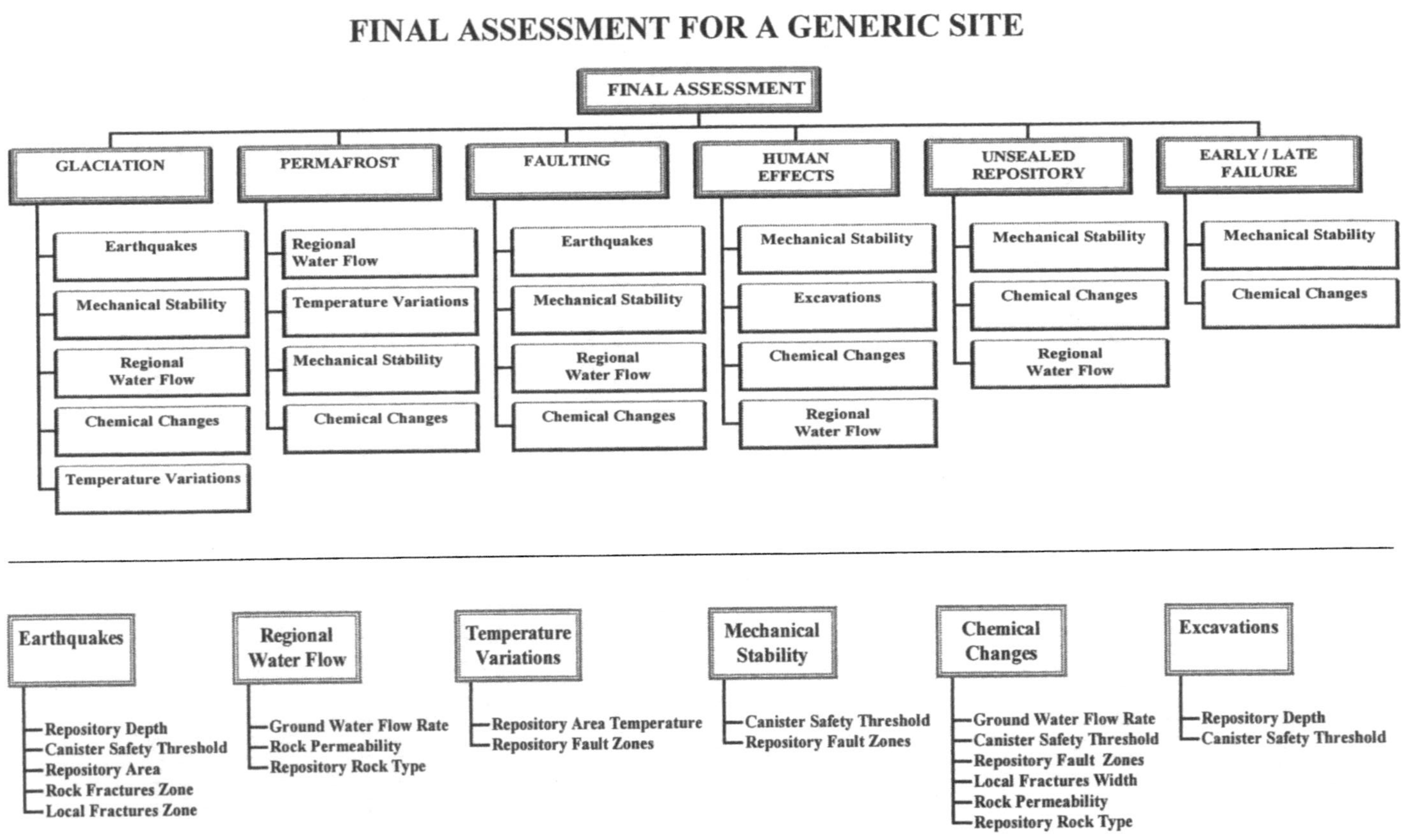

Fig. 1. Scenarios hierarchy

The basic is the mathematical formalism of the theory of evidence [Shafer 1976], and the evidential logic mechanism [Baldwin, Martin and Pilsworth 1995] is used to model the logical structure. The values of the different parameters of the disposal structure can be either single numbers or fuzzy sets. The main conceptual entities, i.e., the hypothetical scenarios on the future state of the repository site, and the geological and mechanical events that are caused by them, are represented using the evidential logic rule.

In the context of this rule, a concept defined by the head of the rule, is a weighted combination of clauses, each one describing the characteristics of that concept, weighted accordingly to their importance. The overall confidence to the rule is indicated by two probability intervals. The first contains the probability for the head being true when the body of the rule is satisfied, and the second for the case when the negation of the body is true.

Given a new instance in the data, the degree to which it matches the prototypical case described by the rule is calculated in three stages:

1. Each characteristic (clause) is matched against the relevant data, giving a partial probability interval (partial support pair) for that specific characteristic.
2. A general support pair for the whole set of clauses, that make the body of the rule, is calculated from the weighted sums of the partial supports.
3. Finally, the two probability intervals in the definition of the rule are combined to the overall body support pairs, resulting to a probability interval for the satisfaction of the rule.

More formally, the rule has the format

$$
\begin{aligned}
&((Object \quad X \quad is \quad f)\\
&(S\ (feature\_1 \quad has\_value \quad a\ )\ w_1\\
&(feature\_2 \quad has\_value \quad b\ )\ w_2\\
&((feature\_3 \quad has\_value \quad c\ )\ w_3\\
&\vdots\\
&(feature\_n \quad has\_value \quad n\ )\ w_n\ ))):((n_1 \quad p_1)((n_2 \quad p_2))
\end{aligned}
$$

where $(Object \quad X \quad is \quad f)$ is the conclusion of the rule, $feature\_i$ denotes a concept characteristic, and

$$\sum_{i=1}^{n} w_i = 1$$

For the evaluation of the rule, each feature is matched against the relevant data giving a partial support pair (α_i, β_i), and in that way the rule body is assigned a probability interval

$$(\alpha,\beta)=\left[S\left(\sum_{i=1}^{n} w_i\alpha_i\right),\quad S\left(\sum_{i=1}^{n} w_i\beta_i\right)\right]$$

The factor S is a function, $S:[0,1]\rightarrow[0,1]$ acting as a filter that allows for the combination of the terms in the body of the rule to take intermediate values between conjunction and disjunction. The final support pair (γ_1, γ_2) containing the probability for the rule to be true is calculated by taking

$$\gamma_1=\begin{cases} n_1\beta+p_1(1-\beta) & \text{if} \quad n_1\le p_1 \\ n_1\alpha+p_1(1-\alpha) & \text{if} \quad n_1> p_1 \end{cases}$$

and

$$\gamma_2=\begin{cases} n_2\alpha+p_2(1-\alpha) & \text{if} \quad n_2\le p_2 \\ n_1\beta+p_1(1-\beta) & \text{if} \quad n_2> p_2 \end{cases}$$

An illustrative case of a rule describing the effect that earthquakes would have on a geological site to be used as a long term waste repository, could be modeled as:

((*Impact_of_earthquakes on_site_X is_significant*)
(**S**(*Repository_Depth site_X has_value* **deep**) 0.2
(*Canister_Safety_threshold site_X has_value* **low**) 0.3
(*Repository_Area site_X has_value* **large**) 0.1
(*Rock_Fractures_Zone site_X has_value* **wide**)0.25
(*Rock_Fractures_Age site_X has_value* **recent**)0.15)):((0.8 1)(0.1 0.3))

where **deep, low, large, wide, recent** are fuzzy sets defined over the appropriate spaces. The overall support pairs following the definition of the rule, indicate that the probability of an important impact of an earthquake for a site, will be within the interval (0.8 1), given that the conditions in the rule body are all satisfied, i.e.,

Pr((*Impact_of_earthquakes on_site_X is_significant*) | (*Body of the Rule*)) $\in$ (0.8 1)

while the probability of the same event given that the conjunction of the body clauses is false, lies in (0.1 0.3), i.e.,

$$\Pr((Impact\_of\_earthquakes \quad on\_site\_X \quad is\_significant) \mid \neg\,(Body\ of\ the\ Rule)) \in (0.1 \quad 0.3)$$

The rule base of the system contains nested evidential logic rules following the hierarchy depicted in Figure 1.

The system is implemented using the FRIL language [Baldwin, Martin and Pilsworth 1988], and an overview of the control flow within it is given in Figure 2.

The interaction with the system can be performed at different levels. At a simple user mode the system can simply be consulted on site evaluation, using information already stored in the knowledge base. An instance of the windows interface supporting this action, is shown in Figure 3 along with a window for a brief explanation facility regarding the stored knowledge. In an interactive mode, a new candidate site can be evaluated after its parameters values are supplied to the system.

At a higher level, the user can act as the designer of the system, modifying or even extending the reasoning mechanism underlying the system. At that level the user is allowed to perform any combination of the following actions:

- Change the value of a site parameter.
- Change the nature of a site parameter value, i.e., to substitute a single number with a fuzzy one.
- Change the importance of a site characteristic, by altering the weights on a rule clause.
- Add a new feature in the rule body using either existing parameter values or defining new ones.
- Modify the probability intervals giving the confidence on the overall rule.
- Introduce a new rule having a conclusion identical to an existing rule, but a different set of features in its conditions.
- Define a completely new rule.

The windows interface for this set of interactions is shown in Figure 4.This level of interaction allows for full control over the system, providing thus for a very flexible and therefore generic assessment model, which can account for the differences in sites and regulations governing the licensing of a radioactive waste repository, but also forming at the same time a user friendly framework for different scenarios and hypotheses testing.

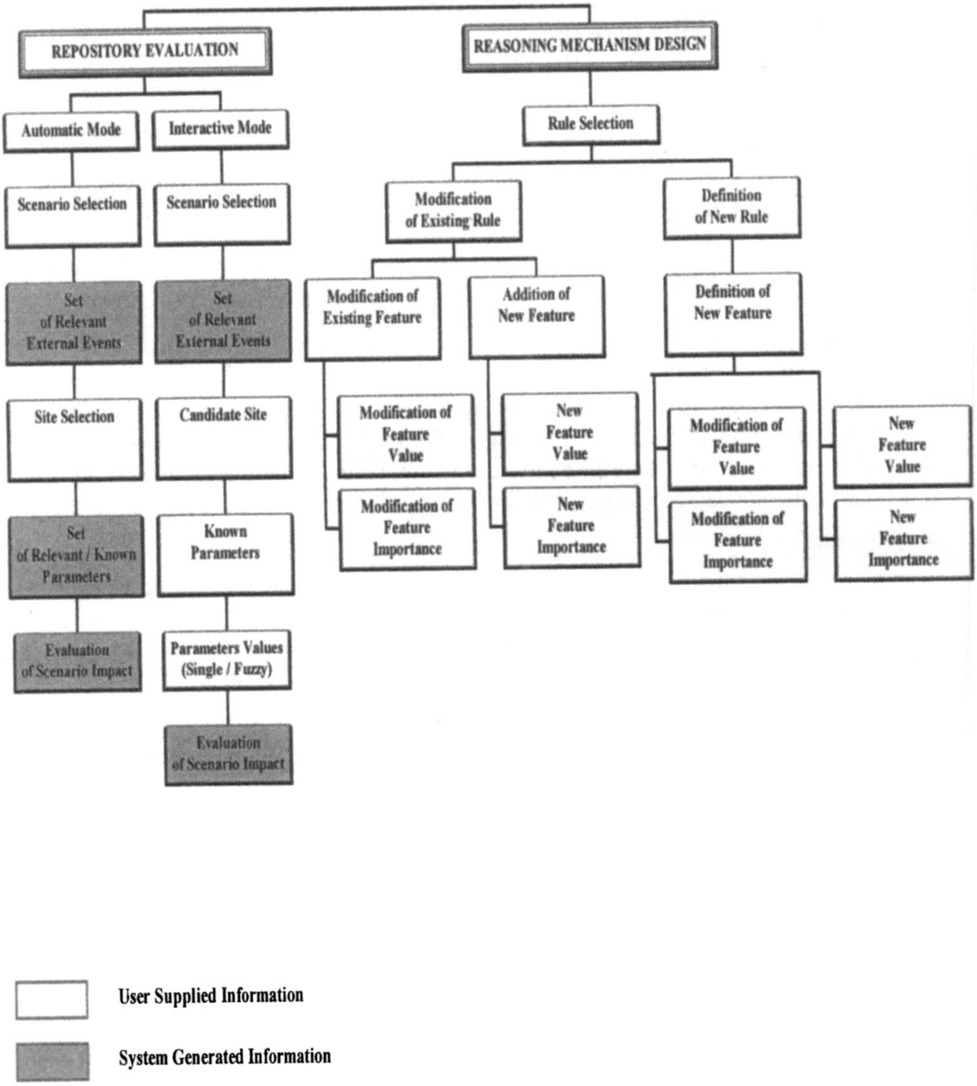

Fig. 2. System control flow diagram

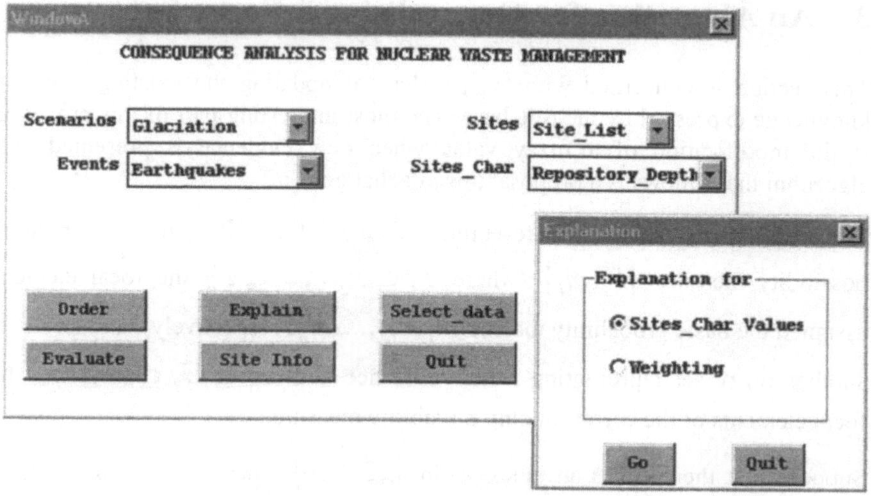

Fig. 3. Interface for repository assessment

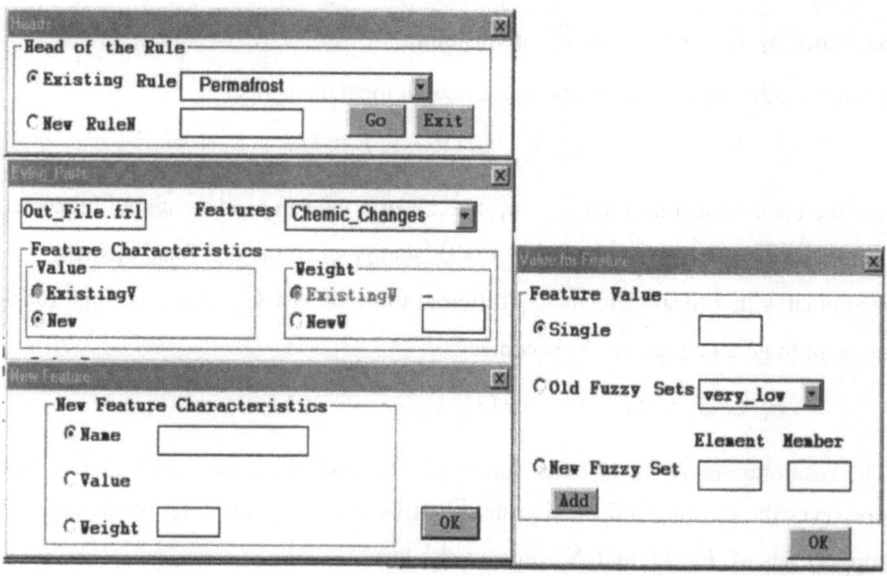

Fig. 4. Interface for the design of the inference mechanism

3 An Algorithm for Generalising a Fuzzy Set

This section is concerned with the problem of updating the existing uncertain knowledge expressed by the rule base. The most interesting part of this procedure is the modification of a fuzzy value when new evidence is presented. The algorithm that follows is a proposal towards that goal.

Consider a fuzzy set F, representing existing information, the corresponding possibility measure $\{F_i, m_i\}$ (where $F_1 \subseteq F_2 \subseteq \ldots \subseteq F_n$ the focal elements assigned the basic probability masses $m_1, m_2, \ldots, m_n$, respectively), and let G be another fuzzy set representing a new evidence and $G_1 \subseteq G_2 \subseteq \ldots \subseteq G_m$ the focal elements of the corresponding possibility measure.

Suppose that there exists an index k in $\{1,2,\ldots,n\}$, such that for some F_k in $F_1, F_2, \ldots, F_n$ the following hold $G_1 \cap F_k = \varnothing$ and $G_1 \cap F_{k+1} \neq \varnothing$.

i.e., the set G_1 containing the elements with full membership in the concept described by G, and the set F_k are disjoint.

Then considering the possibility measure with focal elements

$$F_1, F_2, \ldots, F_k \cup G_1, F_{k+1} \cup G_2, \ldots, F_n \cup G_m$$

and the basic probability masses $m_1, m_2, \ldots, m_n$, we can produce a new fuzzy set S. In the case where $m - n + k - 1 \neq 0$, that is when the last focal element F_n is aligned with a focal element G_j different from the last G_m, then the final focal element to generate the set S becomes

$$F_n \cup G_j \cup G_{j+1} \cup \ldots \cup G_m$$

The resulting set S is an extension of F towards the support set of G, that preserves the existing information in F. If by S_F, S_G, and S_S we denote the support sets of F, G and S respectively, then

$$S_F \subseteq S_S \text{ and } S_G \subseteq S_S$$

Over these sets a probability distribution can be defined, as follows: For each $x \in S_F$ let F_p be a focal element such that

$$\{x\} \cap F_p \neq \varnothing$$

Since $F_p \subseteq \ldots \subseteq F_n$, the value of the probability distribution at the point x can be defined as $$p(x) = \sum_p^n \frac{m_p}{|F_p|}$$

$|F_p|$ being the cardinality of F_p and m_p the mass associated with it.

After the generalization of F, resulting at the set S, the value of the distribution at x will now become

$$p'(x) = \sum_r^n \frac{m_r}{|S_r|}$$

where again S_r is a focal element such that $\{x\} \cap S_r \neq \varnothing$. This defines a uniform probability distribution over each focal element.

The amount $Dp = p(x) - p'(x)$ gives an indication of the degree of the performed generalization.

The above algorithm can be parameterized as follows: Suppose that the fuzzy set F is updated given G producing the set S. If now S is further updated given the same set G as new evidence, then the outcome can be considered as F updated given G with a parameter equal to 2, and this corresponds to a more intense generalization of the initial knowledge. This procedure if repeated converges when the smallest focal element G_1 is unified with F_1 to produce the first focal element generating S. The number of repetitions, leading to the convergence state, constitutes an upper limit for the values that the parameter can take.

In that way is provided a means for gradual updating using a parameter with an intuitive meaning: The integer value of the parameter corresponds to repetition of the new evidence given by the set G, and the higher the number of repetitions the greater will be the generalization of the set F. The procedure reaches a convergence state, where no further generalization is possible.

The number of iterations till the convergence state can also be considered as an index of the similarity between the fuzzy sets F and G [Tocatlidou 1996]. The following example illustrates the algorithm.

Given a fuzzy set

$$F = 10/0.1, 20/0.4, 30/0.8, 40/1, 50/0.7, 60/0.3, 70/0.2$$

considered a piece of existing knowledge, and the set

$$G=10/0.1,20/0.2,30/0.6,40/0.8,50/1,60/0.6,70/0.3$$

representing new evidence, the algorithm will proceed by taking the union of the focal elements:

$$\begin{aligned}
&\{40\}:0.2\\
&[\{40,30\}\cup\{50\}]:0.1\\
&[\{40,30,50\}\cup\{50,40\}]:0.1\\
&[\{40,30,50,20\}\cup\{50,40,30,60\}]:0.1\\
&[\{40,30,50,20,60\}\cup\{50,40,30,60,70\}]:0.1\\
&[\{40,30,50,20,60,70\}\cup\{50,40,30,60,70,20\}]:0.1\\
&[\{40,30,50,20,60,70,10\}\cup\{50,40,30,60,70,20,10\}]:0.1
\end{aligned}$$

This results in the possibility measure

$$\begin{aligned}
&\{40\}:0.2\\
&\{40,30,50\}:0.1\\
&\{40,30,50,60\}:0.3\\
&\{40,30,50,60,20,70\}:0.1\\
&\{40,30,50,60,20,70\}:0.1\\
&\{40,30,50,60,20,70,10\}:0.1\\
&\{40,30,50,60,20,70,10\}:0.1
\end{aligned}$$

or, after probability masses assigned to identical sets are added

$$\begin{aligned}
&\{40,50\}:0.2\\
&\{40,30,50\}:0.1\\
&\{40,30,50,60\}:0.3\\
&\{40,30,50,60,20,70\}:0.2\\
&\{40,30,50,60,20,70,10\}:0.2
\end{aligned}$$

This corresponds to the generalized fuzzy set

$$M1=10/0.1,20/0.4,30/0.8,40/1,50/0.8,60/0.4,70/0.3$$

If now we apply the algorithm with parameter 2 to F the outcome will be the set:

$$M2=10/0.1,20/0.4,30/0.8,40/1,50/1,60/0.4,70/0.3$$

A diagrammatic representation of the previous fuzzy sets is given in Figures 5, 6 and 7.

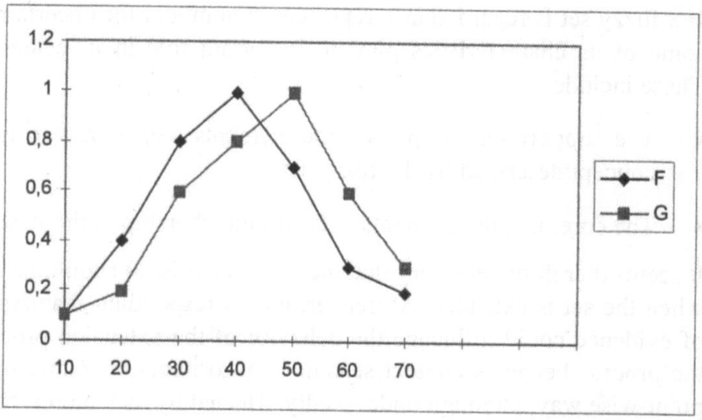

Fig. 5. Fuzzy sets in Rule (F) and Data (G)

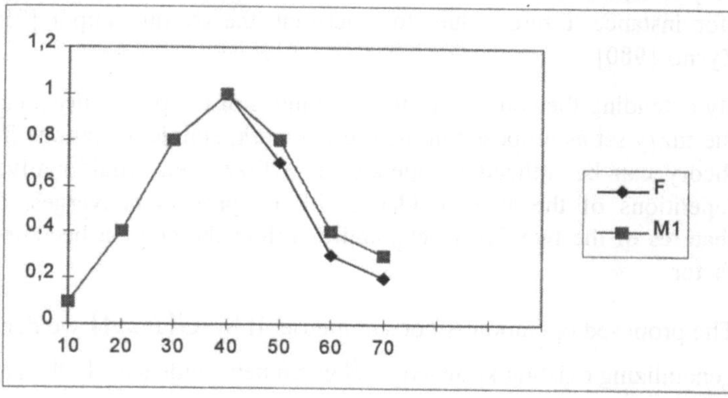

Fig. 6. First generalisation of F

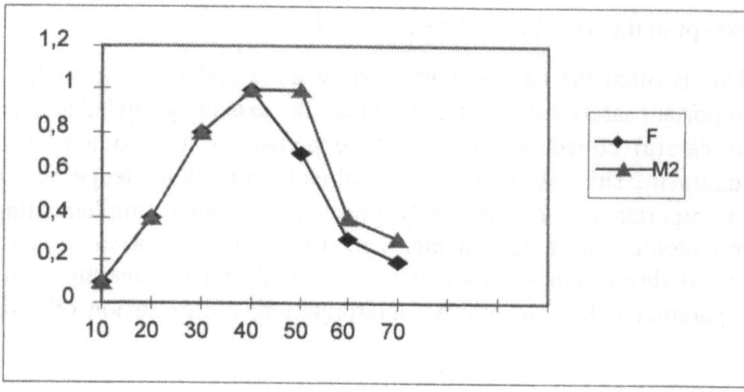

Fig. 7. Second generalisation of F

4 Discussion of the Algorithm Properties

If a fuzzy set is regarded as a representation model for uncertain knowledge, then some of its characteristics play an important role in its semantic interpretation. These include

- The support set comprising the elements with non-zero membership to the concept described by the fuzzy set.
- The core, i.e., the elements with full membership in the fuzzy set.

It seems therefore necessary that these characteristics should be taken into account when the set is extended. Moreover, the corresponding features of the new piece of evidence could influence the behavior of the extension process in which case the process becomes context sensitive. Algorithms operating on a fuzzy set in a point-wise way, although undoubtedly affected by its features, they do not provide for a holistic and intuitive treatment of the fuzzy set. There are even cases where they exhibit a counter intuitive result, such as allowing a single membership value (for instance a zero value) to determine the overall output [Zimmermann and Zysno 1980].

By extending the constituent focal elements of the possibility measure underlying the fuzzy set as proposed in the present work, simple operations from classical set theory can be utilized to operate on a fuzzy set. Additionally the number of repetitions of the same evidence till the process converges, depends on the features of the two fuzzy sets, and therefore the context becomes a determining factor.

The proposed operation is not symmetric. If by $G(B/A)$ we denote the result of generalizing existing knowledge B given new evidence A, then

$$G(B/A) \neq G(A/B)$$

except in the trivial case where $A = B$.

This is often the case in human knowledge updating, especially in situations of important safety considerations, where the existing knowledge may be the product of careful collection of valuable expertise, or the coding of domain experts judgment. This assigns to the data already in the knowledge base a higher degree of importance over the newly presented evidence, implementing in a sense a monotonic reasoning scheme. The use of the repetition of the same piece of knowledge as a means for determining the degree of generalization, and as such as a parameter, has an intuitive interpretation: the extension of a fuzzy set with a parameter higher than one, indicates the presence of persistent evidence, which justifies the greater change in the initial knowledge towards this piece of evidence.

The upper limit of this parameter – the number of repetitions till convergence state – can also be considered an index of the conceptual distance between the entities represented by the two fuzzy sets. According to this interpretation the convergence state produces a new fuzzy set fully comprising the previous two.

Finally, due to the fact that the algorithm is based on simple set operations, it can be applied to fuzzy sets defined over non numerical domains.

5 Conclusions

For the decision making process required in assessing a site to serve as a permanent repository for nuclear waste, a theoretical framework and a prototype system are proposed.

The theoretical modeling uses evidential reasoning as a form for representing and handling the different types of uncertainty faced by the decision maker. Then by considering the consequences of possible future scenarios it assesses the performance of a certain site. The mechanisms of the evidential support logic rule can describe domain knowledge, allowing stochastic or fuzzy uncertainty to be represented, as well as degrees of confidence at different levels of abstraction.

The whole system has a very high degree of flexibility permitting modifications and extensions at every level. Furthermore, a mechanism for updating the existing knowledge in the rule base is proposed.

References

1 Baldwin J.F., (1992) Fuzzy and Probabilistic Uncertainties, in Encyclopedia of Artificial Intelligence, S.C. Shapiro, Ed., Wiley, pp. 528-537.

2 Baldwin J.F., Martin T.P. and Pilsworth B.W., (1988) FRIL Manual, Bristol, FRIL Systems Ltd.

3 Baldwin J.F., Martin T.P. and Pilsworth B.W., (1995) FRIL Fuzzy and Evidential Reasoning in A. I., Research Stusies Press.

4 Baldwin J.F., Martin T.P. and Tocatlidou A., (1995) "Uncertainty Management in Radioactive Waste Repository Site Assessment", Fuzzy Sets and Systems, 74, pp. 81-91.

5 Bonano E.J., Cranwell R.M. and Davis P.A., (1989) "Performance Assessment Methodologies for the Analysis of High Level Nuclear Waste Repositories", Radioactive Waste Management and the Nuclear Fuel Cycle, 13, pp. 229-239.

6 Gallegos D.P. and Bonano E.J., (1993) "Consideration of Uncertainty in Performance Assessment of Radioactive Waste Disposal from an International Regulatory Perspective", Reliability Engineering and Systems Safety, 42, pp. 111-123.

7 Horvitz E.J., Heckerman D.E. and Langlotz C.P., (1986) "A Framework for Comparing Alternative Formalisms for Plausible Reasoning", in 5th National Conference on Artificail Intelligence, Philadelphia, PA: American Association for Artificial Intelligence.

8 Roxburg I.S., (1987) Geology of High Level Nuclear Waste Disposal, London, Chapman and Hall.

9 Shafer G., (1976) A Mathematical Theory of Evidence, Princeton, Princeton University Press.

10 Thorne M.C., (1993) "The Use of Expert Opinion in Formulating Conceptual Models of Underground Disposal Systems and the Treatment of Associated Bias", Reliability Engineering and Systems Safety, 42, pp. 161-180.

11 Tocatlidou A., (1998) "Learning Based Similarity Measurement for Fuzzy Sets", International Journal of Intelligent Systems, 13, pp. 193-220.

12 Voegele M.D., Younker J.L. and Alexander D.H., (1989) "Deriving a Site Characterization Program from the Applicable Regulations", Radioactive Waste Management and the Nuclear Fuel Cycle, 13, pp. 105-120.

13 Walker C.A., (1983) Science and Technology of the Sources and Management of Radioactive Waste, in Too Hot to Hnadle? Social and policy Issues in the Management of Radioactive Wastes, C.A. Walker, L.C. Gould, and E.J. Woodhouse, Editors, Yale University Press, New Haven.

14 Weinberg A.M., (1972) "Social Institutions and Nuclear Energy", Science, 177, pp. 27-34.

15 Wu J.S., Apostolakis G.E. and Okrent D., (1990) "Uncertainties in System Analysis: Probabilistic Versus Nonprobabilistic Theories", Reliability Engineering and System Safety, 30, pp. 163-181.

16 H-J. Zimmermann and P. Zysno, (1980) "Latent Connectives in Human Decision Making", Fuzzy Sets and Systems, 4, pp. 37-51.

A Fuzzy Model of Heavy Metal Loadings in Marine Environment

L.I. Kuncheva[1], J. Wrench[2], L.C. Jain[3], and A. Al-Zaidan[1]

[1]School of Informatics
[2]School of Ocean Sciences
University of Wales
Bangor, Gwynedd LL57 1UT
United Kingdom
e-mail: l.i.kuncheva@bangor.ac.uk
[3]University of South Australia, Adelaide,
Mawson Lakes, S.A.,5095, Australia

Abstract

We design a fuzzy model of the loadings of heavy metals for two coastal areas of the Irish Sea (Liverpool Bay and Morecambe Bay). Each metal concentration is associated with a fuzzy set "contaminated", defined over a set of sampling sites (70 in Liverpool Bay and 203 in Morecambe Bay). The higher the concentration, the higher the degree of membership of the site. Six overall loading indices are calculated using aggregation connectives between fuzzy sets. The loading indices are then interpolated and plotted on a map. A visual inspection shows that: (i) product aggregation is most indicative for the locations of the disposal grounds; (ii) mean aggregation reflects well the sediment movement in the bay; (iii) maximum aggregation indicates all highly contaminated sites. The proposed fuzzy model is easy to implement and the results are directly interpretable.
Keywords: Fuzzy aggregation connectives, environmental modelling, Liverpool Bay, Morecambe Bay, heavy metal concentrations, index of spatial distribution, spatial data analysis.

1 Introduction

Near shore waters have been used as a resource for the disposal of waste material over a prolonged period extending back to the later part of the industrial revolution. In the case of the Irish Sea disposal of sewage material from the heavily populated industrial region of south Lancashire has been carried out since the late 1800's. For most of this time the attitude to disposal may be summarised as 'out of sight out of mind'. With the development of reliable and sensitive methods for measuring trace contaminants, questions have been raised as to the environmental impact of long-term disposal in coastal areas.

Assessing the impact of such long-term activities in areas such as Liverpool Bay and Morecambe Bay (Figure 1) is fundamental to environmental risk management in coastal zones of the Irish Sea.

There are several strands to impact assessment.

1. Assessing the degree of contamination above normal, or background, levels.

2. Defining the pattern, or the geographical distribution, of contamination.

3. Determining the likely changes in contamination with time.

This present work addresses the first two components of risk assessment and management by analysis of data sets on metals in the surface sediments of two distinct coastal environments of the Irish Sea.

The dynamic nature of coastal waters presents a severe challenge to environmental assessment of disposal activities in near shore waters. Surface sediments lying in shallow water act as a primary sink for pollutants delivered through rivers and groundwater flow and are most often studied in an attempt to identify and assess contaminant sources. However, sediments are subjected to strong oscillating tidal forces resulting in bed load transport; storms may result in mixing of contaminated sediment with material from a different source and organisms in sediments may re-work deeper consolidated material to the surface layer. The overall effect of these uncontrollable environmental variables is to obscure spatial data.

The basic problem centers on combining metal data in a way which produces a meaningful distribution pattern given that the metal have different concentration scales. Large data sets are being collected and stored, awaiting processing and analysis. The current study falls into the general (and controversial) category of *statistical spatial data analysis* as defined in (Bailey, 1994). This type of analysis has been recognized as an important research line, gaining speed in the past 10 years and trying to establish its identity. The importance lies with the demand on the part of *Geographical Information Systems* (GIS) for "systems that 'do something' other than display and organize data" (Fotheringham and Rogerson, 1994). In this study we have multiattribute data, i.e., for each location a set of metal concentrations has been measured. The problem is to combine this information for each location and to devise a contamination distribution picture of the whole area of interest. Typical choices for processing such data are principal component analysis (PCA) or cluster analysis, possibly because these techniques are available in

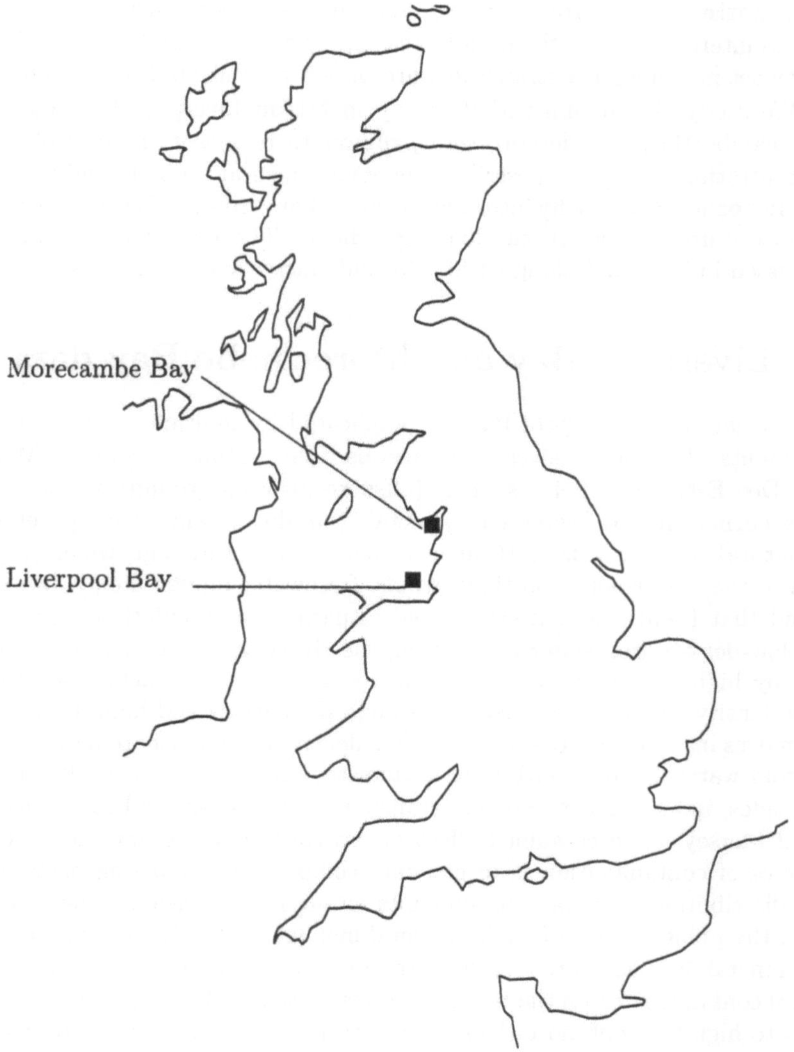

Figure 1: Liverpool Bay and Morecambe Bay location

most statistical software packages. The results of both methods are difficult to interpret, unless the data has favorable structure and characteristics. Fuzzy set modeling is a straightforward option for this kind of problems.

We study the loadings of 10 heavy metals in Liverpool Bay and 7 in Morecambe Bay, and design *loading indices* to represent the overall metal concentration. Section 2 describes the environmental problem and the data set. In Section 3 we briefly introduce fuzzy sets and give the fuzzy aggregation operators used to design the loading indices. The results are shown and discussed in Section 4 along with PCA and cluster analysis results.

2 Liverpool Bay and Morecambe Bay data

The situation in Liverpool Bay is complicated by multiple sources of contaminants. The bay receives heavy metals from continuous sources (Mersey and Dee Estuaries), point sources (offshore disposal ground) and a higher than normal input of some heavy metals (notably arsenic) through erosion of natural mineral sources (Camacho-Ibar, 1992). Although water circulation in the bay depends on tides, winds, freshwater inputs, etc., it has been found that there is a well established estuarine-like circulation induced by the low-density freshwater inputs from the rivers (the Mersey in particular) and by higher density sea water from the Irish Sea (Camacho-Ibar, 1992). Low-density water moves offshore through the surface and high-density water moves inshore near the bottom. This density-driven inshore movement of bottom water, coupled with tidal asymmetry producing stronger flood than ebb tides, induces a net sediment transport directed east and south-east toward Mersey. The combined effect of environmental factors and multiple sources of contamination is to generate complex and changing patterns in the distribution of metal contaminants in surface sediments. The problem from the point of view of environmental management is to develop patterns from metal data which reflect the current status of sediments. The impact of metal contamination on the biota in sediments is complex; exposure of organisms to high levels of more than one metal introduces further environmental stress. It is therefore appropriate that an approach to developing patterns of contaminant distribution should include all metals.

Environmental management in Liverpool Bay is based on yearly analysis of heavy metals in surface sediments from a sampling grid approximately 20 km^2. Figure 2 (lefthand side plot) shows the Liverpool Bay area, the disposal ground and the sampling sites (stations). Interpretation of the sampled data is limited by the factors described above and there is a need to develop a rational protocol whereby a realistic picture may be produced from spatial data.

The data set in this study consists of concentrations of 10 heavy metals relative to aluminium content, measured between 14th and 16th September 1988 at the 70 sites on the sampling grid. The metals are: mercury (Hg),

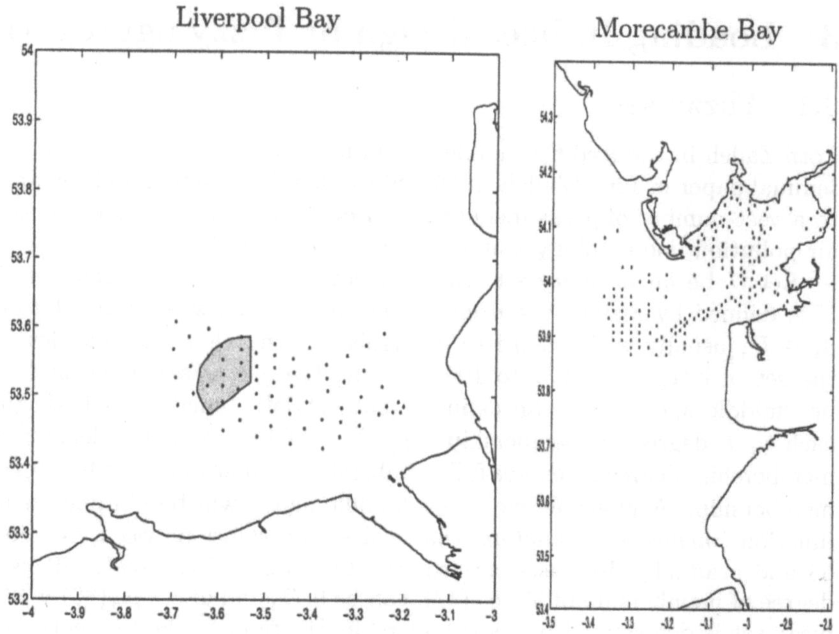

Figure 2: Liverpool Bay, the disposal ground (light gray area), Morecambe Bay, and the sampling sites (70 in Liverpool Bay and 203 in Morecambe Bay).

cadmium (Cd), chromium (Cr), copper (Cu), nickel (Ni), lead (Pb), zinc (Zn), arsenic (As), manganese (Mn), and iron (Fe).

The same stations have been sampled every year. A database has been collected over the period 1986-1993. Our aim in this pilot study was to develop a mathematical tool for analysing metal distributions using one data set. We chose the 1988 data set because a thorough analysis of the processes in Liverpool Bay for year 1988 can be found in (Camacho-Ibar, 1991, 1992).

Morecambe Bay is geographically contiguous with the northern sector of Liverpool Bay (Figure 1) and this area has not been used routinely for disposal of waste organic material. Morecambe Bay sampling stations are shown in Figure 2. A large data set for the late 1980's has been used for the generation of patterns of metals in surface sediments for purposes of comparison with Liverpool Bay.

In this study, Liverpool Bay contamination is the central environmental theme, and Morecambe Bay data is only used as a counter example. The loading indices for Liverpool Bay should be able to verify the low level of contamination of Morecambe Bay.

3 Loading Indices design by fuzzy aggregation

3.1 Fuzzy sets

Lotfi Zadeh introduced the simple and intuitive concept of a fuzzy set in his seminal paper in 1965 (Zadeh, 1965). Since then fuzzy sets have been applied to a vast number of areas including environmental sciences: soil, forest and air pollution, meteorology, water resources, etc. (Bezdek, 1999).

Let U be an ordinary set with elements $u_1, \ldots, u_m$. A *fuzzy set* A on U is defined by assigning a *degree of membership* between 0 and 1 to each $u_i \in U$, usually with regard to a linguistic term. For example, let U be the set of integers from 1 to 100 denoting the age of a person, and let A be "middle aged". We can define a (subjective!) function that assigns to each u_i a degree of membership $\mu_A(u_i) \in [0, 1]$. Degree 0 denotes non-membership, degree 1 denotes full membership, and any other value is partial membership. A plausible model of "middle aged" will be obtained using a function (*membership function*) that yields high values between, say, 40 and 55 and gradually decreases towards the two edges of the scale. Thus, the degree of membership of 37, $\mu_A(37)$, can be 0.75, and of 82, $\mu_A(82) = 0.1$. A fuzzy set is determined by its membership function, so the two notions will be used interchangeably.

Let $S = \{s_1, \ldots, s_{70}\}$ be the set of 70 sites in Liverpool Bay. Let $A_1, \ldots, A_{10}$ be fuzzy sets on S, one for each metal, with membership functions

$$\mu_{A_i} : S \to [0, 1], \quad i = 1, \ldots, 10.$$

The higher the ith metal concentration at site s_j, the higher the degree of membership $\mu_{A_i}(s_j)$. We chose the simple rescaling to devise the 10 membership functions from data: Let $x_i(s_j)$ be the concentration of the ith metal measured at site s_j, and let

$$LB^i_{min} = \min_{k=1}^{70}\{x_i(s_k)\},$$

and

$$LB^i_{max} = \max_{k=1}^{70}\{x_i(s_k)\}.$$

Then

$$\mu_{A_i}(s_j) = \frac{x_i(s_j) - LB^i_{min}}{LB^i_{max} - LB^i_{min}} \tag{1}$$

Figure 3 plots the membership function of mercury over the 2-dimensional space (sampling ground) spanned by the 70 sites in Liverpool Bay. Referring the plot to the original geographical problem, we observe a high concentration of mercury in the area of river estuaries. The contamination with mercury in that area is higher than that at the disposal ground (the second highest peak).

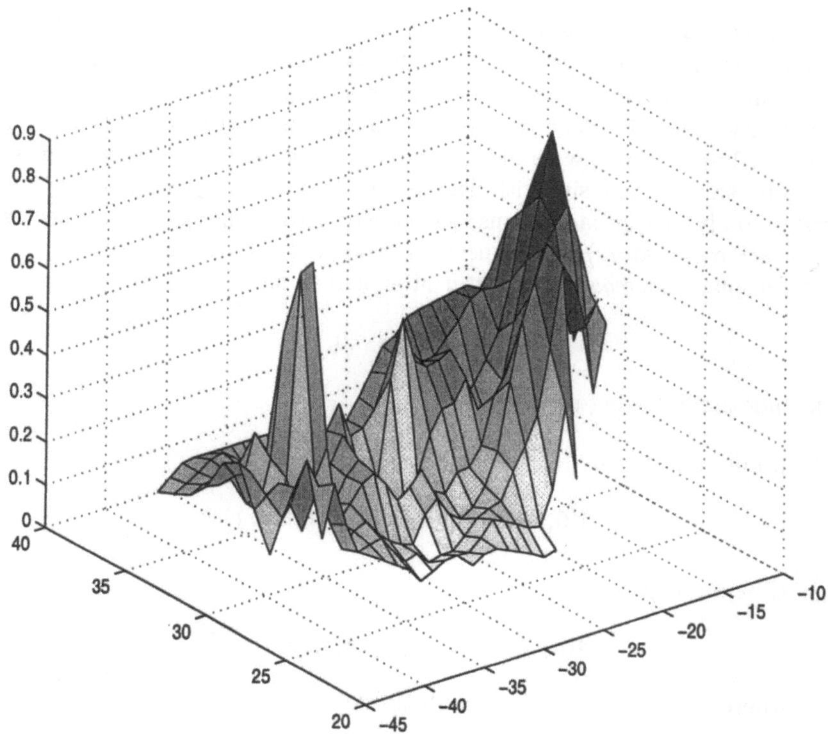

Figure 3: Hg membership function for the fuzzy set "contaminated". Th horizontal axis with the negative numbers is longitude between -4.00 and -3.00 degrees (west of Greenwich) and the one with the positive numbers is 53 degrees and decimal minutes north. The vertical axis is the degree of membership (between 0 and 1).

Another way of representing the membership functions (adopted here) is to use color or contour plot and overlay the scatterplot of the sampling sites.

Clearly, the concentration pattern would be the same if we did not scale the concentration between 0 and 1. Although individual metal distribution is an interesting topic on its own, it was argued above that an index of overall loading (contamination) is needed.

3.2 Fuzzy aggregation connectives

Dubois and Prade (Dubois, 1997) point out that although fuzzy membership functions have numerous possible interpretations, fuzzy mathematics has gone a long way disregarding fuzzy sets semantics: "The risk is to leave the user with no guidelines about how to apply fuzzy set theory..." They distill three main semantics: similarity, preference and uncertainty. The interpretation of the fuzzy sets used here fits best in the second category: preference (in a broad sense) because the membership functions do not measure a similarity to some "ideal" prototypes nor do they express any type of uncertainty. For example, we *prefer* (to call "contaminated") a site with degree of membership 0.7 to a site with degree 0.4. Taking this interpretation, fuzzy decision-analysis approach seems reasonable. Fuzzy aggregation connectives (aggregation operators) will be used to define overall loading indices.

An m-place *aggregation operator* $\mathcal{A}$ is defined as

$$\mathcal{A} : [0,1]^m \rightarrow [0,1],$$

satisfying the following properties:

1. *Limit conditions*:

$$\mathcal{A}(0,0,\dots,0) = 0, \quad \mathcal{A}(1,1,\dots,1) = 1.$$

2. *Commutativity*:

$$\mathcal{A}(a_1, a_2, \dots, a_m) = \mathcal{A}(b_1, b_2, \dots, b_m),$$

   where $b_1, \dots, b_m$ is any permutation of $a_1, a_2, \dots, a_m$, $a_i \in [0,1]$, $i = 1, \dots, m$.

3. *Monotonicity*:

$$\mathcal{A}(a_1, a_2, \dots, a_m) \geq \mathcal{A}(b_1, b_2, \dots, b_m),$$

   for any $a_1, a_2, \dots, a_m, b_1, \dots, b_m \in [0,1]$, such that $a_i \geq b_i, \forall i = 1, \dots, m$.

There are a great variety of fuzzy connectives and aggregation operators (Bloch, 1996, Dubois, 1985, Grabisch, 1995, Yager, 1994). Since this is a pilot study, here we use perhaps the simplest 6 aggregation operators:

1. *Pessimistic-type aggregation*

   - Minimum

$$\mathcal{A}_1(a_1, a_2, \dots, a_m) = \min\{a_1, a_2, \dots, a_m\},$$

   - Product

$$\mathcal{A}_2(a_1, a_2, \dots, a_m) = a_1 \cdot a_2 \cdot \dots \cdot a_m.$$

   - Geometric mean

$$\mathcal{A}_3(a_1, a_2, \dots, a_m) = (a_1 \cdot a_2 \cdot \dots \cdot a_m)^{1/m}.$$

2. *Optimistic-type aggregation*

   - Maximum

$$\mathcal{A}_4(a_1, a_2, \ldots, a_m) = \max\{a_1, a_2, \ldots, a_m\},$$

3. *Indifferent-type aggregation*

   - Arithmetic mean

$$\mathcal{A}_5(a_1, a_2, \ldots, a_m) = \frac{1}{m}(a_1 + \ldots + a_m).$$

   - "Competition jury". This is an operator, where we discard the highest and the lowest values from the set $a_1, a_2, \ldots, a_m$, and average the remaining $m-2$ values.

$$\mathcal{A}_6(a_1, a_2, \ldots, a_m) =$$

$$\frac{1}{m-2}\left(a_1 + \ldots + a_m - \max_k a_k - \min_k a_k\right).$$

Replacing a_i with $\mu_{A_i}(s_j)$, each of these 6 aggregation operators defines a **Loading Index** as a fuzzy set on S. The indices are denoted respectively $\mathrm{LI}_1, \ldots \mathrm{LI}_6$.

4 Results

Table 1 shows the correlation coefficients ($\times$ 100) between the metal concentrations in Liverpool Bay.

Table 1: Correlations between the metal concentrations ($\times$ 100)

	Cd	Cr	Cu	Ni	Pb	Zn	As	Mn	Fe
Hg	0	-16	-23	-34	-31	-23	-27	-31	-27
Cd		55	54	63	71	70	51	55	48
Cr			77	83	76	85	70	64	87
Cu				70	77	80	60	57	78
Ni					78	84	65	73	79
Pb						95	81	82	78
Zn							88	88	88
As								90	87
Mn									76

Cluster analysis is often used in environmental studies to find spatial areas corresponding to values "low", "medium", and "high" of a certain parameter indicative of the concentration (Markus, 1996). When this parameter is based on more than one variable, e.g., contamination with heavy metals, clustering might not be appropriate. Figure 4 shows the difference between "favourable data" (where the clusters correspond to "low" and "high" contamination) and "unfavourable data" (where such correspondence does not exist). Shown on the left plot are two clusters found by the hard c-means (HCM) clustering procedure for zinc (Zn) and lead (Pb) from the Liverpool Bay data. Their concentrations correlate well (0.95), so the two cluster centers can be labeled "low" and "high" as shown. The plot on the right shows the results from the same procedure applied to mercury (Hg) and lead (Pb). The metal concentrations have a negative correlation coefficient (-0.31), so the cluster centers (circled) do not correspond to "low" and "high" in any order.

When more than two components are involved such inconsistencies are difficult to resolve, likely to be obscured, and the result can be misinterpreted.

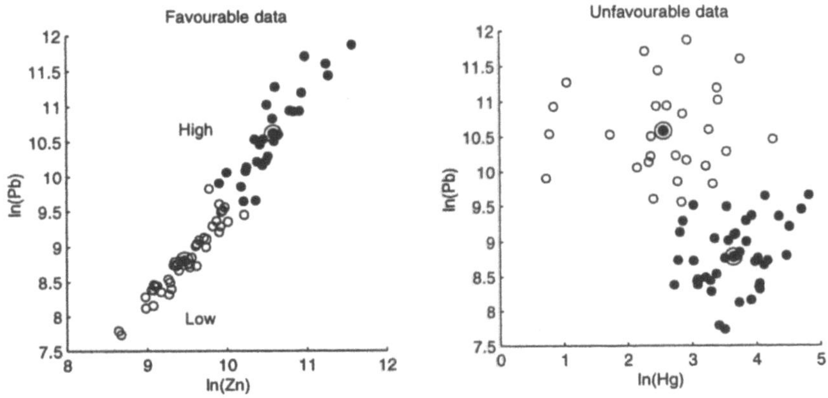

Figure 4: Scatterplot of "favourable" and "unfavourable" data

As Table 1 shows, most of the correlations are positive (due to high concentrations around the waste disposal ground and low elsewhere) except for Hg, and therefore cluster analysis can be expected to produce sensible results. For three suspected clusters the hard c-means algorithm on the scaled data (eqn. (1)) found the centers shown in Table 2.

All center components but the first one (Hg) are ordered so that the three clusters correspond to "high", "low", and "medium", respectively. Figure 5 shows the result as the spatial distribution of loading on the geographical sampling grid. Dark regions correspond to high metal loading, and bright regions, to low loading. Hard c-means clustering uses random initialization, hence different sets of centers can be obtained. The two plots show the results from clustering the original metal concentrations (left) and the scaled data (right). Clustering results are influenced by data transformation (Duda and Hart, 1973). This effect is demonstrated by the differences in the two plots

Table 2: Three cluster centers from one HCM run

	Cluster 1	Cluster 2	Cluster 3
Hg	0.1259	0.3011	0.2448
Cd	0.4159	0.0810	0.1679
Cr	0.7228	0.1739	0.5275
Cu	0.6875	0.0663	0.3147
Ni	0.7823	0.1846	0.4810
Pb	0.5663	0.0322	0.1836
Zn	0.5707	0.0674	0.2413
As	0.4846	0.0252	0.1312
Mn	0.4184	0.0092	0.0863
Fe	0.7141	0.1339	0.3602

and clearly raises the question of whether the data should be scaled. If not, the metals with concentrations which are by orders of magnitude higher than the others will dominate and determine the clustering result. On the other hand, if we decide to scale the data, we need to choose a scaling method (e.g., taking the logarithm, z-normalisation, scaling as in eqn (1), etc.). Each of these methods might lead to a different clustering result.

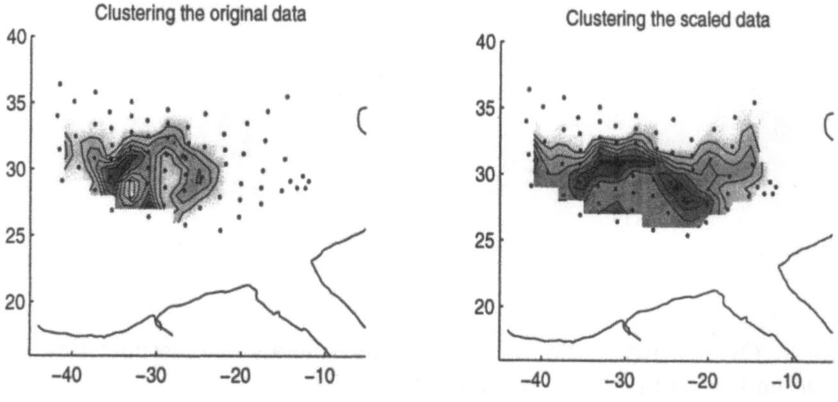

Figure 5: Overall distribution of the 10 heavy metals in Liverpool bay calculated by hard c-means clustering

Fuzzy c-means has also been used for clustering purposes in spatial data analysis (Markus, 1996). It is debatable, however, what the added value of using fuzzy c-means is over that of the hard c-means.

Principle Component Analysis (PCA) gives results that are not easily interpretable in the general case. Here, the variance of most metals is along the same axis: high values at the disposal site, which are an order of magnitude higher than the values elsewhere. In this case the first principal component should follow the pattern found by the HCM clustering. Figure 6 shows the distribution defined by the first and the second principal components. While the first component might correspond to metal loading, the second component is not easy to name. The top two plots show the results with the original data, and the bottom two plots, with the scaled data. As the figure shows, similarly to clustering, PCA is sensitive to the type of data scaling. The difference is especially clear in the second principal component which makes its interpretation even more obscure.

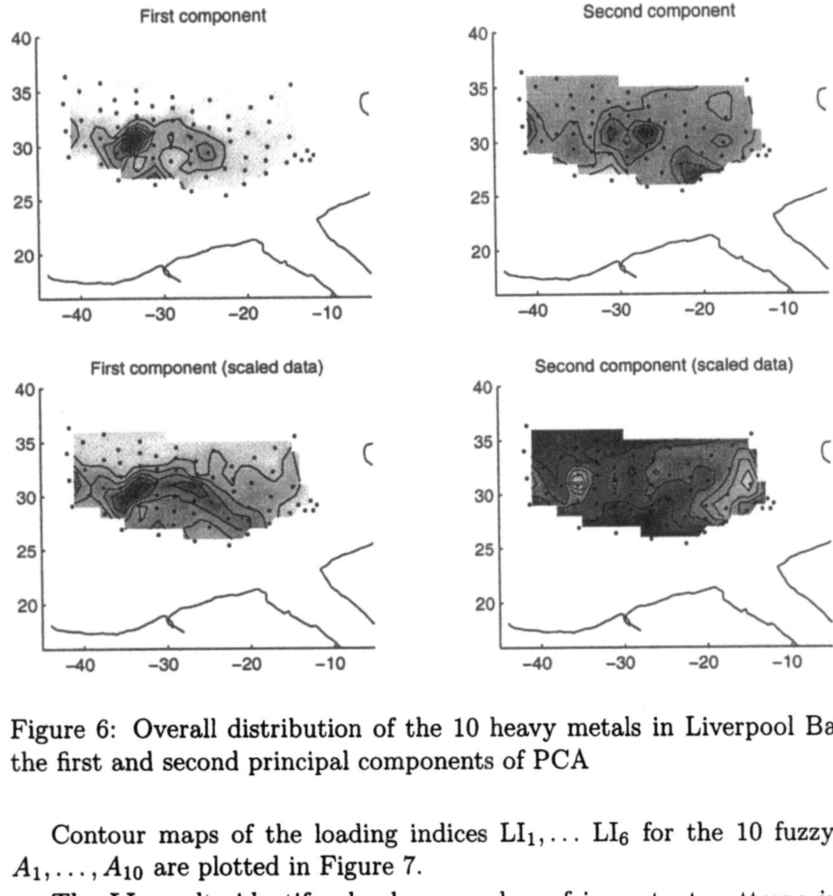

Figure 6: Overall distribution of the 10 heavy metals in Liverpool Bay by the first and second principal components of PCA

Contour maps of the loading indices $LI_1, \ldots LI_6$ for the 10 fuzzy sets $A_1, \ldots, A_{10}$ are plotted in Figure 7.

The LI results identify clearly a number of important patterns in the spatial data. The feature common to all 6 methods is the area of low metal contamination in the north west (top left) section of the sampling grid. The product method of calculation efficiently resolves the disposal ground whilst most of the other methods identify a residual south easterly movement of material from the disposal ground towards the estuary mouth. The smaller high area in the extreme eastern sector (seen most clearly in the Maximum

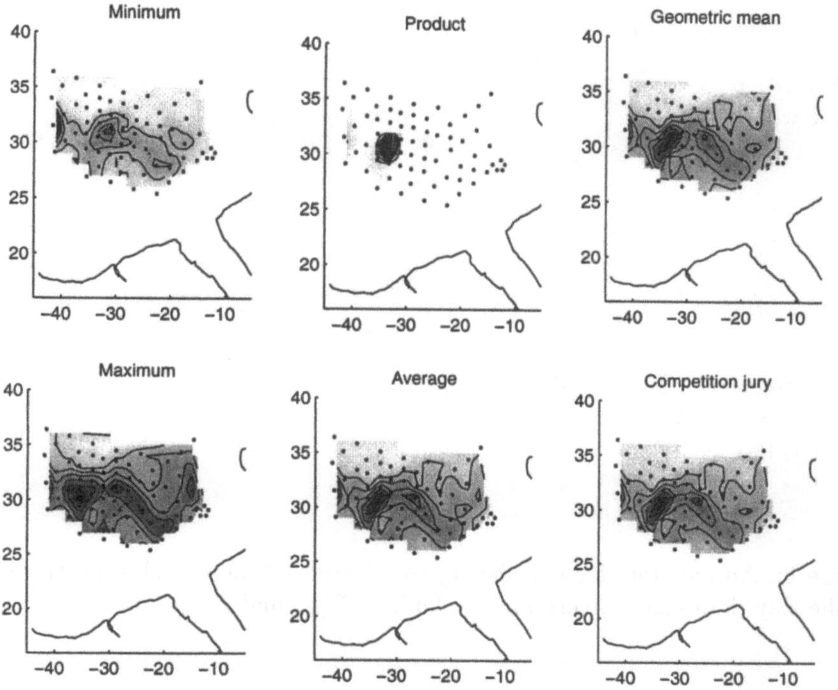

Figure 7: Overall distribution of the 10 heavy metals in Liverpool Bay calculated by the 6 Loading Indices

plot) identifies a widening area at the estuary mouth where there is net deposition of suspended material delivered from industrial sources higher up the River Mersey.

We calculated the six indices with the Morecambe bey data. To make the results comparable, we used formula (1) with the limits LB^i_{min} and LB^i_{max} of metal i as in Liverpool Bay. All negative values were set to 0 and all values greater than 1 were set to 1. Figure 8 gives an example of the scaling of Morecambe data using the liverpool Bay limits.

Because many of the concentrations in Morecambe Bay were lower than LB^i_{min}, a significant number of degrees of membership were set to 0. Thus the product and the minimum loading indices produced entirely flat distributions indicating no significant contamination of Morecambe bay. The maximum loading index is the most sensitive of all six and would label a site as contaminated if even one metal has high concentrations, regardless of the concentrations of the other metals. Figure 9 shows the distributions of maximum and product loading indices for Morecambe Bay. As with Liverpool Bay, the product loading index generates a pattern which is easily interpreted. The featureless distribution in Morecambe Bay indicates, in di-

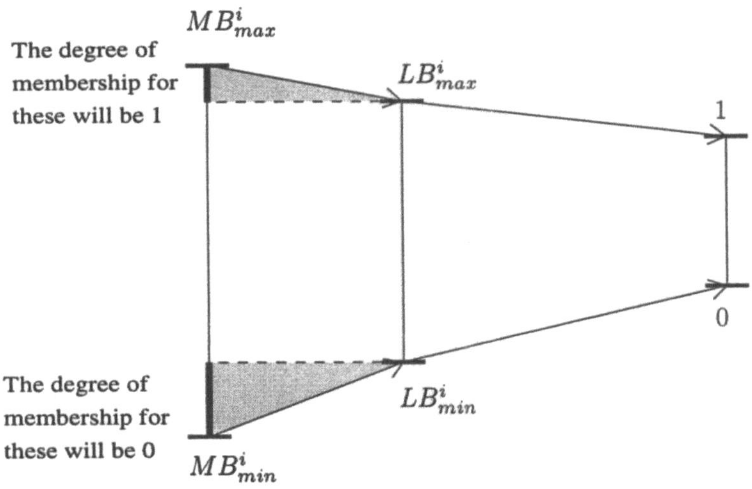

Figure 8: An example for calculating the degrees of membership of Morecambe Bay data using Liverpool Bay limits LB^i_{min} and LB^i_{max}.

rect contrast with Liverpool Bay, no point source for heavy metals. This observation is consistent with the fact that specific areas of Morecambe Bay have not been used for routine disposal.

Initial plots of the maximum loading index showed almost blanket contamination of Morecambe Bay. Stepwise removal of individual metal data sets revealed that this effect was due entirely to lead. The question of lead in Morecambe Bay will be addressed in a future paper. For the purposes of this present work, the maximum loading index plot in Figure 9 has been generated without the lead data. The maximum loading index is a sensitive indicator of possible contamination and the resulting pattern is more complex. There are isolated areas of apparent contamination which require further examination of the data for specific sampling stations.

The difficulty in assessing results such as those in this study comes from the fact that there is no benchmark against which the new solution can be matched. The *plausibility* of the results can be judged only by eye and the interpretation will therefore be subjective. *Product aggregation* clearly indicates where the highest contamination is. This loading index may be favored by the user responsible for the waste disposal. *Maximum aggregation* shows all contaminated sites even if the contamination is due to just one of the components. This loading index may be selected by the user concerned about, say, the fish diversity in the region. The bottom line is that there is no *true* loading distribution nor is there a single one that can be "useful" from

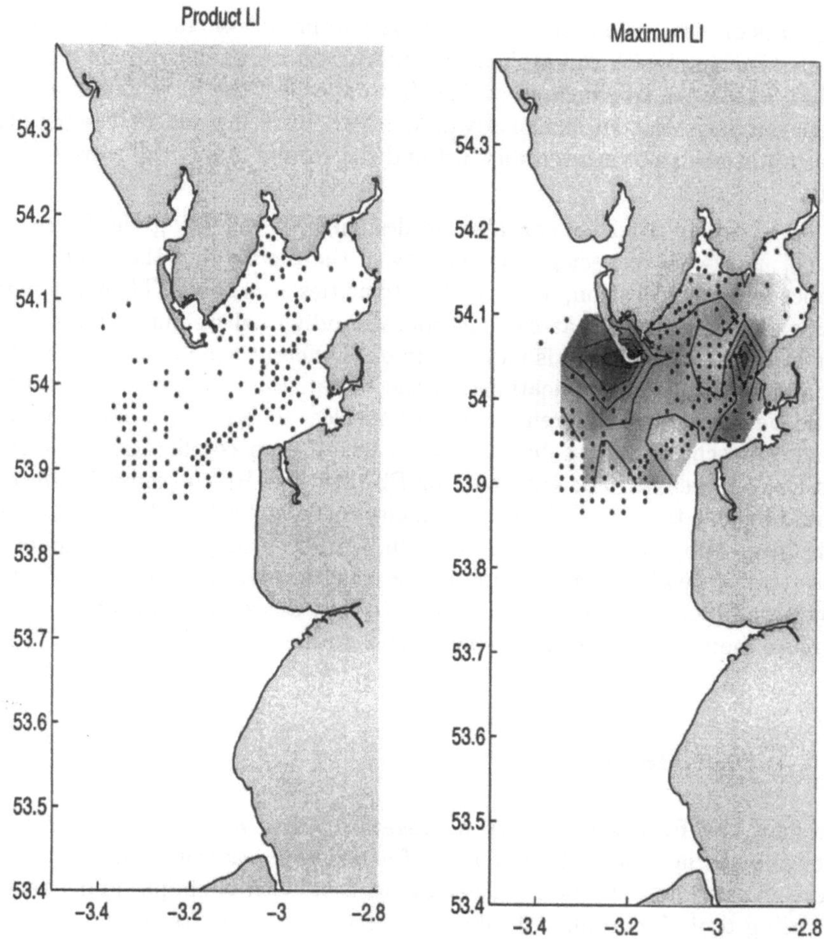

Figure 9: Product and maximum LI for Morecambe Bay

all points of view. What the proposed fuzzy approach offers is a collection of indices, each one with comprehensible interpretation, thereby giving the user a chance to make an educated choice. In this respect the fuzzy sets approach to spatial data analysis has an advantage over clustering and PCA where the interpretation is not straightforward and the results are at the mercy of the data.

An Index of Toxicity can also be designed by weighting the fuzzy sets $A_1, \ldots, A_{10}$ with respect to the toxicity of the metals and then applying a proper fuzzy aggregation, if the relative toxicities are known. There are many fuzzy set connectives that can incorporate individual weights for the fuzzy sets but the problem here is more complex. Different combinations of metals could have different implications on the biota. Besides, these implications could be specific for different groups of species. Therefore, a more complex coefficient scheme has to be considered where each combination of metals has its own weight. For example, for Pb, Mn and Hg the aggregation rule should be able to account for seven toxicity coefficients: for (Pb), (Mn), (Hg), (Pb, Mn), (Pb, Hg), (Mn, Hg), and (Pb, Mn, Hg). An apt fuzzy model for this type of weighted aggregation is the fuzzy integral (cf. Grabisch, 1995). However, determining a set of coefficients that assess the joint effect of two or more metals on a variety of species is not a trivial task.

5 Conclusions

We show how fuzzy set theoretic aggregation operators can be used for modeling the spatial distribution of a set of variables in environmental problems, thereby providing the non-mathematical user with a simple and effective modeling tool. We applied 6 different fuzzy aggregation techniques to a set of heavy metal concentrations sampled from Liverpool Bay. For verification of our findings we used data from Morecambe Bay which is supposed to be low- or non-contaminated. The results were displayed as geographical scatterplots of metal loading and were assessed visually. Morecambe Bay data analysis supported our conclusions about the loading indices. The low contamination appearing on the maximum aggregation loading index plot was not unexpected and we offered a hypothesis about its origin. The other indices such as product and minimum, indicated no contamination across the whole bay, which confirmed the supposed difference of contamination between Liverpool Bay and Morecambe Bay. Unlike PCA and HCM, fuzzy aggregation offers a variety of plots with different information in them. On the two edges of the scale are the product aggregation, which resolves the disposal grounds clearly, and the maximum aggregation, which identifies all possible sites with suspected high contamination. The main advantage of our fuzzy model over PCA and clustering is that the results are directly interpretable in the domain context.

References

[1] T.C. Bailey. A review of statistical spatial analysis in geographical information systems. In A.S. Fotheringham and P.A. Rogerson, editors, *Spatial analysis and GIS*, pages 13–44. Taylor & Francis, London, 1994.

[2] J.C. Bezdek and L.I. Kuncheva. Fuzzy pattern recognition. In J.G. Webster, editor, *Wiley Encyclopedia of Electrical and Electronics Engineering*, pages 173–181. John Wiley and Sons, Inc., Publishers, 1999.

[3] I. Bloch. Information combination operators for data fusion: a comparative review with classification. *IEEE Transactions on Systems, Man, and Cybernetics - Part A: Systems and Humans*, 26:52–67, 1996.

[4] V.F. Camacho-Ibar. *Trace elements and polychlorinated biphenyls (PCB) congeners in Liverpool bay sediments.* PhD thesis, University of Wales, Bangor, 1991.

[5] V.F. Camacho-Ibar, J.J. Wrench, and P.C. Head. Contrasting behaviour of arsenic and mercury in liverpool bay sediments. *Estuarine, Coastal and Shelf Science*, 41(3):241–263, 1992.

[6] D. Dubois and H. Prade. A review of fuzzy set aggregation connectives. *Information Sciences*, 36:85–121, 1985.

[7] D. Dubois and H. Prade. The three semantics of fuzzy sets. *Fuzzy Sets and Systems*, 90:141–150, 1997.

[8] R.O. Duda and P.E. Hart. *Pattern Classification and Scene Analysis.* John Wiley & Sons, NY, 1973.

[9] A.S. Fotheringham and P.A. Rogerson, editors. *Spatial Analysis and GIS.* Taylor & Francis, London, 1994.

[10] M. Grabisch. A new algorithm for identifying fuzzy measures and its application to pattern recognition. In *Proc. FUZZ/IEEE'95*, pages 145–150, Yokohama, Japan, 1995.

[11] M. Grabisch. On equivalence classes of fuzzy connectives - the case of fuzzy integrals. *IEEE Transactions on Fuzzy Systems*, 3(1):96–109, 1995.

[12] J.A. Markus and A.B. McBratney. An urban soil study: Heavy metals in Glebe, Australia. *Australian Journal of Soil Research*, 34:453–465, 1996.

[13] R.R Yager and D.P. Filev. *Essentials of Fuzzy Modeling and Control.* John Wiley & Sons, N.Y., 1994.

[14] L.A. Zadeh. Fuzzy sets. *Information and Control*, 8:338–353, 1965.

5.

RISK IN FINANCE

Capital Budgeting Techniques Using Discounted Fuzzy Cash Flows

Cengiz Kahraman

Department of Industrial Engineering
İstanbul Technical University, Maçka 80680, İstanbul, Turkey
cengiz2@ayasofya.isl.itu.edu.tr

In an uncertain economic decision environment, an expert's knowledge about dicounting cash flows consists of a lot of vagueness instead of randomness. Cash amounts and interest rates are usually estimated by using educated guesses based on expected values or other statistical techniques to obtain them. Fuzzy numbers can capture the difficulties in estimating these parameters. In this chapter, the formulas for the analyses of fuzzy present value, fuzzy equivalent uniform annual value, fuzzy future value, fuzzy benefit-cost ratio, and fuzzy payback period are developed and given some numeric examples. Then the examined cash flows are expanded to geometric and trigonometric cash flows and using these cash flows fuzzy present value, fuzzy future value, and fuzzy annual value formulas are developed for both discrete compounding and continuous compounding.

1 Introduction

The purpose of this chapter is to develop the fuzzy capital budgeting techniques. The analyses of fuzzy future value, fuzzy present value, fuzzy rate of return, fuzzy benefit/cost ratio, fuzzy payback period, and fuzzy equivalent uniform annual value are examined for the case of discrete compounding.

To deal with vagueness of human thought, Zadeh [1] first introduced the fuzzy set theory, which was based on the rationality of uncertainty due to imprecision or vagueness. A major contribution of fuzzy set theory is its capability of representing vague knowledge. The theory also allows mathematical operators and programming to apply to the fuzzy domain.

A fuzzy number is a normal and convex fuzzy set with membership function $\mu_A(x)$ which both satisfies normality: $\mu_A(x)=1$, for at least one $x \in R$ and convexity: $\mu_A(x') \geq \mu_A(x_1) \Lambda \mu_A(x_2)$, where $\mu_A(x) \in [0,1]$ and $\forall x' \in [x_1, x_2]$. '$\Lambda$' stands for the minimization operator.

Quite often in finance future cash amounts and interest rates are estimated. One usually employs educated guesses, based on expected values or other statistical techniques, to obtain future cash flows and interest rates. Statements like *approximately between \$ 12,000 and \$ 16,000* or *approximately between 10% and 15%* must be translated into an exact amount, such as *\$ 14,000* or *12.5%* respectively. Appropriate fuzzy numbers can be used to capture the vagueness of those statements.

A tilde will be placed above a symbol if the symbol represents a fuzzy set. Therefore, $\tilde{P}, \tilde{F}, \tilde{G}, \tilde{A}, \tilde{i}, \tilde{r}$ are all fuzzy sets. The membership functions for these fuzzy sets will be denoted by $\mu(x|\tilde{P}), \mu(x|\tilde{F}), \mu(x|\tilde{G})$, etc. A fuzzy number is a special fuzzy subset of the real numbers. The extended operations of fuzzy numbers are given in the appendix. A triangular fuzzy number (TFN) is shown in Figure 1. The membership function of a TFN ($\tilde{M}$) defined by

$$\mu(x|\tilde{M}) = (m_1, f_1(y|\tilde{M}) / m_2, m_2 / f_2(y|\tilde{M}), m_3) \tag{1}$$

where $m_1 \prec m_2 \prec m_3$, $f_1(y|\tilde{M})$ is a continuous monotone increasing function of y for $0 \le y \le 1$ with $f_1(0|\tilde{M}) = m_1$ and $f_1(1|\tilde{M}) = m_2$ and $f_2(y|\tilde{M})$ is a continuous monotone decreasing function of y for $0 \le y \le 1$ with $f_2(1|\tilde{M}) = m_2$ and $f_2(0|\tilde{M}) = m_3$. $\mu(x|\tilde{M})$ is denoted simply as $(m_1 / m_2, m_2 / m_3)$.

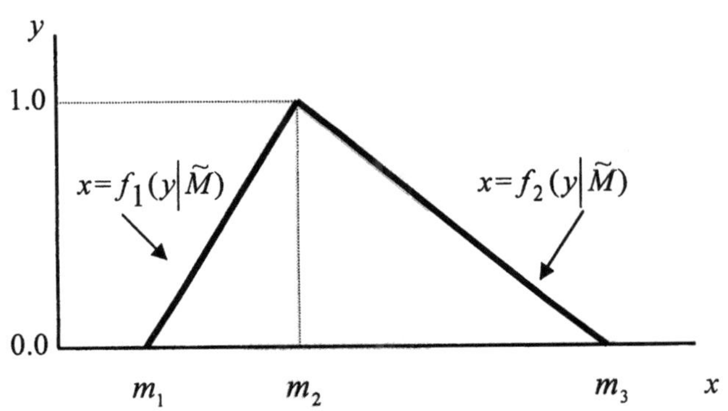

Fig. 1. A triangular fuzzy number, $\tilde{M}$

A flat fuzzy number (FFN) is shown in Figure 2. The membership function of an FFN, $\widetilde{V}$ is defined by

$$\mu(x|\widetilde{V}) = (m_1, f_1(y|\widetilde{V}) / m_2, m_3 / f_2(y|\widetilde{V}), m_4) \qquad (2)$$

where $m_1 \prec m_2 \prec m_3 \prec m_4$, $f_1(y|\widetilde{V})$ is a continuous monotone increasing function of y for $0 \le y \le 1$ with $f_1(0|\widetilde{V}) = m_1$ and $f_1(1|\widetilde{V}) = m_2$ and $f_2(y|\widetilde{V})$ is a continuous monotone decreasing function of y for $0 \le y \le 1$ with $f_2(1|\widetilde{V}) = m_3$ and $f_2(0|\widetilde{V}) = m_4$. $\mu(y|\widetilde{V})$ is denoted simply as $(m_1 / m_2, m_3 / m_4)$.

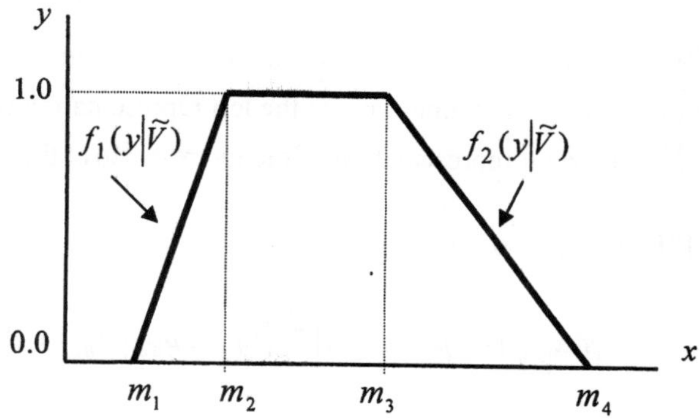

Fig. 2. A flat fuzzy number, $\widetilde{V}$

The fuzzy sets $\widetilde{P}, \widetilde{F}, \widetilde{G}, \widetilde{A}, \widetilde{i}, \widetilde{r}$ are usually fuzzy numbers but n will be discrete positive fuzzy subset of the real numbers [2]. The membership function $\mu(x|\widetilde{n})$ is defined by a collection of positive integers n_i, $1 \le i \le K$, where

$$\mu(x|\widetilde{n}) = \begin{cases} \mu(n_i|\widetilde{n}) = \lambda_i, 0 \le \lambda_i \le 1 \\ 0, otherwise \end{cases} \qquad (3)$$

2 Fuzzy Present-Value Method

The present-value method of alternative evaluation is very popular because future expenditures or receipts are transformed into equivalent dollars now. That is, all of the future cash flows associated with an alternative are converted into present dollars. If the alternatives have different lives, the alternatives must be compared over the same number of years.

Chiu and Park [3] propose a present value formulation of a fuzzy cash flow. The result of the present value is also a fuzzy number with nonlinear membership function. The present value can be approximated by a TFN. Chiu and Park [3]'s formulation is

$$P\widetilde{V} = \left(\sum_{t=0}^{n} \left(\frac{\max(P_t^{l(y)},0)}{\prod_{t'=0}^{t} (1+r_{t'}^{r(y)})} + \frac{\min(P_t^{l(y)},0)}{\prod_{t'=0}^{t} (1+r_{t'}^{l(y)})} \right),\ \sum_{t=0}^{n} \left(\frac{\max(P_t^{r(y)},0)}{\prod_{t'=0}^{t} (1+r_{t'}^{l(y)})} + \frac{\min(P_t^{r(y)},0)}{\prod_{t'=0}^{t} (1+r_{t'}^{r(y)})} \right) \right) \quad (4)$$

where $P_t^{l(y)}$: the left representation of the cash at time t, $P_t^{r(y)}$: the right representation of the cash at time t, $r_t^{l(y)}$: the left representation of the interest rate at time t, $r_t^{r(y)}$: the right representation of the interest rate at time t.

Buckley's [2] membership function for $\widetilde{P}_n$,

$$\mu(x|\widetilde{P}_n) = (p_{n1}, f_{n1}(y|\widetilde{P}_n)/p_{n2}, p_{n2}/f_{n2}(y|\widetilde{P}_n), p_{n3}) \quad (5)$$

is determined by

$$f_{ni}(y|\widetilde{P}_n) = f_i(y|\widetilde{F})(1+f_k(y|\widetilde{r}))^{-n} \quad (6)$$

for $i = 1,2$ where $k=i$ for negative $\widetilde{F}$ and $k=3-i$ for positive $\widetilde{F}$.

Ward [4] gives the fuzzy present-value function as

$$P\widetilde{V} = (1+r)^{-n}(a,b,c,d) \quad (7)$$

where (a, b, c, d) is a flat fuzzy filter function (4F) number.

Example 1. A \$ (-14,000, -12,000, -10,000) investment will return annual benefits for six years with no salvage value at the end of six years. Compute the fuzzy present worth of the cash flow using an interest of (7.12%, 10.25%, 13.42%) per year.

$$f_{6,1}(y|\widetilde{P}) = \sum_{j=0}^{6} f_{j,1}(y|\widetilde{F}_j)(1 + f_{k(j)}(y|\widetilde{r}_f))^{-j} \tag{8}$$

$$f_{6,2}(y|\widetilde{P}) = \sum_{j=0}^{6} f_{j,2}(y|\widetilde{F}_j)(1 + f_{k(j)}(y|\widetilde{r}_f))^{-j} \tag{9}$$

i=1, 2 where $k(j)=i$ for negative $\widetilde{F}_j$ and $k(j) = 3 - i$ for positive $\widetilde{F}_j$.

For $y = 0$, $f_{6,1}(y|\widetilde{P}) = \$ - 3{,}525.57$

For $y = 1$, $f_{6,1}(y|\widetilde{P}) = f_{6,2}(y|\widetilde{P}) = \$ - 24.47$

For $y = 0$, $f_{6,2}(y|\widetilde{P}) = \$ + 3{,}786.34$

The possibility of $NPV = 0$ for this triangular fuzzy number can be calculated using a linear enterpolation :

$x = -3{,}810.81y + 3{,}786.34$

For $x = 0$, $Poss(NPV = 0) = 0.9936$.

3 Fuzzy Capitalized Value Method

A specialized type of cash flow series is a perpetuity, a uniform series of cash flows which continues indefinetely. An infinite cash flow series may be appropriate for such very long-term investment projects as bridges, highways, forest harvesting, or the establishment of endowment funds where the estimated life is 50 years or more.

In the nonfuzzy case, if a present value P is deposited into a fund at interest rate r per period so that a payment of size A may be withdrawn each and every period forever, then the following relation holds between P, A, and r:

$$P = \frac{A}{r} \tag{10}$$

In the fuzzy case, lets assume all the parameters as triangular fuzzy numbers: $\widetilde{P}=(p_1,p_2,p_3)$ or $\widetilde{P}=((p_2-p_1)y+p_1),(p_2-p_3)y+p_3)$ and $\widetilde{A}=(a_1,a_2,a_3)$ or $\widetilde{A}=((a_2-a_1)y+a_1),(a_2-a_3)y+a_3)$ and $\widetilde{r}=(r_1,r_2,r_3)$ or $\widetilde{r}=((r_2-r_1)y+r_1,(r_2-r_3)y+r_3)$, where y is the membership degree of a certain point of A and r axis. If $\widetilde{A}$ and $\widetilde{r}$ are both positive,

$$\widetilde{P}=\widetilde{A}\varnothing\widetilde{r}=(a_1/r_3,a_2/r_2,a_3/r_1) \tag{11}$$

or

$$\widetilde{P}=(((a_2-a_1)y+a_1)/((r_2-r_3)y+r_3),((a_2-a_3)y+a_3)/((r_2-r_1)y+r_1)) \tag{12}$$

If $\widetilde{A}$ is negative and $\widetilde{r}$ is positive,

$$\widetilde{P}=\widetilde{A}\varnothing\widetilde{r}=(a_1/r_1,a_2/r_2,a_3/r_3) \tag{13}$$

or

$$\widetilde{P}=(((a_2-a_1)y+a_1)/((r_2-r_1)y+r_1),((a_2-a_3)y+a_3)/((r_2-r_3)y+r_3)) \tag{14}$$

Now, let $\widetilde{A}$ be an expense every nth period forever, with the first expense occurring at n. For example, an expense of (\$5,000,\$7,000,\$9,000) every third year forever, with the first expense occurring at t=3. In this case, the fuzzy efffective rate $\widetilde{e}$ may be used as in the following:

$$f_i(y|\widetilde{e})=(1+(1/m)f_i(y|\widetilde{r}'))^m-1 \tag{15}$$

where $i=1,2;$ $f_1(y|\widetilde{e})$:a continuous monotone increasing function of y; $f_2(y|\widetilde{e})$:a continuous monotone decreasing function of y; m: the number of compoundings per period; $\widetilde{r}'$: the fuzzy nominal interest rate per period. The membership function of $\widetilde{e}$ may be given as

$$\mu(x|\widetilde{e})=(e_1,f_1(y|\widetilde{e})/e_2,e_2/f_2(y|\widetilde{e}),e_3) \tag{16}$$

If $\widetilde{A}$ and $f_i(y|\widetilde{e})$ are both positive,

$$\widetilde{P}=\widetilde{A}\varnothing\widetilde{e}=(((a_2-a_1)y+a_1)/f_2(y|\widetilde{e}),((a_2-a_3)y+a_3)/f_1(y|\widetilde{e})) \tag{17}$$

If $\widetilde{A}$ is negative and $f_i(y|\widetilde{e})$ is positive,

$$\widetilde{P}=\widetilde{A}\varnothing\widetilde{e}=(((a_2-a_1)y+a_1)/f_1(y|\widetilde{e}),((a_2-a_3)y+a_3/f_2(y|\widetilde{e})) \quad (18)$$

$(a_2-a_1)y+a_1$ and $(a_2-a_3)y+a_3$ can be symbolized as $f_1(y|\widetilde{a})$ and $f_2(y|\widetilde{a})$ respectively.

Example 2: Project ABC consists of the following requirements. Find the capitalized worth of the project if $\widetilde{r}=(12\%,15\%,18\%)$ annually.

1. A ($40,000, $50,000, $60,000) first cost ($F\widetilde{C}$) at t=0,

2. A ($4,000, $5,000, $6,000) expense ($\widetilde{A}_1$) every day,

3. A ($20,000, $25,000, $30,000) expense ($\widetilde{A}_2$) every third year forever, with the first expense occurring at t=3.

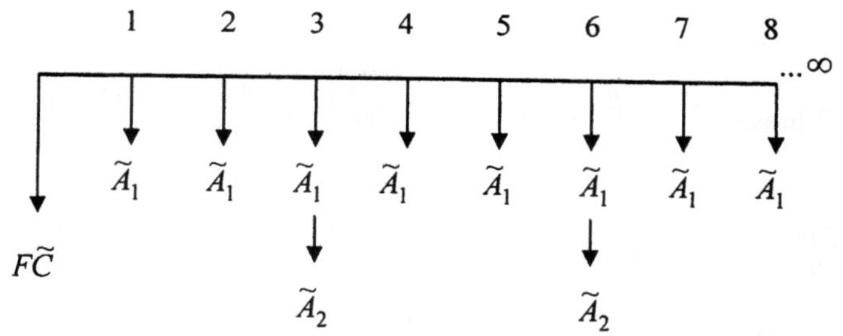

Fig. 3. The cash flow diagram for Example 2

$$\widetilde{P}=F\widetilde{C}\oplus(\widetilde{A}_1\varnothing\widetilde{r})+(\widetilde{A}_2\varnothing\widetilde{e}) \quad (19)$$

$$\widetilde{P}=((10{,}000y+40{,}000),(-10{,}000y+60{,}000))+((1{,}000y+4{,}000)/(-0.03y+0.18),$$
$$(-1{,}000y+6{,}000)/(0.03y+0.12))+((5{,}000y+20{,}000)/((1.18-0.03y)^3-1),$$
$$(-5{,}000y+30{,}000)/((1.12+0.03y)^3-1))$$

For y=0, $f_{\infty,1}(y|\widetilde{P})=\$93{,}326.42$

For y=1, $f_{\infty,1}(y|\widetilde{P})=f_{\infty,2}(y|\widetilde{P})=\$131{,}317.97$

For y=0, $f_{\infty,2}(y|\widetilde{P})=\$184{,}074.07$

4 Fuzzy Future Value Method

The future value (FV) of an investment alternative can be determined by using the relationship

$$FV(r)=\sum_{t=0}^{n}P_t(1+i)^{n-t} \tag{20}$$

where $FV(r)$ is defined as the future value of the investment using a minimum attractive rate of return $(MARR)$ of $r\%$. The future-value method is equivalent to the present-value method and the annual value method.

Chiu and Park's [3] formulation for the fuzzy future value has the same logic of fuzzy present value formulation:

$$\begin{aligned}\widetilde{FV}=\{&\sum_{t=0}^{n-1}[\max(P_t^{l(y)},0)\prod_{t'=t+1}^{n}(1+r_{t'}^{l(y)})+\min(P_t^{l(y)},0)\prod_{t'=t+1}^{n}(1+r_{t'}^{r(y)})]+P_n^{l(y)},\\ &\sum_{t=0}^{n-1}[\max(P_t^{r(y)},0)\prod_{t'=t+1}^{n}(1+r_{t'}^{r(y)})+\min(P_t^{r(y)},0)\prod_{t'=t+1}^{n}(1+r_{t'}^{l(y)})]+P_n^{r(y)}\}\end{aligned} \tag{21}$$

Buckley's [2] membership function $\mu(x|\widetilde{F})$ is determined by

$$f_i(y|\widetilde{F}_n)=f_i(y|\widetilde{P})(1+f_i(y|\widetilde{r}))^n \tag{22}$$

For the uniform cash flow series, $\mu(x|\widetilde{F})$ is determined by

$$f_{ni}(y|\widetilde{F})=f_i(y|\widetilde{A})\beta(n,f_i(y|\widetilde{r})) \tag{23}$$

where i=1,2 and $\beta(n,r)=(((1+r)^n-1)/r)$ and $\widetilde{A}\succ 0$ and $\widetilde{r}\succ 0$.

5 Fuzzy Benefit/Cost Ratio Method

The benefit/cost ratio (BCR) is often used to assess the value of a municipal project in relation to its cost; it is defined as

$$BCR = \frac{B-D}{C} \tag{24}$$

where B represents the equivalent value of the benefits associated with the project, D represents the equivalent value of the disbenefits, and C represents the project's net cost. A BCR greater than 1.0 indicates that the project evaluated is economically advantageous. In BCR analyses, costs are not preceded by a minus sign.

When only one alternative must be selected from two or more mutually exclusive (stand-alone) alternatives, a multiple alternative evaluation is required. In this case, it is necessary to conduct an analysis on the incremental benefits and costs. Suppose that there are two mutually exclusive alterantives. In this case, for the incremental BCR analysis ignoring disbenefits the following ratios must be used:

$$\Delta B_{2-1}/\Delta C_{2-1} = \Delta PV\,B_{2-1}/\Delta PVC_{2-1} \tag{25}$$

where PVB: present value of benefits, PVC: present value of costs. If $\Delta B_{2-1}/\Delta C_{2-1} \geq 1.0$, the alternative 2 is preferred.

In the case of fuzziness, first, it will be assumed that the largest possible value of Alternative 1 for the cash in year t is less than the least possible value of Alternative 2 for the cash in year t. The fuzzy incremental BCR is

$$\Delta\tilde{B}/\Delta\tilde{C} = \left(\frac{\sum_{t=0}^{n}(B_{2t}^{l(y)} - B_{1t}^{r(y)})(1+r^{r(y)})^{-t}}{\sum_{t=0}^{n}(C_{2t}^{r(y)} - C_{1t}^{l(y)})(1+r^{l(y)})^{-t}}, \frac{\sum_{t=0}^{n}(B_{2t}^{r(y)} - B_{1t}^{l(y)})(1+r^{l(y)})^{-t}}{\sum_{t=0}^{n}(C_{2t}^{l(y)} - C_{1t}^{r(y)})(1+r^{r(y)})^{-t}}\right) \tag{26}$$

If $\Delta\tilde{B}/\Delta\tilde{C}$ is equal or greater than (1, 1, 1), Alternative 2 is preferred.

In the case of a regular annuity, the fuzzy $\tilde{B}/\tilde{C}$ ratio of a single investment alternative is

$$\tilde{B}/\tilde{C} = \left(\frac{A^{l(y)}\gamma(n,r^{r(y)})}{C^{r(y)}}, \frac{A^{r(y)}\gamma(n,r^{l(y)})}{C^{l(y)}}\right) \tag{27}$$

where $\tilde{C}$ is the first cost and $\tilde{A}$ is the net annual benefit, and $\gamma(n,r) = ((1+r)^n - 1)/(1+r)^n r)$. The $\Delta\tilde{B}/\Delta\tilde{C}$ ratio in the case of a regular annuity is

$$\Delta\tilde{B}/\Delta\tilde{C} = \left(\frac{(A_2^{l(y)} - A_1^{r(y)})\gamma(n,r^{r(y)})}{C_2^{r(y)} - C_1^{l(y)}}, \frac{(A_2^{r(y)} - A_1^{l(y)})\gamma(n,r^{l(y)})}{C_2^{l(y)} - C_1^{r(y)}}\right) \tag{28}$$

6 Fuzzy Equivalent Uniform Annual Value Method

The equivalent uniform annual value (*EUAV*) means that all incomes and disbursements (irregular and uniform) must be converted into an equivalent uniform annual amount, which is the same each period. The major advantage of this method over all the other methods is that it does not require making the comparison over the least common multiple of years when the alternatives have different lives [5]. The general equation for this method is

$$EUAV = A = NPV\gamma^{-1}(n,r) = NPV\frac{(1+r)^n r}{(1+r)^n - 1} \tag{29}$$

where NPV is the net present value. In the case of fuzziness, $N\widetilde{P}V$ will be calculated and then the fuzzy $EU\widetilde{A}V$ $(\widetilde{A}_n)$ will be found. The membership function $\mu(x|\widetilde{A}_n)$ for $\widetilde{A}_n$ is determined by

$$f_{ni}(y|\widetilde{A}_n) = f_i(y|N\widetilde{P}V)\gamma^{-1}(n, f_i(y|\widetilde{r})) \tag{30}$$

and *TFN(y)* for fuzzy *EUAV* is

$$\widetilde{A}_n(y) = (\frac{NPV^{l(y)}}{\gamma(n,r^{l(y)})}, \frac{NPV^{r(y)}}{\gamma(n,r^{r(y)})}) \tag{31}$$

Example 3. Assume that $N\widetilde{P}V = (-\$3{,}525.57, -\$24.47, +\$3{,}786.34)$ and $\widetilde{r} = (3\%, 5\%, 7\%)$. Calculate the fuzzy *EUAV*.

$$f_{6,1}(y|\widetilde{A}_6) = (3{,}501.1y - 3{,}525.57)\frac{(1.03+0.02y)^6(0.02y+0.03)}{(1.03+0.02y)^6 - 1}$$

$$f_{6,2}(y|\widetilde{A}_6) = (-3{,},810.81y + 3{,}786.34)\frac{(1.07-0.02y)^6(0.07-0.02y)}{(1.07-0.02y)^6 - 1}$$

For y=0, $f_{6,1}(y|\widetilde{A}_6) = -\$650.96$

For y=1, $f_{6,1}(y|\widetilde{A}_6) = f_{6,2}(y|\widetilde{A}_6) = -\$4.82$

For y=0, $f_{6,2}(y|\widetilde{A}_6) = +\$795.13$

7 Fuzzy Payback Period Method

The payback period method involves the determination of the length of time required to recover the initial cost of investment based on a zero interest rate ignoring the time value of money or a certain interest rate recognizing the time value of money. Let C_{j0} denote the initial cost of investment alternative j, and R_{jt} denote the net revenue received from investment j during period t. Assuming no other negative net cash flows occur, the smallest value of m_j ignoring the time value of money such that

$$\sum_{t=1}^{m_j} R_{jt} \geq C_{j0} \tag{32}$$

or the smallest value of m_j recognizing the time value of money such that

$$\sum_{t=1}^{m_j} R_{jt}(1+r)^{-t} \geq C_{j0} \tag{33}$$

defines the payback period for the investment j. The investment alternative having the smallest payback period is the preferred alternative. In the case of fuzziness, the smallest value of m_j ignoring the time value of money such that

$$(\sum_{t=1}^{m_j} r_{1jt}, \sum_{t=1}^{m_j} r_{2jt}, \sum_{t=1}^{m_j} r_{3jt}) \geq (C_{1j0}, C_{2jo}, C_{3j0}) \tag{34}$$

and the smallest value of m_j recognizing the time value of money such that

$$(\sum_{t=1}^{m_J} (R_{jt}^{l(y)} / (1 + r^{r(y)})^t), \sum_{t=1}^{m_J} (R_{jt}^{r(y)} / (1 + r^{l(y)})^t) \geq \tag{35}$$

$$((C_{2j0} - C_{1j0})y + C_{1j0}, (C_{2j0} - C_{3j0})y + C_{3j0})$$

defines the payback period for investment j, where r_{kjt}: the kth parameter of a triangular fuzzy R_{jt}; C_{kj0}: the kth parameter of a triangular fuzzy C_{j0}; $R_{jt}^{l(y)}$: the left representation of a triangular fuzzy R_{jt}; $R_{jt}^{r(y)}$: the right representation of a triangular fuzzy R_{jt}. If it is assumed that the discount rate changes from one period to another, $(1+r^{r(y)})^t$ and $(1+r^{l(y)})^t$ will be $\prod_{t'=1}^{t}(1+r_{t'}^{r(y)})$ and $\prod_{t'=1}^{t}(1+r_{t'}^{l(y)})$ respectively.

It is now necessary to use a ranking method to rank the triangular fuzzy numbers such as Chiu and Park's [3], Chang's [6] method, Dubois and Prade's [7] method, Jain's [8] method, Kaufmann and Gupta's [9] method, Yager's [10] method. These methods may give different ranking results and most methods are tedious in graphic manipulation requiring complex mathematical calculation. In the following, two of the methods which do not require graphical representations are given. Chiu and Park's (1994) weighted method for ranking TFNs with parameters (*a, b, c*) is formulated as

$$((a+b+c)/3)+wb \tag{36}$$

where w is a value determined by the nature and the magnitude of the most promising value. The preference of projects is determined by the magnitude of this sum.

Kaufmann and Gupta (1988) suggest three criteria for ranking TFNs with parameters (a,b,c). The dominance sequence is determined according to priority of:

1. Comparing the ordinary number (a+2b+c)/4
2. Comparing the mode, (the corresponding most promise value), b, of each TFN.
3. Comparing the range, c-a, of each TFN.

The preference of projects is determined by the amount of their ordinary numbers. The project with the larger ordinary number is preferred. If the ordinary numbers are equal, the project with the larger corresponding most promising value is preferred. If projects have the same ordinary number and most promising value, the project with the larger range is preferred.

Example 4. Assume that there are two alternative machines that are under consideration to replace an aging production machine. The associated cash flows are given in the following table. Determine the best alternative by using the payback period method recognizing the time value of money. The fuzzy interest rate is (12%, 15%, 18%) annually.

Table 1. Fuzzy cash flow (x$1,000)

End of Year	0	1	2	3	4
Alt. *A*	(-7, -5, -3)	(2, 3, 4)	(4, 4.5, 5)	(1, 1.5, 2)	(3.5, 4, 4.5)
Alt. *B*	(-12, -10, -8)	(3, 4, 5)	(4.5, 5, 5.5)	(3, 3.5, 4)	(4, 4.5, 5)

If Chiu and Park 's [3] method (*CP*) is used for ranking *TFN*s, it is obtained (w=0.3):

For Alternative *A*,

$$CP_0 = (\frac{C_{1jt} + C_{2jt} + C_{3jt}}{3}) + wC_{2jt} = -6,500$$

$$TFN_1 = (\frac{1,000y + 2,000}{1.18 - 0.03y}, \frac{-1,000y + 4,000}{1.12 + 0.03y}) = (1695, 2608.7, 3571.4)$$

$$CP_1 = 3,407.6$$

$$TFN_2 = (\frac{500y + 4,000}{(1.18 - 0.03y)^2}, \frac{-500y + 5,000}{(1.12 + 0.03y)^2}) = (2872.7, 3402..6, 3986)$$

$$CP_2 = 4,441.2$$

$$\sum_{t=1}^{2} CP_t \succ CP_0 \rightarrow PP_A = 1.656 \text{ years.}$$

For Alternative *B*,

$$CP_0 = -13,000$$

$$TFN_1 = (2542.4, 3478.3, 4464.3)$$

$$CP_1 = 4538.5$$

$$TFN_2 = (3231.8, 3780.7, 4384.6)$$

$$CP_2 = 4,993.2$$

$$TFN_3 = (1825.9, 2301.3, 2847.1)$$

$$CP_3 = 3,015.2$$

$$TFN_4 = (2063.2, 2572.9, 3177.6)$$

$CP_4 = 3{,}376.4$

$\sum_{t=1}^{4} CP_t \succ CP_0 \rightarrow PP_B = 3.3$ years.

Alternative A is the preferred one.

8 Fuzzy Internal Rate of Return Method

The internal rate of return *(IRR)* method is referred to in the economic analysis literature as the discounted cash flow rate of return, internal rate of return, and the true rate of reurn. The internal rate of return on an investment is defined as the rate of interest earned on the unrecovered balance of an investment. Letting r^* denote the rate of return, the equation for obtaining r^* is

$$\sum_{t=1}^{n} P_t (1+r^*)^{-t} - FC = 0 \tag{37}$$

where P_t is the net cash flow at the end of period t.

Assume the cash flow $\tilde{F} = \tilde{F}_0, \tilde{F}_1, \ldots, \tilde{F}_N$ is fuzzy. $\tilde{F}_n$ is a negative fuzzy number and the other $\tilde{F}_t$ may be positive or negative fuzzy numbers. The fuzzy $IRR(\tilde{F}, n)$ is a fuzzy interest rate $\tilde{r}$ that makes the present value of all future cash amounts equal to the initial cash outlay. Therefore, the fuzzy number $\tilde{r}$ satisfies

$$\sum_{t=1}^{n} PV_{k(i)}(\tilde{F}_t, i) = -\tilde{F}_0 \tag{38}$$

where $\sum$ is fuzzy addition, *k(i)*=1 if $\tilde{F}_t$ is negative and *k(i)*=2 if $\tilde{F}_t$ is postive.

Buckley [2] shows that such simple fuzzy cash flows may not have a fuzzy *IRR* and concludes that the *IRR* technique does not extend to fuzzy cash flows. Ward [4] considers Eq. (37) and explains that such a procedure can not be applied for the fuzzy case because the right hand side of Eq. (37) is fuzzy, 0 is crisp, and an equality is impossible.

9 An Expansion to Geometric and Trigonometric Cash Flows

When the value of a given cash flow differs from the value of the previous cash flow by a constant percentage, *j%*, then the series is referred to as a *geometric series*. If the value of a given cash flow differs from the value of the previous cash flow by a sinusoidal wave or a cosinusoidal wave, then the series is referred to as a *trigonometric series*

9.1 Geometric Series-Fuzzy Cash Flows in Discrete Compounding

The present value of a crisp geometric series is given by

$$P = \sum_{n=1}^{N} F_1 (1+g)^{n-1} (1+i)^{-n} = \frac{F_1}{1+g} \sum_{n=1}^{N} \left(\frac{1+g}{1+i}\right)^n \quad (39)$$

where F_1 is the first cash at the end of the first year. When this sum is made, the following present value equation is obtained:

$$P = \begin{cases} F_1 \dfrac{1-(1+g)^N (1+i)^{-N}}{i-g}, i \neq g \\ \\ \dfrac{NF_1}{1+i}, i = g \end{cases} \quad (40)$$

and the future value is

$$F = \begin{cases} F_1 \dfrac{(1+i)^N - (1+g)^N}{i-g}, i \neq g \\ \\ NF_1 (1+i)^{N-1}, i = g \end{cases} \quad (41)$$

In the case of fuzziness, the parameters used in Eq.(4) will be assumed to be fuzzy numbers, except project life. Let $\gamma(i,g,N) = \dfrac{1-(1+g)^N (1+i)^{-N}}{i-g}, i \neq g$. As it is in Figure 1 and Figure 2, when k=1, the left side representation will be depicted and when k=2, the right side representation will be depicted. In this case, for $i \neq g$

$$f_{Nk}(y|\widetilde{P}_N) = f_k(y|\widetilde{F}_1)\gamma(f_{3-k}(y|\widetilde{i}), f_{3-k}(y|\widetilde{g}), N) \tag{42}$$

In Eq. (42), the least possible value is calculated for $k = 1$ and $y = 0$; the largest possible value is calculated for $k = 2$ and $y = 0$; the most promising value is calculated for $k = 1$ or $k = 2$ and $y = 1$.

To calculate the future value of a fuzzy geometric cash flow, let $\zeta(i, g, N) = \frac{(1+i)^N - (1+g)^N}{i-g}, i \neq g$. Then the fuzzy future value is

$$f_{Nk}(y|\widetilde{F}_N) = f_k(y|\widetilde{F}_1)\zeta(f_k(y|\widetilde{i}), f_k(y|\widetilde{g}), N) \tag{43}$$

In Eq. (43), the least possible value is calculated for $k = 1$ and $y = 0$; the largest possible value is calculated for $k = 2$ and $y = 0$; the most promising value is calculated for $k = 1$ or $k = 2$ and $y = 1$. This is also valid for the formulas developed at the rest of the chapter.

The fuzzy uniform equivalent annual value can be calculated by using Eq. (44):

$$f_{Nk}(y|\widetilde{A}) = f_k(y|\widetilde{P}_N)\vartheta(f_k(y|\widetilde{i}), N) \tag{44}$$

where $\vartheta(i, N) = \frac{(1+i)^N i}{(1+i)^N - 1}$ and $f(y|\widetilde{P}_N)$ is the fuzzy present value of the fuzzy geometric cash flows.

9.2 Geometric Series - Fuzzy Cash Flows in Continuous Compounding

In the case of crisp sets, the present and future values of discrete payments are given by Eq. (45) and Eq. (46) respectively:

$$\mathrm{P} = \begin{cases} F_1 \dfrac{1 - e^{(g-r)N}}{e^r - e^g}, r \neq g \\ \dfrac{NF_1}{e^r}, g = e^r - 1 \end{cases} \tag{45}$$

$$F = \begin{cases} F_1 \dfrac{e^{rN} - e^{gN}}{e^r - e^g}, r \neq g \\ NF_1 e^{r(N-1)}, g = e^r - 1 \end{cases} \tag{46}$$

and the present and future values of continuous payments are given by Eq. (47) and Eq. (48) respectively:

$$P = \begin{cases} F_1 \dfrac{1 - e^{N(g-r)}}{r - g}, r \neq g \\ \dfrac{NF_1}{1+r}, r = g \end{cases} \tag{47}$$

$$F = \begin{cases} F_1 \dfrac{e^{rN} - e^{gN}}{r - g}, r \neq g \\ \dfrac{NF_1 e^{rN}}{1+r}, r = g \end{cases} \tag{48}$$

The fuzzy present and future values of the fuzzy geometric discrete cash flows in continuous compounding can be given as in Eq. (49) and Eq. (50) respectively:

$$f_{Nk}(y|\tilde{P}_N) = f_k(y|\tilde{F}_1)\beta(f_{3-k}(y|\tilde{r}), f_{3-k}(y|\tilde{g}), N) \tag{49}$$

$$f_{Nk}(y|\tilde{F}) = f_k(y|\tilde{F}_1)\tau(f_k(y|\tilde{r}), f_k(y|\tilde{g}), N) \tag{50}$$

where $\beta(r,g,N) = F_1 \dfrac{1 - e^{(g-r)N}}{e^r - e^g}, r \neq g$ for present value and $\tau(r,g,N) = F_1 \dfrac{e^{rN} - e^{gN}}{e^r - e^g}, r \neq g$ for future value.

The fuzzy present and future values of the fuzzy geometric continuous cash flows in continuous compounding can be given as in Eq. (51) and Eq. (52) respectively:

$$f_{Nk}(y|\widetilde{P}_N) = f_k(y|\widetilde{F}_1)\eta(f_{3-k}(y|\widetilde{r}), f_{3-k}(y|\widetilde{g}), N) \tag{51}$$

$$f_{Nk}(y|\widetilde{F}_N) = f_k(y|\widetilde{F}_1)\upsilon(f_k(y|\widetilde{r}), f_k(y|\widetilde{g}), N) \tag{52}$$

where $\eta(r,g,N) = F_1 \frac{1-e^{(g-r)N}}{r-g}$, $\upsilon(r,g,N) = F_1 \frac{e^{rN}-e^{gN}}{r-g}, r \neq g$

9.3 Trigonometric Series - Fuzzy Continuous Cash Flows

In Figure 4, the function of the semi-sinusodial wave cash flows is depicted. This function, *h(t),* is given by Eq. (53) in the crisp case:

$$h(t) = \begin{cases} D\sin \pi t, 0 \leq t \leq 1 \\ 0, \quad otherwise \end{cases} \tag{53}$$

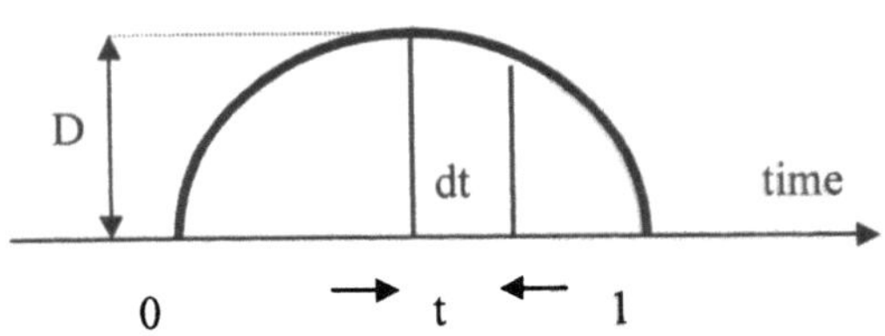

Fig. 4. Semi-sinusoidal wave cash flow function

The future value of a semi-sinusoidal cash flow for T=1 and g is defined by Eq. (54) :

$$V(g,1) = D\int_0^1 e^{r(1-t)} \sin \pi t dt = D\frac{\pi(2+g)}{r^2+\pi^2} \tag{54}$$

Figure 5 shows the function of a cosinusoidal wave cash flow. This function, *h(t),* is given by Eq. (55):

$$h(t) = \begin{cases} D(Cos 2\pi t + 1), 0 \leq t \leq 1 \\ 0, \; otherwise \end{cases} \tag{55}$$

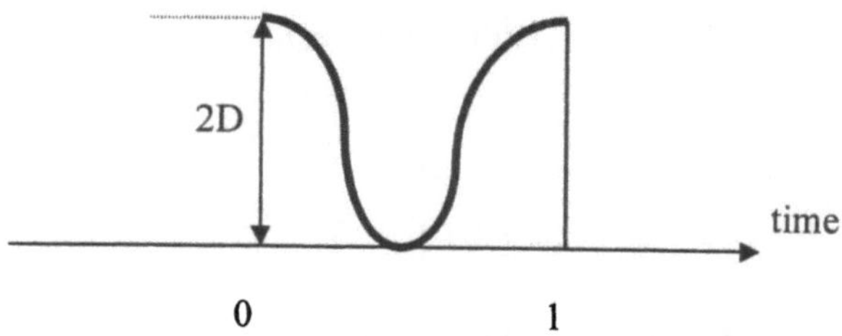

Fig. 5. Cosinusoidal wave cash flow function

The future value of a cosinusoidal cash flow for T=1 and g is defined as

$$V(g,1) = D\int_0^1 e^{r(t-1)}(\cos 2\pi.t + 1)dt = D(\frac{gr}{r^2 + 4\pi^2} + \frac{g}{r}) \tag{56}$$

Let the parameters in Eq. (54), r and g, be fuzzy numbers. The future value of the semi-sinusoidal cash flows as in Figure 6 is given by

$$f_{Nk}(y|\widetilde{F}_N) = f_k(y|\widetilde{D})\phi(f_{3-k}(y|\widetilde{r}), f_k(y|\widetilde{g}))\varphi(f_k(y|\widetilde{r}), N) \tag{57}$$

where $\phi(r,g) = \pi(2+g)/(r^2 + \pi^2)$, $\varphi(r,N) = (e^{rN} - 1)/(e^r - 1)$.
The present value of the semi-sinusoidal cash flows is given by Eq. (58):

$$f_{Nk}(y|\widetilde{P}_N) = f_k(y|\widetilde{D})\phi(f_{3-k}(y|\widetilde{r}), f_k(y|\widetilde{g}))\psi(f_{3-k}(y|\widetilde{r}), N) \tag{58}$$

where $\psi(r,N) = (e^{rN} - 1)/((e^r - 1)e^{rN})$.

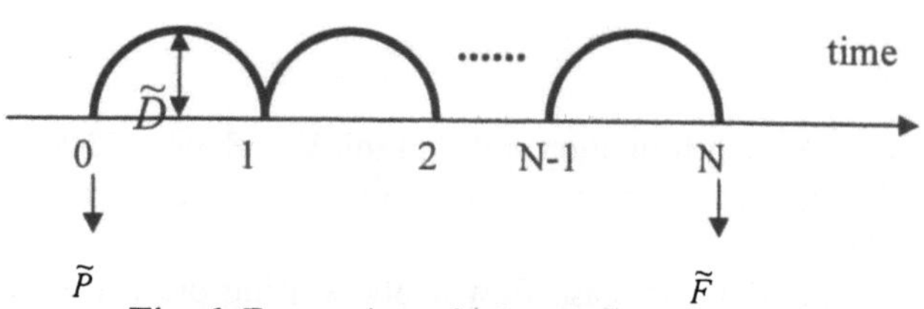

Fig. 6. Fuzzy sinusoidal cash flow diagram

The present and future values of the fuzzy cosinusoidal cash flows can be given by Eq. (59) and Eq. (60) respectively:

$$f_{Nk}(y|\widetilde{P}_N) = f_k(y|\widetilde{D})\xi(f_{3-k}(y|\widetilde{r}), f_k(y|\widetilde{g}))\psi(f_{3-k}(y|\widetilde{r}), N) \tag{59}$$

where $\xi(r,g) = (\frac{gr}{r^2 + 4\pi^2} + \frac{g}{r})$ and the fuzzy future value is

$$f_{Nk}(y|\widetilde{F}_N) = f_k(y|\widetilde{D})\xi(f_{3-k}(y|\widetilde{r}), f_k(y|\widetilde{g}))\varphi(f_k(y|\widetilde{r}), N) \qquad (60)$$

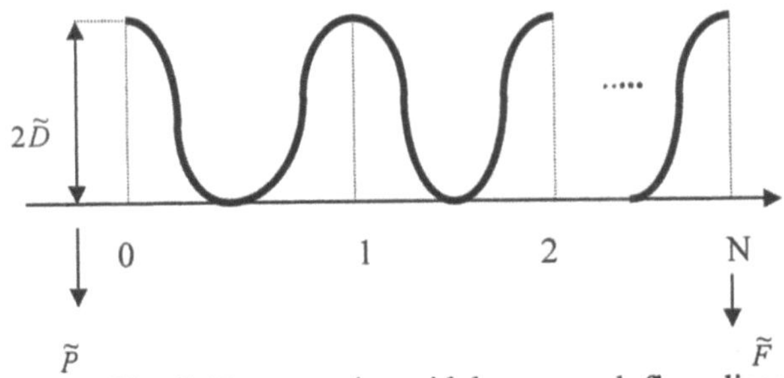

Fig. 7. Fuzzy cosinusoidal wave cash flow diagram

10 Conclusions

In this chapter, capital budgeting techniques in the case of fuzziness and discrete compounding have been studied. The cash flow profile of some investments projects may be geometric or trigonometric. For these kind of projects, the fuzzy present, future, and annual value formulas have been also developed under discrete and continuous compounding in this chapter. Fuzzy set theory is a powerful tool in the area of management when sufficient objective data has not been obtained. Appropriate fuzzy numbers can capture the vagueness of knowledge. The other financial subjects such as replacement analysis, income tax considerations, continuous compounding in the case of fuzziness can be also applied [11,12]. Comparing projects with unequal lives have not been considered in this chapter. This will also be a new area for a further study.

References

[1] Zadeh, L.A. (1965) Fuzzy Sets, Information and Control, Vol. 8, pp. 338-353.
[2] Buckley, J.U. (1987) The fuzzy mathematics of finance, Fuzzy Sets and Systems, Vol. 21, pp. 257-273.
[3] Chiu, C.Y., Park, C.S. (1994) Fuzzy cash flow analysis using present worth criterion, The Engineering Economist, Vol. 39, No. 2, pp. 113-138.
[4] Ward, T.L. (1985) Discounted fuzzy cash flow analysis, in 1985 Fall Industrial Engineering Conference Proceedings, pp. 476-481.
[5] Blank, L.T., Tarquin, J.A. (1987) Engineering Economy, Third Edition, McGraw-Hill, Inc.
[6] Chang, W. (1981) Ranking of fuzzy utilities with triangular membership functions, Proc. Int. Conf. of Policy Anal. and Inf. Systems, pp. 263-272.

[7] Dubois, D., Prade, H. (1983) Ranking fuzzy numbers in the setting of possibility theory, Information Sciences, Vol. 30, pp. 183-224.
[8] Jain, R. (1976) Decision making in the presence of fuzzy variables, IEEE Trans. on Systems Man Cybernet, Vol. 6, pp. 698-703.
[9] Kaufmann, A., Gupta, M. M. (1988) Fuzzy Mathematical Models in Engineering and Management Science, Elsevier Science Publishers B.V.
[10] Yager, R.R. (1980) On choosing between fuzzy subsets, Kybernetes, Vol. 9, pp. 151-154.
[11] Kahraman, C., Ulukan, Z. (1997) Continuous compounding in capital budgeting using fuzzy concept, in the Proceedings of 6th IEEE International Conference on Fuzzy Systems (*FUZZ-IEEE'97)*, Bellaterra-Spain, pp. 1451-1455.
[12] Kahraman, C., Ulukan, Z. (1997) Fuzzy cash flows under inflation, in the Proceedings of Seventh International Fuzzy Systems Association World Congress (IFSA'97), University of Economics, Prague, Czech Republic, Vol. IV, pp. 104-108.
[13] Zimmermann, H.-J. (1994) Fuzzy Set Theory and Its Applications, Kluwer Academic Publishers.

Appendix

One of the most basic concepts of fuzzy set theory which can be used to generalize crisp mathematical concepts to fuzzy sets is the extension principle. Let X be a Cartesian product of universes $X=X_1 \ldots X_r$, and $\tilde{A}_1,\ldots,\tilde{A}_r$ be r fuzzy sets in $X_1,\ldots,X_r$, respectively. f is a mapping from X to a universe Y, $y=f(x_1,\ldots,x_r)$. Then the extension principle allows us to define a fuzzy set $\tilde{B}$ in Y by [13]

$$\tilde{B} = \{(y, \mu_{\tilde{B}}(y)) \mid y = f(x_1,\ldots,x_r), (x_1,\ldots,x_r) \in X\} \tag{A.1}$$

where

$$\mu_{\tilde{B}}(y) = \begin{cases} \sup\limits_{(x_1,\ldots,x_r) \in f^{-1}(y)} \min\{\mu_{\tilde{A}_1}(x_1),\ldots,\mu_{\tilde{A}_r}(x_r)\} & , \text{if } f^{-1}(y) \neq \varnothing \\ 0 & , \text{otherwise} \end{cases} \tag{A.2}$$

where f^{-1} is the inverse of f.

Assume $\widetilde{P}=(a, b, c)$ and $\widetilde{Q}=(d, e, f)$. a, b, c, d, e, f are all positive numbers. With this notation and by the extension principle, some of the extended algebraic operations of triangular fuzzy numbers are expressed in the following.

Changing Sign

$$-(a, b, c) = (-c, -b, -a) \tag{A.3}$$

or

$$-(d, e, f) = (-f, -e, -d) \tag{A.4}$$

Addition

$$\widetilde{P} \oplus \widetilde{Q} = (a+d, b+e, c+f) \tag{A.5}$$

and

$$k \oplus (a, b, c) = (k+a, k+b, k+c) \tag{A.6}$$

or

$$k \oplus (d, e, f) = (k+d, k+e, k+f) \tag{A.7}$$

if k is an ordinary number (a constant).

Subtraction

$$\widetilde{P} - \widetilde{Q} = (a-f, b-e, c-d) \tag{A.8}$$

and

$$(a, b, c) - k = (a-k, b-k, c-k) \tag{A.9}$$

or

$$(d, e, f) - k = (d-k, e-k, f-k) \tag{A.10}$$

if k is an ordinary number.

Multiplication

$$\widetilde{P} \otimes \widetilde{Q} = (ad, be, cf) \tag{A.11}$$

and

$$k \otimes (a, b, c) = (ka, kb, kc) \tag{A.12}$$

or

$$k \otimes (d, e, f) = (kd, ke, kf) \tag{A.13}$$

if k is an ordinary number.

Division

$$\widetilde{P} \oslash \widetilde{Q} = (a/f, b/e, c/d) \tag{A.14}$$

Survival Risk and Project Evaluation

Francesca Beccacece, Paolo Gallo** and Lorenzo Peccati**

* IMQ – Bocconi University
Via Sarfatti 25, 20136 Milan, Italy
E-mail: {francesca.beccacece, lorenzo.peccati}@uni-bocconi.it

** SAA – Torino University
Via Ventimiglia 115, 10100 Torino, Italy
E-mail: faolg@tin.it

1 Introduction

This paper sketches a procedure for the evaluation of industrial projects under serious survival risk which at present is under testing in the business sector of energy.

The survival risk mainly stems both from the action of natural variables which characterize the sites where the investment should be made and from the effects of high social, political, juridical instability of the site countries.

The strong requirement we have been forced to meet since the start-up of this study was that of constructing a model simple enough to be understood by managers without specific experience in the energy sector, but sufficiently rich in order to cope with special aspects of those types of investments which cannot be thought of as automatically incorporated in standard financial parameters.

The structure of the paper is the following. In Section 2 we present the model for the evaluation of industrial projects under the assumption of deterministic cash-flows; Section 3 and Section 4 are respectively devoted to the study of the variation of the value of the project in time and to the analysis of a simple but effective numerical example. In Section 5, the case of random cash-flows is considered. Section 6 contains the description of an investment made by FIAT-Avio, which has been evaluated via this procedure.

2 The model

The investment is developed in two time periods of n and k years respectively. The first one of n years is the *cost period* whose the amounts (in absolute value) are $c_1, c_2, \ldots, c_n$, payed at the dates $1, 2, \ldots, n$. The second period, which lasts till k years, should provide revenues of amounts $r_1, r_2, \ldots, r_k$ at the dates $n+1, n+2, \ldots, n+k$.

We can denote the cash flows by $f_1, f_2, \ldots, f_n, f_{n+1}, \ldots, f_{n+k}$ occurring at the various milestones by:

$$\begin{cases} f_s = -c_s \text{ for } s = 1, 2, \ldots, n \\ f_s = r_s \text{ for } s = n+1, n+2, \ldots, n+k \end{cases}$$

It can be convenient to complete the flows' sequence with a null one at 0: $f_0 = 0$.

The project can be stopped at any epoch of its life for different reasons, briefly sketched in the introduction. This can happen during the year 1, before the payment of the first cost, during the year 2, after the first cost but before the second one, and so on. When we consider the case in which the project arrives to its conclusion, we refer to it as death of the project during the year $n + k + 1$. The tools we are going to use are the ones standard in actuarial mathematics. See, for instance, the reference text by Gerber [3].

An assessment of the probabilities of extinction of the project during any year $j = 1, 2, 3, \ldots, n+k, n+k+1$ is available. We denote these probabilities by $Q_j = \Pr(\text{death during the year } j)$. Obviously

$$\sum_{j=1}^{n+k+1} Q_j = 1.$$

Let j be the year of the extinction of the project ($j = 1, 2, 3, \ldots, n+k+1$). The discounted cash flow (**DCF**) of the project under this assumption is

$$\begin{aligned} G_j(x) &= f_0 + f_1(1+x)^{-1} + f_2(1+x)^{-2} + \cdots \\ &\quad \cdots + f_{j-1}(1+x)^{-j+1} \\ &= \sum_{h=0}^{j-1} \frac{f_h}{(1+x)^h}. \end{aligned}$$

The expected **DCF** is

$$\begin{aligned} \overline{G}(x) &= G_1(x)Q_1 + G_2(x)Q_2 + \cdots + G_{n+k+1}(x)Q_{n+k+1} \\ &= \sum_{j=1}^{n+k+1} G_j(x)Q_j. \end{aligned}$$

Denoting by i the opportunity cost of the capital which supports the project, the expected net present value (**NPV**) of the project turns out to be $\overline{G}(i)$.

The model allows for the valuation of the project at any date $t \geq 1$ during its life.

When we arrive at t we know that the project is still alive at that moment. The unitary mass of probability, previously distributed over $n + k + 1$ years, since each of them was a "possible" year of death, must be now divided among the remaining $n+k+1-t$ years. We denote by $Q_j^{(t)}$ the probability of death during the year j, $j=t+1,t+2,\ldots,n+k+1$; it is estimated at t, *i.e.*, it is conditional on the survival of the project in t. We have

$$Q_j^{(t)} = \frac{Q_j}{\sum_{s=t+1}^{n+k+1} Q_s} = \frac{Q_j}{Q_{t+1}+Q_{t+2}+\cdots+Q_{n+k+1}}.$$

We remark that $Q_j = Q_j^{(0)}$.

The value V_t of the project at t can be determined through an equation whose financial and economic meaning is very simple. The basic idea is to compare the final value of the project in the two cases of selling or not selling it. First, let us assume that at t, just after the maturity of f_t, the project is sold: in this case, at the last maturity $n+k$, the final wealth will be modified, from one hand, for an amount equal to the (certain) final value of the previous flows, from the other hand, for the final value of V_t. Let us consider now the not selling hypothesis: in this case the final value is random since the project would be die in any future year. If the decision maker is risk neutral, the certainty equivalent at t of the project is the price V_t which makes the final value deriving from the selling equal to the expectation of the stochastic final value in the case of project retention. See, for related approaches, [2] and [1]. Denoting by y_t the final value at $n+k$ of the previous flows

$$y_t = \sum_{h=0}^{t} f_h\,(1+i)^{n+k-h}\,,$$

we can state the following "evaluation equation"[1]

$$y_t + V_t\,(1+i)^{n+k-t} = y_t + \sum_{j=t+2}^{n+k+1}\left[\sum_{h=t+1}^{j-1} f_h\,(1+i)^{n+k-h}\right] Q_j^{(t)}.$$

Moreover, defining the survival probability of the project at the date h, on the basis of the information available at t, as

$$P_h^{(t)} = \sum_{j=h+1}^{n+k+1} Q_j^{(t)}$$

after some manipulations, we obtain

$$V_t = f_{t+1}\,(1+i)^{-1}\,P_{t+1}^{(t)} + f_{t+2}\,(1+i)^{-2}\,P_{t+2}^{(t)} + \cdots$$
$$\cdots + f_{n+k}\,(1+i)^{-(n+k-t)}\,P_{n+k}^{(t)}$$

[1] In the right hand sides the sum starts from $j=t+2$, as at $j=t+1$ there is no cash-movements.

or

$$V_t = \sum_{h=t+1}^{n+k} f_h\,(1+i)^{t-h}\,P_h^{(t)} \qquad (1)$$

Formula (1) provides us with a very simple interpretation of the value of the project at t: it is the sum of the values at t of the future flows, each of them weighted with the correspondent survival probability (that is the probability which the project is alive almost until the maturity of the flow).

3 The temporal dynamics of the value

In this section we investigate how V_t varies in time. It can result interesting for several reasons, first of all to single out in which date(s) the value of the project reaches a very high and positive value: it would represent the optimal time to dispose of it.

In order to describe the evolution of V_t, we calculate the variation of the value of the project between t and $t+1$. It is not difficult, even if a bit boring, to obtain the difference equation

$$V_{t+1} - V_t = -f_{t+1}\,(1+i)^{-1}\,P_{t+1}^{(t)} + \sum_{h=t+2}^{n+k} f_h\,(1+i)^{t-h}\left[(1+i)\,P_h^{(t+1)} - P_h^{(t)}\right]$$

The right hand side of the equation is the sum of two components: the first addendum is the contribution of the flow maturing at t, the second one depends on the future flows according to the modified information state passing from t to $t+1$. We want to investigate the sign of both the components.

First addendum — It is the maturing flow discounted for one year and multiplied by the survival probability of one year: We can reasonably assume that it is no far from f_{t+1}, *but with the opposite sign*. Hence, during the cost period it is always positive and, in particular, at the end of the cost period, on the bearing of the most costs, the positive drift will be relevant. On the contrary, during the revenue period, the drift will be negative. From an economic viewpoint, this fact is absolutely clear and coherent: during the investment phases of a financial project, its value increases, while, when it becomes profitable, its value decreases.

Second addendum — It is composed by the future flows and the spread of survival probabilities evaluated at t and at $t+1$. Since

$$P_h^{(t+1)} > P_h^{(t)}$$

then we can conclude that the sign of the spread is always positive

$$(1+i)\,P_h^{(t+1)} - P_h^{(t)} > 0 \text{ for any } h \geq t+2.$$

Really, considering standard situation of low interest rates and absence of jumps in the information state, the value of this quantity will be not relevant. As it is obvious such conclusions are strictly related to the analysis of realistic data.

4 A numerical example

Let us consider an industrial project characterized by a high risk with $n = 4$ and $k = 3$. The costs to be (eventually) supported are constantly increasing of 20 and the first one is 10, hence $c_1 = 10$, $c_2 = 30$, $c_3 = 50$, $c_4 = 70$. Also the revenues are increasing and assessed as follows: $r_5 = 600$, $r_6 = 750$ and $r_7 = 900$.

The following table collects the assessed extinction probabilities of the project:

year of death	probability
1	19.231%
2	17.308%
3	15.385%
4	13.462%
5	11.538%
6	9.615%
7	7.692%
8	5.769%

The opportunity cost of the capital is $i = 8\%$.

We can calculate in correspondence of any possible date of death for the project the NPV. We present them in the following table:

year of death	values of NPV = G
1	0
2	−9.259
3	−34.979
4	−74.671
5	−126.123
6	282.227
7	754.854
8	1279.995

The expectation of the NPV is

$$\mathrm{E}(\mathbf{G}) = 127.46$$

Since it is positive, the projects turns out to be convenient for a risk neutral decision maker. We can also calculate the *survival probabilities* $P_h^{(t)}$ where t is the evaluation date and h is any future date, that is $t < h$; the following table shows them

$t\backslash h$	1	2	3	4	5	6	7
1	1	0.786	0.595	0.429	0.286	0.167	0.071
2	1	1	0.758	0.545	0.364	0.212	0.091
3	1	1	1	0.720	0.480	0.280	0.12
4	1	1	1	1	0.667	0.389	0.167
5	1	1	1	1	1	0.583	0.25
6	1	1	1	1	1	1	0.429

Using the equation (1) we obtain the values V_t at the various maturities $t = 1, 2, 3, \ldots, 7$. Obviously, the last value V_7 will be equal to 0, since after the year 7 there are no flows originated by the project. We collect the values for V_t obtained above in the following table

t	V_t
1	180.433
2	278.013
3	446.335
4	739.502
5	597.994
6	357.143
7	0

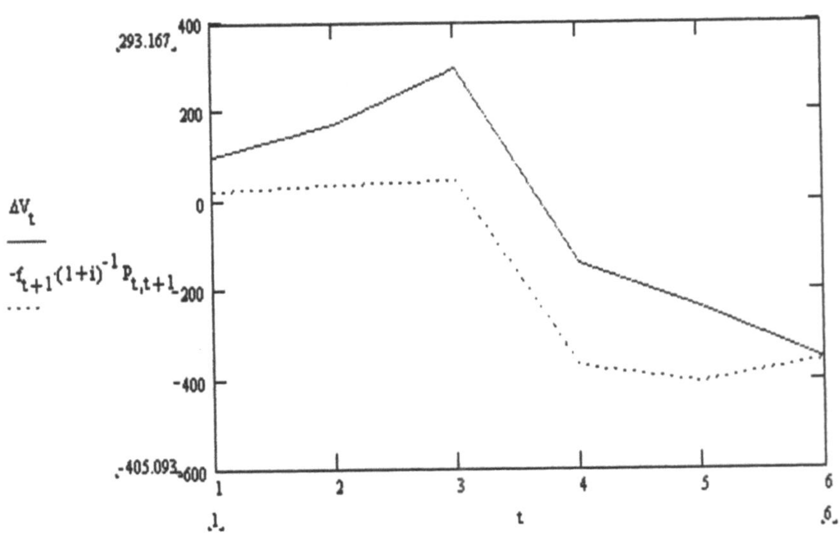

Fig. 1.

It is interesting to study the variation over the time of the value of the project and, in particular, the behavior of the first addendum: the component deriving from the maturing flow. In Figure 4 there are shown both the graphs; it is immediate to notice that the first component is *magna pars* of the variation of the value.

5 The case of random cash flows

The model presented in Section 2 assumes that all the flows deriving from the project (costs and revenues) are certain. We now remove such an assumption in order to construct an evaluation procedure which is more flexible and better to fit the context we are considering. We will obtain expressions for V_t which are nothing but the natural transposition of formula (1) in the stochastic case.

Even if uncertainty, in the most cases, is confined to the revenues, there is no difficulty in considering that all flows, positive and negative, are stochastic. In order to avoid excessive complexities, we assume stochastic independence of the amounts: the cash-flow (cost or revenue) with maturity s, $s = 1, 2, \ldots, n+k$, is a random variable $\widetilde{f}_s$ with expectation m_s and standard deviation σ_s. All the $\widetilde{f}_s$ have the same distribution but are characterized by different values of the parameters. Uncertainty of the flows is the unique modification we introduce in the model: all the other elements are preserved, first of all the formalization of the extinction probabilities.

Let t the date of the evaluation. In the case of retention, the value of the project is composed by two parts: the final value at $n+k$ of the past flows, which have maturity $h \leq t$, and the final value at $n+k$ of the future flows, which have maturity $h > t$. While the first component is deterministic and given by y_t, the final value evaluated at t of the future flows is a mixture of random variables for which we can assess the (conditional) probability distribution. We denote such random variable with $\widetilde{M}_t$. For each possible expiration date $j = t+1, t+2, \ldots, n+k+1$ the final conditional value of the project $\widetilde{M}_t^j$ is the random variable final value at $n+k$ of the random flows which will take place from $t+1$ till j. Hence we have

value of the final amount: $\widetilde{M}_t^j$	with probability $Q_j^{(t)}$
0	$Q_{t+1}^{(t)}$
$\widetilde{f}_{t+1}(1+i)^{n+k-t-1}$	$Q_{t+2}^{(t)}$
$\widetilde{f}_{t+1}(1+i)^{n+k-t-1} + \widetilde{f}_{t+2}(1+i)^{n+k-t-2}$	$Q_{t+3}^{(t)}$
...	...
$\sum_{h=t+1}^{j-1} \widetilde{f}_h (1+i)^{n+k-h}$	$Q_j^{(t)}$
...	...
$\sum_{h=t+1}^{n+k} \widetilde{f}_h (1+i)^{n+k-h}$	$Q_{n+k+1}^{(t)}$

For a risk-neutral decision maker looking at the final wealth the certainty equivalent V_t at t turns out to be the price x satisfying the evaluation equation

$$y_t + x\,(1+i)^{n+k-t} = y_t + \sum_{j=t+2}^{n+k+1} \left[\sum_{h=t+1}^{j-1} m_h\,(1+i)^{n+k-h} \right] Q_j^{(t)}.$$

After some manipulations, we have

$$V_t = \sum_{h=t+1}^{n+k} m_h\,(1+i)^{t-h}\,P_h^{(t)} \tag{2}$$

It is immediate to notice that formula (2) is the same that (1) where each (random) cash-flow $\tilde{f}_h$ has been replaced with its expectation m_h.

Another approach deserves to be considered. Consider the case of a decision maker who uses the evaluation principle of the comparison of the final wealth, but who is risk averse. The certainty equivalent V_t is determined by stating the evaluation equation in terms of expected utility, through a concave utility function

$$u\left[y_t + V_t\,(1+i)^{n+k-t}\right] = \mathrm{E}\left[\mathrm{u}(\mathrm{y_t} + \widetilde{\mathrm{M}}_\mathrm{t})\right]$$

In particular, assuming exponential utility functions and normal distribution of the flows, the expression of the certainty equivalent V_t becomes simpler and characterized by a very natural financial interpretation.

Therefore, let us assume that the decision maker evaluates final possible wealths via the exponential utility function

$$u(x) = -\exp\{-\alpha x\} \text{ with } \alpha > 0$$

The evaluation equation turns out to be

$$-\exp\{-\alpha\,(y_t + V_t(1+i)^{n+k-t})\} = \mathrm{E}\left[-\exp\{-\alpha(\mathrm{y_t} + \widetilde{\mathrm{M}}_\mathrm{t}\}\right]$$

leading to

$$\exp\{-\alpha \cdot V_t(1+i)^{n+k-t}\} = \mathrm{E}\left[\exp\{-\alpha \cdot \widetilde{\mathrm{M}}_\mathrm{t}\}\right] \tag{3}$$

The right had side of (3) is $\mathrm{E}\left[\exp\{-\alpha \cdot \widetilde{\mathrm{M}}_\mathrm{t}\}\right]$, and it is given by

$$\mathrm{E}\left[\exp\{-\alpha \cdot \widetilde{\mathrm{M}}_\mathrm{t}\}\right] = \mathrm{Q}_{\mathrm{t+1}}^{(\mathrm{t})} + \sum_{\mathrm{j=t+2}}^{\mathrm{n+k+1}} \mathrm{Q}_\mathrm{j}^{(\mathrm{t})} \cdot \mathrm{g_j}(-\alpha)$$

where g_j is the moment generating function of $\widetilde{M}_t^j$.

Since $\widetilde{M}_t^{t+1} = 0$, we can denote with $g_{t+1}(-\alpha) = \exp\{-\alpha \cdot 0\} = 1$ the moment generating function of $\widetilde{M}_t^{t+1}$ even if it is a degenerate random variable. Using such a convention, we obtain

$$\mathrm{E}\left[\exp\{-\alpha \cdot \mathrm{M_t}\}\right] = \sum_{\mathrm{j=t+1}}^{\mathrm{n+k+1}} \mathrm{Q_j^{(t)}} \cdot \mathrm{g_j}(-\alpha)$$

and substituting in (3)

$$\exp\{-\alpha V_t(1+i)^{n+k-t}\} = \sum_{j=t+1}^{n+k+1} Q_j^{(t)} \cdot g_j(-\alpha)$$

Consequently, the certainty equivalent V_t is given by

$$V_t = \frac{\ln\left(\sum_{j=t+1}^{n+k+1} Q_j^{(t)} \cdot g_j(-\alpha)\right)}{-\alpha\,(1+i)^{n+k-t}} \tag{4}$$

Let us consider now a more special case, which is of remarkable interest in the practice: cash-flows $\widetilde{f}_h \sim \mathrm{N}(\mathrm{m_h}, \sigma_\mathrm{h}^2)$ for $h = t+1, t+2, \ldots, t+k$ are mutually stochastically independent.

The final value at any possible date of extinction $j = t+2, t+3, \ldots, t+k+1$ is a normal random variable

$$M_t^j \sim \mathrm{N}\left(\sum_{\mathrm{h=t+1}}^{\mathrm{j-1}} \mathrm{m_h}(1+\mathrm{i})^{\mathrm{n+k-h}}, \sum_{\mathrm{h=t+1}}^{\mathrm{j-1}} \sigma_\mathrm{h}^2(1+\mathrm{i})^{2(\mathrm{n+k-h})}\right)$$

recalling that M_t^{t+1} is equal to 0 with probability 1.

The expression (4) can be also obtained via a different approach which directly stems from the basic von Neumann-Morgenstern paradigm. As it is well known, a lottery A composed by random variables is equivalent to the lottery composed by the certainty equivalents of the random variables of A, with the same probability distribution. Applying this principle to our context, we can construct the random variable which, for any date of extinction (i.e. with probability distribution given by the $Q_j^{(t)}$) presents the correspondent certainty equivalent of the stochastic final value deriving from the project. If z_j is the certainty equivalent of M_t^j for $j = t+1, t+2, \ldots, n+k+1$, the random variable D_t equivalent to M_t is given by

0	$Q_{t+1}^{(t)}$
z_{t+1}	$Q_{t+2}^{(t)}$
...	...
z_{n+k}	$Q_{n+k+1}^{(t)}$

According to the von Neumann–Morgenstern expected utility paradigm D_t is equivalent to M_t, and consequently

$$\mathrm{E}\left[\mathrm{u}(\mathrm{D_t})\right] = \mathrm{E}\left[\mathrm{u}(\mathrm{M_t})\right]$$

In particular, using exponential utility function we have

$$\mathrm{E}\left[\exp\{-\alpha \cdot \mathrm{D_t}\}\right] = \mathrm{E}\left[\exp\{-\alpha \cdot \mathrm{M_t}\}\right]$$

and, by substituting in (3)

$$\mathrm{E}\left[\exp\{-\alpha \cdot \mathrm{M_t}\}\right] = \exp\{-\alpha \cdot \mathrm{V_t}(1+\mathrm{i})^{\mathrm{n+k-t}}\}$$

Finally we obtain

$$V_t = \frac{\ln(\sum_{j=t+1}^{n+k+1} Q_j^{(t)} \exp\{-\alpha z_j\})}{-\alpha(1+i)^{n+k-t}}$$

Since $g_j(-\alpha) = \mathrm{E}(\exp\{-\alpha \cdot \mathrm{M_j^t}\} = \exp\{-\alpha \mathrm{z_j}\}$, we have obtained for V_t the same expression given by (4).

6 An application

The model presented in this paper is currently used by FIAT-Avio in the evaluation of investments in the energy sector.

In order to show how the theoretical apparatus described above can be used effectively to help management decisions, we provide some details on a recent practical case where the methodology has been used.

The project consists in the development, finance, construction and operation of 200 MW gas fired power plant in combined cycle, based on two industrial (heavy-duty) gas turbines. It can be labelled as a *greenfield-project*.

As far as the technical features are concerned, the following table provides us with the necessary information

		Remarks
Gas Turbine Power	70 MW	Gross power
Steam Turbine Power	65 MW	Gross power
Auxiliary consumption	5 MW	
Net Output Power	200 MW	
Net Efficiency	53%	
Fuel	Gas	Natural gas, no dual fuel capability

The development phase of the project can be lengthy (many years) and requires a significant budget (several million dollars). It consists in structuring the project, in making it viable up to its financial closing.

The development phase starts from a so-called *(1) pre-feasibility study*, aiming to identify needs for new power generation capacity and to outline a preliminary *pro-forma* for the major economics. The period dedicated to this study is limited (1-3 months) and so is the expenditure (around $ 100,000). Its conclusion may lead to drop the project idea or to move into a more detailed study.

If the project turns out to be interesting, the *(2) feasibility study* is started.

The initial objectives of this phase are:

- assessment of potential options to determine the project economic viability
- identify major issues, such as fuel, permits, tax and legal,
- develop a risk mitigation approach,
- develop a preliminary engineering approach,
- develop financing options.

These initial activities will end up with a more detailed economic *pro-forma* to clearly identify the potential terms of the project. Once the project concepts have been developed, the promoters will start structuring it through Letters of Intent (LoI) with customers and contractors (such as fuel suppliers and construction). This period lasts for 12 months with cost expenses of $ 1,000,000 (both internal resources and external advisors).

The signature of the LoI relevant to power supply represents the first milestone achieved by the project, followed by a LoI with the fuel supplier and the *(3) negotiations of the main contracts*: Power Purchase Agreement (PPA), Fuel Supply Agreement (FSA), Engineering, Procurement and Construction contract (EPC), Operation and Maintenance contract (O&M).

During the negotiations the promoters are assisted by appointed financial and legal advisors. A fiscal advisor is also appointed to solve all the issues relevant to tax and duties. The financial advisor is paid on a monthly retainer fee plus a success fee (percentage on total project cost) at financial close. The other advisor are paid on an hourly fee.

This phase is the most expensive ($ 4,000,000) and (internal and external) costs are directly funded by the promoters (not having reached yet the financial close): the negotiations of all contracts plus the clearance of all permits normally takes at least 18 months.

Once the strategic contracts are ready and the necessary permits are available, the project moves into its *(4) financial agreements*: arrangements and syndication. These are the steps necessary to achieve the financial close, 3 – 6 months are needed an the promoters need to find an additional $ 500,000 – 1,000,000 (financial and legal advisors), before receiving the financing.

The total cost is financed through a 70/30 debt equity ratio. The debt component is divided between two loans: 40% of the total project cost is financed by an ECA (i.e. Export Credit Agency) loan, the remaining 30% by a standard commercial loan.

At financial close (drawdown of the debt), the equity ($ 6,000,000) is partially recovered through development and promoter success fees and the financial advisor receives the success fee, reduced by the monthly retainer fees (already paid by the promoters). Thereafter the EPC Contractor receives the so-called "final notice to proceed" and starts the construction activities (receiving a 10% advance payment).

The *(5) construction phase* represents the physical construction of the Power Plant, carried out accordingly to the EPC Contract terms and conditions. A 24 months period has been foreseen for the construction activities. During this period payments to the EPC Contractor are consistent with the construction progress: payments are made when the milestones, agreed in the contract, are achieved. When the plant is completed, then the Take Over Certificate is issued by the owner and the commercial operation starts: 90% of the total contract price has been paid to the EPC Contractor and the remaining 10% is kept by the owner for the guarantee period (2 years) and aid in two installments, at the end of the first year of commercial operation (5%) and at the end of the guarantee period (5%).

If the power plant does not perform as guaranteed the so-called Liquidated Damages are applied both at the Take Over date and at the end of the Guarantee Period.

During this phase the lenders do not require any repayment (period of grace) and the debt generates the so-called interests during construction (IDC) that become part of the capital to be repaid.

We arrive to the *(6) operation phase.* A 15 years period has been considered for the operation of the plant and it is equal to the PPA validity. During this period the plant will operate as a base load plant for 7,500 hour per year and the energy produced is sold according to the terms and conditions previously agreed between the owner and the customers (utility, industrial customers, traders,...) and according to a tariff structured as follows:

- **Fuel Charge** — fuel is a pas through, the owner of the power plant guarantees on the heat rate,
- **Debt Service Charge** — it provides the annual coverage of the project debt (principal repayment and interests),
- **Equity Service Charge** — it represents the agreed ROE recognized by the customer to the plant power owner,
- **Other Charges** — they are supposed to cover other operational costs such as O&M, insurance, management, income tax, ...

References

1. P. Gallo, L. Peccati (1993): 'The appraisal of industrial investments: a new method and a case study', *International Journal of Production Economics*, **30–31**, 465–476.
2. P. Gallo, L. Peccati (1995): 'A model for the Evaluation of a business plan', Working paper pres. at the EWGFM XVII Meeting (Bergamo, Italy).
3. H.U. Gerber (1990): *Life Insurance Mathematics*, Springer–Verlag, Swiss Association of Actuaries. Berlin.

Coherent Risk Measures and Linear Programming

Stefano Benati
Dipartimento di Informatica e Studi Aziendali
Universitá di Trento
Via Inama 5
38100 Trento, Italy

In recent years, the need for risk measures in highly complex systems, brought a general reconsideration about what features a sound definition of risk should satisfy. Most of the time, the measure of risk must be applied to many different situations, for example the numbers involved are used to compare and to control portfolio investments, as well as trading activities of a department, the exposure to default risk of a class of debts, and so on. It is a natural requirement that those numbers can be compared. But it is necessary that the definition of the risk measure satisfies some mathematical properties, that are given in the form of axioms. This idea forms the core of the theory of coherent risk measures, that was put forward by four scholars in a recent work, [1].

In this paper, the theory is revised by linear programming methodology. By this way, some results, that were already known, are proved by different methods and some new results can be obtained, such as the duality relation between the 'model-free' risk measure and the 'scenario-dependent' risk measure. Moreover, it can be easily discussed the extension of the risk measure to the case in which many financial instruments are available to hedge against risks. Finally, it is discussed how to build a coherent risk measure from the definition of an acceptable set and then how to adjust the popular Value at Risk measure to obtain a coherent measure.
Keywords: Risk measures, linear programming, value at risk.

1 Introduction

Risk measure is source of concern for many financial institutions. The large use of complex financial instruments, such as options, callable bonds, mortgage-backed securities and so on, and the growing development of financial markets, made financial institutions aware that they are exposed to risks that are difficult to control and to forecast.

In the past, it was a common practice to model future revenues as random variables with given distributions. The risk measure was the operational problem of computing a number, that was able to represent the 'negative' results of the probability distribution. In this way, firm management could read it as a measure of possible losses. The simplest model assumes that the random variable is normal. In this case, all the information is summarized by the expected value and the standard deviation of the distribution. Therefore, the standard deviation is considered the appropriate measure of risk. In financial markets, nobody knows the real parameters underlying the model, therefore the sample variance is used to estimate the true parameters of the model. This hypothesis is introduced in the early definition and usage of risk, as in the Markowitz model and the definition of the Sharpe Ratio, the Treynor ratio and so on, see [5].

In this framework, some estimation problems can emerge. Even for simple investments, such as portfolios of stocks, the normal assumption is not verified, so that the sample variance looses its efficiency property. An alternative estimator can be used, such as the mean absolute deviation, see [3], that is a more robust statistic. An application to optimal portfolios is available in [4].

There are theoretical problems, too. Some authors pointed that the risk measure should consider unfavorable results of an investment only, and that it can be erroneous to weight positive and negative deviation from the expected value by the same way. This is specially true when one has to deal with an asymmetric distribution of future results and moments higher than second are needed to describe the probability function. But the market evolution leads to complex portfolios with exactly this feature, due to the fact that options are asymmetric instruments and that decision makers try to exploit this feature. In this case variance is nor a representative measure of the distribution. About this problem, the vast literature about the existence of the third order moment can be cited. For recent results, one can see [2] and [10]. It is worth noting that in operational environment, e.g. in the J.P.Morgan trading system, a new measure has been developed that potentially takes into account the necessity of considering asymmetric distributions. This measure is the Value at Risk (VaR), for which there is a fast growing literature. Basically, it is the estimation of the lowest quantile of the probability distribution. If no assumption is made about the distribution, it is a risk measure that considers negative events only. See [8] and [9].

Nowadays, scholars and practitioners can choose among many risk definitions. Pros and cons of the methodologies are not always clear, therefore a strong contribution on the debate can come from the recent work of Artzner, Delbaen, Eber, Heath, [1] (from now on: ADEH). Their starting point is an axiomatic treatment of risk measure, that is, they start defining what features must a risk measure supports, to be a reliable number to control and to compare risks. Therefore the problem is not focused on statistic estimation, as often happens in the literature, but on mathematical properties. In their framework, some properties are defined first and then it is verified whether they are satisfied by a risk measure. If so, the risk measure is called coherent. As it will be seen, many definitions currently adopted are not coherent.

In this paper, results available in [1] are considered and their links to linear programming are outlined. By this way, some of the theorems given by ADEH are shown by different methods. Some original proof is available too, such as the duality relation between model-free and scenario-dependent measures of risk. Moreover, it can be discussed whether coherent risk measures can be built in the operational practice, when there are multiple financial instruments to hedge against risk. Finally, it will be discussed how to compute coherent risk measures in practice, e.g. by the definition of an acceptable set and by the generation of scenarios. Finally, it will be shown how to adjust the popular Value at Risk measure to obtain a coherent measure.

2 Axioms for the risk measure and for the acceptable set

Suppose a future date T is given to evaluate a set of financial assets. Suppose that, in time T, one and only one of the events ω, belonging to a discrete set Ω of states of nature, can happen. Suppose that the set of the states of nature is finite with cardinality n. For example, Ω can stand for all the prices of a stock, or all the values of an interest rate, and so on.

We call *risk* the vector $x \in \Re^n$, whose entries $x_i, i = 1, ..., n$ stand for the final net value of wealth of a portfolio of financial activities, under the assumption that event ω_i took place.

If we let I_{ω_i} be the total income under ω_i and C_{ω_i} the total incurred cost, we have the budget equation $x_i = I_{\omega_i} - C_{\omega_i}$.

The vector $x \in \Re^n$ is composed by positive and negative value, since in some state of the world costs can be greater than gains. A decision maker, in front of an investment that is represented by a risk, can decide whether it is acceptable or not. That is, he makes a 'yes or no' decision. Clearly, he can accept risks that under some circumstances will give negative values, if those bad events are sufficiently compensated by positive wealth under other events. So he is able to define a set $\mathcal{A} \subset \Re^n$, through which his choices are modeled. A risk, such that $x \in \mathcal{A}$, is an acceptable risk, if $x \notin \mathcal{A}$, the risk is not acceptable. Set $\mathcal{A}$ is the set of the acceptable risks, or acceptable set.

If the decision maker is consistent, then the acceptable set $\mathcal{A}$ must satisfy some logical properties. The decision maker accepts any activity that guarantees positive wealth under any state of the world. Let C^+ be the positive cone of $\Re^n$. The first property to be satisfied is:

Axiom 1: The acceptable set $\mathcal{A}$ contains C^+, that is: $C^+ \subset \mathcal{A}$.

Conversely, the decision maker does not take risks that always provide negative results. Let C^{--} be the set of the strictly negative arrays.

Axiom 2: The intersection between C^{--} and the acceptable set is empty. That is: $\mathcal{A} \cap C^{--} = \emptyset$.

Axiom 2 can be given in tighter form. Let C^- the negative cone of $\Re^n$.

Axiom 2(a): The only intersection between $\mathcal{A}$ and C^- is the array of zeros: : $\mathcal{A} \cap C^- = 0$.

Axiom 2(a) defines acceptable a solution in which the wealth of the decision maker does not change.

The third axiom establishes that the decision maker is risk adverse.

Axiom 3: The acceptable set $\mathcal{A}$ is convex.

Finally, expositions are not bounded.

Axiom 4: The acceptable set $\mathcal{A}$ is a positively homogeneous cone.

The fifth axiom is not necessary and it is not given by ADEH, but it simplifies some technical detail. Since we are going to use linear programming, we'd rather work with the min and max operator instead of sup and inf. Therefore it is simply required that the definition of $\mathcal{A}$ is given by defining what are the acceptable risk, rather than defining the risk that are not. That is, $\mathcal{A}$ contains its frontier.

Axiom 5: The acceptable set $\mathcal{A}$ is closed.

The geometric representation of an acceptable set is drawn in Figure 1. There are only two states of nature, that support risks $x = [x_1; x_2]$. The area where the vector of the frontier are pointing to is the set $\mathcal{A}$ of the acceptable risks.

By the definition of $\mathcal{A}$, risks can be measured. The measure of risk relies over the distance between x and the frontier of the acceptable set. Suppose that we have a not acceptable risk x: the budget equation $x_i = I_{\omega_i} - C_{\omega_i}$ can be modified by adding a positive number $\lambda_{\omega_i} = \lambda$, constant for every i, to obtain:

$$x_i = I_{\omega_i} - C_{\omega_i} + \lambda$$

There will be a value of λ that will make the risk acceptable. The minimum value λ is the measure of risk. The same procedure can be applied to an acceptable risk x. In this case, revenues can decrease of a value λ in every state of the world. For a given λ, a risk on the frontier of the acceptable set can obtained. This number is the risk measure of an acceptable position.

In other words, the risk measure is how much we are willing to gain or to loose in every state of the world, to obtain or to maintain an acceptable risk.

We can describe the same operation by assuming that there is a reference financial instrument, that is represented by an array $u \in \Re^n$, where $u_i = 1$ for every i. This financial instrument is the pure liquidity. Pure liquidity provides the same results under any state of the world and, at least from an ideal point

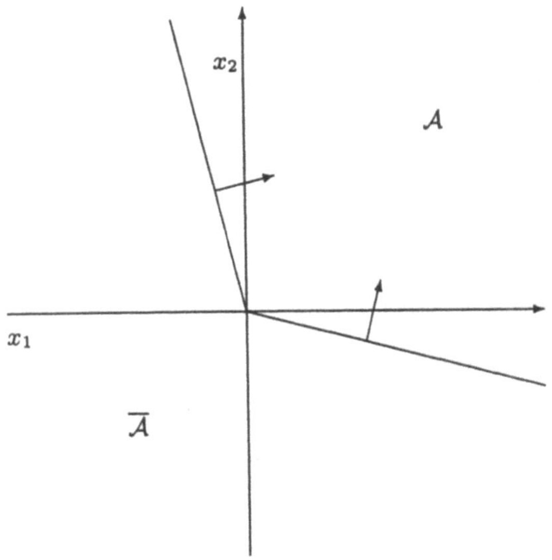

Fig. 1. The geometric representation of the acceptable set $\mathcal{A}$

of view, it can be added to or withdrawn from a risk by an external office or by the hand of fate.

Definition 1: *Risk measure over an acceptable set:* Risk x, the reference instrument u and the acceptable set $\mathcal{A}$ are given, then the risk measure over $\mathcal{A}$, is the function $\rho_{\mathcal{A}} : \Re^n \to \Re$:

$$\rho_{\mathcal{A}}(x) = \min\{\lambda | x + \lambda u \in \mathcal{A}\}$$

The graphic representation of the definition is given in Figure 2.

Suppose that we can compute the risk measure of an investment, without the explicit knowledge of $\mathcal{A}$. That is, we have a 'black box', that computes $\rho_{\mathcal{A}}(x)$ for every x. Still, we can have informations about $\mathcal{A}$. As we will see later, if $\mathcal{A}$ satisfies axioms 1-4, then the risk measure $\rho_{\mathcal{A}}$ satisfies another set of axioms. These axioms are the following. Let $\rho : \Re^n \to \Re$ be a function by which risk is measured. Then ρ must satisfy the following axioms.

Axiom MS: Measurability: for every risk x:

$$|\rho(x)| < +\infty$$

This axiom is not given by ADEH, but it is necessary to prevent incongruity in the definition of $\mathcal{A}$ and of u.

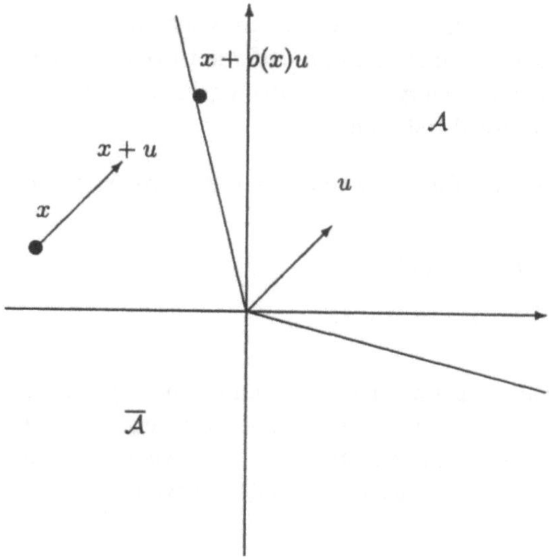

Fig. 2. Geometric representation of the non acceptable risk x, the reference instruments u and the acceptable risk $x + \rho(x)u$

Axiom T: Translation invariance: for every risk x and every real number α:

$$\rho(x + \alpha u) = \rho(x) - \alpha$$

Axiom T requires that if revenues increases (or decreases) of α units, then the risk measure decreases (increases) by α units. It can be observed that, by axiom T, $\rho(x + \rho(x)u) = 0$ for every x.

Axiom S: Subadditivity: for any two risks x_1 and x_2:

$$\rho(x_1 + x_2) \leq \rho(x_1) + \rho(x_2)$$

This axiom has the following meaning. Suppose that the risk from highly complex positions must be computed, e.g. the risk of a portfolio that contains a large number of assets. Then a strategy of valuation is to divide the portfolio in homogeneous clusters and then to compute the risk measure for every cluster. The total risk is the sum of the risks of every cluster. To obtain a reliable number, it is necessary that risks are not hidden by this procedure. The worst thing that can happen is to compute low risks for two clusters, and then, by repeating the computation at aggregate level, to discover that risk is very high. Therefore, what is wanted here is that, if we merge clusters and measure risk at aggregate level, we do not obtain an unexpected number. One way to guarantee that disaggregate measures are safe, is to require that the risk measure is subadditive, so that, every time we disaggregate and sum, we obtain an overestimation of the total risk. That is:

Axiom PH: Positive homogeneity: for every real $\lambda \geq 0$ and every risk x:

$$\rho(\lambda x) = \lambda \rho(x).$$

Axiom PH establishes that the measure of risk depends linearly on the value of a position. This assumption prevents scale effects both negative, e.g. when the value of the loss is too large to be accepted, and positive, e.g. moral hazard: the loss is so large that a central institution will intervene to cover it.

Axiom M Monotonicity: for every risk pair x_1 e x_2, such that $x_1 < x_2$:

$$\rho(x_1) > \rho(x_2).$$

Axiom M ensures that the risk of x_1 is greater than the risk of x_2, if x_1 gives worse results than x_2 in every state of nature. It is worth noting that mean-variance based criteria of choice do not satisfy this assumption, unless other hypotheses are introduced..

Axiom R: Relevance (in weak sense): for every risk x such that $x \leq 0$, but $x \neq 0$, then $\rho(x) > 0$.

Axiom R prevents that there are risks that are not correctly measured. It is the axiom that ADEH used. However a stronger version can be given, that will be useful in the following sections.

Axiom R(a): Relevance (in strong sense): for every risk $x \notin \mathcal{A}$:

$$\rho(x) > 0.$$

Definition 2: Coherency: a risk measure that satisfies the axioms of measurability, translation invariance, subadditivity, positive homogeneity, monotonicity and relevance, is said coherent.

We delay to the last chapter the discussion about the coherency of the Value at Risk measure of risk. In the following, the link between coherent measures and acceptable sets is proved.

3 Correspondence between coherent risk measures and acceptable sets

The axioms for the acceptable set $\mathcal{A}$ and the coherent risk measures are linked, as the following theorems show.

Theorem 1: If $\mathcal{A}$ satisfies axioms 1-4, then $\rho_{\mathcal{A}}(x)$is a coherent risk measure.

Proof:

1) Measurability: let $x \notin \mathcal{A}$. Let $\lambda = \min\{x_i\}$ the worst event of risk x. Since $C^+ \subset \mathcal{A}$, then $x + \lambda u \in \mathcal{A}$. Therefore $\rho_{\mathcal{A}}(x) < \lambda < +\infty$. If $x \in \mathcal{A}$ the proof is similar.

2) Translation invariance:

$$\begin{aligned}\rho_{\mathcal{A}}(x + \alpha u) &= \min_p\{p|x + (\alpha + p)u \in \mathcal{A}\}\\ &= \min_{p,\alpha}\{p + \alpha|x + (\alpha + p)u \in \mathcal{A}\} - \alpha\\ &= \rho_{\mathcal{A}}(x) - \alpha\end{aligned}$$

Note the use of the property of monotonicity, that ensures that the risk on the frontier is the same for both the minimum problems.

3) Subadditivity:

$$\begin{aligned}\rho_{\mathcal{A}}(x + y) &= \min_{p,q}\{p + q|x + y + (p + q)u \in \mathcal{A}\} \leq\\ &\leq \min_p\{p|x + pu \in \mathcal{A}\} + \min_q\{q|y + qu \in \mathcal{A}\} =\\ &= \rho_{\mathcal{A}}(x) + \rho_{\mathcal{A}}(y).\end{aligned}$$

The inequality is correct, since $\mathcal{A}$ is a convex cone. Therefore, if $y+qu \in \mathcal{A}$ and $x + pu \in \mathcal{A}$, then $x + y + (p + q)u \in \mathcal{A}$.

3) Positive homogeneity:
Instead of solving the minimum problem with respect to a variable, solve it with respect to the product of a variable p for a constant λ. A change of scale of the underlying variable is obtained, but the optimal value does not change:

$$\begin{aligned}\rho_{\mathcal{A}}(\lambda x) &= \min_p\{\lambda p|\lambda x + (\lambda p)u \in \mathcal{A}\} =\\ &= \min_{\lambda p}\{\lambda p|\lambda(x + pu) \in \mathcal{A}\}\end{aligned}$$

Since $\mathcal{A}$ is a cone, $\lambda(x + pu) \in \mathcal{A}$ if and only if $(x + pu) \in \mathcal{A}$. Therefore:

$$\begin{aligned}\rho_{\mathcal{A}}(\lambda x) &= \min_p\{\lambda p|x + pu \in \mathcal{A}\} =\\ &= \lambda \min_p\{p|x + pu \in \mathcal{A}\} =\\ &= \lambda\rho_{\mathcal{A}}(x)\end{aligned}$$

where the last inequality is true, since λ is fixed.

5) Relevance (both in strong and weak sense): If $x \notin \mathcal{A}$ and $\rho_{\mathcal{A}}(x) \leq 0$, then $x + \rho_{\mathcal{A}}(x)u \leq x$, but at the same time $x + \rho_{\mathcal{A}}(x)u \in \mathcal{A}$, a contradiction of the assumption that $\mathcal{A}$ is a positive cone. ◇

The opposite direction of the statement of theorem 1 can be proved, too:

Theorem 2: If $\rho_{\mathcal{A}}(x)$ is coherent, then the acceptable set $\mathcal{A}$ satisfies axioms 1-4.

Proof

Axioms 1 and 2: Observe that $|\rho_{\mathcal{A}}(x)| < +\infty$ by the measurability axiom. Then, if $\rho_{\mathcal{A}}(x)$ is positively homogeneous, it follows that $\rho_{\mathcal{A}}(0) = \rho_{\mathcal{A}}(0x) = 0\rho_{\mathcal{A}}(x) = 0$. If $x < 0$ then, for monotonicity, $\rho_{\mathcal{A}}(x) > \rho_{\mathcal{A}}(0) = 0$, and axiom 2 follows. If $x > 0$, then $\rho_{\mathcal{A}}(x) < \rho_{\mathcal{A}}(0) = 0$, and axiom 1 follows.

Axiom 3: If $\rho_{\mathcal{A}}(x)$ is relevant, then $\rho_{\mathcal{A}}(x) < 0$ implies that $x \in \mathcal{A}$. Suppose that $\rho_{\mathcal{A}}(x) < 0$, $\rho_{\mathcal{A}}(y) < 0$. Then, for $\lambda \leq 1$, it follows:

$$\begin{aligned} 0 &\geq \rho_{\mathcal{A}}(\lambda x) + \rho_{\mathcal{A}}((1-\lambda)y) = \\ &= \lambda\rho_{\mathcal{A}}(x) + (1-\lambda)\rho_{\mathcal{A}}(y) \geq \\ &\geq \rho_{\mathcal{A}}(\lambda x + (1-\lambda)y). \end{aligned}$$

Therefore, if $x, y \in \mathcal{A}$, we have that $(1-\lambda)x + \lambda y \in \mathcal{A}$, so that $\mathcal{A}$ is convex.

Axiom 4. If $\rho_{\mathcal{A}}(x)$ is positive homogeneous and $x \in \mathcal{A}$, then $\rho_{\mathcal{A}}(x) \geq 0$, therefore $\rho_{\mathcal{A}}(\lambda x) \geq 0$, if $\lambda \geq 0$. By which $\lambda x \in \mathcal{A}$, so that $\mathcal{A}$ is a cone. ◇

4 Scenario dependent risk measures

A risk measure that relies on the definition of an acceptable set, is also called model-free by ADEH, since it does not require any assumption on the probabilities of the states of nature. Conversely, let's suppose that a finite set $\mathcal{P}$ of probability distributions $p^k \in \Re^n, k = 1, ..., K$; over the events in Ω are available to the decision maker. We will refer to a probability distribution p_k as a scenario.

Suppose that $\mathcal{P}$ is consistent with the decision maker analysis. That is, in Ω there is not an impossible event under every scenario and it is not verified that there is some ω_i such that $p_i^k(\omega_i) = 0$ for every $p^k \in \mathcal{P}$. Clearly, it is not binding assumption, for if such an event exists, then it can be discarded from Ω without any consequence.

Definition 3: Scenario dependent risk measure: A set $\mathcal{P}$ of probability distribution is given. A scenario dependent risk measure is the value:

$$\rho_{\mathcal{P}}(x) = \max\{E_{p^k}(-x) | p^k \in \mathcal{P}\}.$$

To measure risks, the decision maker has available a set of scenarios, for example provided by a pool of experts, instead of an acceptable set $\mathcal{A}$. He computes the expected result for every scenario, then the worst of these results is the risk measure. It can be wondered what are the links between this measure and the system of axioms that were introduced before. This is discussed in the following theorems.

Theorem 3: The scenario dependent risk measure $\rho_{\mathcal{P}}(x)$ is coherent.

Proof: Let $\rho_k(x) = -x^T p^k$ the expected risk under the condition that scenario k takes place .

Translation invariance:

$$\begin{aligned}\rho_{\mathcal{P}}(x+\alpha u) &= \max\{\rho_1(x+\alpha u), \rho_2(x+\alpha u), ..., \rho_n(x+\alpha u)\} = \\ &= \max\{\rho_1(x), \rho_2(x), ..., \rho_n(x)\} - \alpha = \\ &= \rho_{\mathcal{P}}(x) - \alpha.\end{aligned}$$

Subadditivity:

$$\begin{aligned}\rho_{\mathcal{P}}(x+y) &= \max\{\rho_1(x+y), \rho_2(x+y), ..., \rho_n(x+y)\} \leq \\ &\leq \max\{\rho_1(x), \rho_2(x), ..., \rho_n(x)\} + \max\{\rho_1(y), \rho_2(y), ..., \rho_n(y)\} = \\ &= \rho_{\mathcal{P}}(x) + \rho_{\mathcal{P}}(y).\end{aligned}$$

Positive homogeneity:

$$\begin{aligned}\rho_{\mathcal{P}}\{\alpha x\} &= \max\{\rho_1(\alpha x), \rho_2(\alpha x), ..., \rho_n(\alpha x)\} = \\ &= \alpha \max\{\rho_1(x), \rho_2(x), ..., \rho_n(x)\} = \\ &= \alpha \rho(x)_{\mathcal{P}}\end{aligned}$$

Relevance (in weak sense): If there are not impossible events under any scenario, if $x < 0$, then $\rho(x) > 0$. $\diamond$.

We are not in the position to prove strict relevance, since the acceptable set is not defined yet. However, it is expected that, if $\mathcal{A}$ is defined by the property that $\rho_{\mathcal{P}}(x) \leq 0$, then it must satisfy axioms 1-4. It is the corollary of the following theorem.

Theorem 4: The risk measure over the acceptable set $\rho_{\mathcal{A}}$ and the scenario dependent risk measure $\rho_{\mathcal{P}}$ are dual to each other.

Proof
Consider $\rho_{\mathcal{A}}$: Since $\mathcal{A}$ is a convex cone, then it can be described by convex combinations of rays $r^j : j = 1, ..., J$, see [7]. Therefore $x \in \mathcal{A}$, if and only if

there are $\beta_j \in \Re, \beta_j \geq 0$ such that: $x = \sum_j \beta_j r^j$. Therefore the risk measure can be obtained by solving the following linear programming problem:

$$\begin{aligned}
&\rho_{\mathcal{A}}(x) = \min_{\alpha,\beta} \alpha \\
&\text{s.t.} \\
&x + \alpha u \geq \textstyle\sum_j \beta_j r^j \\
&\beta_j \geq 0 \text{ for every } j; \\
&\alpha \text{ free}
\end{aligned}$$

In matrix form, let $\beta = [\beta_1, ..., \beta_J]^T \in \Re^J$, $R = [r^1; ...; r^J] \in \Re^{n \times J}$:

$$\begin{aligned}
&\min[1;0]^T[\alpha;\beta] \\
&s.t. \\
&[u; -R][\alpha; \beta^T]^T \geq -x \\
&\beta \geq 0, \alpha \text{ free}
\end{aligned}$$

The dual problem is:

$$\begin{aligned}
&\max_y y^T x \\
&\text{s.t.} \\
&y^T u = 1 \\
&-y^T R \leq 0 \\
&y \geq 0
\end{aligned}$$

That is, in extensive form:

$$\begin{aligned}
&\max \textstyle\sum_i -x_i y_i \\
&\text{s.t.} \\
&\textstyle\sum_i y_i = 1 \\
&\textstyle\sum_i r_i^1 y_i \geq 0 \\
&\textstyle\sum_i r_i^2 y_i \geq 0 \\
&\dots \\
&\textstyle\sum_i r_i^J y_i \geq 0 \\
&y_i \geq 0
\end{aligned}$$

We can observe that the feasible region is a polyhedron contained in the set $\mathcal{Q} = \{y | \sum_i y_i = 1; y_i \geq 0\}$. Therefore every vector y that satisfies the constraints is a probability distribution. Moreover, since the feasible region is a finite polyhedron, the problem attains a maximum on a vertex. Let $y^k : k = 1, ..., K$ all the vertices of the polyhedron, since they are probability distributions, let $\mathcal{P} = [y^1, ..., y^K]$. The dual problem can be written as:

$$\begin{aligned}\rho_{\mathcal{A}}(x) &= \max\{-x^T y^1, -x^T y^2, ..., -x^T y^K\} = \\ &= \max\{E_y(-x) | y \in \mathcal{P}\} = \\ &= \rho_{\mathcal{P}}(x)\end{aligned}$$

Then the duality relation is proved, provided that the minimum and the maximum problems have solution. But it is always the case, since it assumed that $|\rho(x)| \leq +\infty$ by the axiom of measurability. Moreover the existence of the impossible event in any scenario is prevented, because the primal problem admits always finite solution, equal to the dual.

5 Risk measure with multiple reference instruments

To define $\rho_{\mathcal{A}}$, we have introduced an acceptable set $\mathcal{A}$ and we have assumed that gains can be improved or worsen by a fixed amount in any state of the world, but no operational justification has been attempted. As in the framework of ADEH, it can be assumed that an external institution can intervene in the risk management by adding liquidity to a position with positive risk. The liquidity is invested in a fixed rate asset, e.g. government bonds, to obtain a safe income under any state of the world. If this amount is λ and r is the fixed rate, the risk becomes: $x + \lambda(1 + r)u$. The discussion in [1] used this assumption. From the technical point of view, the factor $(1 + r)$ does not make a large difference, since in their framework the operation has an ideal meaning, as it is only the instrument, e.g. the direction, by which it is possible to modify positions and reach the acceptable set.

But in practice decision maker must measure and hedge against risks and in this case the investment on fixed rate bonds is just one of the opportunities. For example, a stock exchange office may hedge against a potential loss by purchasing currency options. A bank branch may sell part of credits of firms that operate on the same sector, in order to diversify risk. It is not to be take for granted that fixed rate bonds are the best instruments to hedge against risk.

It can be noted that the methodology to solve the optimal hedge problem can be the same as to solve the risk measure problem: first of all, an acceptable set $\mathcal{A}$ must be defined and it must satisfy some axioms. Then, given a risk that is not contained in the acceptable set, a given amounts of financial instruments must be chosen, to obtain an acceptable risk. Clearly there is a link between the two problem. One may wonder whether the risk measure of definition 1 can be generalized, both to define a coherent risk measure and to solve the optimal hedge problem by the same linear programming problem. From the operational point of view, two problems are solved as one.

To justify our attempt, we can see that sometimes misleading results can emerge from treating the two problems separately. Consider the following example (Figure 3). There are only two states of the world, over that two results x_1 and x_2 are defined. Cone $\mathcal{A}$ is represented by the part of the region were the vectors are pointing to. There are two financial instruments available to modify a position. The first one, u_1, is the pure liquidity, that is something that has the same unitary future value for a unit of capital today (there is no interest, but conceptually nothing changes) and an insurance, u_2, that guarantees 3 units of capital if the event ω_1 takes place, 0 unit if ω_2 happens, and the cost of the insurance is 1 unit today. Two risks are given: A=(-3,1) and B=(3/4, -3/4). It is easy to see that $\rho_{\mathcal{A}}(A) > \rho_{\mathcal{A}}(B)$, according to definition 1. But the cost of hedging is different. In the B case, the only possible hedge is pure liquidity, so that it needs $\frac{3}{4}\sqrt{2}$ units, while in the A case, the cheaper opportunity is to buy an insurance for a cost of $\frac{2}{3}$ unit. In spite of the fact that $\rho_{\mathcal{A}}(A) > \rho_{\mathcal{A}}(B)$, to hedge A is more expensive than hedging B.

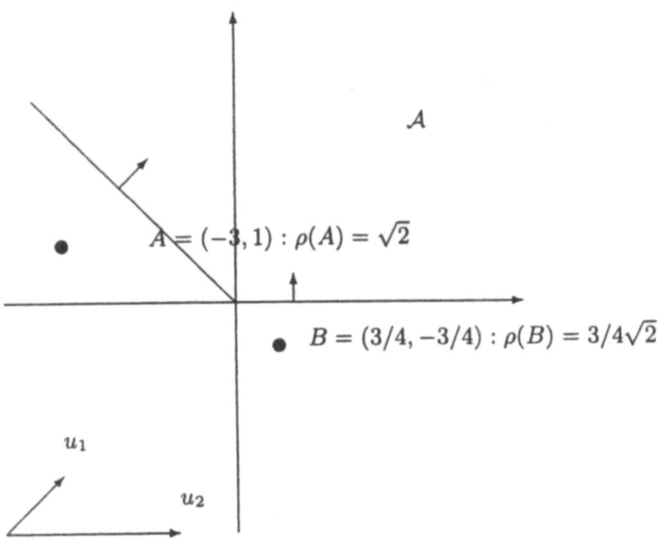

Fig. 3. Reppresentation of risks A and B. If it were possible to use the reference instrument u_2, the cost to hedge A would be equal to 2/3, less than the cost of hedging B

Consider the problem of the optimal hedge against risk. In the market, a set of financial instruments $\mathcal{U} = \{u^k \in \Re^n, i = 1, ..., K\}$ is available, each of unitary cost. Every instrument is characterized by an array u^k, whose element i, u_i^k, is the economic result at time T, under the assumption that event ω_i will take place. We suppose that $\mathcal{U}$ is composed by *contracts of pure insurance*: that is $u_i^k \geq 0$. Suppose that among the financial instruments there is 'pure liquidity', 1 unit today for 1 unit tomorrow (if we consider 1 unit plus interest rates tomorrow, we have to change the notation slightly, but the conclusion are the same). Let $u^1 = 1$ be the corresponding array. Let $\alpha_k \in \Re$ be the variable associated to the quantity of instrument k that is sold or purchased by the decision maker. Let $\Theta \in \Re^K$ be the set of the feasible strategies that he can choose: that is $\alpha \in \Theta$ if it stands for a set of quantities that can be actually be sold or purchased. The following assumption is necessary, that is, a limited set of strategies are available to the decision maker: he can buy contracts of pure insurance only, he cannot sell them. In some occurrence, it is a realistic feature, a trader can buy options, but he cannot write them. Then the following constraints define Θ: $\alpha \in \Theta$ if and only if $\alpha_k \geq 0$ for every $k \geq 2$ and α_1 free. The following definition can be given.

Definition 4: *Risk measure with multiple instruments:* Risk x, a set of reference instruments $\mathcal{U}$, a set of feasible strategies Θ and an acceptable set $\mathcal{A}$ are given, then the risk measure with multiple instruments is:

$$\rho_{\mathcal{U}}(x) = \{\min_{\alpha \in \Theta} \sum \alpha_i | \sum \alpha_i u^i + x \in \mathcal{A}\}$$

Observe that $\rho_{\mathcal{U}}(x)$ is the minimum cost hedge of risk x. The definition is equivalent to the solution of the following linear programming problem. Let $U = [u^1; ...; u^K]$ be the matrix whose columns are the reference instruments:

$$\begin{aligned} &\rho_{\mathcal{U}}(x) = \min_{\alpha,\beta} \textstyle\sum_i \alpha_i \\ &s.t. \\ &x + \textstyle\sum u^i \alpha_i \geq \sum \beta_j r^j \; \alpha_i, \beta_j \geq 0 \text{ for every } i \text{ and } j \end{aligned}$$

We can observe that the definition of Θ makes 'asymmetric' the risk measure: if an acceptable risk $x \in \mathcal{A}$ must be evaluated, then the only reference instruments is liquidity. If we are measuring a not acceptable risk, then multiple instruments can be used. Therefore we need to modify the axiom of translation invariance.

Axiom T(a): Translation invariance: Let $\alpha^* = \arg\min_\alpha \{\sum \alpha_i | x + \sum \alpha_i u^i \in \mathcal{A}\}$. Then α^* stands for the optimal mix of contracts of pure insurance to be purchased to hedge $x \notin \mathcal{A}$, or the potential loss of revenues that can occur without making the risk non acceptable. A risk measure with multiple instruments is said translation invariant, if:

$$\rho_U(x + \lambda\alpha^*) = \rho_U(x) - \lambda \sum \alpha_i^*, \text{ for } \lambda \in \Re, 0 \leq \lambda \leq 1$$

Axiom T(a) is an extension of axiom T, because it introduces an optimal reference direction α^* and a limited step λ. They were not necessary to define T, because there was only one reference instrument, so that only one direction was feasible. For the hypotheses on Θ, the following theorem can be obtained:

Theorem 5: $\rho_{\mathcal{U}}(x)$ is a coherent risk measure.

Proof:

Relevance: If $x \notin \mathcal{A}$ and if $\rho_{\mathcal{U}}(x) \leq 0$, then it implies that $x + \sum \alpha_i^* u^i \in \mathcal{A}$, but, by the constraints on $\mathcal{U}$ and on Θ, it is also $x \geq x + \sum \alpha_i^* u^i$, a contradiction of the hypothesis that $\mathcal{A}$ is a cone.

Measurability: If $x \notin \mathcal{A}$, then it follows that $\rho_{\mathcal{U}}(x) > -\infty$, because the array u_1 of pure liquidity is available. But, from relevance, it is also $\rho_{\mathcal{U}}(x) < 0 < +\infty$. The proof in the case that $x \in \mathcal{A}$ is similar.

Subadditivity: Observe that $\rho_{\mathcal{U}}(x) = \sum \beta_i^*$ implies that $x + \sum \beta_i^* u^i \in calA$, and observe that $\rho_{\mathcal{U}}(y) = \alpha_i^*$ implies that $y + \sum \alpha_i^* u^i \in \mathcal{A}$. Since $\mathcal{A}$ is a convex cone, then $(x+y) + \sum(\beta_i^* + \lambda_i^*)u^i \in \mathcal{A}$, by which $\rho_{\mathcal{U}}(x+y) \leq \rho_{\mathcal{U}}(x) + \rho_{\mathcal{U}}(y)$.

Positive homogeneity: Consider $\lambda^* \in \Re^k$, such that $\lambda_i^* = \lambda$ for every i. Then $\rho_{\mathcal{U}}(\lambda x) = \min_{\lambda\alpha}\{\sum \lambda\alpha_i | \lambda x + \sum \lambda\alpha_i u^i \in \mathcal{A}\}$. But, if $\mathcal{A}$ is a cone, then $\lambda x + \sum \lambda\alpha_i^* u^i \in \mathcal{A}$ implies that $\sum \alpha_i^* + x \in \mathcal{A}$. The reverse proposition also holds, by which $\rho_{\mathcal{U}}(\lambda x) = \lambda\rho_{\mathcal{U}}(x)$.

Translation invariance: If $x \notin \mathcal{A}$,

$$\begin{aligned}\rho_{\mathcal{U}}(x+\lambda\alpha^*) &= \min_\beta\{\textstyle\sum \beta_i | \sum \beta_i u^i + \lambda\alpha_i^* u^i + x \in \mathcal{A}\} = \\ &= \min_\beta\{\textstyle\sum \beta_i + \lambda\alpha_i^* | \sum(\beta_i + \lambda\alpha_i^*)u^i + x \in \mathcal{A}\} - \lambda\sum\alpha_j^* = \\ &= \rho_{\mathcal{U}}(x) - \lambda\textstyle\sum\alpha_i.\end{aligned}$$

If $x \in \mathcal{A}$, axiom T(a) is the same as axiom T. The same proof that was provided for theorem 1 is also true. $\diamond$

If we are in a practical situation in which the hedge strategies of the decision maker consist in purchasing contracts of pure insurance only, then the same linear programming problem can be used to determine the optimal risk hedge and the coherent risk measure. The duality relation stated by theorem 4 holds too, but note that $\rho_{\mathcal{U}}(x) \leq \rho_{\mathcal{A}}(x)$.

One may wonder whether the condition on Θ can be relaxed to include purchasing and selling contracts of pure insurances. In the linear programming problem, it corresponds to enlarging Θ, by letting α_i free, for every $i \geq 2$. Unfortunately, the risk measure is not coherent anymore: it does not satisfy the relevance axiom. Consider the following example (Figure 4). There are only 2 future events, there are three reference instruments: liquidity $u^1 = [1;1]$ and 2 contracts of insurance: $u^2 = [0;10]$, $u_3 = [2;0]$. Set $\mathcal{A}$ is the area towards which the vectors are pointing to. Consider risk $x = [-1;-1]$. Since $x < 0$, we must have $\rho_U(x) > 0$ by the relevance axiom. It is not so: if $\alpha_1 = 0, \alpha_2 = -\alpha_3 = 1$, it results that $x + \alpha_2 u^2 + \alpha_3 u^3 \in \mathcal{A}$ e $\alpha_2 + \alpha_3 = 0$, by which $\rho_U(x) \leq 0$. In this case there are arbitrage opportunities that the decision maker can exploit. Therefore definition 4 is a useful tool to establish such opportunities, but it does not correspond to a coherent risk measure.

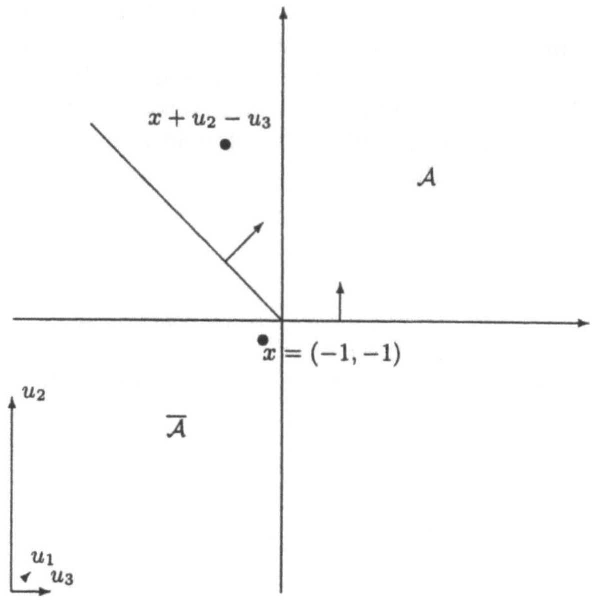

Fig. 4. Risk with multiple instruments. By buying u_2 and selling u_3, the decision maker can obtain an acceptable risk with zero cost

6 The construction of coherent risk measures

In the previous sections, it was assumed that the acceptable set $\mathcal{A}$ or the family of probability functions $\mathcal{P}$ were readily available to the decision-maker to measure risk. But the definition of $\mathcal{A}$ requires that the rays r^j, describing the cone, are determined by the agent. A task that sometimes can be difficult. Conversely, as far as the family of probabilities $\mathcal{P}$ is concerned, we can observe that economic agents usually describe future events by one probability function. Therefore $\mathcal{P}$ contains only one element and the risk measure is the expected value of the investment. Clearly, an unsatisfactory result. In the following section, it will be discussed how to deal with these arguments.

6.1 The construction of the risk measure by "standard" risks

Instead of requiring the definition of $\mathcal{A}$, an undirect definition is obtained by the measure of "standard", e.g. simple, risks and by stating a linear programming problem, that is a straightforward modification of the previous example.

The procedure is the following. Consider a set $\mathcal{B} = \{b^i | b^i \in \Re^n, i = 1, ..., I\}$ of representative risks: the decision-maker is required to give a measure $\rho_i \in \Re$ for every risk. For example, the 'standard' risk b^i can be defined as one having entries $b^i_j = 0$, if $j \neq i$, $b^i_i = 1$ otherwise. Then the decision-maker is required to compute the amount of liquidity that he is willing to receive for safe to make b^i acceptable. This amount is the risk measure ρ_i of a standard risk and it is a useful definition to measure the 'complex' risk $x \in \Re^n$. In fact, x can be compared to a portfolio composed by "standard" risks b^i and pure liquidity u. Let $y \in \Re^{I+1}$ be the array describing the amount of every risk and liquidity of the portfolio, if the inequality:

$$x \geq \sum_{i=1}^{I} y_i b^i + y_{I+1} u$$

is satisfied, then x is better than the portfolio in every state of nature. It is not necessary to know explicitly $\mathcal{A}$ to derive:

$$\rho_{\mathcal{A}}(x) \leq \rho_{\mathcal{A}}(\sum_{i=1}^{I} y_i b^i + y_{I+1} u)$$

Therefore, to compute $\rho_{\mathcal{A}}(x)$, it is sufficient to find the linear combination of standard risks and liquidity that supports x with minimum risk. That is, one has to solve:

$$\begin{aligned} \rho_{\mathcal{A}}(x) \;&=\; \min_y \textstyle\sum_{i=1}^{I} y_i \rho_i - y_{I+1} \\ s.t.& \\ & \textstyle\sum_{i=1}^{I} y_i b^i + y_{I+1} u \leq x \\ & y_i \text{ libere} \end{aligned}$$

It can be proved by linear programming properties that $\rho_{\mathcal{A}}(x)$, computed by this method, is a coherent risk measure. It is sufficient to observe that x is the right hand side of the constraint inequalities. Moreover, the underlying acceptable set $\mathcal{A}$ can be obtained by elementary row operations on the objective function and the constraints matrix. We need only to obtain an objective function where there is only one element different from zero, corresponding to $-y_{I+1}$. Then the columns of the matrix that correspond to the zero coefficients are the rays of $\mathcal{A}$.

6.2 Value at Risk and coherent risk measures

Value at Risk (VaR) has recently become a popular risk measure that is used by several banks and financial institutions. But, as it will be seen, it is not a coherent risk measure. Its drawbacks can be serious, since it underestimates the potential losses of investments. Due to its diffusion, it can be convenient to try to obtain a coherent risk measure by the same methodology that makes Var successful. That is, by the estimation of numbers that represent extremal quantiles of the probability distribution.

Some preliminary definitions are necessary. Remember that we are supposing that events are represented by a finite set Ω, $\omega_i \in \Omega$ and they are mutually exclusive. Let p_i be their respective probability. The economic result $x_i \in \Re$ correspond to event ω_i. Let $F(x) = \Pr[\xi \leq x]$ be the resulting cumulative distribution function. Observe that, since it is defined over a discrete set of events, $F(x)$ is not a continuous function.

Definition 5: *α-quantile* A real number α, such that $\alpha \in (0, 1)$, is given. Then $q \in \Re$ is an α-quantile if:

$$\Pr(\xi < q) \leq \alpha \leq \Pr(\xi \leq q)$$

There are other equivalent definitions of α-quantiles, but definition 5 can be easily drawn: see Figure 5.

Note that the cumulative distribution function is expressed by a set of pairs (q_i, α_i), such that: $\Pr(\xi \leq q_i) = \alpha_i = \Pr(\xi < q_{i+1})$. By definition 5, two cases result. If $\alpha_i < \alpha < \alpha_{i+1}$, then the '$\alpha$-quantile is the single number q_{i+1}. If $\alpha = \alpha_i$ for some i, then all the numbers between q_i and q_{i+1} are α-quantiles. This discussion shows that an α-quantile is a closed interval. Let q_α^- be the infimum value of the interval, let q_α^+ be the supremum.

Definition 6. *Value at Risk, VaR* A real number $\alpha \in (0, 1)$ is given, a random variable X and its cumulative distribution function $F(x)$ are given, then the Value at Risk is the negative of the α-quantile q_α^-. That is:

$$VaR_\alpha = -\min\{x | \Pr(X \leq x) \geq \alpha\}$$

In ADEH, Value at Risk is defined by q_α^+. There are no theoretical justifications to prefer one definition against the other, but, in our case, the graphic representation is more intuitive and the following analysis is simplified. First of all it is obtained:

Lemma 1: $\Pr(x \leq VaR_\alpha) \geq \alpha$.

According to the definition of ADEH only the strict inequality $\Pr(x \leq Var_\alpha) > \alpha$ can be obtained, but this provide some further complications.

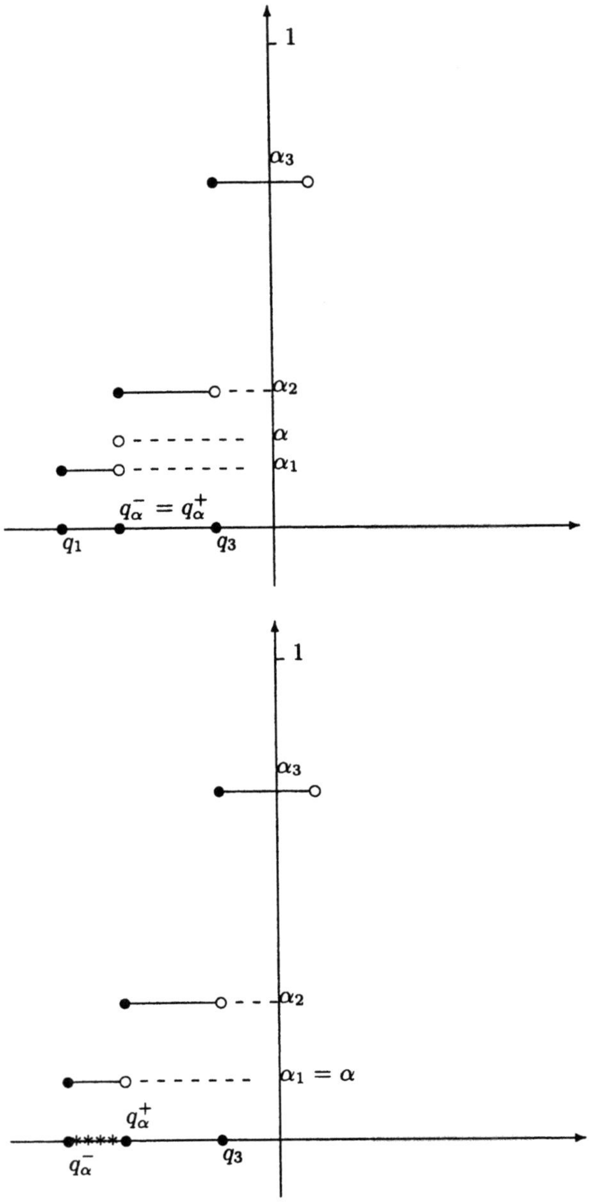

Fig. 5. Representation of α-quantiles. In the fisrt case $q_\alpha^- = q_\alpha^+$. In the second case, the gross line stand for the close interval (q_α^-, q_α^+)

As it was discussed, Value at Risk is not a coherent risk measure. The following example is taken from [1], but it is useful to explain the problems emerging from non-coherent measures.

Consider two binary options, both with the same expiring date T. The first option, denoted by A, for the price of u, will pays 1000 if the stock price in time T is greater than the price U and it will not pay anything otherwise. The second option, called B, for the price of l, will pay 1000 if the price of the underlying asset in time T is less than L, with $L < U$, it will pay nothing otherwise. Choose L and U such that: $\Pr(S < L) = \Pr(S > U) = 0.008$. Consider the case that an institution must write these options and must evaluate their Value at Risk, with $\alpha = 0.01$ If it decides once at the time, then it can compute the VaR of two A options, that is $-2u$. There is no risk, it writes the options. Then it computes the VaR of two B options, it is $-2l$ and there is no risk. But if the institution consider the emission of one A option and one B option, (that is the convex combination of two acceptable investments) the value at risk becomes very high: $1000 - l - u$. There is a big amount of risk that was not recovered by analyzing the single position.

The Value at Risk can be adjusted, in order to obtain a coherent risk measure.

Let $P(\omega)$ be the probability of the event $\omega \subset \Omega$. A method to generate a family of probabilities is by conditioning $P(\omega)$ to events A, to obtain $P(\omega|A)$, where A is a subset of events. It can be required that A can be restricted to satisfy some property, therefore $A \in \mathcal{A} \subset \mathcal{P}(\Omega)$, where $\mathcal{P}(\Omega)$ is the power set of Ω.

Since Ω is supposed to be a finite and discrete set, the following definition is correct:

Definition 5: *Worst conditional expected value.* A probability measure $P(\omega)$ over $\omega \in \Omega$, a family $\mathcal{A} \subset \mathcal{P}(\Omega)$ of events and a real value $\alpha \in (0,1)$ are given, then the worst conditional expected value is:

$$WCE_\alpha(X) = -\min\{E_P(X|A)|P(A) \geq \alpha, A \in \mathcal{A} \subset \mathcal{P}(\Omega)\}$$

Since $WCE_\alpha(X)$ is the minimum over a finite set of probability distribution, the following corollary can be obtained:

Corollary 1: $WCE_\alpha(X)$ is a coherent risk measure.

Observe that $WCE_\alpha(X)$ is defined without any mention to the cumulative distribution function. But most of the time, it is easier to work with probability function $F(x)$, rather than elementary probabilities $P(\omega)$.

Definition 8: *Tail expected value.* A probability distribution function $F(x)$ and a measure of Value at Risk, VaR_α, are given. Then, the tail expected value is:

$$TCE_\alpha(X) = -E_F\{X|X \leq VaR_\alpha\}$$

An interesting result, for its operational consequences, is the following:

Theorem 6: $TCE_\alpha(X)$ is a coherent risk measure.

Proof Let $A_l = \{\omega | x(\omega) \leq l, l \in \Re\}$.
We define $\mathcal{A}$ in such a way that, for every $A_l, A_m \in \mathcal{A}$, $l \neq m$, we have $A_l - A_m \neq \emptyset$ (note that $\mathcal{A}$ is a sequence of nested sets). Moreover, $\mathcal{A}$ is a finite set. By this restriction on $\mathcal{A}$, we have:

$$\begin{aligned} WCE_\alpha &= -\min\{E_P(X|A)|P(A) \geq \alpha, A \in \mathcal{A}\} \\ &= \min_l -E_P(x|x \leq l)|P(x \leq l) \geq \alpha\} \end{aligned}$$

By lemma 1, the minimum is taken exactly if $l = q_\alpha^-$. Then:

$$WCE_\alpha = -E_F\{x|x \leq VaR_\alpha\} = TCE_\alpha$$

◇

(In ADEH, only the relation $TCE_\alpha \leq WCE_\alpha$ was shown, for the different definition of VaR). From the operational point of view, a coherent risk measure can be quickly obtained by determining VaR_α first, and then by computing the conditional expected value of the inferior results. Clearly, estimation problems are left open.

7 Concluding remarks

In this paper, coherent risk measures are discussed in the framework of linear programming theory. Linear programming provides a general method to prove that a risk measure is coherent and it provides a method to compute it, therefore it can be claimed that it is the more appropriate methodology. We think that two research issues are interesting for future work. On one side some theoretical as well as empirical insights are needed to establish how scenarios must be constructed. On the other side, it is interesting to formulate portfolio problems that maximize returns under risk constrained by a coherent measure. The performance of such portfolios can be compared with other methods, such as MAD models, [4], or other stochastic programming models, [6].

Acknowledgements. The author wishes to acknowledge prof. Luca Erzegovesi, prof. Franco Molinari and dr. Enrico Moretto for their useful comments.

This research has been partly supported by M.U.R.S.T., under grants of the project "Problemi ed opzioni di politica economica in regimi monetari alternativi, con particolare riferimento all'Italia nella U.M.E.". This support gratefully acknowledged.

References

1. Artzner P., F.Delbaen, J.M. Eber, D. Heath "Coherent measures of risk", Working Paper, University of Zurich, 1998.
2. Gamba A., F.Rossi "Mean-variance-skewness analysis in Portfolio choice and capital markets" *Ricerca Operativa, the Journal of the Italian Society of Operations Research*, 81, 5-46, (1997)
3. Huber P. *"Robust statistics"*, Wiley, New York, 1981.
4. Konno H., H. Yamazaki, "Mean-absolute deviation portfolio optimization model and its application to Tokyo stock market", *Management Science*, **37**, 519-531, 1991.
5. Levy H.,M. Sarnat *"Portfolio and investment selection: Theory and Practice"* Prentice-Hall, New York, 1984.
6. Mulvey J.M., H. Vladimirou "Stochastic network programming for financial planning models" *Management Science* 38 (1992) 1643-1664.
7. Nemhauser G., L. Wolsey *"Integer and Combinatorial Optimization"*, John Wiley, New York, 1988.
8. Risk Special Supplement, *Value at Risk*, Risk Magazine, June 1996
9. RiskMetrics, *Technical Document*, Morgan Guarantee Trust Company, Global Research, New York (1995)
10. Simaan Y. "Portfolio selection and asset pricing: a three parameter framework" *Management Science*, **39**, 586-577, 1993.

Fuzzy Risk Management for Project Financing

Piera Mazzoleni

Institute of Econometrics and Mathematics for Economic Decisions
Catholic University
Largo A. Gemelli, 1
20123 Milan, Italy

Abstract. The new risk measures, economic in nature, allow us to deal with fuzziness in dynamic asset - liability models and give a powerful tool to manage Project Financing techniques.

The extended cash flow models are endowed with real options, representing the variety of prudential clauses, enriching the contracts: vagueness can therefore be introduced explicitly, by adding fuzzy risk parameters.

Key words: risk measures, project financing, fuzzy sets and orderings

1 Introduction

The traditional analysis of Project Financing is based on technical aspects and only recently attention has been devoted to economic and financial problems.

Then risk has become a crucial element to state the convenience of separate management under a Project Society (Shah - Thakor, 1987). But the complexity of the several phases and the very long time horizon add a vague and imprecise character and the risk parameter is made more flexible under fuzziness (Section 2).

The recent development of risk measures has introduced a new class of economic rather than statistical indices (Section 4)): fuzziness can be added in the optimization process to state both the objective function and the feasibility constraints and allows us to cope with the high uncertainty level, when measuring risk for very large and lasting projects (Section 5).

An example of the contribution of fuzzy set theory to project management is described in Section 6. The new risk measures allow us to overcome this approach and Section 7 presents a management model to control active and passive operations under solvability constraints both for the project life horizon and for every period, according to the decomposition procedure developed in Section 3.

But to reach an effective prudent management, which is typical of the public and social sectors, usually developed under the Project Financing procedure, it is worth adding suitable guarantees to protect the variety of contracts linked with the project.

Section 8 applies the real option theory in a fuzzy environment and offers an example from the Italian market to emphasize their powerful contribution.

2 Risk level and project financing

The essential nature of Project Financing is the either no recourse or the limited recourse to sponsors to overcome capital deficiencies and it may be read as a risk management process which transfers and insurers risks according to the economic environment.

The level of risk is the decision variable to state the possible separation of a project from the whole firm and its definition as a project financing autonomous unit (Mazzoleni, 1997). A risk measure is defined for long-term operations giving the potential effects of the default scenarios in the financial markets. The risk parameter θ associated with a project depends on the degree of information and the possible development of costly procedures to evaluate the project solvability. It depends also on the possible interaction of the project with the already existing activities for the firm. Then θ has to be introduced with an imprecise definition, that is, as a fuzzy variable and we will add a tilde, $\widetilde{\theta}$. Due to the very long period characterizing projects under examination, the probability concerning prices, quantities, returns and so on, are highly uncertain because there are limited data and experts disagree about the states of nature possibly occurring.

Suppose there are only two points in time and a single project. it requires an investment of I units at the beginning of the period $t = t_0$ and yields a gross return of R at the end of the period, $t = t_1$. The random variable R has a known probability density: it is assumed that the effects of vagueness and imprecision can be represented by introducing a fuzzy risk parameter $\widetilde{\theta}$, affecting the probability distribution, $f\left(R, \widetilde{\theta}\right)$.

Let αI be the amount of initial investment which is financed with debt, $(1 - \alpha) I$ the amount which is financed with shareholders'equity, K the payment in $t = t_1$ for every unit of project, the creditors lend in $t = t_0$, ρ the rate of return expected by creditors, τ the wealth tax rate.

Under conditions of imperfect information, the project sponsors within the firm know both the expected return $\overline{R}$ and the risk parameter $\widetilde{\theta}$ for their project, creditors know $\overline{R}$ but only the possible values for the fuzzy risk parameter $\widetilde{\theta}$.

Competition induces each creditor to design its offer in order to maximize the fuzzy utility of its borrower project and it is realized through a Pareto optimal equilibrium among creditors,

$$\max \quad E_S\left(\widetilde{\theta}\right) = I\alpha\left(\widetilde{\theta}\right)\rho\tau$$

under the fuzzy constraint

$$\begin{aligned} \int I\left[R-L\left(\widetilde{\theta}\right)\right] f\left(R,\widetilde{\theta}\right) dR + IL\left(\widetilde{\theta}\right) &\leq I\alpha\left(\widetilde{\theta}\right)\rho, \\ L\left(\widetilde{\theta}\right) &= \alpha\left(\widetilde{\theta}\right) K\left(\widetilde{\theta}\right) \\ 0 &\leq \alpha\left(\widetilde{\theta}\right) \leq 1 \end{aligned}$$

The objective function $E_S\left(\widetilde{\theta}\right)$ and the budget constraint are fuzzy sets with a corresponding menmbership function inherited by the risk parameter: then the optimal decision $\alpha^*\left(\widetilde{\theta}\right)$ is a fuzzy set with the same membership function, that can be interpreted as the degree to which each of the alternative indebtness quotas satisfies the goal and the constraint.

The density decomposition $f\left(R,\widetilde{\theta}\right) = f^0(R) + \widetilde{\theta} \otimes \left[f^1(R) - f^0(R)\right]$ emphasizes the role of parameter $\widetilde{\theta}$ and makes easier the construction of the overall membership function.

The two strongest implications of this result are that the riskiest investments will always use project financing and the indebtness is related with the imprecision of the project future development. Indeed, it is possible to find a threshold for $\widetilde{\theta}$ to change project management to Project Financing.

Project Financing is advantageous whenever it increases the project's net equity value.

If the new project is conventionally financed and incorporated within the firm, so that the firm's portfolio has $J+1$ projects after such incorporation, the total risk parameter $\theta_{J+1}(\delta)$ reacts to the new entry signal and the portfolio return is $P_{J+1} = P_J + IR$.

Let us denote $E\left[\theta_{J+1}(\delta), \widetilde{\theta}\right]$ the incremental expected net cash flow from the new project and suppose it is fuzzy.

Now suppose the project is set up as a legally separate entity. Then only the project's risk parameter matters and the net end-of-period expected cash flow from the new project is $E_S\left(\widetilde{\theta}\right) = I\alpha^*\left(\widetilde{\theta}\right)\rho_T$.

We can now compare $E_S\left(\widetilde{\theta}\right)$ and $E\left[\theta_{J+1}(\delta), \widetilde{\theta}\right]$.

Let us briefly present a possible criterion to order L-fuzzy numbers (Furukawa, 1997).

Let L be a function from $\Re$ to $\Re$ satisfying conditions: $L(x) = L(-x)$ for any $x \in \Re$; $L(x) = 1$ if and only if $x = 0$; $L(\cdot)$ is nondecreasing on $[0, +\infty)$.

If we set $x_0 = \inf\{x > 0 : L(x) = 0\}$, we have $x_0 \in [0, +\infty)$ and L defines the shape function with zero point x_0. Introduce now any real number m and any positive number α :

DEFINITION 1. We call *L- fuzzy number* a fuzzy number $\widetilde{X}$ whose membership function is given by

$$\mu_{\widetilde{X}} = L\left(\frac{x-m}{\alpha}\right) \vee 0, x \in \Re$$

where $a \vee b = \max\{a, b\}$: therefore, m is the center, α the deviation parameter for $\widetilde{X}$.

A *parametric order relation* is stated for two L-fuzzy numbers $\widetilde{X} = (m, \alpha)_L$, $\widetilde{Y} = (n, b)_L$ as function of parameter $\lambda \in [0, 1]$ by setting

$$\widetilde{X} \preceq_\lambda \widetilde{Y} \Leftrightarrow \begin{array}{c} x_0 \, |\alpha - b| \leq n - m \text{ or} \\ \lambda x_0 \, |\alpha - b| \leq n - m < x_0 \, |\alpha - b| \quad \text{or} \\ |n - m| < \lambda x_0 \, |\alpha - b| \text{ and } b > \alpha \end{array} :$$

for instance

$$\begin{aligned} \widetilde{X} &\preceq {}_0\widetilde{Y} \Leftrightarrow m \leq n \\ \widetilde{X} &\preceq {}_1\widetilde{Y} \Leftrightarrow a_0 \, |\alpha - b| \leq n - m \end{aligned}$$

For every shape function L and each $\lambda \in [0, 1]$ the relation $\preceq_\lambda$ induces a total ordering on the family of all $L-$fuzzy symmetric numbers.

Such a parametric total order relation allows us to compare the net end-of-period expected cash flows accruing to the shareholders, for the single project $E_S\left(\widetilde{\theta}\right) = (m_\theta, a_\theta)$ and for the incorporated project.$E\left[\theta_{J+1}(\delta), \widetilde{\theta}\right] = (n_\theta, b_\theta)$, as L-fuzzy numbers.

Take for instance $\lambda = 0$. Then we suppose suitable assumptions are verified so that a level of the fuzzy risk parameter $\widetilde{\theta}^*$ exists, such that $m_{\widetilde{\theta}^*} = n_{\widetilde{\theta}^*}$ and it results to be convenient to separate the project for $\widetilde{\theta} > \widetilde{\theta}^*$.

If the rate of increase of the expected cash flow from the project is sufficiently higher than the incorporated one, according to the kind of comparison chosen with parameter λ, we have a possible intersection of the two fuzzy functions of the risk parameter $\widetilde{\theta}$ and this enables us to choose the most efficeint strategy, according to the membership function stating the degree of risk evaluation by the Project Manager.

3 The economic-financial evaluation of the project and period decomposition

A very risky project cannot be evaluated in a precise way, especially because the capability to reimburse debts relies only on the possible future returns. Moreover, the leverage decisions have to be taken in advance, only on the basis of the Project Manager feeling of the future returns, in an imprecise and subjective way. Both cash flows and interest rates have then to be treated as fuzzy numbers.

Intervals are the basic representation of fuzzy numbers and for them it is immediate to define center and deviation to give the corresponding L-function.

Let us refer to a fundamental result, that allows us to suitably transform a capital into its future and present value: any continuous and strictly monotonic transformation of invertible fuzzy intervals is also an invertible fuzzy interval, keeping nice properties.

Suppose that $\widetilde{C}$ and $\widetilde{i}$ be compact fuzzy intervals and let the discount factor $g\left(t,\widetilde{i}\right)$ be a continuous and strictly decreasing function. The present value $V = \widetilde{C} \otimes g\left(t,\widetilde{i}\right)$ is an invertible fuzzy interval (Li Calzi, 1990).

The model is easily generalized to multiple flows for dynamic problems. Let us introduce a collection of compact fuzzy intervals $\widetilde{\mathcal{I}}=\left(\widetilde{i}_1, ..., \widetilde{i}_n\right)$ for the interest rates: the present value of the income flows' vector:
$\widetilde{C} = ((\widetilde{c}_s, t_s), s = 1, ..., n)$,
$\widetilde{V} = \sum_{s=1}^{n} \widetilde{c}_s \otimes g_s\left(t,\widetilde{i}_s\right)$,
is an invertible fuzzy number.

Such a value measures the rentability on a market basis rather than by comparing debt versus equity capital and we better refer to an internal indicator. If we use an annual rate $\widetilde{x}^*$, it has to solve equation $V(\widetilde{x}^*) = 0$; otherwise we apply an actualization factor

$$V\left(\widetilde{\Psi}\right) = \sum_{s=1}^{n} \widetilde{c}_s \otimes \widetilde{\Psi}(t_s, 0) = 0$$

and $\widetilde{\Psi}$ measures the project financial law. This approach allows us to compare the project with the market standard, not only on the final horizon, but also period by period.

Rentability and solvability problems can be treated in a global setting on the whole planning period, but useful information can be obtained by a period risk-return evaluation. Then it is worth decomposing the project present value and calculate the periods' outstanding (Peccati, 1991).

If we take simultaneously into account the income flows from the project and the loan service, it is possible to define the outstanding w_s for each period, $s = 0, 1, ...n$, in order to control the degree of leverage and the mutual influence of new investments and loans.

Denote $\rho_1, ...\rho_n$ the opportunity cost for the investors' own capital. The return on capital is given by

$$G_1\left(\widetilde{\Psi}\right) = \sum_{s=0}^{n} \widetilde{\gamma}_s, \widetilde{\gamma}_s = w_s \otimes \widetilde{\Psi}(s, 0) \otimes (\widetilde{x}_s^* - \rho_s)$$

Suppose now the project has been financed with debt capital and reimbursed by flows $C_1, ..., C_n$ under the debt rates $\delta_1, ..., \delta_n$. The recurrence relation on debt exposure is set according to the recursive relation

$$D_s = D_{s-1}(1+\delta_s) - C_s$$

The ratios $\alpha_{s+1} = \frac{D_s}{w_s}$ measure the quota of the project financed with the debt capital and give the leverage ratio in any period. Amount $(1-\alpha_s)\,w_{s-1}$ is the outstanding quota which has been financed with internal capital, $\alpha_s w_{s-1}$ is the one financed with the debt capital.

The period flow contribution can be split into

$$\widetilde{\gamma}_s = w_{s-1} \otimes \widetilde{\Psi}(s,0) \otimes \left[(1-\alpha_s)\left(\widetilde{x}_s^* - \rho_s\right) \oplus \alpha_s\left(\widetilde{x}_s^* - \rho_s\right)\right]$$

where $\widetilde{c}_s = w_{s-1}\,(1 \oplus \widetilde{x}_s^*)$.

For any possible leverage decision α_s, $\widetilde{\gamma}_s$ is a fuzzy interval endowed with its membership function, allowing us to read the period outstanding in a vague and imprecise way.

4 From technical risks to economic and financial risks

Any project faces several kinds of risks, different in nature, ranging from construction to market risks. Some risks are of actuarial type, due to the different kinds of damage during the development of the project; others are political, legal, technological and environmental, due to the country and to the technical characteristics of the project.

First of all, we need *classify* the different kinds according to the operational sector. We might recognize the default risk for which we cannot apply the diversification approach; the market risk, due to interest and exchange rates volatility, affecting both indebtness and rentability; the solvability risk, referring to the possible mismatching for realization phases and payments, unexpected costs, running defects and, what is even more relevant, credit risk.

As a second step, we need *measure* the risks facing adverse events: indeed, we can provide the ruin probability and we can evaluate the sensitivity of the estimated probability distributions for adverse events to the market dynamics.

An indirect estimate can also be obtained by the rating tables, according to the possible transition from lower to higher quality classes, by a suitable improvement of risk management.

From a financial viewpoint, interest rate and credit risks are fundamental. In the traditional approach, interest rate risk is measured by the duration analysis, which takes the economic value of the activities explicitly into account, and computes the exposure of the net income to the risk coming from differential changes in the rate conditions applied to active and passive operations.

The credit risk mainly concerns the banking side. The interest rate risk heavily influences the credit risk itself and the banks evaluate particular economic and financial indices in order to verify not only the profitability, but also the riskiness of the project.

Denote r the minimum return on capital, expected by the investors, τ the tax rate of the Project Society, CP the project's capital.

Suppose that the project's cash flows PCF_k for any year $k = 1, ..., N$ be known with certainty, while the annual payment varies randomly according to the interest rates. Denote D_0 the original amount of debt and let C_k be the periodic capital repayment. Whenever the Project Society would like to guarantee both the repayment service and the minimum return on capital, the maximum annual rate on debt is given by ratio

$$J^*_{k-1} = \max J_{k-1} = \frac{(PCF_k - C_k)(1-\tau) - rCP}{D_{k-1}(1-\tau)}$$

and in correspondence the maximum variable rate sustainable by the project is given by $TV_{k-1} = J^*_{k-1}$−spread. The Project Society compares this affordable rate and the market ones, in order to state its repayment capability.

As a third step, we are finally led to the risk control. We could find a coverage in the insurance market for physical risk of damages to machineries or plants, during the building and inspection processes and for responsibility damages to workers. Therefore, a very accurate reliability analysis has to be developed both in the starting and in the operating periods in order to properly estimate the project reliability. The current literature provides a very deep insight in reliability techniques, but the practice usually emphasizes their impact in terms of functioning costs of the system, rather than embedding risk analysis in the project evolution, explicitly.

We are now going to present a possible approach to the risk measures, leaving the optimization approch for the project control to Section 7.

5 Fuzzy risk measures

The banking system is very sensitive to the credit risk and a variety of more or less sophisticated measures has been developed. One line of development is of statistical type: a very known example is given by the Value-at-Risk. Nevertheless it exhibits some drawbacks and the particular success encountered has promoted a deep analysis of its coherence.

Let us briefly remind the axioms proposed by (Artzner et al., 1998) to characterize the minimal set of properties that a risk measure has to verify:

A1. A risk measure $\rho(X)$ should verify inequalities,
$E[X] \leq \rho(X) \leq \max(X)$;

A2. Scale and transition invariance, $\rho(aX + b) = a\rho(X) + b, a \geq 0, b \geq 0$;

A3. If $X = b$ with probability 1, then $\rho(X) = b, b \geq 0$;

A4. Sub-additivity: $\rho(X+Y) \leq \rho(X) + \rho(Y)$

In correspondence several new risk measures have been proposed. But it is worth introducing a different point of view, based on the project economic evaluation. Indeed, a very basic definition of risk measure is: the risk of an economic setting is the degree with which the future values of a project belong to a risk freasible set, assigned by a supervisor. This approach is particularly useful to deal with the uncertainty in the risk definition.

Denote X the risk under examination, r the reference return of a prudential investment.

The risk measure is defined as the *expected losses*, that is, as the supremum of the negative expected value for the final wealth under the examined scenarios,

$$\rho(X) = \sup_{P \in \mathcal{P}} E_{\mathcal{P}}\left(-\frac{X}{r}\right)$$

$\rho(X)$ might be read as the worst case approach within the scenarios' collection provided by the decision maker.

Another way to set this measure is the *effort* to bring the current situation within the feasible set $\mathcal{G}$ for risks,

$$\inf\{m : mr + X \in \mathcal{G}\}$$

This approach allows us to deal with the vague and imprecise description of feasibility and with the dynamic setting, which characterize the development of Project Financing.

During the early starting phase of the project, a target level is assigned to every development step and we can say that there is a failure, if a certain phase is not completed within the assigned standard and an additional investment is required.

It is worth defining suitable reliability measures, both for technical and for economic purposes. In the Project Financing Literature the reliability measures are usually translated into non functioning costs for the system. But the theory of martingales and semimartingales allows us to give a risk-neutral measure for the project completion relative to the information process. From an operational viewpoint this can be managed by allowing suitable risk equivalent premia as fuzzy intervals, according to the reliability level. Then the feasible set $\mathcal{G}$ of risks can be thought of as a collection of fuzzy quantities, measuring the economic impact of the reliability constraints.

The analysis developed in section 2 has shown the decomposition approach, which allows us to recognize the solvability constraint over the single periods, by comparing the current outstanding with a period prudential investment.

We can conclude that:

- the new risk measures operate in monetary rather than in statistical terms, thus leading to a fuzzy formalization of the solvability constraints;
- the period decomposition of the outstanding allows us to take solvability constraints both for the whole project life,

$$\inf\left\{\widetilde{m}_w : \widetilde{m}_w r + \widetilde{X}_w \in \mathcal{G}\right\}$$

and period by period

$$\inf\left\{\widetilde{m}_p : \widetilde{m}_p r + \widetilde{X}_p \in \mathcal{G}\right\}$$

6 The contribution of fuzzy graph theory

Graph theory offers a powerful tool to plan, coordinate and control project's development. The activity set can be represented by an acyclic network and priorities are given as precedence relations.

Classical techniques as CPM and PERT are applied under certainty conditions, but complete information is available only when activities are actually completed and their characteristics can often be presented only in qualitative terms.

Indeed, the description of the early phase and of the development of a Project Financing operation has shown the variety of aspects to be considered, the objectives' interaction and the environment influence, the evolution of the economic conditions, in which the project is placed.

A substantial help to deal with all these problems simultaneously and to formalize the imprecise and varying conditions, is given by the fuzzy set theory, leading to an effective solvability.

A mathematical programming model for project scheduling is found in (Hapke et al., 1994):

$\mathcal{Z}$ is the set of n activities $(\mathcal{Z}_1, ..., \mathcal{Z}_n)$, requiring discrete amounts of resources $(r_{1j}, ..., r_{pj})$. For each activity $\mathcal{Z}_j$ fuzzy numbers give duration $\widetilde{p}_j$, ready time $\widetilde{a}_j$, due date $\widetilde{d}_j$; $B_j$denotes the collection of all the immediate predecessors of activity $\mathcal{Z}_j$.

The decision variables are zero–one:

$$x_{jt} = \{1 \text{ if activity } \mathcal{Z}_j \text{ is completed in period } t; 0 \text{ otherwise}\}$$

The aim of the scheduling problem is either to minimize the project completion time

$$\min \widetilde{C}_{\max} = \sum_{t=\widetilde{e}_n}^{\widetilde{l}_n} t \cdot x_{nt}$$

or to minimize the maximum lateness

$$\min \widetilde{\mathcal{L}}_{\max} = \max_{j} \left\{ 0, \sum_{t=\widetilde{e}_n}^{\widetilde{\ell}_n} \left(t \cdot x_{nt} - \widetilde{d}_j \right) \right\}$$

where $\widetilde{e}_j$ denotes the early finish completion period and $\widetilde{\ell}_j$ is the late finish completion period.

Constraints are set:

- on the precedence relationships

$$\sum_{t=\widetilde{e}_j}^{\widetilde{\ell}_j} (t - \widetilde{p}_j) \cdot x_{jt} - \sum_{t=\widetilde{e}_f}^{\widetilde{\ell}_{fn}} t \cdot x_{ft} \geq 0 \text{ for any } f \in B;$$

- on renewable resource availability,

$$\sum_{j=1}^{n} \sum_{q=t}^{t+\widetilde{p}_j-1} r_{jk} \cdot x_{jq} \leq N_{kt}, k = 1, ..., p, t = 1, ..., T$$

N_k being the available resource.

The solution procedure is an approximate one and leads back to a deterministic problem to which known algorithms can be applied.

Indeed, a decision-maker can very rarely describe the fuzzy numbers precisely. For any cut-level ξ a lower and an upper bounds are given, so that the corresponding problems are no longer fuzzy. For the activity duration, the lower bound represents the optimistic estimate, the upper bound represents the pessimistic one; for the programmed dates, the upper bound states the optimistic level, the lower bound gives the pessimistic one. Any fuzzy number is then assigned by six real numbers.

But the risk management for Project Financing requires an overall control for the active and passive operations by applying an asset - liability management model in a dynamic setting.

7 Asset liability management in a fuzzy environment

The recent development of the analysis on risk measures and their coherence has led to introduce new tools for the economic evaluation of projects.

Let the risk X/r be represented as the " Accumulated Penalized Shortfalls with respect to a Prudential Target", no longer function of the subjective attitude of the decision - maker, but depending on the market scenarios and on the technical sector of the project.

Moreover, it would be worth adding the influence of all the penalties on the final project result to the global penalty, and the objective function is set equal to

$$\max E \left[\text{Final Wealth - } \frac{\text{Accumulated Penalized Shortfalls-Final Penalty}}{\text{Risk Tolerance}} \right]$$

One of the crucial problems is the scenarios' generation: indeed, several exogeneous parameters affect the project realization and development. Think for instance of wage and plant costs, inflation rate, bonds return and loans' rates.

The dynamic development has to be flexible enough to adjust itself to the economic changes and the multistage dynamic programming will fit it.

We will endow any variable with two indices, one for the time period, t, the other one for the state, s.

The tree structure for the states will represent the time evolution of information and states' interaction and all the nodes will have only one predecessor, but several successors.

States will characterize one of the two possible systems:

- increase/decrease,
- pessimism/optimism/highest probability.

The discrete distribution of the future states will associate to each of them a degree of reliability, that is the corresponding value of the membership function.

The Project Society starts with a bounded capital exposition Y_0,

$$Y_0^{\ell} \leq Y_0 \leq Y_0^{u}$$

At time t=0, the debt capital will be allocated between project quotas L_0, such as plants, workers, structures, and intermediate investments $X(I)$, under the current level I of information,

$$Y_0 - L_0 = \sum_{i} X_{i0}(I_0)$$

subject to suitable bounds

$$x_{i0}^{\ell}(Y_0 - L_0) \leq X_{i0} \leq x_{i0}^{u}(Y_0 - L_0)$$

At the beginning of each period t the Project Society updates decisions: the returns on investments are added

$$A_t(I_t) = \sum_{i} h_{it} X_{i,t-1}(I_{t-1})$$

The debt fraction $L_t = \alpha Y_t$, $0 < \alpha < 1$ is devoted to the project development and additional costs $Z_t(I_t)$ are paid in order to maintain the assigned project standard

$$A_t(I_t) + Y_t - L_t - Z_t(I_t) = \sum_i X_{it}(I_t)$$

Future solvability will be guaranteed by a constraint on all the future obligations,

$$\sum_{k=t+1}^{n} \left(A_k(I_k) + Y_k - L_k - Z_k(I_k)\right) \geq 0$$

The decomposition approach, described in section 2, allows us to guarantee not only the global solvability on the planning horizon, but also a period solvability for a more prudent management, when a debt capital is added.

From one hand the overall objective function guarantees the global planning evaluation, on the other hand, the recursive setting allow the management to adapt the decisions to the information improvement.

The typical backward approach of dynamic programming gives the basis for the solution procedure (Ziemba - Mulvey, 1998).

8 Project guarantees for fuzzy contracts

The expected discounted cash flow, denoted briefly DCF, is no longer an acceptable evaluation tool, when projects allow the management to add discretionary elements depending on future decisions: the giving up of the project, the choice of the starting date, the change of the production rate. Several guarantee clauses are introduced to manage the complexity of the contracts' structure and to bound the possible losses from below. We recognize several kinds of behaviour that can be formalized as real options.

Indeed, any bounding process can be formalized by applying real options, with no reference to the existing underlying asset. Suppose X be a random variable, taking any value: if we constrain it to fall below threshold A, we can value the "insured" amount as

$$V = \min\{X, A\} = X - \max\{0, X - A\}$$

that is, as the observed value, net of a call option with strike price A. Such a constraint is typical of the price level.

Analogously, for a lower bound B, often used to fix the quality level, the "insured" amount can be set as

$$V = \max\{X, B\} = B + \max\{0, X - B\}$$

and the addition of the constraint can be read as a call option with strike price B.

The corresponding DCF's distribution becomes clearly asymmetric,

$$\text{extended DCF} = \text{basic DCF} + \text{REAL OPTIONS}$$

and nonlinearities are admitted in the evaluation process.

Risk management techniques are mainly devoted to recognize and control the several risk sources. The main guarantee for the Project Financing is given by the cash flows, but banks require a formal definition of contracts, endowed with several implicit and explicit guarantees. Indeed, Project Financing itself can be defined as a coordination of contracts and real options give a suitable evaluation model. Typical are the concession contract, the construction contract and the trading contract.

In this paper we will analyze the construction contract, covering construction, planning and production capacity risks. A guaranteed form, the so called " Lump Sum Turn Key" contract, states the constraints for the time to deliver, for the product quality and for the maximum price.

The construction cost C is very difficult to be estimated, summing up a range of very different costs, from workmanship costs to raw material costs.

The underlying asset is represented by the construction cost which is determined by a complex and uncertain process and which is therefore described by a functional dependence of the distribution on a fuzzy risk parameter $\widetilde{\theta}$, $C\left(\widetilde{\theta}\right)$. If X denotes the price cap on the construction cost, fixed by the contract, the option value is

$$\max\left\{C\left(\widetilde{\theta}\right) - X, 0\right\}$$

and it inherits the representative power of the fuzzy risk parameter $\widetilde{\theta}$, described in section 1, according to the future perception of the Project Manager.

An example from the Italian market

We consider a project to transform waste goods into electric power and analyze the influence of the key variables in the guarantees on the financial ratios.

In the construction contract, we recognize a call option with strike price equal to the cost fixed by the contract and an the underlying asset given by the actual cost to implement the project. The Black-Scholes formula is applied to the real option, assuming variance $\sigma^2 = 0.03447$, risk-free rate $r = 0.03$, underlying asset $S = 93.500$, strike price $X = S$ and residual life $\tau = 3.25$,

$$C(S, \tau, X) = 28,389.75972$$

and the cost to add the guarantee, compared with the initial investment is around 30%.

But the evaluation of the construction costs is highly uncertain. We then suppose that the fuzzy risk influence may arrive up to a 20% increase in the construction cost.

In correspondence, we face an increase in the internal rate of debt cost from 8,51% for the basic project up to 11.80%, a decrease in the internal rate of return from 12.92% down to 7.14%.

In correspondence, the Loan Life Cove Ratio, as the ratio of the present value of the project cash flows over the outstanding debt, falls below the safety threshold 1.10:1.

This example shows the role of the manager perception of uncertainty, which allows us to keep or not to keep the project within the feasibility bounds.

(*) This paper has been partially financed by the Italian Ministry of University and Scientific Research and by the National Research Council

References

P.Artzner - F.Delbaen - J.M.Eber - D.Heath, Coherent measures of risk, *Mathematical Finance*, vol.9/3, july 1999, pp.203-228

M.Li Calzi, Towards a general setting for the fuzzy mathematics of finance, *Fuzzy Sets and Systems*, vol.35, 1990, pp.265-280

N.Furukawa, Parametric orders on fuzzy numbers and their roles in fuzzy optimization problems, *Optimization*, vol.40, 1997, 171-192

M.Hapke - A.Jaszkiewicz - R.Slowinski, Fuzzy project scheduling system for software development, *Fuzzy Sets and Systems*, vol.67, 1994, pp.101-117

P.Mazzoleni, Sulla gestione del rischio nel Project Financing, *Ingegneria Economica*, nn.62-63, 1997

P.Mazzoleni, Rischi economici e finanziari e loro copertura, in Proceed. of the Workshop on "Risk analysis - a global approach, Bocconi University, april 17, 2000

L.Peccati, Valutazione analitica e sintetica di attività finanziarie, *Rivista Milanese di Economia, vol.21, 1991*

S.Shah - A.V.Thakor, Optimal capital structure and Project Financing, *Journal of Economic Theory*, vol.42, 1987, pp.209-243

W.T.Ziemba - J.M.Mulvey, World Wide Asset Liability Modeling, Cambridge University Press, 1998

Fuzzy Logic Based Risk Management in Financial Intermediation

Konstantin Korolev, Kai D. Leifert and Heinrich Rommelfanger

Institut für Statistik und Mathematik
Fachbereich Wirtschaftswissenschaften
Johann Wolfgang Goethe Universität
Merton-Str. 17–23
D-60054 Frankfurt, Germany
E-mail: `rommelfanger@wiwi.uni-frankfurt.de`

Abstract. In an environment of uncertainty the financial intermediary facilitates an optimal allocation of resources in the national economy by conducting the efficient risk-pooling and risk-sharing. Our analysis explores the aspect of the risk management and control of the intermediary's portfolio, which provides the nation with a saver, more flexible and more stable financial system. In the presence of asymmetrical information the exact valuation of the financial intermediary's portfolio becomes a highly complicated and expensive process for the supervisory institution. We point out that, the risk-management of a financial intermediary is strongly correlated and often explained by the pricing of financial assets and contingent claims. In the present paper the foundations of option pricing theory are extended by using the theory of fuzzy sets. The introduction of fuzzy asset prices weakens the assumption, that the expectations of all possible future payoffs have to correspond for every investor on the one hand and furthermore allows us, to know less of the underlying ex ante. However, we show that the application of fuzzy set theory in finance can be handled. In addition, fuzzy set theory enables to manage the risk more sophisticated, the different sources of risk can be subtly differentiated.

Keywords: Fuzzy Valued Assets, Fuzzy-Logic, Arbitrage Theory, Options, Brownian Motion.

1 Introduction

Nowadays the term "financial intermediary" is used in a very broad sense. According to this, a financial intermediary is an institution, which produces and provides informations on financial markets, so that this term includes such companies as rating agencies or broker houses. We concentrate our analysis on a specific economic function rather than an institutional structure. In the present paper, we develop a continuous-time model for controlling and supervising the risk-pooling and the risk-sharing functions of the intermediaries in an environment of uncertainty, which facilitates an optimal allocation of resources and provides the national economy with a save, stable and flexible financial system.

The rapidly growing financial, commodity and catastrophe risk markets casted doubt upon the stability of the financial system as a whole. It became possible for the market participants to take positions in derivative securities and, with that, to change the risk exposure of equity without holding and investing money in underlying assets. Since the great disaster of the German Metallgesellschaft in hedging oil futures contracts or the surprising bankruptcy of the British Barings bank due to speculation losses of Nick Leeson, voices became louder to improve the monitoring abilities of the supervising institutions, in order to determine the real portfolio risk of financial intermediaries. In this paper we analyse difficulties of risk management and control of the intermediary's portfolio.

On the one hand, we believe, that the insufficient theoretical financial models are the reason of mistrading and unperfect hedging. Fischer Black and Myron Scholes [2] have - without question - revolutionized finance theory and developed some very powerful methods of pricing and hedging options. Nevertheless they could never foresee, that the trading volume on the markets would increase so rapidly in a few years and therefore could not anticipate problems resulting from that. Additionally, today's financial experts use these models in a way, that does not reveal the original meaning and is interpreted theoretically uncorrect. This leads to one of our major critics, namely that the market participants use the Black-Scholes model without discussing the assumptions of this model. Therefore, it is not surprising, that their results are often unreasonable or unsatisfying.

In the well known approach of Fischer Black and Myron Scholes the whole uncertainty of economy is reduced to only one parameter, the volatility. It should be estimated as a constant value from the available data. This procedure has led to inconsistencies in option pricing known as smile effect or term structure phenomenon. Recently, different authors investigate models with stochastic volatility or other parameters to explain such observable patterns.

On the other hand, the presence of information asymmetry makes the exact valuation of the financial intermediary's portfolio to a highly complicated and expensive process for the supervisory institution. Because of the information asymmetry there exist a principal-agent relationship between the financial intermediary and the supervising institution. The important implications of these relationship are welfare reductions, because the estimates of the assets values are biased. The significant errors of estimation result in additional risks for the financial intermediary as well as for the supervising institution. The traditional models do not take those risks into consideration, because the markets are assumed to be perfect.

Describing implied risks more exactly would - without question - increase the stability of the whole financial sector. Moreover it becomes possible to obtain one instrument, which could fit to a wide variety of different users just by selecting their own optimal preferences. Since uncertainty is only described by different states of nature in almost every scientific research nowadays, it remains uncertain in reality, which exact consequences one state would obey. The approach in this paper includes this uncertainty also.

In the first part of this paper we develop an alternative in handling uncertainty. In particular, for managing the risk more efficiently we distinguish two sources of risk. Firstly, we assume volatility itself as a given observable constant. Furthermore, we suggest the use of fuzzy future expectations of market participants to explain additional riskiness in options, which couldn't be explained by changes of volatility yet. In contradiction to common models of option pricing theory the assumption of homogeneous expectations of future payoffs is neglected. Instead we assume that every market participant contributes to a set of possible future payoffs. This modification weakens the necessary condition in neoclassical financial models to a considerable extent and makes the financial theory more realistic.

An implementation of fuzzy set theory in finance was established by few authors yet. A good overview of fuzzy applications in the modern financial theory is given in the collected volume, edited by Ribeiro et al. [14], where different fuzzy models are presented to support portfolio selection, creditability checks of small business firms and risk classification.

Since principles of the present article can be extended to other similar aspects of financial theory easily, those ones selected for this paper describe the foundations of fuzzy set theory in finance – going back to [10] – as well as the logical economic background and the explanation of the use of such new concepts. First we will show, how the modern finance theory can be extended by using the fuzzy logic and explain the most important definitions, necessary to understand the further path of science. Afterwards we will turn to the supervising problems of a financial intermediary. We will see that fuzzy option pricing allows the monitoring institution to evaluate the risk-management ability of the market participant and therefore can e.g. guarantee liquidity of the bank more exactly. Furthermore, we will explain that the existence of an underlying described by a fuzzy set allows a more realistic view of the world.

While a wide discussion takes place in the literature, if either stochastic calculus or fuzzy set theory is more efficient to model risk, we do not restrict ourselves to one concept but combine both notions in a single framework. After a brief presentation of basic definitions we discuss time continuous option pricing in section 3. The relevance of this chapter becomes obvious, if we turn our sight to more complex financial problems, just like the rate of return on deposits. This is especially discussed in section 4 afterwards. The paper closes with a short conclusion.

2 Preliminaries

In neoclassical financial theory an asset price is a crisp expected value of future payoffs. Therefore, everyone in the economy has to agree to the same set of crisp prices. Uncertain is only which payoff will be realized, depending on the observed states of financial markets. As described above, we propose to use Fuzzy Valued Assets (FVA) here, because it is a very strict and unrealistic assumption that all participants have the same information and consequently the same crisp expected values of future payoffs. Keep in mind, that this was the most critical argument against Merton's theory [12] of the optimal rate of return on banks' deposits. Therefore we will describe the future payoffs by fuzzy sets.

Definition 1. A FVA $\tilde{S}_t$, i.e. a future payoff dependent on a state of nature, is defined as a fuzzy set on the crisp payoff scala.

Remark. For simplicity we assume in this paper, that $\tilde{S}_t$ is a fuzzy number of the LR-type [15], i.e. $\tilde{S}_t$ can be described by $\tilde{S}_t = (S_t; \underline{S}; \overline{S})_{LR}$. Furthermore from Definition 1 follows, that everyone in economy has to agree to a certain set of fuzzy membership functions of future payoffs, but not to real numbers.

On the one hand, risk is now modeled by the uncertainty that different states may occur and on the other hand by fuzzy values presenting the future payoffs. In comparison to neoclassical option pricing theory we are able to manage uncertainty, which covers a wider range of possible outcomes.

In early papers on option pricing and modern finance theory, authors have shown the relevance of completeness and arbitrage on markets. So it is easy to see, that we have to extend these main definitions for FVA's, too.

Definition 2. [10] If there exists for every asset $j \in \{1, 2, \dots, K\}$ a duplicating portfolio, s.t.

$$\sum_{k=1}^{K} \tilde{x}_{sk} n_k = \tilde{x}_{s,j} \quad s = 1, 2, \dots, S;\ j \neq k,$$

where S is the number of states of financial markets,
K is the number of assets in the economy,
$\tilde{x}_{sk}$ is the fuzzy payoff of asset k if the state s occurs,
n_k is the number of asset k in portfolio.

Moreover, for a given ρ-level, $\rho \in [0, 1]$, the ρ-level sets of the payoffs $\tilde{x}_{sk}$ are disjunct sets. This means

$$\tilde{x}_{ik} \succ_\rho \tilde{x}_{sk} \text{ or } \tilde{x}_{ik} \prec_\rho \tilde{x}_{sk} \text{ with } i, s \in \{1, 2, \dots, S\} \text{ and } i \neq s. \tag{1}$$

Then we call a market to be "complete on level ρ".

Definition 3. [10] Let $\mathbf{p} = (\tilde{p}_1, \tilde{p}_2, \ldots, \tilde{p}_K)$ be a known vector of asset prices and let $\tilde{x}_{s,K+1}$ with

$$\tilde{x}_{s,K+1} = \sum_{k=1}^{K} \tilde{x}_{sk} n_k \text{ for } s = 1, 2, \ldots, S. \tag{2}$$

be a duplicating portfolio with an asset price of $\tilde{p}_{K+1} = \sum_{k=1}^{K} n_k \tilde{p}_k$.
Then we call a portfolio "fuzzy arbitrage free", if

$$\sup \operatorname{supp}(\tilde{p}_{K+1}) > 0 \text{ or } \inf \operatorname{supp}(\tilde{x}_{s,K+1}) < 0 \text{ for all } s = 1, 2, \ldots, S.$$

Being aware of those definitions, we now turn towards the pricing of FVA's. The problem we face here is the question, how to handle fuzziness in our framework such that prices will not become fuzzier in every step of the mathematical proof. This can be achieved by the introduction of an asset without spreads. It is easy to show, that this asset allows us, to avoid further fuzzyness during normalization of asset processes. Therefore, we assume

Assumption 1 *There exists a risk-free interest rate without fuzziness, which is constant through time and known with certainty*[1].

In order to develop a model we furthermore consider:

Assumption 2 *Trading in securities takes place continuously in time. The securities and exchange markets are "sufficiently perfect", i.e. there exist no transaction costs, no limits in short selling.*

3 Continuous Time Models for Fuzzy Valued Assets

In modern finance theory incremental changes of an asset are usually assumed to be normally distributed and are defined, i.e. [1], as:

Definition 4. The process $W = (W_t : t \geq 0)$ is a P-Brownian motion if and only if

(i) W_t is continuous, and $W_0 = 0$,
(ii) the value of W_t is distributed, under P, as a normal random variable $N(0, t)$,
(iii) the increment $W_{s+t} - W_s$ is distributed as a normal $N(0, t)$ under P, and is independent of $\mathcal{F}_f$, the history of what the process did up to time s.

---

[1] Assumption 1 is used for simplicity. In general, the existence of any (also risky) asset without fuzziness is sufficient. see [10].

Definition 5. The process of prices for the FVA is described by a Brownian motion, s.t.

$$d(S;\underline{S};\overline{S})_{LR} = (\mu S;\underline{S};\overline{S})_{LR}dt + (\sigma S;0;0)_{LR}dW. \quad (3)$$

From Definition 5 follows that the process of prices for the FVA possesses two parts for market risk description. The first term is a deterministic drift of the FVA and the second term describes a random risk source, which is not fuzzy but identical to neoclassical financial models. Due to fuzziness of the first deterministic term we are able to model some additional uncertainty and to manage market risk more detailed.

While the present price of an underlying asset can be observed on the market as a crisp value, there still arises the question, if it is reasonable to assume, that the value of future payoffs of an asset are ex-ante known as a crisp real number. The problems involved are obvious. The market participants often have different expectations of future payoffs due to imperfect market conditions. This becomes clear when looking at option prices for example. Even though every option may be written on the same underlying they still imply different volatilities[2]. On the one hand, there exists a significant discrepancy of know-how between the professional traders and the general public. On the other hand, the presence of asymmetrical information on the markets, some difficulties in valuation of assets-in-place and future growth opportunities make an exact, crisp valuation of the underlying asset nearly impossible. We will have a closer look on this problem in chapter 4.1.

Furthermore, according to Definition 5 the fuzziness of FVA is not dependent on time passed, in other words the spreads are constant through time. This is a simplifying condition, which should be changed in further research. For predicting the spreads in the following time periods some additional information about the fuzziness of asset prices in the future is required. Such information can be provided by experts, such as financial analysts in market research divisions or technical analysts. In absence of expert knowledge we assume that, once fixed, the spreads of FVA stay constant until more information is available. This information will be given by market observations, i.e. option prices.

The second term of Brownian motion in Definition 6 describes the volatility of asset prices. As already mentioned, it remains crisp in this framework. The market participants can calculate the necessary input value as the historical average volatility. It seems to be rather unrealistic to obtain future volatilities from historical average volatility. Permanent announcements of additional information and an enormous number of activities taken by market participants during a trading period imply random movements of asset prices. Nevertheless the information obtained in the past is the best estimate for future data. In comparison to former models we can rely on the given

[2] for the concept of implied volatility see i.e. [4]

volatility because we especially have an eye on what market participants think of the current asset price, but not on the volatility.

Before we may take a closer look at differential equations based upon Definition 5 we have to discuss briefly some differentiation rules for fuzzy numbers. Such rules are required to derive partial differential equations for any process of asset prices. General differentiation rules for fuzzy numbers do not exist in literature. Hence we examine already existing solutions and check their validity in our framework.

In opposite to classical functions, in which only a single function value changes on variation of a variable, the fuzziness of the function may vary. That means, that not only the main value but also the spreads of a function may depend on one or more variables. In other words, while the solution for Newtonian differentials is a crisp real number or function, the solution for fuzzy differentials is a set of real numbers, which is described by a membership function.

In terms of Newtonian calculus, for a differentiable function with more than one variable there exists a unique solution:

$$\frac{\partial F(x,y)}{\partial x} = f(x,y), \tag{4}$$

with y as a vector of some independent variables, $y = (y_1, y_2, \ldots, y_n)$.

To formalize our discussion of differentiation rules for fuzzy numbers, consider a function $F(\tilde{x}, y)$. Its derivative is a fuzzy-number $\tilde{f}$, where the main value and the spreads depend on $F(\tilde{x}, y)$:

$$\frac{\partial F(\tilde{x},y)}{\partial x} = \tilde{f}(\tilde{x},y) \text{ with } \tilde{f}(\tilde{x},y) = \left(f(\tilde{x},y); \underline{f}(\tilde{x},y); \overline{f}(\tilde{x},y)\right)_{LR} \tag{5}$$

and for the membership function we have

$$\mu_{\tilde{f}(\tilde{x},y)}[f(x,y)] = \tag{6}$$
$$\begin{cases} L\left(\frac{f(\tilde{x},y)-f(x,y)}{\underline{f}(\tilde{x},y)}\right) & \text{for } f(x,y) \le f(\tilde{x},y), \quad \underline{f}(\tilde{x},y) > 0 \\ R\left(\frac{f(x,y)-f(\tilde{x},y)}{\overline{f}(\tilde{x},y)}\right) & \text{for } f(\tilde{x},y) > f(x,y), \quad \overline{f}(\tilde{x},y) > 0 \end{cases}$$

The term $f(x,y)$ describes the value of the derivative for each particular x. Therefore we can conclude that we have to analyze for each derivative the characteristics of $\underline{f}(\tilde{x},y)$ and $\overline{f}(\tilde{x},y)$ in order to calculate the value of a derivative on a function with fuzzy variables.

The solution for the problem mentioned above can be achieved in two different ways. Either we look for some explicit solutions for the derivative under consideration of fuzzy mathematics, as in [6], or we search for an appropriate fuzzy metric to measure the space between the two different states of a fuzzy number.

Now consider a metric, which can be calculated in three steps. First we select an appropriate but subjective α level cut. Then we reduce the fuzzy set on the already given α level to one particular value Ψ. In the last step we solve a Newtonian differential for a function, which depends only on crisp variables:

$$\frac{\partial F\left({}^{\Psi}x^{\alpha}, y\right)}{\partial\, {}^{\Psi}x^{\alpha}} = f({}^{\Psi}x^{\alpha}, y). \tag{7}$$

Furthermore we have to implement the fuzzy derivatives in stochastic calculus, i.e. Ito's Lemma:

Lemma 6. *Let $\tilde{C}(\tilde{S}, t)$ be a twice-differentiable function of t and $\tilde{S}$ defined as in equation 3.*
Then we have:

$$d\tilde{C} = [\Delta^F(\mu S; \underline{S}; \overline{S})_{LR} + \tfrac{\partial \tilde{C}}{\partial t} + \tfrac{1}{2} \cdot \tilde{\Gamma}(S^2; 0; 0)_{LR}]dt + [(\Delta^F \sigma S; 0; 0)_{LR}]dW. \tag{8}$$

Proof. The chain rule for $\tilde{C}(\tilde{S}, t)$ can be solved by approximating the stochastic differential equation with the Taylor expansion. This gives:

$$\begin{aligned}\Delta\tilde{C} = {} & \Delta^F \cdot \Delta(S; \underline{S}; \overline{S})_{LR} + \tfrac{\partial (C;\underline{C};\overline{C})_{LR}}{\partial t}\Delta t + \\ & \tfrac{1}{2}(\Gamma; \underline{\Gamma}; \overline{\Gamma})_{LR} \cdot \Delta[(S; \underline{S}; \overline{S})_{LR}]^2 + \\ & \tfrac{\partial^2 (C;\underline{C};\overline{C})_{LR}}{\partial t (S;\underline{S};\overline{S})_{LR}} \cdot \Delta(S; \underline{S}; \overline{S})_{LR} \cdot \Delta t + \tfrac{1}{2}\tfrac{\partial^2 (C;\underline{C};\overline{C})_{LR}}{\partial t^2} \cdot \Delta t^2 + \\ & R(\Delta S, \Delta t)\end{aligned} \tag{9}$$

With $\Delta t \to 0$ terms of order greater than 1 in the above equation can be neglected. Therefore, we have

$$\begin{aligned}d\tilde{C} = {} & \Delta^F \cdot d(S; \underline{S}; \overline{S})_{LR} + \frac{\partial (C; \underline{C}; \overline{C})_{LR}}{\partial t} \cdot dt \\ & + \frac{1}{2}(\Gamma; \underline{\Gamma}; \overline{\Gamma})_{LR} \cdot d[(S; \underline{S}; \overline{S})_{LR}]^2.\end{aligned} \tag{10}$$

Since $d[(S; \underline{S}; \overline{S})_{LR}]^2$ can be considered to be $(S^2; 0; 0)_{LR}\sigma^2 dt$ one substitutes equation 3 into equation 10 which gives equation 8. □

Discussing differentiation rules for fuzzy numbers the question arises, which derivatives are important for pricing options. In analogy to the procedure of Fischer Black and Myron Scholes in [2], we also create a hedge portfolio depending on FVA's for proving the following theorem:

Theorem 7. *Any claim $\tilde{C}$ described by a stochastic differential equation like 3 in a market with FVA's can be valued through a partial differential equation given by:*

$$\frac{\partial \tilde{C}}{\partial t} + r\Delta^F \tilde{S} + \frac{1}{2}\sigma^2 (S^2; 0; 0)_{LR} \cdot \tilde{\Gamma} - r\tilde{C} \approx 0. \tag{11}$$

Proof. To eliminate the risk source , a hedge portfolio is constructed, s.t.

$$P = -\tilde{C} + \Delta^F \cdot \tilde{S}. \tag{12}$$

For

$$dP = -d\tilde{C} + \Delta^F \cdot d\tilde{S} \tag{13}$$

the implementation of equations 3 and 8 gives

$$\begin{aligned} dP = &-[\Delta^F \cdot (\mu S; \underline{S}; \overline{S})_{LR} + \frac{\partial \tilde{C}}{\partial t} + \frac{1}{2}\tilde{\Gamma}(S^2;0;0)_{LR}\sigma^2]dt \quad (14) \\ &-(\Delta^F \sigma S; 0; 0)_{LR} dW + \Delta^F \cdot (\mu S; \underline{S}; \overline{S})_{LR} dt \\ &-(\Delta^F \sigma S; 0; 0)_{LR} dW \\ \Leftrightarrow dP = &-\left[\left(\frac{\partial \tilde{C}}{\partial t}\right) + \frac{1}{2}\tilde{\Gamma} \cdot (S^2;0;0)_{LR} \cdot \sigma^2\right] dt. \quad (15) \end{aligned}$$

Since

$$\frac{dP}{P} = \frac{-\left[\left(\frac{\partial \tilde{C}}{\partial t}\right) + \frac{1}{2}\tilde{\Gamma} \cdot (S^2;0;0)_{LR} \cdot \sigma^2\right] dt}{-\tilde{C} + \Delta^F \cdot \tilde{S}} \approx r\, dt. \tag{16}$$

Hence,

$$\frac{\partial \tilde{C}}{\partial t} + r\Delta^F \tilde{S} + \frac{1}{2}\sigma^2 (S^2;0;0)_{LR} \cdot \tilde{\Gamma} - r\tilde{C} \approx 0. \tag{17}$$

□

The above fuzzy PDE contains three partial derivatives which have to be discussed further:

1. The first partial derivative of $\tilde{C}$ with respect to $\tilde{S}$,
2. the second partial derivative of $\tilde{C}$ with respect to $\tilde{S}$ and
3. the first partial derivative of $\tilde{C}$ with respect to t.

Usually yields are considered to be normally distributed, implying asset prices to be distributed log-normally, which gives us the price process of $\tilde{C} = \ln \tilde{S}$. Definition 5 implies, that the fuzziness of FVA is not dependent on time passed; thus any derivative of an asset or its spreads will equal zero. The asset price $\tilde{C}$ also depends on the value of $\tilde{S}$, but since the spreads of $\tilde{S}$ were assumed to be constant through time and for every realization of S, the spreads are not dependent on the function itself[3]. This gives a solution which equals the classical Newtonian differentials:

$$\frac{\partial C}{\partial \tilde{S}} = \frac{1}{S} \wedge \frac{\partial^2 C}{\partial \tilde{S}^2} = -\frac{1}{S^2}.$$

[3] the independancy of spreads is a simplification, which of course matters our evaluation process. Since we like to hedge though, it is negligable when considering a short time period only.

Now, a time continuous option pricing model could be derived, that can be solved by numerical methods or the martingale evaluation. We use the latter to price a plain vanilla option to round off the discussion of the fuzzy valued asset pricing in continuous time .

The payoffs of an European Call on a FVA is determined by

$$\tilde{C}_T = \max\{(S_T; \underline{S}; \overline{S})_{LR} - \kappa; 0\}.$$

Since the maximum operator can be substituted by an indicator function I, we get

$$\tilde{C}_T = (S_T; \underline{S}; \overline{S})_{LR} \cdot I_{\{(S_T;\underline{S};\overline{S})_{LR} \succ \kappa\}} - \kappa \cdot I_{\{(S_T;\underline{S};\overline{S})_{LR} \succ \kappa\}}. \tag{18}$$

Both parts of the subtraction 18 can be evaluated separately by the well-known martingale method. To make the approach clear, we will describe the method more detailed for the second term of the above equation. Therefore we define $\tilde{y}_T$ to be

$$\tilde{y}_T = \kappa \cdot I_{\{(S_T;\underline{S};\overline{S})_{LR} \succ \kappa\}} \tag{19}$$

and normalize it by

$$\begin{aligned} \frac{\tilde{y}_0}{K_0} &= \hat{E}_0\left[\frac{\tilde{y}_T}{K_T}\right] = \frac{\kappa}{K_T} \cdot \hat{E}_0\left[I_{\{(S_T;\underline{S};\overline{S})_{LR} \succ \kappa\}}\right] \\ &= \frac{\kappa}{K_T} \cdot Prob[(S_T; \underline{S}; \overline{S})_{LR} \succ \kappa] \end{aligned} \tag{20}$$

with K_t to be the money market account for time t. To calculate the probability for the event that the asset price $\tilde{S}$ lies "in-the-money" we have to take a closer look on the dynamics of our Brownian motion. Furthermore, we have to determine the dynamic equation for the price process when the underlying is log-normally distributed.

Hence,

$$\begin{aligned} d\tilde{S}_t &= (rS_t; \underline{S}; \overline{S})_{LR} dt + \sigma S_t d\hat{W}_t \\ \text{with } d\hat{W}_t &= dW_t + \frac{\mu - r}{\sigma} dt. \end{aligned} \tag{21}$$

Applying Lemma 6 to the price process $C(\tilde{S}, t) = \ln(S; \underline{S}; \overline{S})_{LR}$ and making use of the derivatives already discussed, we receive the dynamic equation

$$\begin{aligned} d(\ln \tilde{S}) &= \frac{1}{S}\left[(rS; \underline{S}; \overline{S})_{LR} dt + \sigma S d\hat{W}_t\right] + \frac{1}{2}\left(-\frac{1}{S^2}\right)\sigma^2 S^2 dt \\ &= \left[(r; \frac{\underline{S}}{S}; \frac{\overline{S}}{S})_{LR} - \frac{1}{2}\sigma^2\right] + \sigma d\hat{W}_t. \end{aligned} \tag{22}$$

The Brownian motion $\hat{W}_t$ in equation 22 obeys the following condition

$$\hat{W}_t(\tilde{S}_t \succ \kappa) = \hat{W}_t(\ln \tilde{S}_t \succ \ln \kappa) \tag{23}$$

and $\ln \tilde{S}_T$ has therefore to be

$$\ln \tilde{S}_T = \ln \tilde{S}_0 + \int_0^T d \ln \tilde{S}_T \tag{24}$$

$$= \ln \tilde{S}_0 + \left[(r; \frac{\underline{S}}{S}; \frac{\overline{S}}{S})_{LR} - \frac{1}{2}\sigma^2 \right] T + \sigma \hat{W}_T. \tag{25}$$

Substituting 25 into 23 gives

$$\ln \tilde{S}_T \succ \ln \kappa$$

$$\Leftrightarrow \ln \tilde{S}_0 + \left[(r; \frac{\underline{S}}{S}; \frac{\overline{S}}{S})_{LR} - \frac{1}{2}\sigma^2 \right] T + \sigma \hat{W}_T \succ \ln \kappa. \tag{26}$$

From Definition 4 we know, that $\sigma \hat{W}_T$ equals an error term $\epsilon \sigma \sqrt{T}$ with $\epsilon \sim N(0,1)$, leading to the term

$$\frac{-\ln \tilde{S}_0 + \ln \kappa - \left[(r; \frac{\underline{S}}{S}; \frac{\overline{S}}{S})_{LR} - \frac{1}{2}\sigma^2 \right] T}{\sigma \sqrt{T}} \prec \epsilon. \tag{27}$$

Since the condition in equation 23 holds and we know from [2] that $\hat{W}(\epsilon \succ -d_2) \Rightarrow N(d_2)$, we obtain

$$N(\tilde{d_2}) = \frac{\ln \tilde{S}_0 - \ln \kappa + \left[(r; \frac{\underline{S}}{S}; \frac{\overline{S}}{S})_{LR} - \frac{1}{2}\sigma^2 \right] T}{\sigma \sqrt{T}}. \tag{28}$$

Last, we normalize the process again to receive a price under the original probability measure P:

$$\tilde{y}_0 = \kappa e^{-rT} N(\tilde{d_2}). \tag{29}$$

Evaluating the first term in equation 18, the measures must be changed with the FVA instead of the money market account. Therefore we define $\tilde{y}_T$ as

$$\tilde{y}_T = \tilde{S}_T \cdot I_{\{(S_T; \underline{S}; \overline{S})_{LR} \succ \kappa\}} \tag{30}$$

and normalize it by

$$\frac{\tilde{y}_0}{\tilde{S}_0} = \hat{E}_0 \left[\frac{\tilde{y}_T}{\tilde{S}_T} \right] = \hat{E}_0 \left[I_{\{(S_T; \underline{S}; \overline{S})_{LR} \succ \kappa\}} \right] \tag{31}$$

$$= Prob[(S_T; \underline{S}; \overline{S})_{LR} \succ \kappa].$$

The obvious problem is the division by the FVA, since the fuzziness will increase according to the definition of extended arithmetic operations with fuzzy numbers. In this example increasing fuzziness would lead to economically senseless results, since the fuzziness of $\tilde{y}_0$ is directly determined by the fuzziness of $\tilde{S}_0$, which is assumed to be constant through time. Therefore we suggest to neglect this problem here. Furthermore the applied change of measure for the Brownian motion modifies from equation 21 to

$$d\tilde{S}_t = (rS_t; \underline{S}; \overline{S})_{LR} dt + \sigma S_t d\hat{W}_t \qquad (32)$$
$$\text{with } d\hat{W}_t = dW_t + \frac{\mu - r + \sigma^2}{\sigma} dt.$$

Now we follow the same argumentation, which results into

$$N(\tilde{d}_1) = \frac{\ln \tilde{S}_0 - \ln \kappa + \left[(r; \frac{\underline{S}}{S}; \frac{\overline{S}}{S})_{LR} + \frac{1}{2}\sigma^2 \right] T}{\sigma\sqrt{T}}, \qquad (33)$$

and the price of a European call option on the FVA $\tilde{S}$ equals:

$$\tilde{C}_0 = \tilde{S}_0 \cdot N(\tilde{d}_1) - \kappa \cdot e^{-rT} N(\tilde{d}_2). \qquad (34)$$

The equivalent valuation of a European put option is given in the appendix. There are some astonishing differences to the former Black-Scholes formulae. The influence of the risk free rate on the options price is most interesting. Especially the return on assets can become even more negative than in the standard option pricing model. It is easy to show, that this influence becomes relatively more important, when volatility decreases. This gives us important insight into the correlation of volatility and the risk free return rate.

Furthermore the question arises, how the Fuzzy Valued Underlying is found. As already mentioned, the parameter $\tilde{S}$ describes different expectations of market participants towards the future payoffs of the underlying in different states. It seems to be rather complicated to observe this parameter on the financial markets. Nevertheless we give a brief presentation on how it can be put into practice.

4 Financial Intermediation and Fuzzy Valued Assets

4.1 The Role of Deposit Insurance

The management of deposits is one of the traditional aims for private banks. The bank receives the savings of individuals and lends money to firms, so that the investments are financed. For that reason, the banks are financial intermediaries, which facilitate the essential transformation function in the national economy. The common advantages for depositors to go to a bank rather than to purchase the fixed-income securities directly on the market are economies of scale, transactions costs, liquidity, and convenience. However, if the deposits are not considered as riskless, the depositor has to analyze the bank's balance sheet, the quality of its management, and the market conditions to determine the risks. In addition, after the risk of the bank is determined the depositor has to diversify his holdings across many banks, so that his savings face the minimum of risk. If the conditions change, the depositor has to repeat his analysis in order to adjust the portfolio of holdings. Therefore, the advantages to keep the savings in a bank disappear, if the deposits are risky.

This shows, that an institutional structure is useful, which saves large information and surveillance costs and makes the private deposits riskless. This institution provides a guarantee for a specified amount of deposits, so that the savings up to this amount become save for the individuals. It prevents also from the panic reactions of depositors, such as bank runs. In general, an institutional structure is required, which provides the nation with a saver, more flexible and more stable financial system. Two general institutional forms can be distinguished - private insurance of deposits or an insurance organization funded by the government or one of its agencies. For instance, the Federal Deposit Insurance Company (FDIC) for commercial banks and the Federal Savings and Loan Insurance Corporation (FSLIC) for savings and loan associations were founded for that purpose in the USA. In Germany the Bundesaufsichtsamt fuer das Kreditwesen (BAKred) supervises the banks and other financial intermediaries with regard to their assets, liabilities, liquidity and their business risks. In return for the insurance the guarantor charges the banks with a certain premium at the beginning of the term of the guarantee. In many countries this payment is constant for all insurers, independent of how risky their assets and liabilities are. Another approach is used in the USA, where a risk-based premium is computed.

The insurance of deposits plays a central role for the management of a bank, because the selection of an investment portfolio depends on the tradeoff between the profit and the risk. The management tends to choose a more risky investment alternative, if the guarantor promises to make a certain payment in the case of default. In consequence of the information asymmetry the probability of default and the "shortfall" (the difference between the payment on the guarantee and the price received from the sale of the assets) increases. Therefore, the aim of the insurance company is the monitoring of banks in order to control the shortfall losses.

This chapter presents a fuzzy logic approach to determine a risk-based premium for deposits. Merton has already pointed out that the model for pricing deposit premiums has its foundation in the isomorphic relationship between deposit insurances and common stock put options. We extend the model of Merton and show that the fuzzy logic approach has a great advantage in comparison to the common models, because it allows to describe vague data, which is directly observable for the principal, as a fuzzy number.

We start the discussion of the fuzzy logic approach with the description of the relationship between the agent (the management of a bank) and the principal (the deposit insurer) in the case of deposits insurance. We continue then with the question, how the contracts between the agent and the principal should be arranged, so that the agency costs are minimized. The chapter closes with the presentation of the fuzzy logic based model for pricing the deposit insurance.

4.2 Principal-Agent Relationship and the Deposit Insurance

The principal-agent relationship results from the information asymmetry between the deposit insurer and the bank and has two important implications - the phenomena of moral hazard and adverse selection. In general, there are welfare reductions, if in order to maximize his benefits the agent performs or refrains from some actions, which are not optimal for the principal. The principal is not able to observe or control those actions. The cost to avoid these welfare reductions are called agency costs. These are residual costs for deviation from the optimal solution for the principal (First Best Solution) and choosing the Second Best Solution instead. They can be classified as following:

1. Surveillance costs and cost for controlling the agent (monitoring costs).
2. Cost for binding the agent to a specified contract.

According to the phenomenon of moral hazard it is optimal for the management of a bank (agent) to increase the expected value of the portfolio by choosing riskier investment alternatives on the conclusion of the insurance contract. Therefore, the insurance contract has the opposite effect as one would expect. The taking over of default risk by a private bank implies that the bank increases the risk of its assets. And as a result of that, the probability of a financial collapse in the national economy increases also. We conclude, that the insurance contract has to include certain specifications, which do not allow the bank to change the assets and liabilities without the permission of the deposit insurers.

The problem of adverse selection occurs, if the agent has some hidden information, which is not directly observable for the agent. Firstly, the deposit insurer is not able to distinguish the quality of the banks in regard to the risk they are facing. If the insurer offers a contract with an insurance premium, which is calculated with the average default risk, the high risk banks are going to accept it. In opposite to that, the low risk banks will consider the average premium as too high and refuse the insurance. Then, the aim of the insurer is to prevent this, from his point of view, adverse selection process and to separate the banks in different risk classes. At best, the bank discloses the quality of his assets and chooses the contract specifications, which are suitable for its risk class during the negotiations. Therefore, the selection of contract specifications may implicitly reveals the risk of the held portfolio. This process is called signaling. Another opportunity to determine the risk of a bank is to surveil its assets and liabilities and afterwards to offer a risk-based premium. This process is called screening.

4.3 Managing the Guarantee Business

The success of the guarantor depends on the proper arrangement of guarantees contracts. It can be achieved by a correct mixture of risk-based premiums, control of operating costs, and control of the frequency and the severity of shortfall losses.

The shortfall losses can be minimized, if the guarantor has a right to audit the value of assets continuously. The key elements of an efficient surveillance and control system are [13]:

1. the guarantor has possession of the collateral;
2. the value of the assets is recomputed frequently and are marked to market;
3. the guarantor has the right to sell the collateral immediately in the case of default.

As already mentioned above, it is optimal for a bank to cheat the insurer and to value its assets higher then the actual market price. To avoid these mispricings, some rules have to be implemented, which provide exact results and are acceptable for both parties. In addition, the opportunity to manipulate the results must be limited. If the collateral assets are traded on the organized and liquid markets such as stock exchanges or securities markets, the prices can be observed directly. But even under this perfect conditions the estimates of the assets prices from the market are often biased, because the trade on the markets has different regulations or contract specifications in comparison to the collateral assets. The significant errors of estimation result in risks for both parties, the guarantor and the bank. If the errors exceed the values, the guarantor recognizes the default too late and the proceeds of selling the collateral will be less than expected. If the errors fall short of the values, the bank will be liquidated, even if it is solvent actually. The marking to market is even more difficult, if the assets are traded infrequently or are illiquid.

We may conclude that, the insurance company is interested in calculating fair values. But the obtaining of information about the true values of collateral assets and the controlling of the intermediary causes large costs for the guarantor. Thus, from the point of view of the insurer there exists a tradeoff between minimizing the probability of default and controlling surveillance costs.

An alternative approach to the frequent controlling is the creation of a capital cushion for the insured intermediary. The greater the capital of the intermediary, the smaller are deposits relative to the own capital. Thus, the frequency of surveillance and the intensity of audit can be reduced, which consequently saves the costs of controlling. Unfortunately, these costs savings do not mean, that the capital cushion approach is more efficient than the monitoring of banks. In the financial literature of recent years a wide discussion about the optimal capital structure of a corporate enterprise. To summarize the major idea, we can state, that there exists a principal-agent relationship between the equity holders (principals) and the management of the company (agent). But the position of the shareholders is even weaker than those of the insurance company. The only way for the equity holder to control the company's value and the dividend payments is to elect the management. In the case of default the equity holder does not receive any payments. The agency and tax costs between the shareholders and the management become larger, if the capital cushion grows. Therefore, the capital cushion is a cost shifting from the insurance company to the equity holders, with the only effect that the equity holders carry the controlling and surveillance.

A different approach to the capital cushion is the inclusion of some regulations of asset restrictions for the financial intermediary in the insurance contract. The aim of such asset restrictions is to reduce the variability between the debits and the credits of the bank. This type of regulations is called "matching assets to liabilities". For instance, a bank with long-term and money fixed liabilities could be restricted to hold securities with similar specifications. Matching assets to liabilities reduces the risk of a intermediary considerably, even if a full matching is not achieved.

While the concept is simple, the implementation of matching in practice is rather complex. The financial intermediaries face the problem, that most of them (especially in the USA and increasingly in Europe) have limited business fields, which determine their circle of creditors and depositors. If the bank offers short-term and fixed-income conditions for the deposits but holds long-term bonds, the deposit insurer has to deal with an interest rate risk, even though the bonds are default free. The interest rate risks as much as the exchange rate risks can be managed successfully by using the derivative securities markets, which are growing rapidly. For instance, by trading swaps the bank is able to match its assets and liabilities very effectively. In addition, there are other trends like securitization of mortgages, which allow the intermediaries to reduce their risk exposure without changing the investment conditions for a customer, shifting to new types of costumers.

The presence of the organized markets for swaps and other derivatives does not facilitate the insurance company to refrain from monitoring completely. These markets make it possible to take an uncovered position as easily as to trade covered positions. Hence, the intermediary can even increase instead of decreasing the risk exposure by buying a speculative strategy on the derivatives markets, which makes its assets unhedged. Consequently, the insurance company has to monitor assets restrictions regularly.

4.4 Determination of a Risk-Based Premium Based on Fuzzy Logic

Robert Merton[12] developed a model of pricing the risk-based premium for deposit insurance. His approach has its foundation on the isomorphic relation between deposit insurance and common-stock put options. Therefore, the model for pricing the premium is based on the Black-Scholes formulae[2]. One of the advantages of this procedure is that the Black-Scholes formulae is widely known and accepted. On the other hand, the empirical research of options markets has shown that in this approach there exist some inconsistencies known as smile effect or term structure phenomenon. One explanation of these phenomenons is the assumption of the constant volatility, which reduces the whole uncertainty to one parameter. We may conclude that even in the case of a correct valuation of the underlying assets and their volatility, the theoretical price of the put option will differ in comparison to a similar option, which is traded on the derivatives markets. In addition, we have to take into account all problems associated with the controlling, surveillance and marking to market of assets and liabilities, mentioned in the above section. Because of information asymmetry, the estimation of true values for the underlying assets contains significant errors.

Therefore, the model for pricing the risk-based premium based on the standard Black-Scholes approach contains errors of two types. Firstly, the errors can be made as a result of using different mathematical instruments compared to the ones used by the market, on which the assets will be sold in the case of default. Secondly, the model estimates wrong results due to mispricing of the underlying assets. Under these circumstances the fuzzy logic approach for pricing the options seems to be more appropriate, because it minimizes the errors of both types.

As already discussed in the section above, the errors in the valuation of assets result in an additional risk source for both parties, the insurer and the intermediary. We suggest to value the assets of the intermediary by modeling these risks with the help of fuzzy numbers. On the one hand this procedure saves the controlling and the surveillance costs for the insurance company, because approximate values are sufficient for the calculations. On the other hand in some cases it is not possible to estimate correct market values of the underlying assets, for instance because they are illiquid. Here, the insurer can use the expert knowledge or its experience gained in similar cases to price the illiquid assets as an approximate value.

With this background, let's consider a simple model. A company borrows money by issuing a zero-bond, which has no coupon payments during the lifetime. According to the contract specifications of the debt issue the company promises to pay a total of B dollars at the maturity date. In the event of default the company promises a payment in the amount of all the firm's assets. In the following we assume, that at the issue date the value of the underlying assets at the maturity date is known only approximately and is described as a fuzzy number. On the maturity date the value of assets is known for sure, and therefore the assets become crisp.

In other words, if the value V of the assets is larger than the payment on the bond issue at maturity, it is optimal for the management of the company to make the payment, even by selling the assets if necessary. Hence, in this case the value of the company's equity will be $V - B$. If the value of the firm's assets is less than the promised payment at maturity, the management is not able to pay the total of B. Accordingly, the value of the equity will be zero and and the value of the bond issue will be V. Therefore, the value of the bond at the maturity date can be expressed as $min(V, B)$ and the value of the equity as $max(0, V - B)$.

The argumentation up to this point is straightforward and is similar to the common approach. Now, we have to consider, that today's values of the underlying assets are known as fuzzy numbers. Thus, at the maturity date the bondholders are receiving $min(\tilde{V}, B)$ and the equityholders are possessing $max(0, \tilde{V} - B)$.

Let's consider the impact of a third-party guarantee for the payment to the bondholders. From point of view of the bondholders the guarantee converts a risky debt into a riskless, because due to the conditions of the guarantee the insurer is obliged to pay the promised amount of at least B dollars to the bondholders in the case of default. Therefore, the guarantee can be compared to a traditional insurance policy. The insurer has to hedge the value of the guarantee, which causes cost. Hence, the company is expected to pay the insurer for the guarantee an amount at least equal to his cost, which is called a fair premium.

To determine the fair premium, we have to analyze the payments on the maturity date, if a guarantee contract is concluded. On the maturity date the value of the equity is independent of the guarantee, $max(0, \tilde{V} - B)$. The value of the debt is riskless and always B and the guarantee value is $min(0, \tilde{V} - B)$. A fair premium for the company would be $-min(0, \tilde{V} - B)$ dollars, which can be reexpressed as $max(0, B - \tilde{V})$[4].

If we denote $\tilde{G}(T)$ as the value of the guarantee to the firm at the maturity date, then

$$\tilde{G}(0) = max(0, B - \tilde{V}), \tag{35}$$

which corresponds to the payoff structure of the common put option. Hence, we can use the results of the previous chapter3 to derive the formula for the value of the guarantee. The explicit derivation of a put option on fuzzy valued assets is shown in appendix, because there exist some important differences compared to call options on fuzzy valued assets. and write:

$$\tilde{G}(T) = B \cdot e^{(-rT)} N(-\tilde{d}_2) - \tilde{V} \cdot N(-\tilde{d}_1) \tag{36}$$

where

$$N(-\tilde{d}_1) = \frac{\ln \tilde{V} - \ln B + \left[(r; \frac{\underline{V}}{\overline{V}}; \frac{\overline{V}}{V})_{LR} + \frac{1}{2}\sigma^2 \right] T}{-\sigma\sqrt{T}}$$

$$\text{and } N(-\tilde{d}_2) = \frac{\ln \tilde{V} - \ln B + \left[(r; \frac{\underline{V}}{\overline{V}}; \frac{\overline{V}}{V})_{LR} - \frac{1}{2}\sigma^2 \right] T}{-\sigma\sqrt{T}}$$

[4] Note that these operations do not change the fuzziness of the expression.

The reader has to notice, that by using the formulae 36 the guarantee premium becomes more fuzzy than the underlying assets. The fuzziness of guarantee premium is caused not only by the fuzziness of underlying assets but also by an additional uncertainty of the stochastic process. The term $\left[(r; \frac{\underline{S}}{S}; \frac{\overline{S}}{S})_{LR} - \frac{1}{2}\sigma^2\right]$ is often called the market-price-of-risk process, which can take some positive as well as negative values and is obviously fuzzy in our model.

The guarantee premium formulae 36 offers a great advantage of flexibility for users and can help to determine an optimal portfolio strategy. We will explain the general procedure with the help of our example above. Any other guarantee could be formed in the same manner by putting an eye on the economics behind the different insurance strategies. The only task we have to perform is, to determine, which price of the underlying contains the lowest risk. To speak in formal specifications, we have to determine the parameters Ψ and α only. In detail, first the guarantor has to decide, which α-level cut is relevant for him. To interprete this decision, imagine the fact, that if α equals 1, the underlying assets are marked-to-market perfectly. In this case the contract specifications for the securities traded on the market and for the companies assets are identical. There is also no fuzzy risk in the market-price-of-risk process, because the market participants have the same expectations. If α tends to zero, then the underlying assets are illiquid, their exact value in case of default is not known. The expectations of market participants differ significantly. If the guarantor chooses $\alpha < 1$, then he also has to decide, which element of the bounded interval on the level α has to be hedged. This can be done by choosing the parameter Ψ. If Ψ equals 1, then the guarantor is rather pessimistic about the future development of the underlying and expects a lower price of the underlying in case of default. In opposite to this, rather positive or optimistic expectations can be represented by $\Psi \to 0$.

For instance, let's behave very pessimistic and moreover think of the most illiquid assets we could model. This means we assume the worst case possible in our world. Then, we consider parameter α to be zero, because it gives us the possibility to select the very left boundary $\underline{V}$ for the equity together with the parameter $\Psi = 1$.

With this knowledge the equation 36 can be simplified to

$$^{\Psi}\tilde{G}^{\alpha} = B \cdot e^{-rT} \cdot {}^{\Psi}N^{\alpha}(-\tilde{d}_2) - {}^{\Psi}\tilde{V}^{\alpha} \cdot {}^{\Psi}N^{\alpha}(\tilde{d}_1) \tag{37}$$

with

$$^{\Psi}N^{\alpha}(-\tilde{d}_1) = \frac{\ln {}^{\Psi}\tilde{V}^{\alpha} - \ln B + \left[{}^{\Psi}(r; \frac{\underline{V}}{V}; \frac{\overline{V}}{V})^{\alpha}_{LR} + \frac{1}{2}\sigma^2\right] T}{-\sigma\sqrt{T}}$$

$$^{\Psi}N^{\alpha}(-\tilde{d}_2) = \frac{\ln {}^{\Psi}\tilde{V}^{\alpha} - \ln B + \left[{}^{\Psi}(r; \frac{\underline{V}}{V}; \frac{\overline{V}}{V})^{\alpha}_{LR} - \frac{1}{2}\sigma^2\right] T}{-\sigma\sqrt{T}}$$

The guarantor is also able to choose, on which risk source he should apply the (Ψ, α)-combination approach. If the underlying assets are marked-to-market but there are significant differences in market expectations, the (Ψ, α)-combination approach will be applied on the term $\tilde{V}$ in the equation 36. On the other hand in case of illiquid assets and identical market expectations, the (Ψ, α)-combination approach will be applied on the terms $N(\tilde{d}_1)$ and $N(\tilde{d}_2)$ in the same equation. We conclude, that the fuzzy price of a guarantee is an equilibrium solution. However, the (Ψ, α)-combination approach facilitates to handle the fuzziness of the guarantee premium very flexible.

As Merton and Bodie [13] already pointed out, there exists a fundamental identity for the cost of loan guarantees:

$$\text{Risky Loan} + \text{Loan Guarantee} \equiv \text{Default Free Loan}$$

If the premium of the guarantee is calculated fairly, the savings in the yield payments should be exactly the same as the cost of the guarantee from point of view of the issuing company. It is well known that the price of a zero-bond corresponds to its yield payment. If the market value of common zero-bond is $Be^{(-R(T)T)}$, with $R(T)$ as promised yield, then the market value of a bond with guarantee will be

$$\tilde{G}(T) + Be^{(-R(T)T)} = Be^{(-\tilde{r}T)}. \tag{38}$$

From definition 38 follows, that the promised yield on the default-free loan will be less than the yield on the risky loan. But we are not able to determine the exact amount of savings. The value depends on the equity of the issuing company. If the assets are easy to surveil and control, the default-free interest rate can be computed exactly. If, however, the information about the value of underlying assets is only imprecise, the default-free interest rate still face some risk source. Therefore the default-free interest rates of different companies can vary significantly.

Now let's consider the deposit insurance. In comparison to the above model, we suppose the company to be a bank and let the debt issue correspond to the deposits. This is possible, if we interpret the time interval until the maturity of debt as the time period up to the next audit of the banks assets and the bank pays interest on the savings, which equals the default-free rate. Therefore, from point of view of the insurance company the deposits can be treated as debt issue, or better to say, zero-coupon bonds. If the insurance company guarantees the deposits, the deposits become riskless and their current value can be described as:

$$D = B \cdot e^{(-rT)}. \tag{39}$$

Let $\tilde{g}$ be the cost of the guarantee per dollar of insured deposits, $\frac{\tilde{G}(T)}{D}$. Then we can modify the formulae 36 to:

$$\tilde{g}(\tilde{d}, \tau) = N(\tilde{x}_2) - \frac{\tilde{V}}{D} \cdot N(\tilde{x}_1) \tag{40}$$

where

$$\tilde{x}_1 = \frac{\ln \tilde{V} - \ln D + \tau/2}{\tau^{1/2}}$$

and

$$\tilde{x}_2 = \frac{\ln \tilde{V} - \ln D - \tau/2}{\tau^{1/2}},$$

with $\tilde{d} = \frac{D}{\tilde{V}}$ as the current deposit-to-asset value ratio, and $\tau = \sigma^2 T$ as the variance of the logarithmic change in the value of the assets during the term of the deposits.

5 Conclusion

Modern finance theory is based on a certain number of assumptions. Especially, option pricing theory is a powerful instrument in financial intermediation also. Merton has shown, that pricing deposit insurances for instance is fundamentally based upon the pricing of options. Nevertheless, it is incontestable that some assumptions are too strict to be realistic.

Therefore, in this article, a new approach in pricing options was suggested. The modeling of price processes is based upon the random walk models of neoclassical finance theory on the one hand. One the other hand the fuzzy set theory is implemented to handle the market risk more efficiently. We have shown, that the valuation of financial claims is possible, even if the underlying is only described in terms of LR-fuzzy numbers, which weakens some necessary conditions of the modern financial theory. Moreover, differentiation rules for fuzzy numbers, as a major problem of this approach, were discussed. Based on linguistic examination of the basic economic background in financial markets our argumentation was straightforward.

Time continuous evaluation led to a set of prices, described by membership functions. It is important to notice, that a fuzzy price for a claim is a, not preference related, equilibrium solution when markets are complete. However, any market participant can get a crisp option price, when his personal preference structure is taken into account.

With the help of those results, principles in pricing deposit insurances were extented by a fuzzy approach. We have shown further that it is possible to describe risks more exactly and even more realistic by using fuzzy sets instead of crisp real numbers. Now, it depends on the insurer, which type of fuzzy number is selected and which spreads are chosen. The obvious advantage is the exact evaluation of risks, even if it is not possible to fix deterministic values for the underlying assets, i.e. the firm's value.

The reader may notice, that further research is necessary to determine most realistic fuzzy numbers for different problem sets. It was not our aim, to solve every risk situation, but to supply a mathematical framework to handle such problems.

Appendix A

Lemma A. *The price of a put option is given by* $\tilde{P}_0 = -\tilde{S}_0 \cdot N(-\tilde{d}_1) + \kappa \cdot e^{-rT} N(-\tilde{d}_2)$ *with*

$$N(-\tilde{d}_1) = \frac{\ln \tilde{S}_0 - \ln \kappa + \left[(r; \frac{\underline{S}}{S}; \frac{\overline{S}}{S})_{LR} + \frac{1}{2}\sigma^2\right] T}{-\sigma\sqrt{T}}$$

$$\text{and } N(-\tilde{d}_2) = \frac{\ln \tilde{S}_0 - \ln \kappa + \left[(r; \frac{\underline{S}}{S}; \frac{\overline{S}}{S})_{LR} - \frac{1}{2}\sigma^2\right] T}{-\sigma\sqrt{T}}$$

Proof. The payoffs of a European Put on a FVA is determined by $\tilde{P}_T = \max\{\kappa - (S_T; \underline{S}; \overline{S})_{LR}; 0\}$. Since the maximum operator can be substituted by an indicator function I, we get

$$\tilde{P}_T = \kappa \cdot I_{\{\kappa \succ (S_T;\underline{S};\overline{S})_{LR}\}} - (S_T; \underline{S}; \overline{S})_{LR} \cdot I_{\{\kappa \succ (S_T;\underline{S};\overline{S})_{LR}\}}. \tag{41}$$

We assume

$$\tilde{y}_T = \kappa \cdot I_{\{\kappa \succ (S_T;\underline{S};\overline{S})_{LR}\}} \tag{42}$$

and use the normalization

$$\begin{aligned} \frac{\tilde{y}_0}{K_0} &= \hat{E}_0\left[\frac{\tilde{y}_T}{K_T}\right] = \frac{\kappa}{K_T} \cdot \hat{E}_0\left[I_{\{\kappa \succ (S_T;\underline{S};\overline{S})_{LR}\}}\right] \\ &= \frac{\kappa}{K_T} \cdot Prob[\kappa \succ (S_T; \underline{S}; \overline{S})_{LR}] \end{aligned} \tag{43}$$

where K_t is the money market account for time t. Then we obtain

$$\begin{aligned} d\tilde{S}_t &= (rS_t; \underline{S}; \overline{S})_{LR} dt + \sigma S_t d\hat{W}_t \\ \text{with } d\hat{W}_t &= dW_t + \frac{\mu - r}{\sigma} dt. \end{aligned} \tag{44}$$

Using the same argumentation as we did for evaluating a European call option in section 3 we receive:

$$\begin{aligned} & \ln \tilde{S}_T \prec \ln \kappa \\ \Leftrightarrow \ln \tilde{S}_0 + \left[(r; \frac{\underline{S}}{S}; \frac{\overline{S}}{S})_{LR} - \frac{1}{2}\sigma^2\right] T + \sigma \hat{W}_T & \prec \ln \kappa \end{aligned} \tag{45}$$

and, as $\sigma \hat{W}_T$ equals an error term $\epsilon\sigma\sqrt{T}$ with $\epsilon \sim N(0,1)$, we have

$$\frac{-\ln \tilde{S}_0 + \ln \kappa - \left[(r; \frac{\underline{S}}{S}; \frac{\overline{S}}{S})_{LR} - \frac{1}{2}\sigma^2\right] T}{\sigma\sqrt{T}} \succ \epsilon. \tag{46}$$

When we compare this result to equation 27, it is obvious, that - just like in the standard Black Scholes model - this value is directly correlated to $N(\tilde{d}_2)$. We receive

$$N(\tilde{e}_2) = \frac{-\ln \tilde{S}_0 + \ln \kappa - \left[(r; \frac{S}{S}; \frac{\overline{S}}{S})_{LR} - \frac{1}{2}\sigma^2\right] T}{\sigma\sqrt{T}} \tag{47}$$

first. Please note, that this does not equal $N(-\tilde{d}_2)$ for now, since fuzzy subtraction rules are not commutative. On the other hand, we do not like the put option to have different spreads than an equivalent call option would obey. So we would like to compare the normal density functions for puts to those of calls and use a slightly different argumentation here. Since

$$\begin{aligned} & \ln \tilde{S}_T \prec \ln \kappa \\ \Leftrightarrow \ln \tilde{S}_0 + & \left[(r; \frac{S}{S}; \frac{\overline{S}}{S})_{LR} - \frac{1}{2}\sigma^2\right] T + \sigma \hat{W}_T \prec \ln \kappa \end{aligned} \tag{48}$$

and $\sigma \hat{W}_T$ equals an error term $\epsilon\sigma\sqrt{T}$ with $\epsilon \sim N(0,1)$, we get

$$\begin{aligned} -\ln \tilde{S}_0 + \ln \kappa - \left[(r; \frac{S}{S}; \frac{\overline{S}}{S})_{LR} - \frac{1}{2}\sigma^2\right] T \succ \epsilon \cdot \sigma\sqrt{T} \\ \Leftrightarrow \frac{-\ln \tilde{S}_0 + \ln \kappa - \left[(r; \frac{S}{S}; \frac{\overline{S}}{S})_{LR} - \frac{1}{2}\sigma^2\right] T}{-\sigma\sqrt{T}} \prec \epsilon \\ \Leftrightarrow \frac{\ln \tilde{S}_0 - \ln \kappa + \left[(r; \frac{S}{S}; \frac{\overline{S}}{S})_{LR} - \frac{1}{2}\sigma^2\right] T}{-\sigma\sqrt{T}} \succ \epsilon. \end{aligned} \tag{49}$$

Then, we obtain

$$N(-\tilde{d}_2) = \frac{\ln \tilde{S}_0 - \ln \kappa + \left[(r; \frac{S}{S}; \frac{\overline{S}}{S})_{LR} - \frac{1}{2}\sigma^2\right] T}{-\sigma\sqrt{T}}. \tag{50}$$

In common financial theory the difference to $N(\tilde{e}_2)$ is trivial but we have to stress our attention to it here. In Fuzzy Set Theory both equations would lead to different results. With equation 50 we nevertheless obtain a formulae, comparable to the approach of Black and Scholes, still consistent with the theory of put-call-parities.

Last, we normalize the process again to receive a price under the original probability measure P:

$$\tilde{y}_0 = \kappa e^{-rT} N(-\tilde{d}_2). \tag{51}$$

Analogous we value the second term in the put pricing formulae with the help of $\tilde{y}_T$, which becomes

$$\tilde{y}_T = \tilde{S}_T \cdot I_{\{\kappa \succ (S_T; \underline{S}; \overline{S})_{LR}\}} \tag{52}$$

and normalize it by

$$\frac{\tilde{y}_0}{\tilde{S}_0} = \hat{E}_0\left[\frac{\tilde{y}_T}{\tilde{S}_T}\right] = \hat{E}_0\left[I_{\{\kappa\succ(S_T;\underline{S};\overline{S})_{LR}\}}\right] \tag{53}$$
$$= Prob[\kappa \succ (S_T;\underline{S};\overline{S})_{LR}].$$

As we discussed before, the normalization through our asset $\tilde{S}_0$ does not increase the fuzziness and we apply the change of measure for the Brownian motion , s.t.

$$d\tilde{S}_t = (rS_t;\underline{S};\overline{S})_{LR}dt + \sigma S_t d\hat{W}_t \tag{54}$$
$$\text{with } d\hat{W}_t = dW_t + \frac{\mu - r + \sigma^2}{\sigma}dt.$$

This results in the discussion of the above probabilities and we obtain, together we the assumption, that the spreads of calls and puts should be comparable

$$Prob(\ln\kappa \succ \ln(S_T;\underline{S};\overline{S})_{LR}) \tag{55}$$

$$\Leftrightarrow\ Prob(\gamma\cdot\sigma\sqrt{T} \prec \ln\kappa - \ln\tilde{S}_0 - \left[(r;\frac{\underline{S}}{S};\frac{\overline{S}}{S})_{LR} + \frac{1}{2}\sigma^2\right]T \tag{56}$$

$$Prob(\gamma \succ \frac{\ln\kappa - \ln\tilde{S}_0 - \left[(r;\frac{\underline{S}}{S};\frac{\overline{S}}{S})_{LR} + \frac{1}{2}\sigma^2\right]T}{-\sigma\sqrt{T}} \tag{57}$$

$$Prob(\gamma \prec \frac{\ln\tilde{S}_0 - \ln\kappa + \left[(r;\frac{\underline{S}}{S};\frac{\overline{S}}{S})_{LR} + \frac{1}{2}\sigma^2\right]T}{-\sigma\sqrt{T}} \tag{58}$$

Hence,

$$N(-\tilde{d}_1) = \frac{\ln\tilde{S}_0 - \ln\kappa + \left[(r;\frac{\underline{S}}{S};\frac{\overline{S}}{S})_{LR} + \frac{1}{2}\sigma^2\right]T}{-\sigma\sqrt{T}}, \tag{59}$$

and – after normalizing the process $\tilde{y}_T$ again – the price of a European call option on the FVA $\tilde{S}$ equals

$$\tilde{P}_0 = -\tilde{S}_0\cdot N(-\tilde{d}_1) + \kappa\cdot e^{-rT}N(-\tilde{d}_2). \tag{60}$$

□

References

1. M. Baxter and A. Rennie (1996): Financial Calculus, Cambridge University Press, Cambridge.

2. F. Black and M. Scholes (1973): The Pricing of options and corporate liabilities, Journal of Political Economy, 637–654.
3. J. Cox, S. Ross, and M. Rubinstein (1979): Option pricing, a simplified approach, Journal of Financial Economics 7, 229–263.
4. J. Cox, and M. Rubinstein (1979): Options Markets,Prentice Hall.
5. P. Diamond and P. Kloeden (1994): Metric Spaces of Fuzzy Sets. Theory and Applications, World Scientific.
6. D. Dubois and H. Prade (1980): Fuzzy Sets and Systems. Theory and Applications, Academic Press.
7. C.C.-J. Fu, and E.S. Lee (1999):Neuro-Fuzzy in the Valuation of Stock Price Based on Fundamental Analysis, Proceedings of the Eight International Fuzzy Systems Association World Congress, Vol.1, 57–62, Taipei, Taiwan ROC.
8. Div. Authors (1998-2000): The Handbooks of Fuzzy Sets, Vol. 1-7, Kluwer Academic Publishers.
9. T. Kaino, and K. Hirota (1999): Derivative of Fuzzy Relations and its Applications to Capital Investment Decision Making Problem, Proceedings of the Eight International Fuzzy Systems Association World Congress, Vol.1, 995–998, Taipei, Taiwan ROC.
10. K. Korolev, K. D. Leifert, and H. Rommelfanger (1999): Arbitragebewertung bei vagen erwartungen der marktteilnehmer, Working Paper Series: Finance and Accounting, Department of Economics, University of Frankfurt, ISSN 1434-3401, 44.
11. K. Korolev, K. D. Leifert, and H. Rommelfanger (1999): Optionspreistheorie bei vagen Daten, Working Paper Series: Finance and Accounting, Department of Economics, University of Frankfurt, ISSN 1434-3401,42.
12. R.C. Merton (1977): An analytic derivation of the cost of deposit insurance and loan guarantees, Journal of Banking and Finance,1.
13. R.C. Merton and Z. Bodie (1992): On the management of financial guarantees, Financial Management,21.
14. R.A. Ribeiro, H. J Zimmermann, R. R. Yager, and J. Kacprzyk, editors (1999): Soft Computing and Financial Engineering, Physica Verlag.
15. H. Rommelfanger (1994), Fuzzy Decision Support Systeme, Springer Verlag, Berlin.
16. L. A. Zadeh (1965): Fuzzy Sets,Information and Control, 8, 338–353.

Bank Creditworthiness Using Fuzzy Systems: A Comparison with a Classical Analysis Approach[1]

G.Facchinetti, S.Bordoni and G.Mastroleo

Department of Political Economics
Faculty of Economics
University of Modena and Reggio Emilia
Via Berengario 51
41100 Modena, Italy
E-mail: facchinetti@unimo.it

Abstract

In the last decades, the problem of measuring credit risk has been the object of analysis by researchers. Financial organisations, banks, credit institutions, clients, suppliers, etc., need prediction of failure for firms in which they have any kind of interest. The most widely used methods are based on econometric analysis, which estimates the probability of clients' insolvency (default). Often they do not show a satisfactory ability to discriminate between creditworthy and non-creditworthy clients. This situation is often due to the use of unrealistic assumptions of statistical hypotheses and to the complete lack of communication with the decision makers. The aim of this research is to overcome these limitations. In this paper a fuzzy logic approach, an alternative method, is used to provide a system able to evaluate bank creditworthiness. The financial data of 400 clients, offered by the Bank of Sardinia[2] and relating to small businesses are used to compare the econometric and fuzzy approaches[3]. The results are really interesting and show how the fuzzy system may offer better solutions to the problem of bank creditworthiness.

1. Introduction

In the recent years there has been an increase of bank competition in the activity of credit grants. This induces banks to assume higher risks in order not to lose clients and money. It thus becomes all the more necessary to use subtler tools as decision supports. These problems are typical not only of banks but also of

---

[1] This research work is financed by Modena University (project " Fuzzy logic prototype of a scoring for bank creditworthiness" 1999).
[2] We thank the Bank of Sardinia for the data offered and for all the time their experts dedicated us. In particular, we thank again the Bank's President, Prof. Sebastiano Brusco, who believed in this research and supported us.
[3] The fuzzy system is built with the support of "fuzzyTECH" by INFORM.

financial organisations, credit institutions, clients, suppliers, insurance, which need predictions of credibility for firms or persons in which they have any kind of interest.

In the last decades, the problem of measuring of credit risk has been the object of analysis by researchers and credit operators. The more widely used methods are based on econometric models, which estimate the probability of clients' insolvency (default) and are used in the screening and monitoring phases. They produce a score and are based on accounts and financial data of the client requesting money. The analysis results in a positive or negative client judgement. Among these methods are "linear probability" analysis, logit [8] and probit analysis [14] and discriminant analysis [1]. In particular, Altman's Z-score has, in the past, given some good performances. Recent studies have shown the disadvantages of this approach, [4], [6] and to provide a higher level of prediction accuracy new methods have been proposed. They use methodologies like neural networks [2], expert system [7], and fuzzy logic. [9],[12]. These methodologies fall under the rubric of "soft computing".

In this paper a fuzzy logic system is built for evaluation of bank creditworthiness, and the results are compared with what is obtained by classic methods such as a multivariate approach. This is the first paper in which two such different methods are compared on the same data and even though not all the power of the fuzzy inference system is exploited, the results obtained show that this new approach offers a better performance than the classic one.

2. The Soft Computing Approach

Lofti A. Zadeh in the foreword of [9] writes: "The heart of the applications of quantitative methods to business and finance is decision analysis. The pioneer searchers of this field are mathematicians like von Neumann, Morghenstern, Wald. They thought that it is possible to construct a theory of decision-making that has an axiomatic foundation. With the passage of time and accumulation of experience, it is becoming increasingly clear that the real-world decision-making is much too complex, too uncertain, and too imprecise to lend itself to precise, prescriptive analysis. It is this realization that underlies the rapidly growing shift from conventional techniques of decision analysis to techniques based on fuzzy logic, neurocomputing, genetic computing, more in general "Soft Computing". Soft Computing is a consortium of methodologies that are tolerant of imprecision, uncertainty, and partial truth."

These methodologies are complementary rather than competitive. It is often useful to employ them in combination, giving rise to "hybrid systems", like neurofuzzy computing. They have become very widespread, for two principal reasons.

The first is that they are more effective in dealing with problems in which the dependencies between variables are too complex or too ill defined to admit a characterization by models, which are linear or quadratic. Such problems are the norm in economics and financial fields. The second is the progress in electronic data processing. Till a few decades ago with the first batch processing systems, users had to wait a long time before they got the output of their computations. The introduction of time-sharing systems has given users the benefit of accessing their data in real time.

Although, engineering applications of soft computing have gained much more public interest than business and financial applications, an even larger potential exists here.

The development in computer and information sciences has made the world of business and finance a magnet for methodologies that can exploit the ability of modern information systems to process huge volumes of data at high speed and with high reliability. Fuzzy logic has the benefit of enabling software to make human-like decisions. Zadeh himself, founder of fuzzy logic, contends that a computer cannot solve problems as well as human experts unless it is able to think in the characteristic manner of human beings [9]. Enabling software to make human-like decisions yields many benefits:

- For decisions that need to be made in large quantities, such as buy or sell decision in a stock-trading system, automation of decision-making greatly expands capacity at low cost.
- Complex decision-making processes become transparent and can thus be explicitly evaluated and optimized.
- The experience of more than one single person can be agglomerated into one system.

Fuzzy logic, introduced by Lofti Zadeh [10], has attracted the attention of many researchers, who have contributed to its development and applications to language, automata theory, and learning systems. In the early seventies, however, the introduction of linguistic variable and fuzzy if-then rules opened the door to many other applications like control theory. In the absence of complete and precise information, fuzzy logic and fuzzy sets are effective tools for modelling complex business, finance and management systems. The subjective judgement of experts who have used fuzzy logic techniques produces better results than the objective manipulation of inexact data [3]. Fuzzy logic exceeds the inability of classical logic to capture the vague language, common sense reasoning used by people every day. It deals with objects that have all the possible grades of truth between "yes" and "no".

A fuzzy expert system applied to decision-making problems is defined in the same way as ordinary expert systems, but here the methods of fuzzy logic are applied. Fuzzy expert systems use fuzzy data, fuzzy rules, and fuzzy inference, in addition to the standard ones implemented in the ordinary expert systems.

The following are the main stages of a fuzzy system design [5]:

1. Identifying the problem and choosing the type of fuzzy system which best suits the requirements of the problem. A modular system can be designed consisting of several fuzzy modules linked together. A modular approach, if applicable, may greatly simplify the design of the whole system, dramatically reduce its complexity and make it more comprehensible.
2. Defining the input and output variables, their linguistic attributes (fuzzy values) and their membership function (fuzzification of input and output).
3. Articulating the set of heuristic fuzzy rules. (IF-THEN rules).
4. Choosing the fuzzy inference method (selection of aggregation operators for precondition and conclusion).
5. Translation of the fuzzy output in a crisp value (defuzzification methods).
6. Experimenting with the fuzzy system prototype; drawing the goal function between input and output fuzzy variables; changing membership functions and fuzzy rules if necessary; tuning the fuzzy system; validation of the results.

3. The Creditworthiness Expert System

The building of a fuzzy expert system has two main problems: the fuzzification and the construction of blocks of fuzzy rules. These two steps are obtained in different ways. One approach is the interview with experts of the problem. Another is the use of the methods of machine-learning, neural networks and genetic algorithms to learn membership functions and fuzzy rules. The two approaches are quite different. The first approach does not use the past history of the problem and allows a real contact with the experts who may permit all the experience matured in years of work in that field to enter the study. The second is based only on the past data and transfers for the future the same structure as the past.

The aim of this paper is to produce a method to evaluate the creditworthiness of a client and to forecast its good performance. Usually the historical data offered by the bank are relative only to credits granted (we have no information about clients who have not been granted credit) and those granted several years previously. For to have complete historical data it is necessary to analyse positions that have started and ended. For these reasons, we have preferred the first approach. We can formalize the steps in the following manner.

For each linguistic variable, input x_i (i=1...m) and output y, we must fix the one's range of variability U_i and V.

$\forall$ i, (i=1...m), if n_i is the number of the linguistic attributes of the variable x_i and $\hat{n} = \max_{i\in[1,m]} n_i$, we define the set

$$A^i = \{A_1^i, A_2^i, ..., A_{j_i}^i, ..., A_{n_i}^i\}$$

where $\forall\ j_i \in [1,\ n_i]$, $\forall\ n_i \in [1, \hat{n}]$ $A_{j_i}^i$ are the fuzzy numbers that describe the linguistic attributes of the input variable x_i..

In the same way we define the set

$$B = \{B_1, B_2, ..., B_k, ..., B_r\}$$

where $\forall k \in [1, r]$ B_k are the fuzzy numbers which describe the linguistic attributes the output variable y.

With every elements of A^i and B a membership function

$$\mu_{A_{j_i}^i}(x): U_i \to [0,1] \qquad \text{and} \qquad \mu_{B_k}: V \to [0,1]$$

is associated.

The elements of A^i and B are overlapped in some "grey" zones in which they are not easy to characterize precisely. Most things in the world do not fall clearly into one crisp category or another. Experts who use abstraction as a way of simplifying the problem identify the location of these "grey" zones.

The choice of the slopes of the elements of A^i and B is a mathematical translation of what the experts think about the single terms.

The second step is the construction of block rules.

We define the set Ψ of h fuzzy rules, where $h \le \prod_1^m n_i$, $\forall\ n_i \in [1, \hat{n}]$ of this type:

$\forall$ i$\in$ [1,m], $\forall\ j_i \in [1, n_i]$, $\forall k \in [1, r]$

IF (x_1 is $A_{j_1}^1$) $\otimes$ (x_2 is $A_{j_2}^2$) $\otimes \ldots \otimes$ (x_m is $A_{j_m}^m$) (5)

THEN (y is B_k), (6)

The relation (5) is called "precondition" and the symbol $\otimes$ represents one of the possible aggregation operators. The ones most widely used in practical applications are the MIN and the MAX operators, or a convex combination of them

$$\gamma MIN + (1-\gamma) MAX \qquad \text{with } \gamma \in [0,1]$$

where $\gamma \in [0,1]$ could be interpreted as the "degree of positive compensation".

If, instead of Min and Max, other t-norms or conorms are used, "negative", or "positive" compensation will occur for different value of γ. One can also choose different ways of linking the "and" with the "or". If we start with a multiplicative coupling, we define:

$$\mu_{A\Theta B} = \mu_{A\cap B}^{1-\gamma} * \mu_{A\cup B}^{\gamma} \qquad \text{with } \gamma \in [0,1]$$

where $\mu_{A\Theta B}$ is the membership of the aggregated set $A\Theta B$, and $\mu_{A\cap B}$ and $\mu_{A\cup B}$ are general memberships of the intersection and the union.

If the intersection and the union are algebraically represented by the product and the algebraic sum, we obtain the γ-operator [13]

$$\mu = \left(\prod_{1}^{n} \mu_i\right)^{(1-\gamma)} * \left(1 - \prod_{1}^{n} (1-\mu_i)\right)^{\gamma} \qquad (7)$$

If $\gamma = 0$, then:

$$\mu_{\cap} = \prod_{1}^{n} \mu_i$$

if $\gamma = 1$, then:

$$\mu_{\cup} = 1 - \prod_{1}^{n} (1-\mu_i)$$

The γ-operator is pointwise injective, continuous, monotonous and commutative. In later empirical studies it was shown that this aggregator concept represents the human decision process more accurately than the others. [11]

When we have decided what type of precondition aggregator $\otimes$ to use, we begin with the introduction of the inputs values $x_i^0 \;\; \forall\, i \in [1,m]$. This fact produces that only a subset Ω of the rules set Ψ is involved. The elements of this subset are the rules in which are present in the IF part the fuzzy numbers $A_{j_i}^i$ that have x_i^0 in its domain. Every element of Ω produces a numerical values of this type:

$$\alpha_{j_1, j_2, \ldots, j_m} = \mu_{A^1_{j_1}}(x_1^0) \otimes \mu_{A^2_{j_2}}(x_2^0) \otimes \ldots \mu_{A^m_{j_m}}(x_m^0) \quad (8)$$

This value produces the strength of the rule involved and gives its level of "firing" (activation). At this step we must check, in Ω, what are the B_k involved.

The "conclusion" (6) is obtained by:

$$[\mu_{A^1_{j_1}}(x_1^0) \otimes \mu_{A^2_{j_2}}(x_2^0) \otimes \ldots \mu_{A^m_{j_m}}(x_m^0)] \wedge \mu_{B_k}(y) \quad (9)$$

The result of (9) is a fuzzy set $\tilde{B}_k$ obtained by B_k cut off at the level $\alpha_{j_1, j_2, \ldots, j_m}$.

When the same B_k is affected by different rules, we must choose what will be the level of cut off to consider. We have several methods available. The choice depends of the type of application. The most used are the MAX and the BSUM method. The MAX has the meaning of keeping as "winner" the strongest rule, in the sense that if a rule is "firing" more then one time, the result is the maximum level of firing. In the BSUM case, all the firing degree is considered and the final result is the sum of the different level of activation (not more than one). The aggregation of the control outputs produces a fuzzy set, with membership function $\mu_{agg}(y)$, obtained by the union of the different $\tilde{B}_k$. We now have a result of the fuzzy inference system, which is a fuzzy replay. We need to return to a "crisp" value, a step that is called "defuzzification". This operation produces a "crisp" action $\bar{y}$ that adequately represents the output fuzzy set defined by its membership function $\mu_{agg}(y)$. There is no single way to perform this operation. To select the proper method, it's necessary to understand the linguistic meaning that underlies the defuzzification process. Two of these different linguistic meanings are of practical importance: the "*best compromise*" and the "*most plausible result*". One of the first type methods is the Centre of Area (CoA) that produces the abscissa of the centre of gravity of the fuzzy output set

$$\bar{y} = \frac{\int_V y\mu_{agg}(y)dy}{\int_V \mu_{agg}(y)dy}$$

One of the methods of the second type is the "Mean of Maximum" (MoM) that, rather then balancing out the different inference results, selects the typical value of the terms that is most valid. [9]

4. Classical Econometric Methods

At the end of 1998, the Bank of Sardinia ordered a study of a system of Small Business Scoring from a well-known software house. This product should enable the criteria of evaluation for the screening moment to be automatic. Here, we present the results of their study. Before passing to the fuzzy approach, we used the same approach as theirs, in order to be sure that the initial data, offered by the bank, were the same.

This approach uses a classical discriminant and logistic analysis, but we were requested by the Bank not to specify the exact model used. The study begins with the analysis of the bank data. This entails reducing the number of clients and identifying some statistical indicators that led to a reduction of the variables. A scorecard is constructed in which, for every selected variable, the range of variation is decided. This range is divided into some intervals and a score value is assigned to each interval.

The final score is obtained by the sum of the values that every client obtained using the selected variables. As a last step, some measurements of the discriminant power of the model are presented.

The target is the type of Small Business that, for the Bank of Sardinia, is:

- Individual firms, personal society or capital society with a turnover of less than 1,5 million Ecu.
- Credit up to 100.000 ECU

4.1. Classical Environment Analysis

- **Population of reference.**

After selection of the 8020 initial cases, 2557 positions are analysed.

- **Performance definition**

The clients' performance at the end of 1998 is the following:

Situation	Definition
"Bad"Customers	nonperforming, nonaccrual, bad loans,
"Good" Customers	good
Bad rate	100*Bad/(Bad+Good)

- **Selection of the sample**

From the population they chose by lot a random sample of 26% of the total population which comprised 330 good and 330 bad clients.

The documents offered by bank were not completed for all the clients identified by sampling, so they considered only 284 good and 199 bad clients.

4.2. Data analysis report

The data analysis report studies the trend of the risk of the variables that are eligible to enter the score grid.

For every variable used by the bank for clients' creditworthiness (about 130), a table was constructed in which two statistics are produced:

- Bad rate: measures the risk and represents the percentage of bad clients out of the total.
- Power statistic: measures the level of separation between bad and good clients. Large values of the power statistic identify the variables that are predictive in the identification of the bad clients.

Some variables are presented in a collected form. These clusters are made by using their experience and to satisfy the following objectives:

- To collect classes with similar bad rate.
- To smooth, where possible, the risk trend and remove statistical anomalies.
- To use intuition and logic in the clustering of the variables.

4.3. The evaluation of the variables

The 130 variables that the bank uses for client creditworthiness were analyzed. Using the two indexes previously described, 22 most predictive ones were obtained. Among these, nine were considered most interesting. There are no explanations for this choice. The selected variables are:

- Legal structure (type of legal structure of the Company)
- Branch economic activity (line or branch of economic activity)
- Available assets (total available assets: commercial credits + other activities at short run + partner's withdrawal)
- Total equity (economic wealth + year-end profit and loss picture)
- Number of employees
- Debts at short (debt with suppliers + other short term debts + short-term debts to banks)

- Variation of turnover ({[turnover(t) - turnover(t-1)]/ turnover(t-1)}*100)
- Profit and loss (year-end profit and loss picture)
- Risky data base registration (possible position: yes or no)

5. The Fuzzy Inference System

The goal of this research is to show the different improvements that a fuzzy inference system may offer in building a score for bank creditworthiness, as compared to the use of a classical scorecard. To reach this result, we built a model for every phase of the development.

The first phase: we used a neuro-fuzzy approach to reproduce the classic scorecard results. For the training of the network we used all the disposable knowledge, as it is typical of the hybrid approach.

The experts have decided the inputs aggregation schema and the numbers of the linguistic attributes. As we have nine input variables, we have built three aggregation blocks, with three variables each. The system produces three intermediate variables i_1, i_2, i_3 that are outputs for the first rule-blocks and inputs for the next rule-block. The result obtained is a fuzzy algorithm that approximates the scorecard results but has at its core some expert knowledge.

The second phase: we optimised the first system erasing the rules with Degree of Support (DoS) near zero. The back testing on the data with this second system produced better results in comparison to the scorecard results.

The third phase: we built a new fuzzy system, adding four new inputs. This decision was due to the participation of the experts. They decided that the nine variables proposed by the classical approach were not sufficient to describe the problem. They suggested four new variables, their aggregation, their fuzzification and the relative rule block.

- Cashflow (year-end profit and loss + depreciation + allocations)
- Financial costs / Turnover (financial charges / turnover)
- Used credit / Granted credits
- ROI ([turnover ± stocks – raw materials costs – operating costs – wage costs – depreciation – allocations] / Total Assets)

This addition produced the new intermediate output i_4, which with the input ROI was been added to i_1, i_2, i_3 to build the five new inputs of the system, which produces the final valuation.

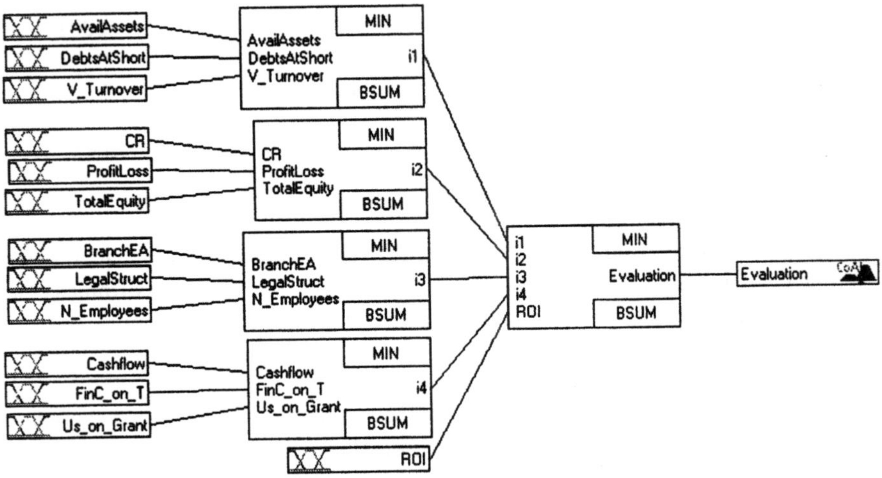

Figure 1

In Figure 1 we have the schema of the system as described in the foregoing explanations. This project editor is built with the software "fuzzyTECH" by Inform. On the left we have four groups of three variables each. The first three groups generate the intermediate variables i_1, i_2, i_3, the fourth group is made with three of the four new variables and generates the intermediate variable i_4. The last variable ROI and the others produce the output Valuation. We have used the MIN operator for the aggregation of the precondition, the BSUM for the conclusion, and CoA for the defuzzification.

6. Comparison Result

In order to compare the classical and the fuzzy logic approach, we present the two scores and two indexes of discriminant power of the models, as Maximum Spread and Span Area. The two scores give an evaluation of creditworthiness and offer different levels of "cut-off", that are the thresholds of the scoring under which the bank decide to refuse requests. In the first column are present the score values; in the "Good" Customers column is present the cumulate percentage of Good; the same for the column of "Bad" Customers. The Maximum Spread measures the maximum distance between the cumulated distributions of the "Good" and the "Bad" Customers. Higher this value is and higher the model discriminant capacity is. The Span Area is the area between the same distributions. Even in this case higher values correspond higher model discriminant capacity.

In Table 1 there is the classical score and the corresponding Maximum Spread and Span Area. The two values are:

Maximum Spread = 48,20
Span Area = 19,3%

Table 1

Score	"Good" Customers	"Bad" Customers
29	0,00%	0,00%
31	0,00%	0,50%
41	0,00%	1,00%
51	0,00%	2,00%
61	0,00%	4,50%
71	0,00%	7,50%
81	0,00%	12,10%
91	0,40%	19,10%
101	1,80%	26,60%
111	4,20%	34,20%
121	6,00%	42,20%
131	10,60%	51,80%
141	16,50%	61,80%
151	24,30%	72,50%
161	31,30%	78,40%
171	41,20%	86,90%
181	50,70%	92,00%
191	59,50%	94,00%
201	71,10%	98,00%
211	80,60%	99,50%
221	86,60%	100,00%
231	91,50%	100,00%
241	93,30%	100,00%
251	95,40%	100,00%
261	97,50%	100,00%
271	98,60%	100,00%
281	98,60%	100,00%
291	98,60%	100,00%
301	99,60%	100,00%
303	100,00%	100,00%

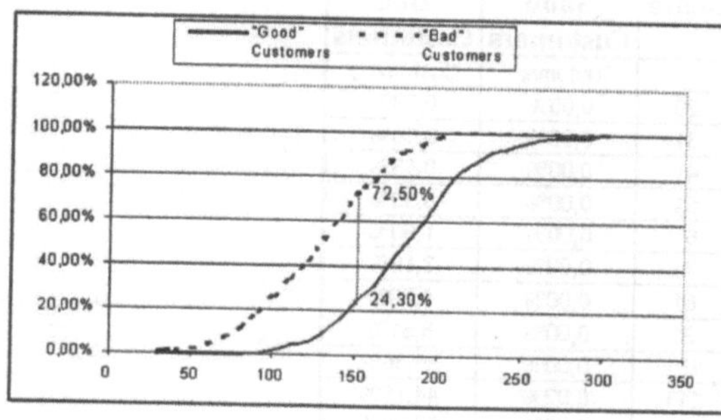

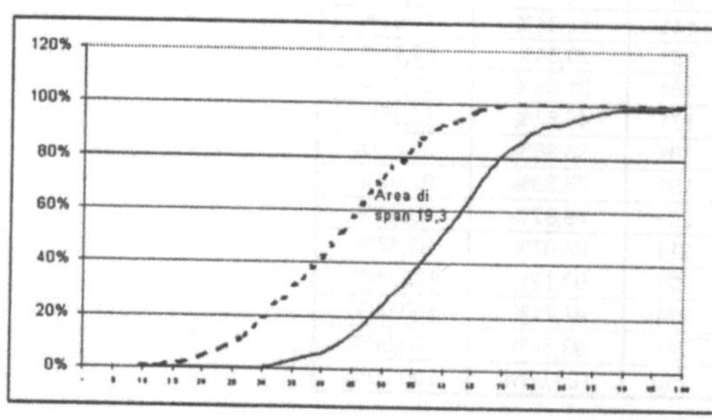

In Table 2 there is the score obtained with the fuzzy approach. In this case the two indexes are:

Maximum Spread=61,81
Span Area=43,5%

Table 2

Score	"Good" Customers	"Bad" Customers
	Customers	Customers
29	0,00%	0,00%
31	0,00%	0,00%
41	0,00%	0,53%
51	0,00%	0,53%
61	0,00%	1,60%
71	0,00%	3,19%
81	0,00%	4,79%
91	0,00%	8,51%
101	0,00%	39,36%
111	0,92%	44,15%
121	2,29%	56,38%
131	7,34%	63,30%
141	12,84%	71,81%
151	20,64%	82,45%
161	31,65%	86,70%
171	45,87%	90,96%
181	56,88%	97,87%
191	73,39%	97,87%
201	79,82%	98,94%
211	88,07%	99,47%
221	93,12%	100,00%
231	97,71%	100,00%
241	99,54%	100,00%
250	100,00%	100,00%

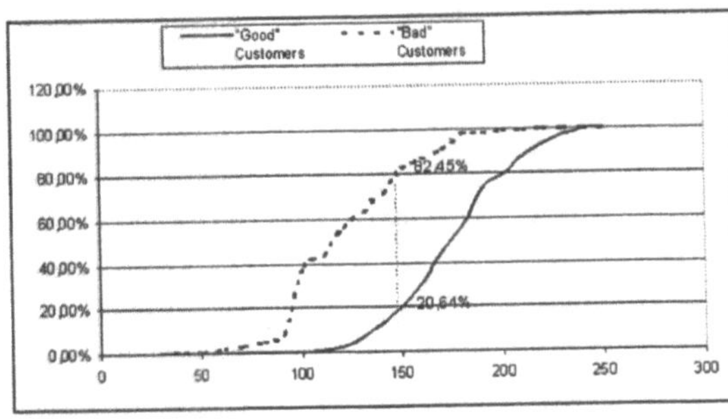

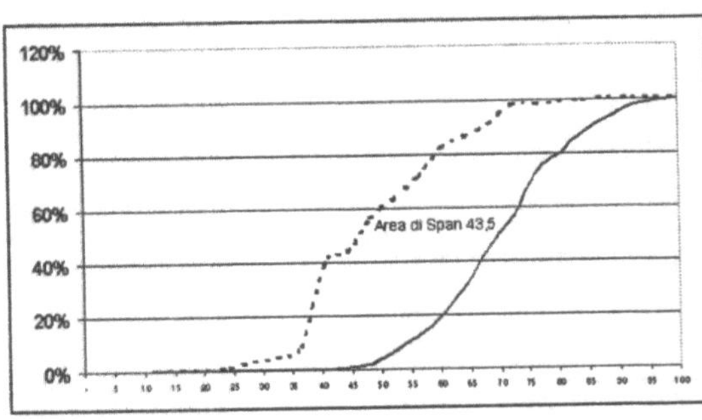

7. Conclusion and Remarks

In this paper, we propose a fuzzy expert system for bank creditworthiness. Comparison and implementation are performed with classical econometric techniques. The classical approaches and results were proposed to the Bank of Sardinia by a well-known software house. The results obtained show that this fuzzy expert system offers a better measurement of the discriminant power of the model. This happens even if we have not used the real capability of a fuzzy inference system that does not need to reduce the number of variables. Owing to its nature, it has the possibility to work with a great number of inputs. We think that the combination of the strong structure of a fuzzy system with the background knowledge of the decision-makers is a correct way to obtain good results. In addition, this type of method offers other advantages:

1. It offers transparency of classification decisions.
2. It does not use past data for the construction of the scoring.

These two facts are really interesting tools. For the variable fuzzification and the rule blocks are completely readable at any moment by the users of the system. They may decide to change some of those without destroying the system, but in a very simple way. The past data are used only for tuning of the system, not for building it. In addition, this fact enables a good method of gauging creditworthiness to be constructed even in the absence of sufficient historical data, a situation that not infrequently occurs.

Bibliography

[1] Altman, E.I.(1968). *"Financial ratios, discriminannt analysis and the prediction of corporate bankruptcy"* The Journal of Finance 23,589-609.

[2] Altman, E.I.-Marco, G:- Varetto, F. (1994). *"Corporate distress diagnosis: Comparison using discriminant analysis and neural network (the italian experience)"*. Journal of Banking and Finance 18,505-529.

[3] G.Bojadziev, M.Bojadziev (1997). *"Fuzzy Logic for Business, Finance and Management"*, (World Scientific Publishing co, Singapore), 1997.

[4] Eisembeis, R.A. (1977). *"Pitfalls in the application of discriminant analysis in business and economics"*. The Journal of Finance 32, 875-900..

[5] Kasabov, N.K.(1996) *"Foundation of Neural Networks,Fuzzy Systems, and Knoledge"* Engineering.(MIT Press)

[6] Malecot,J.F. (1981) .*"Limites des modèles de prévision de défaillances"* Finance 2,4, pag. 291-315

[7] Messier, W.F.; Hansen, J.V. (1988). .*"Including rules for expert systems development: An example using default and bankruptcy data"*. Management Sciences 34, 12, p.1403-1415.

[8] Ohlson, J.A. (1980) *"Financial ratios and probabilistic prediction of bankruptcy"* Journal of Accounting Research , Spring, p. 109-131.

[9] C. von Altrock (1997). .*"Fuzzy Logic and neurofuzzy applications in business and finance."* (Prentice Hall),

[10] Zadeh, L.A. (1965).*"Fuzzy sets"* Information and Control, 8, p.338-353

[11] Zimmermann, H. J. Zysno P.(1980) .*"Latent Connectives in Human Decision Making"*, Fuzzy Sets and Systems, 4, 37-51.

[12] Zimmermann H. J. (1996).*"Fuzzy Sets Theory and its Applications."* 3rd

revised edition, Kluver Academic Publisher, Boston and Dordrecht.

[13] Zimmermann H. J (1997)"*Operators in models of decision making*". In "Fuzzy Information Engineering" p. 471-496.

[14] Zmijewski,M.E. (1984)." *Methodological Issues related to the estimation of financial distress prediction models*". Studies on Current Econometric Issues in Accounting Research, p. 59-82

Risk of Options – Impact of Volatility Parameter

Krzysztof Jajuga and Katarzyna Kuziak

Department of Financial Investments and Insurance
Wrocław University of Economics
ul. Komandorska 118/120
53-345 Wrocław
Poland
Email: {jajuga, kakuziak}@manager.ae.wroc.pl

1. Derivative Instruments – Introduction

A derivative instrument is a financial instrument whose value depends on the value on the underlying index, for example on the price of another financial instrument (e.g. currency, stock). Derivative instruments gained a lot of attention of financial theoreticians and practitioners in the last decade. The main reason is the increasing risk both on the financial markets and in the whole economy. On the one hand, the derivative instruments proved to be very useful instruments in risk management. This refers to the following types of risk:

- market risk (interest rate risk, exchange rate risk, stock price risk, commodity price risk);
- credit risk;
- enterprise-wide risk;
- weather risk;
- insurance (catastrophe) risk.

On the other hand, the theory of derivative instruments (especially option pricing theory) proved to be very important tool in the development of modern financial theory.

There are two basic groups of derivative instruments – here the main difference is relation between two parties with respect to the distribution of risk. These are:

- option-like instruments, being the "asymmetric" contracts, here classical ("plain-vanilla") options and exotic options can be distinguished;
- forward-like instruments, being the "symmetric" contracts, here forwards, futures and swaps can be distinguished.

In the option-like instruments one party of the contract (called option holder) transfers risk to the other party of the contract (called option writer). In the forward-like instruments both parties of the contract exchange (part of) different types of risk, so that the total risk of both parties decreases.

In this paper we consider classical options. The pricing models for these instruments are well known, relatively advanced and widely used.

The easiest application of options in risk management is very straightforward:

- the purchase of call option is the insurance against the increase of the value of underlying instrument;
- the purchase of put option is the insurance against the decrease of the value of underlying instrument.

There are however, more advanced strategies for risk management, called hedging strategies. We will discuss them later.

2. Risk of Derivative Instruments

Options are financial instruments. Similarly as other financial instruments, including the derivative instruments, options are subject to financial risk. There are different types of financial risk, including:

- market risk, this is the risk caused by the changes of prices in the financial market (e.g. interest rates, exchange rates, stock prices, etc.);
- credit risk, this is the risk of a loss arising from the failure of a counterparty to make a contractual payment – in the case of derivative instruments is present in OTC derivatives;
- operational risk, this is risk caused by malfunctioning of different systems, including information systems, reporting systems, management systems,

settlement systems etc as well as the risk caused by the errors made by people.

Market risk is the most quantifiable type of financial risk. There are three main groups of market risk measures:

- volatility measures;
- sensitivity measures;
- downside risk measures.

Volatility measures aim at determining how large is the fluctuation (volatility) of financial prices or financial returns. Of course, the higher volatility, the higher risk. Clearly, volatility is understood as the dispersion of prices. Therefore the common volatility measures are the spread (scale, scatter) parameters obtained either for theoretical distribution or for the set of observations, for example standard deviation of returns.

Sensitivity measures reflect the influence of some variables (called risk factors) on the financial prices or returns. The more sensitive is the financial price (or return) to the changes of the factors, the higher is market risk of financial instrument. To apply this type of measure one has to derive the risk model, where the price of financial instrument is the function of suitable risk factors. The sensitivity measure is usually defined as a partial derivative of the price with respect to the risk factor.

The third group of market risk measures contains **downside risk measures**. They put the emphasis on the adverse deviations from expectations. One of the most common and recommended measures is **Value at Risk (VaR)**.

Value at Risk (VaR) is such an amount of money that the likelihood of experiencing a loss in the market value of a financial instrument in excess of that amount, due to the adverse change in market risk factors over a specified risk horizon, is less than a specified tolerance level.

It is easy to notice that VaR can be defined through the quantile of the distributions of returns.

Out of these three groups of measures, two groups are particularly useful in the case of options. These are: Greek coefficients (sensitivity measures) and Value at Risk (downside risk measure). The so-called Greek coefficients are sensitivity measures therefore market risk is understood as the sensitivity of the option price with respect to different factors influencing the option price. From the point of view of risk management the most important Greek coefficients are:

- delta (δ) – it is the sensitivity of the option price to the value of the underlying index;
- gamma (γ) – it is the sensitivity of the option price to the delta coefficient (the measure of the convexity of the relationship between the value of the option and the value of the underlying instrument);
- vega (denoted by λ) – it is the sensitivity of the option price value to the volatility of the underlying index.

In this paper we consider mainly the so-called model risk. It is different from the mentioned types of risk, since it does not come from the financial market. It is understood as the risk due to the errors in the model derived in theory and used in real applications (for example the risk model).

Two types of errors are common when the model risk is considered:

- errors resulting from taking not proper form of the model (specification errors);
- errors of the estimation of the parameters of the model (estimation errors).

3. Model Risk in the Black-Scholes Option Pricing Model

In this paper we consider the famous option pricing model, belonging to the so-called Black-Scholes-Merton methodology (Black, Scholes (1973), Merton (1973)). These models are based on the risk-free arbitrage pricing. The considered model is Black-Scholes model for pricing European option on the non-dividend-paying stock given as:

$$c = SN(d_1) - Xe^{-rT}N(d_2)$$

$$p = Xe^{-rT}N(-d_2) - SN(-d_1)$$

$$d_1 = \frac{\ln(\frac{S}{X}) + (r + \frac{\sigma^2}{2})T}{\sigma\sqrt{T}}$$

$$d_2 = \frac{\ln(\frac{S}{X}) + (r - \frac{\sigma^2}{2})T}{\sigma\sqrt{T}} = d_1 - \sigma\sqrt{T}$$

where:

c – the value of European call option,
p – the value of European put option,
S – the value of the underlying stock,
X – the exercise price of the option,
T – time to the expiration of the option,
σ – the volatility of the underlying stock (given as standard deviation of returns),
r – risk-free rate of return,
$N(d)$ – the cumulative normal distribution function.

Under the Black-Scholes model, the Greek coefficients are given as:

- delta coefficient:

$$\delta_{call} = N(d_1)$$
$$\delta_{put} = N(d_1) - 1$$

- gamma coefficient:

$$\gamma_{call} = \gamma_{put} = \frac{n(d_1)}{S\sigma\sqrt{T}}$$

- vega coefficient:

$$\lambda_{call} = \lambda_{put} = Sn(d_1)\sqrt{T}$$

where:

$n(d)$ – the value of the density function for standardized normal distribution at d.

It can be also proved (e.g. Jorion (1997)) that Value at Risk for the individual option, under Black-Scholes approach is equal to:

$$VaR = c\sqrt{\delta^2 S^2 \sigma^2 + 0.5(\gamma S^2 \sigma^2)^2}$$

where:

c – value depending on the probability,
for 0.95 we take $c = 1.65$; for 0.99 we take $c = 2.33$.

So, Value at Risk is a function of delta coefficient and gamma coefficient.

Although the Black-Scholes model is a pricing model, it can be considered as the model used for hedging purposes. It was derived by the application of risk-free arbitrage argument. This argument means that the investor cannot make extra profit by doing risk-free and no initial investment transactions.

Greek coefficients are used to design the so-called delta-gamma-vega hedging strategy. Greek coefficients determine the number of stocks and options in the portfolio so that this portfolio is not sensitive to the changes of price and volatility of the stock. Therefore they are used for hedging purposes.

The Black-Scholes option pricing model is subject to model risk. Both types of the mentioned errors can occur. Firstly, we discuss specification errors.

It is well known that the Black-Scholes option pricing model is based on the assumption that the process of the underlying stock prices is geometric Brownian motion, given as:

$$\frac{dS}{S} = \mu dt + \sigma dz$$
$$dz = \varepsilon\sqrt{dt}$$

where:

ε – the stochastic process where random variables are independent and normally (N(0;1)) distributed.

Geometric Brownian motion is a very convenient from the theoretical point of view. However in practice the stock price process very often departs from geometric Brownian motion. First of all, the random process ε can have non-normal distribution. Secondly, quite different process can be suitable for the prices of the underlying instrument, e.g.:

1. Ornstein-Uhlenbeck process (type of mean-reverting process), given as:

$$\frac{dS}{S} = \eta(M - S)dt + \sigma dz$$
$$dz = \varepsilon\sqrt{dt}$$

where:

M – the parameter – the mean level of the process,
η – the parameter – "speed" of the mean reversion.

2. Jump diffusion process, given as:

$$\frac{dS}{S} = (\mu - \lambda k)dt + \sigma dz + dq$$
$$dz = \varepsilon\sqrt{dt}$$

where:

dq – Poisson process.

The second component of the model risk contains estimation errors. In the Black-Scholes there are five parameters. Three of them (the price of the underlying stock, the exercise price of the option, time to the expiration of the option) are known, the fourth parameter, risk-free rate of return is easily estimated. The only parameter, which is subject to the estimation error, is the volatility parameter, given as the standard deviation of returns of the underlying stock.

The most common methods to estimate this parameter are:

- historical volatility – here historical stock returns are used in the assumed volatility model, like GARCH model (see e.g. Hull (2000));
- implied volatility – here past option prices are used to calculate volatility directly from the Black-Scholes model.

Both types of errors have impact on the application of Black-Scholes model for financial risk management, namely for the strategies used in risk management. In particular, the following strategies are subject to model risk:

- delta–gamma–vega hedging; here the portfolio of financial instruments (particularly options) is constructed that it is not sensitive (in short period) to the changes of stock price and stock volatility;
- option trading strategies; here different combination of options are constructed in such a way that the desired risk profile in certain date is achieved;
- more advanced strategies – it is worth to mention Hindy optimization model (see Hindy (1995)).

4. Some Empirical Studies on the Impact of Volatility Parameter

We present the results of two empirical studies, where the impact of the volatility parameter is shown. In both studies the Black-Scholes model is considered. In these studies we considered only call option.

Study 1

Here we studied the behavior of market risk measures of the options for different values of volatility parameter. This reflects the sensitivity of market risk measures on the values of volatility parameter, which (as we argued) is not known in practice. Four market risk measures were studied: delta (δ), gamma (γ), vega (λ), and Value at Risk (for the tolerance level equal to 0.01). The following values of the other parameters were assumed:

- stock price (S): 100,
- exercise price (X): 90, 95, 100, 105, 110,
- time to the expiration (T) – in days: 180, 90, 30, 15, 3,
- risk-free rate (r): 5%, 10%.

The values of stock price and exercise price define the so-called:

- deep-in-the-money option: S=100, X=90;
- in-the-money option: S=100, X=95;
- at-the-money option: S=100, X=100;
- out-of-the-money option: S=100, X=105;
- deep-out-of-the-money option: S=100, X=110.

Then for each set of values of these parameters the behavior of four market risk measures for the different values of volatility (between 0.01 to 0.5) was studied. This gives the idea how the error in the estimation of volatility parameter reflects the values of risk measures, which are then used for risk management purposes.

The following conclusions can be derived on the base of the obtained results.

1. Impact of increasing volatility parameter on delta coefficient.

1A. First no changes, then the decrease of delta:
the case of in-the-money and deep-in-the-money option;

1B. First no changes, then the increase of delta:
the case of out-of-the-money and deep-out-of-the-money option:

- if simultaneously risk-free rate increases and time to expiration (30, 15, 3 days) decreases;
- or if time to expiration decreases;

1C. First the increase of delta, then no changes:
the case of out-of-the-money and deep-out-of-the-money option;

1D. First the decrease of delta, then the increase of delta:
the case of out-of-the-money option – if risk-free rate increases;

1E. First the decrease of delta, then no changes:
the case of at-the-money option.

2. Impact of increasing volatility parameter on gamma coefficient.

2A. First no changes, then the increase of gamma and then the decrease of gamma:
the case of in-the-money and deep-in-the-money option;
the case of out-of-the-money and deep-out-of-the-money option:

- if simultaneously risk-free rate increases and time to expiration (30, 15, 3 days) decreases;
- or if time to expiration decreases;

2B. First no changes, then the increase of gamma:

the case of in-the-money, out-of-the-money and deep-out-of-the-money option:

- if simultaneously risk-free rate increases and time to expiration equal to 3 days;
- or if time to expiration decreases;

the case of deep-in-the-money option:

- if simultaneously risk-free rate increases and time to expiration equal to 15 and 3 days;
- or if time to expiration decreases;

2C. First the increase of gamma, then the decrease of gamma:

the case of out-of-the-money, deep-out-of-the-money and at-the-money option;

2D. First the decrease of gamma, then no changes:

the case of out-of-the-money option – if risk-free rate increases;

the case of at-the-money option

- if simultaneously risk-free rate increases and time to expiration equal to 30, 15, 3 days;
- or if time to expiration decreases (15, 3 days).

3. Impact of increasing volatility parameter on vega coefficient.

3A. First the increase of vega, then no changes:

the case of out-of-the-money, deep-out-of-the-money and at-the-money option;

3B. First no changes, then the increase of vega:

the case of in-the-money and deep-in-the-money option:

the case of out-of-the-money and deep-out-of-the-money option:

- if simultaneously risk-free rate increases and time to expiration decreases;
- or if time to expiration decreases;

3C. First the increase of vega, then the decrease of vega:

the case of out-of-the-money – if risk-free rate increases.

4. Impact of increasing volatility parameter on Value at Risk.

4A. Approximately linear increase of VaR:

the case of all options (except for the cases mentioned below);

4B. First no changes, then approximately linear increase of VaR:

the case of out-of-the-money and deep-out-of-the-money option:

- if simultaneously risk-free rate increases and time to expiration decreases;
- or if time to expiration decreases;

4C. Nonlinear increase of VaR:

the case of in-the-money option – if the time to expiration decreases.

All these relationships can be helpful for the institution, which uses hedging strategy but expects some changes in the volatility of underlying stock.

Study 2

Here we show the results of simple simulation studies. We simulated the time series of 250 daily stock prices, where the geometric Brownian motion was assumed with the values of parameters equal to:

$\mu=0$, $\sigma=2$.

The initial stock price was equal to 100 and the risk-free rate was assumed to be equal to 5%. The different values of exercise price and time to expiration were assumed:

- exercise price: 97, 100, 103;
- time to expiration (in days): 180, 90, 45.

Therefore we had 9 sets of parameters. For each set of parameters 500 repetitions were done.

Table 1. Mean values of risk measures in simulation studies – daily data.

Exercise price	Time (in days)	Delta	Gamma	Vega	Value at Risk
97	180	0.5755	0.0244	1.9587	3.0307
100	180	0.5233	0.0164	1.5519	2.8100
103	180	0.4644	0.0221	2.1896	2.5570
97	90	0.5494	0.0159	0.7162	2.9442
100	90	0.5028	0.0211	1.0192	2.7386
103	90	0.4373	0.0216	1.0975	2.4391
97	45	0.5400	0.0113	0.2557	2.8926
100	45	0.4896	0.0238	0.5778	2.7012
103	45	0.4240	0.0236	0.6048	2.3967

Then the volatility parameter was estimated based on daily data and weekly data (49 weeks), using the historical volatility estimate (standard deviation of logarithmic returns) given as:

$$s = \sqrt{\frac{1}{n-1}(r_t - \bar{r})^2}$$

where:

$$r_t = \ln S_t - \ln S_{t-1}$$

Then the values of delta, gamma, vega and VaR were calculated. Tables 1 and 2 present the mean values calculated across 500 repetitions.

The analysis of these tables leads to the following conclusions:

1. The estimates of risk measures show significantly higher values when the weekly data are used than in the case of daily data. The only exception is gamma, for which the difference is smaller.
2. The higher exercise price, the smaller delta – this is well known relationship.
3. The higher exercise price, the larger Value at Risk.
4. Gamma and vega show no relationship to exercise price – this is due to the relatively long time to the expiration.
5. The shorter time to expiration, the lower (slightly) delta.
6. The shorter time to expiration, the lower (slightly) Value at Risk.
7. The shorter time to expiration, the lower (significantly) vega.
8. Gamma shows no relationship to the time to expiration.

Table 2. Mean values of risk measures in simulation studies – weekly data.

Exercise price	Time (in days)	Delta	Gamma	Vega	Value at Risk
97	180	0.6154	0.0210	4.0662	7.3150
100	180	0.5582	0.0189	3.8860	6.7377
103	180	0.5032	0.0193	4.2044	6.1788
97	90	0.5900	0.0202	1.9940	7.1188
100	90	0.5357	0.0173	1.8473	6.5422
103	90	0.4802	0.0205	2.2751	6.0152
97	45	0.5767	0.0198	0.9974	7.0494
100	45	0.5261	0.0162	0.8803	6.4633
103	45	0.4677	0.0216	1.2088	5.9741

Although this simulation study is just one example, it shows some relationships that are worth to be explored.

References

Black F., Scholes M. (1973). The pricing of options and corporate liabilities. Journal of Political Economy. 81. pp. 637-654.

Hindy A. (1995). Elements of quantitative risk management. w: Beaver W.H.. Parker G.. Risk management. problems and solutions. pp. 107-138. McGraw-Hill. New York.

Hull J.C. (2000). Options. futures and other derivatives. Prentice Hall. Upper Saddle River.

Jorion P. (1997). Value at Risk. Irwin. Chicago.

Merton R.C. (1973). Theory of rational option pricing. Bell Journal of Economics and Management Science. 4. pp. 141-183.

Managing Portfolio of Options

Dariusz Gątarek

BRE Bank SA
00-950 Warszawa, ul. Senatorska 18, Poland

Systems Research Institute. Polish Academy of Sciences
01-447 Warszawa, ul. Newelska 6, Poland

Email:dgatarek@bresa.com.pl

The aim of this note is to show that using sophisticate option pricing models is necessary not only for banks with big portfolios of exotic options. It is necessary also for portfolio management of plain vanilla options.

1. Introduction

Option pricing models are used for three purposes. One is to price illiquid (exotic) options. Second is to manage portfolios of options, both liquid and illiquid. Third is risk management. In order to make a model applicable in practice it has to be calibrated. Calibration is the following procedure:

We are given a portfolio of n options with given (by the market) prices $C=\{c_1,c_2,...,c_n\}$. A model is here a function $f(U,P,V)$, where U is the underlying (stock price, currency rate, interest rate, etc.), P – is the set of parameters (strike price, maturity, etc.) and V – is the market volatility – a parameter describing uncertainty of the market, may be multidimensional or even functional. We say that the model f is calibrated to the portfolio C if we have determined volatility V such that $c_i = f(U,P_i,V)$ for any $i=1,2,...,n$. So determined V is called implied volatility. Black-Merton-Scholes, Garman-Kohlhagen and Black'76 formulae are the most popular models for pricing European options. In practise assumptions of above models are only approximately satisfied (in the best case). If two options with the same maturity have different implied volatilities we observe the smile (smirk) phenomenon.

In this paper we do not discuss suitability of respective models, this was done in many other papers. We deal with the following problem:

Suppose we manage a portfolio of options. Assume that they are sufficiently liquid and there are no arbitrage opportunities between them. We do not expect option prices to satisfy any theoretic assumptions. The dynamics of the market prices is not exactly known, although we are given a model in which, to some extend, we believe.

We have the following quality criteria for models:

1. A model is **academic** if it can be calibrated only to small number of options (one, two, three) in terms of implied volatility.
2. A model is **acceptable** if it can be calibrated in any instant of time to any number of options with different maturities.
3. A model is **acceptable with smile** if it can be calibrated in any instant of time to any number of options.
4. A model is **good** if it can be calibrated to any number of options with different maturities. Moreover the calibration is stable, i.e. new calibration is relatively rarely necessary, certainly not for all portfolio evaluations.
5. A model is **good with smile** if it can be calibrated to any number of options in a stable manner.
6. A model is **perfect** if it describes exactly dynamics of the market.

We do not expect any known model to be *perfect*, since perfection is something we have to tend to, not to reach. If we want to manage a portfolio of options we need a model which is at least *acceptable*. Otherwise the notion of portfolio of options is not controllable, it is not possible to compute its risk exposure, if volatility (also an important risk factor!) is not uniquely determined. A *good* model is obviously better then an *acceptable* one, not only because we are lazy. A new calibration of the model changes Greek letters of the model, in consequence also risk exposure changes (sometimes radically) violating self-financing of the portfolio. A radically wrong (but *acceptable* in our sense) model is misleading, anyway some claim that it is better to have false solutions then to know nothing. A model *with smile* is better then a model without smile, although a model without smile is not worthless, since the industry knows some dirty but practical tricks how to handle with smile. One more remark should be made. The notions: *acceptable* and *acceptable with smile* are precise. The notion *good* is not well defined, since the notion *rare* is also not precise. It must be so anyway, since it is not possible to have one model forever or even for longer time: market conditions change. Common sense is a clue also here.

We may imagine another policy: take a model, which describes actual dynamics of prices (in our opinion) and do not care about option prices. If they are overpriced in that model, sell them, if they are underpriced, buy them. That portfolio can be also hedged with underlying securities: stocks, bonds, currency, etc. If we were able to predict dynamics of the market, theory of arbitrage pricing claims that such strategy would bring profits at maturity of options. Such policy is not acceptable: no bank would risk short time profit for, maybe bigger, profits in further future. Predicting future market dynamics has no more sense then predicting future stock prices and maturity of some interest rate and swap options exceeds 5 years. One would state that banks believe rather market then analysts. They are certainly right.

We will discuss quality of various popular models in next sections.

2. Stock options

This sections can be considered as an introduction to the next one dealing with interest rate options, what. The basic stock option model in the celebrated Black-Merton-Scholes model [BS]. In this model we assume that the stock price X(t) satisfies the equation under the risk-neutral measure:

(1)
$$dX(t) = rX(t)dt + \sigma X(t)dW(t),$$
$$X(0) = X_0,$$

where r is the short rate, σ - volatility and W(t) – one-dimensional Brownian motion. The stock price is given by:

(2)
$$X(t) = X(0)\exp\left\{rt + \sigma W(t) - \frac{\sigma^2}{2}t\right\}.$$

Let c(t) be the price of the European call stock option at time t, maturing at time T with strike price K. Its price is given by the famous Black-Merton-Scholes formula

(3)
$$c(0) = e^{-rT} E(X(T) - K)^+,$$

where E is the expected value under risk-neutral measure and Z^+ is the positive part of Z. For constant volatility (3) becomes

(4)
$$c(0) = X(0)N(d_1) - Ke^{-rT} N(d_2),$$

where

$$d_1 = \frac{\ln(X(0)/K) + (r + \sigma^2/2)T}{\sigma\sqrt{T}},$$

$$d_2 = d_1 - \sigma\sqrt{T},$$

where N is a cumulated normal distribution. So defined Black-Merton-Scholes model is a seminal case of an *academic* model, it can be calibrated to exactly one option, because there is only one free parameter in the option pricing formula – namely the volatility σ. Introduction of variable volatility can make it *acceptable* or *good* and introduction of stochastic volatility make it *good with smile*.

The standard generalisation is to introduce variable volatility $\sigma(t)$. It can be done in three ways:

1. Time-dependent volatility $\sigma(t)$.
2. State-dependent volatility $\sigma(t,X(t))$.
3. Stochastic volatility $\sigma(t)$.

Introducing deterministic volatility makes the model *acceptable*. In the model with deterministic volatility the stock price satisfies the equation:

$$dX(t) = rX(t)dt + \sigma(t)X(t)dW(t),$$
$$X(0) = X_0$$

and its solution is given by

$$X(t) = X(0)\exp\left\{rt + \int_0^T \sigma(s)dW(s) - \frac{1}{2}\int_0^T \sigma^2(s)ds\right\}.$$

To prove that the model is *acceptable*, it suffices to notice that the option price c(0) is given by the formula (4), where

$$d_1 = \frac{\ln(X(0)/K) + (r + \Sigma(T))}{\sqrt{\Sigma(T)}},$$

$$d_2 = d_1 - \sqrt{\Sigma(T)},$$

$$\Sigma(T) = \int_0^T \sigma^2(s)ds.$$

It is easy to check that the price (4) can be calibrated to any number of options with different maturities.

In the model with state-dependent volatility the stock price satisfies the equation:

$$dX(t) = rX(t)dt + \sigma(t, X(t))X(t)dW(t),$$

$$X(0) = X_0 ,$$

In this cases pricing of options reduces to solving the following partial differential equation:

$$\frac{\partial c}{\partial t} + \frac{1}{2} x^2 \sigma^2 (t,x) \frac{\partial^2 c}{\partial x^2} + rx \frac{\partial c}{\partial x} = rc \ .$$

We refer to [C] for the proof that the model with state-dependent volatility is *acceptable with smile.*

Introducing stochastic volatility makes the model *good* or even *good with smile.* In the model with stochastic volatility the stock price satisfies the equation:

$$dX(t) = rX(t)dt + \sigma(t)X(t)dW(t) ,$$
$$X(0) = X_0 ,$$

where $\sigma(t)$ is a stochastic process, its solution is given by

$$X(t) = X(0) \exp\left\{ rt + \int_0^T \sigma(s)dW(s) - \frac{1}{2} \int_0^T \sigma^2(s)ds \right\}$$

and the option price is given by (3). Since $\sigma(t)$ is a stochastic process, computing (3) is not easy, moreover, there are theoretical problems related to completeness of the market. For detailed study of the problem we refer to [AZ].

3. Interest rate options

Situation in the option market on: stocks, stock indices, currency and futures is clear and nice. We use Black-Merton-Scholes formula for stock and stock index options, Garman-Kohlhagen for currency options and Black'76 for futures options. All of them are versions of the basic Black-Merton-Scholes model. All the work to be done is to generalise that model to make it work if option prices are perturbed. In the interest rate options models, situation is more complex and even chaotic. Number of equilibrium or arbitrage-free interest rate option pricing models is quite large. Most of them do not leave universities but some are quite popular in the industry. We will examine most popular models as a management tool for a portfolio of interest rate options. Introduce some notations. Let B(t,T) be the price at time t of zero-coupon bond maturing at time T. Define forward instantaneous rates as

$$f(t,x) = -\frac{\partial}{\partial x} \ln(B(t,x))$$

and the short rate by $f(t) = f(t,t)$. Let $\delta > 0$ be a given tenor. Define the LIBOR rate L(t,T) by

$$\delta L(t,T) = \frac{B(t,T)}{B(t,T+\delta)} - 1.$$

Let c(t) be the price of the European call option on LIBOR rate L at time t, maturing at time T with strike price K. The most popular way is to use Black'76 formula:

$$c(t) = \delta B(t,T+\delta)\big(L(t,T)N(d_1) - KN(d_2)\big),$$

where

$$d_1 = \frac{\ln L(t,T) - \ln K + \frac{1}{2}\sigma^2 (T-t)}{\sigma\sqrt{T-t}},$$

$$d_2 = d_1 - \sigma\sqrt{T-t}.$$

The Black model is not a model in our sense, since it is useless as a portfolio management tools. For $S \neq T$ LIBOR rates, L(t,S) and L(t,T) are considered as completely separate instruments without any correlation and this is crucial in portfolio management. When we refer to smile effect in interest rate models we usually mean that two options with the same maturity have different implied Black volatilities, this is because of popularity of the Black'76 formula.

3.1. Short rate models

Short rate models are of the form:

$$df(t) = a(t, f(t))dt + \sigma(t, f(t))dW(t), \tag{6}$$

where a and σ are some deterministic functions. In the most popular Vasicek model [V] we assume that the short rate satisfies the equation:

$$df(t) = a(b - f(t))dt + \sigma dW(t) \tag{6}$$

Let c(t) be the price of the European call option, maturing at time S with strike price K, on zero coupon bond maturing at time T. In such case the option price is equal:

$$c(t) = B(t,T)N(h) - KB(t,S)N(h-\gamma),$$

where:

$$h = \frac{1}{\gamma}\ln\left(\frac{B(t,T)}{B(t,S)K}\right) + \frac{\gamma}{2},$$

$$C(t,T) = \frac{e^{-a(T-t)} - 1}{a},$$

$$\gamma = \lambda(t,S)C(S,T),$$

$$\lambda(t,S)^2 = \frac{\sigma^2\left(1 - e^{-2a(T-t)}\right)}{2a}.$$

If a = 0, then

$$\lambda(t,S) = \sigma\sqrt{S-t}$$

and

$$\gamma = \sigma(T-S)\sqrt{S-t}.$$

The Vasicek model is another seminal *academic* model, it can be calibrated to exactly two option, because there are two free parameters in the option pricing formula – volatility σ and drift factor a. Another *academic* model is the Cox-Ingersoll-Ross [CIR]. In that model we assume that the short rate satisfies the equation:

$$df(t) = a(b - f(t))dt + \sigma\sqrt{f(t)}dW(t). \quad (7)$$

It can be also calibrated to exactly two option, because there are two free parameters in the option pricing formula – volatility σ and drift factor a. For exact formulae we refer to [CIR]. As the case of stock options the standard generalisation is to introduce variable volatility σ(t). It can be done in two ways:

1. Time-dependent volatility σ(t).
2. Multi-dimensional models.

The very popular Hull-White [HW] model can be considered as the Vasicek model with non-constant volatility

$$df(t) = a(t)(b(t) - f(t))dt + \sigma(t)dW(t).$$

Another model with non-constant volatility is the Black-Karasinski [BK] lognormal model:

$$f(t) = e^{r(t)},$$

where r(t) satisfies:

$$dr(t) = a(t)(b(t) - r(t))dt + \sigma(t)dW(t).$$

It was proven [HW], [BK] that both the Hull-White and the Black-Karasinski models are *acceptable* and it was done exactly the same way as for the Black-Merton-Scholes model. These models are certainly not *acceptable with smile*. We may ask the following question: why do not we just take the general model (5) and calibrate it to market prices? This would be the same trick as state-dependent volatility in the Black-Merton-Scholes model. The reason is that the short rate is rarely an underlying for the option. In most cases options are written on LIBOR rates or bonds. Bond prices are functions of the short rate in one-factor models, this function is usually not easy to determine.

Important exceptions are the Vasicek and the Cox-Ingersoll-Ross models, where bond prices can be represented as an exponent of an affine function of the short rate:

(8)
$$B(t,T) = e^{A(t,T)+C(t,T)f(t)}.$$

Notice the following disadvantage of all short rate models. Stock prices, stock indices and currency rates are well defined single digits. Interest rate are not. Neither is 3-months LIBOR completely independent of 6-months LIBOR (Black model) nor are they deterministic functions of one risk factor (short rate models). As statistical analyses show term structure movements are driven by several factors: shift, rotation, butterfly, etc. Contribution of respective factors to the term structure dynamics is shown below.

We can see that most of the term structure movements are described by 4 risk factors.

Tab. 1. [JZ]

Number of factors	**1**	**2**	**3**	**4**
EURO	67.7%	83.9%	93.7%	97.1%
USD	75.8%	85.1%	93%	98.9%

Calibration of one factor models to a portfolio of options depending on more factors is artificial, even if possible. There are several ways to overcome this difficulty.

The Vasicek and Cox-Ingersoll-Ross models can be generalised to a multidimensional case and called affine models [DK]. In an n-dimensional affine model, $f(t)=(f_1(t), f_2(t),..., f_n(t))$ satisfies the equation (6) (Gaussian process) or (7) (positive process), where a and σ are n × n matrices, b is a vector of n elements, $\sqrt{f(t)} = diag\left(\sqrt{\upsilon_1(f(t))}, \sqrt{\upsilon_2(f(t))},..., \sqrt{\upsilon_n(f(t))}\right)$ and finally $\upsilon_i(x) = \alpha_i + \beta_i \cdot x$.

The affine property (8) is kept in both cases for an n-dimensional function C(t,T). As multidimensional f(t) is no more the short rate. The affine models can be calibrated to $2n^2$ options, hence they are *acceptable* or even *good* for large n even if the volatility is not time-dependent, especially if n is large. For more detailed study we refer to [DK].

Notice that we did not investigate the problem of "fitting the initial yield curve" because it is not important for management of options portfolio.

3.2. Whole yield curve models

All the short rate models are based on the following idea: All the bond prices are determined by the dynamics of the short rate f(t) under the risk-free probability measure E, by the formula

$$B(0,T) = E\left\{\exp\left(-\int_0^T f(s)ds\right)\right\}.$$

As it was already mentioned, the short rate is rarely an underlying for the option. It plays an important role in theory, but its importance in practice is limited. In most cases options are written on LIBOR rates or bonds. The Heath-Jarrow-Morton [HJM] model describes the evolution not of the short rate but of the whole yield curve. Bond prices satisfy the following equation:

$$dB(t,x) = B(t,x)\left(f(t)dt + \sigma(t,x)dW(t)\right),$$

where σ(t,x) is the bond volatility. When we shift out attention to forward instantaneous rates f(t,x), they satisfy the following Heath-Jarrow-Morton equation:

$$df(t,x) = \frac{\partial\sigma(t,x)}{\partial x}\cdot\sigma(t,x)dt + \frac{\partial\sigma(t,x)}{\partial x}\cdot dW(t) = \frac{1}{2}\frac{\partial\sigma(t,x)^2}{\partial x}dt + \frac{\partial\sigma(t,x)}{\partial x}\cdot dW(t) \tag{9}$$

We may ask: how many risk factors drive the yield curve dynamics? The answer is: infinity, even if the Brownian motion W(t) is one-dimensional, since whole the term structure $B(t,T)_{T\geq t}$ moves. Introducing more Brownian motions does not mean introducing more risk factor, it is only a way of making calibration easier.

The Heath-Jarrow-Morton approach is a general framework for arbitrage-free interest rate dynamics, when we speak about the Heath-Jarrow-Morton model we usually assume the volatility σ(t,x) to be a deterministic function. Let c(t) be the price of the European call option, maturing at time S with strike price K, on zero coupon bond maturing at time T. In such case the option price is equal:

$$c(t) = B(t,T)N(h) - KB(t,S)N(h-\gamma)$$

where:

$$h = \frac{\ln B(t,T) - \ln B(t,S) - \ln K + \frac{1}{2}\gamma^2}{\gamma},$$

$$\gamma^2 = \int_t^S \left(\frac{\partial\sigma(u,S)}{\partial S} - \frac{\partial\sigma(u,T)}{\partial T}\right)^2 du .$$

Notice that the volatility structure in the Heath-Jarrow-Morton model is rich: σ is a function depending on two parameters, so the Heath-Jarrow-Morton model is the first candidate to be a *good* model. And what about smile? Recall the method of state-dependent volatility. We may assume the volatility σ to be a function of the yield curve $\sigma(t,T,B(t,\cdot))$. Consider a special case of such dependence. First let us derive the Heath-Jarrow-Morton equation for LIBOR rate L(t,T). It can be easily shown that L satisfies:

$$dL(t,x) = \left(L(t,x) + \delta^{-1}\right)\left(\sigma(t,x+\delta) - \sigma(t,x)\right)\left(\sigma(t,x+\delta)dt + dW(t)\right).$$

Now if we set

$$\left(L(t,x) + \delta^{-1}\right)\left(\sigma(t,x+\delta) - \sigma(t,x)\right) = L(t,x)\gamma(t,x) \tag{10}$$

for certain deterministic function γ(t,x) we get the following equation for L:

(11) $$dL(t,x) = L(t,x)\gamma(t,x)\big(\sigma(t,x+\delta)dt + dW(t)\big),$$

where

$$\sigma(t,T) = \sum_{k=0}^{\left[\frac{T-t}{\delta}\right]} \gamma(t,T-k\delta)\frac{\delta L(t,T-k\delta)}{1+\delta L(t,T-k\delta)}.$$

We have derived a version of the Heath-Jarrow-Morton model with state-dependent volatility. This model is usually called a Brace-Gątarek-Musiela [BGM] model or forward LIBOR model. Let c(t) be the price of the European call option on LIBOR rate L at time t, maturing at time T with strike price K. Its price is equal

(12) $$c(t) = \delta B(t,T+\delta)\big(L(t,T)N(h) - KN(h-\zeta)\big)$$

where

$$h = \frac{\ln L(t,T) - \ln K + \frac{1}{2}\zeta^2}{\zeta},$$

$$\zeta^2 = \int_t^T \gamma^2(s,T)ds .$$

hence the volatility structure in the forward LIBOR model is also rich, since γ is a function depending on two parameters, and the forward LIBOR model is also a candidate to be a *good* model. Why did introduce state-dependent volatility by formula (10) and not otherwise? The reason was market practice is quoting OTC interest rate options in terms of Black volatility. Comparing the Black formula and formula (12) we get a perfect prescription how to calibrate the Brace-Gątarek-Musiela model:

$$\sigma_{Black}(t,T)^2(T-t) = \int_t^T \gamma^2(s,T)ds$$

Moreover the model is easy to calibrate to swaption prices [BGM], [R]. There exists also a swap conterparty of the forward LIBOR model called forward swap model [J], [R]. Smile can be treated by the forward LIBOR model exactly the same way as the Black-Merton-Scholes model. The simplest way is to introduce state-dependent volatility γ. Assume that $\gamma(t,T,L(t,T))$ is a function of L. Now we get the following equation for L:

(11) $$dL(t,x) = L(t,x)\gamma(t,x,L(t,x))\big(\sigma(t,x+\delta)dt + dW(t)\big),$$

where

$$\sigma(t,T) = \sum_{k=0}^{\left[\frac{T-t}{\delta}\right]} \gamma(t,T-k\delta, L(t,T-k\delta))\frac{\delta L(t,T-k\delta)}{1+\delta L(t,T-k\delta)} .$$

In this cases pricing of options reduces to solving the following partial differential equation:

$$\frac{\partial c}{\partial t}+\frac{1}{2}x^2\gamma^2(t,T,x)\frac{\partial^2 c}{\partial x^2}=rc .$$

For details we refer to [AA].

4. Conclusions

In this paper we examined several (most important in the author's personal view) option pricing models as tools of management of option portfolio. We do not claim that our list is complete. We did not examine their theoretical correctness, absence of arbitrage and mathematical properties of respective equations – it was done by the authors. We did not also examine their efficiency in pricing exotic derivatives – there is a number of articles and books dealing with that problem. We studied only a very basic problem: We are given a portfolio of options. We expect this portfolio to be arbitrage-free, at least we do no see any easy free lunches. On the hand we have a mathematical model. How to match both of them? In order to answer the question "how?" we must first answer the question "why?". Model calibration is just determining (joint!) volatility of all options in the portfolio. It is not possible to compute its risk exposure, if volatility is not uniquely determined, only certain gap analysis can be made. Value-at-Risk requires calibration. Most models can be relatively easily calibrated to a portfolio of options with different maturities. If the model has less risk factors then options in our portfolio have, value of such calibration becomes questionable. When maturities of options are not necessarily different, we may observe the smile (smirk) effect. In such case known models are less efficient.

We can conclude the following:

- Only models with rich volatility structure can be calibrated to nontrivial portfolios of options.
- Number of risk factors in the model should reflect the number of risk factors in the portfolio.
- There is no one universal model, various models can be well calibrated to various portfolios.
- Smile effect can be treated only with whole yield curve models.

Model calibration is easiest for models with rich volatility structure. The opposite holds when we price exotic option: it is easiest for simplest models, only maybe less accurate (if accuracy has any meaning). Both problems, calibration and pricing, are to be solved simultaneously. We may imagine the following picture: the market quotes not only security prices but also a model driving option prices. Therefore the forward LIBOR and forward swap models are also called implied models. In order to work with *good* models rather then with *acceptable* ones we have to perform back-testing: to calibrate models not only is present time, but also for some data in the past. And this is the general conclusion of the paper.

References

[AA] L. Andersen and J. Andreasen, Volatility skews and extensions of the LIBOR market model, General Re Financial Products working paper (1997), http://www.hkkk.fi/efa99/papers/248.pdf

[AZ] M. Avellaneda and Y. Zhu, A risk-neutral stochastic volatility model, International Journal of Theoretical and Applied Finance 2 (April 1998), 289-310

[B] F. Black, The pricing of commodity contracts, Journal of Financial Economy 3 (1976), 167-179

[BDT] F. Black, E. Derman and W. Toy, A one-factor model of interest rates and its application to treasury bond options, Financial Analysts Journal (1990), I-II, 33-39

[BK] F. Black and P. Karasinski, Bond and option pricing when short rates are lognormal, Financial Analysts Journal (July-August 1991), 52-59

[BS] F. Black and M. Scholes, The pricing of options and corporate liabilities, Journal of Political Economy 81 (1973), 637-653

[BGM] A. Brace, D. Gatarek and M. Musiela, The market model of interest rate dynamics, Mathematical Finance 7 (April 1997), 127-156

[BM] A. Brace and M. Musiela, A multifactor Gauss Markov implementation of Heath, Jarrow, Morton, Mathematical Finance 4 (1994), 259-283

[C] T.F. Coleman, Determining volatility via (inverse) optimization, Financial Engineering News 13 (December 1999), http://fenews.com/1999/Issue13/12189905.htm

[CIR] J.C. Cox, J.E. Ingersoll and S.A. Ross, A theory of the term structure of interest rates, Econometrica 53 (1985), 385-407

[DK] D. Duffie and R. Kan, A yield-factor model of interest rates, Mathematical Finance 6 (1996), 379-406.

[FV] H.G. Fong and O. Vasicek, Fixed-income volatility management, Journal of Portfolio Management (Summer 1991), 41-46

[HJM] D. Heath, R. Jarrow and A. Morton, Bond pricing and the term structure of interest rates: a new methodology, Econometrica 61 (1992), 77-105

[HW] J. Hull and A, White, Pricing interest-rate-derivative securities, Review of Financial Studies 3 (1990), 573-592.

[J] F. Jamshidian, Libor and swap market models and measures, Finance and Stochastics 1 (1997), 293-330.

[JZ] F. Jamshidian, Y. Zhu, Scenario simulation, Finance and Stochastic 1, (1997), 25-42

[MSS] K. Miltersen, K. Sandmann and D. Sondermann, Closed form solutions for term structure derivatives with log-normal interest rates, Journal of Finance LII (1997), 409-430

[R] R. Rebonato, Interest-rate option models, 2nd Edition, Wiley, Chichester 1998

[R1] R. Rebonato, Volatility and correlation in the pricing of equity, FX, and interest-rate options, Wiley, Chichester 1999

[V] O. Vasicek, An equilibrium characterisation of the term structure, Journal of Financial Economy 5 (1977), 177-188

MIX
Papier aus verantwortungsvollen Quellen
Paper from responsible sources
FSC® C105338
FSC
www.fsc.org

If you have any concerns about our products,
you can contact us on
ProductSafety@springernature.com

In case Publisher is established outside the EU,
the EU authorized representative is:
Springer Nature Customer Service Center GmbH
Europaplatz 3, 69115 Heidelberg, Germany

Printed by Libri Plureos GmbH
in Hamburg, Germany